Springer Series in

SOLID-STATE SCIENCES 142

Dear MyCopy Customer,

This Springer book is a monochrome print version of the eBook to which your library gives you access via SpringerLink. It is available to you at a subsidized price since your library subscribes to at least one Springer eBook subject collection.

Please note that MyCopy books are only offered to library patrons with access to at least one Springer eBook subject collection. MyCopy books are strictly for individual use only.

You may cite this book by referencing the bibliographic data and/or the DOI (Digital Object Identifier) found in the front matter. This book is an exact but monochrome copy of the print version of the eBook on SpringerLink.

Springer-Verlag Berlin Heidelberg GmbH

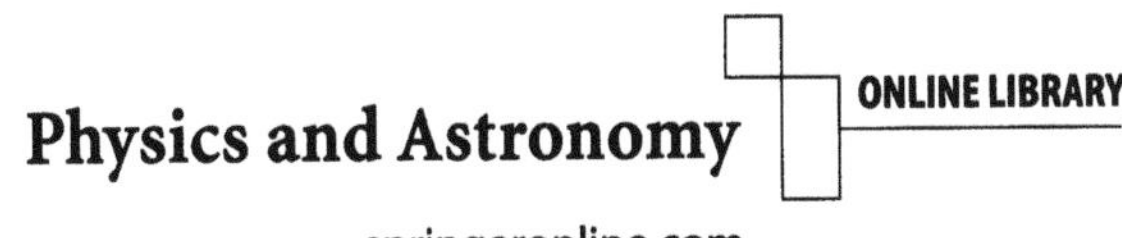

springeronline.com

Springer Series in

SOLID-STATE SCIENCES

Series Editors:
M. Cardona P. Fulde K. von Klitzing R. Merlin H.-J. Queisser H. Störmer

The Springer Series in Solid-State Sciences consists of fundamental scientific books prepared by leading researchers in the field. They strive to communicate, in a systematic and comprehensive way, the basic principles as well as new developments in theoretical and experimental solid-state physics.

126 **Physical Properties of Quasicrystals**
Editor: Z.M. Stadnik

127 **Positron Annihilation in Semiconductors**
Defect Studies
By R. Krause-Rehberg and H.S. Leipner

128 **Magneto-Optics**
Editors: S. Sugano and N. Kojima

129 **Computational Materials Science**
From Ab Initio to Monte Carlo Methods
By K. Ohno, K. Esfarjani, and Y. Kawazoe

130 **Contact, Adhesion and Rupture of Elastic Solids**
By D. Maugis

131 **Field Theories for Low-Dimensional Condensed Matter Systems**
Spin Systems and Strongly Correlated Electrons
By G. Morandi, P. Sodano, A. Tagliacozzo, and V. Tognetti

132 **Vortices in Unconventional Superconductors and Superfluids**
Editors: R.P. Huebener, N. Schopohl, and G.E. Volovik

133 **The Quantum Hall Effect**
By D. Yoshioka

134 **Magnetism in the Solid State**
By P. Mohn

135 **Electrodynamics of Magnetoactive Media**
By I. Vagner, B.I. Lembrikov, and P. Wyder

136 **Nanoscale Phase Separation and Colossal Magnetoresistance**
The Physics of Manganites and Related Compounds
By E. Dagotto

137 **Quantum Transport in Submicron Devices**
A Theoretical Introduction
By W. Magnus and W. Schoenmaker

138 **Phase Separation in Soft Matter Physics**
Micellar Solutions, Microemulsions, Critical Phenomena
By P.K. Khabibullaev and A.A. Saidov

139 **Optical Response of Nanostructures**
Microscopic Nonlocal Theory
By K. Cho

140 **Fractal Concepts in Condensed Matter Physics**
By T. Nakayama and K. Yakubo

141 **Excitons in Low-Dimensional Semiconductors**
Theory, Numerical Methods, Applications
By S. Glutsch

142 **Two-Dimensional Coulomb Liquids and Solids**
By Y. Monarkha and K. Kono

Volumes 1–125 are listed at the end of the book.

Yuriy Monarkha Kimitoshi Kono

Two-Dimensional Coulomb Liquids and Solids

With 109 Figures

Springer

Professor Yuriy Monarkha
Institute for Low Temperature Physics
and Engineering
National Academy of Sciences of Ukraine
47 Lenin Ave.
Kharkov 61103
Ukraine

Professor Kimitoshi Kono
Low Temperature Physics Laboratory
RIKEN
Hirosawa 2-1
Wako-shi 351-0198
Japan

Series Editors:

Professor Dr., Dres. h. c. Manuel Cardona
Professor Dr., Dres. h. c. Peter Fulde*
Professor Dr., Dres. h. c. Klaus von Klitzing
Professor Dr., Dres. h. c. Hans-Joachim Queisser

Max-Planck-Institut für Festkörperforschung, Heisenbergstrasse 1, D-70569 Stuttgart, Germany
* Max-Planck-Institut für Physik komplexer Systeme, Nöthnitzer Strasse 38
D-01187 Dresden, Germany

Professor Dr. Roberto Merlin

Department of Physics, 5000 East University, University of Michigan
Ann Arbor, MI 48109-1120, USA

Professor Dr. Horst Störmer

Dept. Phys. and Dept. Appl. Physics, Columbia University, New York, NY 10027 and
Bell Labs., Lucent Technologies, Murray Hill, NJ 07974, USA

ISSN 0171-1873
DOI 10.1007/978-3-662-10639-6

Library of Congress Cataloging-in-Publication Data applied for.
Bibliographic information published by Die Deutsche Bibliothek
Die Deutsche Bibliothek lists this publication in the Deutsche Nationalbibliografie;
detailed bibliographic data is available in the Internet at http://dnb.ddb.de.

springeronline.com

Originally published by Springer-Verlag Berlin Heidelberg New York in 2004
MyCopy version of the original edition 2004

Typesetting by the authors
Final processing by Frank Herweg, Leutershausen
Cover concept: eStudio Calamar Steinen
Cover production: *design & production* GmbH, Heidelberg

Printed on acid-free paper SPIN: 10976418 57/3141/tr - 5 4 3 2 1 0
www.springer.com/mycopy

Preface

This book is about quantum phenomena in two-dimensional (2D) electron systems with extremely strong internal interactions. The central objects of interest are Coulomb liquids, in which the average Coulomb interaction energy per electron is much higher than the mean kinetic energy, and Wigner solids. The main themes are quantum transport in two dimensions and the dynamics of highly correlated electrons in the regime of strong coupling with medium excitations.

In typical solids, the mutual interaction energy of charge carriers is of the same order of magnitude as their kinetic energy, and the Fermi-liquid approach appears to be quite satisfactory. However, in 1970, a broad research began to investigate a remarkable model 2D electron system formed on the free surface of superfluid helium. In this system, complementary to the 2D electronic systems formed in semiconductor interface structures, the ratio of the mean Coulomb energy of electrons to their kinetic energy can reach approximately a hundred before it undergoes the Wigner solid (WS) transition. Under such conditions, the Fermi-liquid description is doubtful and one needs to introduce alternative treatments. Similar interface electron systems form on other cryogenic substrates like neon and solid hydrogen.

It might be concluded that the Coulomb liquid is a pure classical system because its Fermi energy is much smaller than the interaction energy and (often) even the temperature. But this is not true, especially if the system is subject to a strong quantizing magnetic field. For example, the Coulomb liquid reveals an unconventional Hall effect with behavior quite opposite to the classical picture and complementary to the quantum Hall effect observed in semiconductor 2D electron systems.

This book has been written to provide a new and comprehensive review of the remarkable properties of interface Coulomb liquids and solids. In spite of the broad range covered by the reviewed properties (from free 2D electron gases and single-electron polarons to electron crystals), it was not difficult to maintain the uniformity and coherence required in a single book because of the emphasis placed on the quantum transport framework formulated for electrons with extremely strong mutual interactions. This transport framework combines the memory function formulation of electron conductivity and the momentum-balance equation method. It is used throughout this book and provides a good description of the unconventional Hall effect, quantum cyclotron resonance, WS phonon–medium excitation coupling and conductivity of the electron solid. The universal conductivity form is obtained as a quantum extension of the Drude form with a relaxation kernel containing the real and imaginary parts. The simplest approximation for the relaxation kernel establishes its relationship with the electron dynamical structure factor (DSF), which is proven to provide quite accurate conductivity equations in the whole frequency range, if electron–electron collisions dominate the momentum distribution within the electron layer. For practical purposes, the main advantage of this approach is that one can avoid summing an infinite series of Feynman diagrams, which is impossible in the conventional conductivity treatment, or solving the kinetic equations. The final equations for the effective collision frequency (the imaginary part of the conductivity relaxation kernel) have a very simple integral form applicable to any state of the electron system, liquid or solid, which is very important for Coulomb liquids.

In a 2D Coulomb liquid subject to a magnetic field, mutual interactions are so strong that it is impossible to introduce a universal energy excitation spectrum in a single reference frame. Therefore, to go beyond the usual Fermi-liquid approach, instead of the energy excitation spectrum, one has to consider more global properties of the strongly interacting electrons, namely the quantum correlation functions, such as the dynamical structure factor. We pay special attention to this particular property of Coulomb liquids and solids because it plays an essential role in the quantum transport theory. For a 2D electron gas subject to a magnetic field, it is the single-electron density of states, broadened due to interaction with scatterers, which plays the key role in quantum transport phenomena. For the Coulomb liquid, it is more convenient and useful to operate with the DSF of the whole liquid. As a function of frequency it has a series of maxima which are broadened owing to both the interaction with scatterers and the electron–electron interaction. In terms of density–density correlation functions it is possible to obtain remarkable similarities between an ideal 2D electron gas and the strongly interacting electrons. For example, in some instances the DSF of the Coulomb liquid and even the Wigner crystal can be found as a simple double integral of the single-electron self-correlation function, or the DSF of an ideal gas.

This text is intended for graduate students and research physicists working in the field of quantum liquids, electronic properties of 2D electron systems, and solid-state physics. It has different levels of sophistication and will hopefully be useful for both theorists and experimentalists. The text contains some instructive examples of general interest, showing that a careless utilization of conventional methods and well-known formulae from solid-state physics can sometimes lead to conclusions that are quite opposite to the correct answers. This especially concerns electron interactions with interface irregularities and the behavior of the electron crystal under normal magnetic fields.

We start this book with a broad overview (Chap. 1) of the single-electron properties of the 2D electron system formed on the surface of quantum liquids (^{4}He and ^{3}He) and on the solid cryogenic substrates (H_2 and Ne). In Chap. 2, we discuss those properties of the 2D electron liquid which originate from the strong mutual Coulomb interaction. Among new and interesting themes presented here, we would like to emphasize the electric field effect on the single-electron density of states, the Coulomb narrowing of Landau levels and the Coulomb broadening of the electron DSF.

Chapter 3 concerns the quantum transport framework for practical evaluations applying to the highly correlated Coulomb liquid and Wigner solid. The general relations obtained here are then applied to a description of the unconventional Hall effect (Chap. 4) and quantum cyclotron resonance (Chap. 5). Theoretical results are compared with a large body of experimental data displaying many-electron effects on quantum magnetotransport.

In Chap. 6, we discuss the polaronic effect for 2D electrons strongly interacting with interface displacements. Besides an interesting relationship with the general problem of polaron self-trapping, this text introduces important physical ideas needed to understand the behavior of the 2D electron crystal on 'soft' interfaces. Many different problems related to the non-dissipative dynamics of the 2D Wigner solid are discussed in Chap. 7. The quantum transport framework helps here to describe the properties of the WS which originate from the strong electron coupling to interface excitations. The remarkable transport properties of the Wigner solid formed on the free surfaces of quantum liquids are discussed in Chap. 8.

One of us wishes to express special thanks to the Institute of Physical and Chemical Research (RIKEN, Japan) for its international invitation program (RESIP), visiting professorship and hospitality. We have benefited from collaboration with our colleagues R.W. van der Heijden, P. Leiderer, V.B. Shikin, E. Teske, W.F. Vinen, F.I.B. Williams, and P. Wyder.

Tokyo, Japan,
December 2003

Yuriy Monarkha
Kimitoshi Kono

Contents

1 Two-Dimensional Interface Electron Systems

1.1 Introduction

The electrostatic charging of dielectrics with surface-bound electrons is an everyday occurrence of our life. The essence of the phenomenon is very simple: a dielectric body attracts free electrons and gathers them at its surface, forming a simple capacitor. For a long while, such interface electrons were of minor interest for fundamental physics because small surface irregularities trap them in the plane of the interface owing to the strong polarization forces. Even a single atom at a flat surface can capture an electron, forming a motionless negative ion. Only at the end of the 1960s was it understood that liquid helium, as a dielectric substrate, provides unique possibilities for studying highly correlated two-dimensional (2D) electron systems.

Surface electron (SE) states on the free surface of liquid helium were introduced theoretically by Cole and Cohen [1] and by Shikin [2]. The basic research on SEs above superfluid helium had initially started as a 'raised to the surface' study of bulk negative ions conventionally described as very small charged bubbles of radius $R_{-} \sim 17\,\text{Å}$. Later it was understood [3] that SEs represent a remarkable model system for the experimental realization of the solid state of strongly interacting electrons on the uniform positive background predicted by Wigner in 1934 [4]. Even in its liquid state, when the system of nondegenerate SEs was subjected to a strong normally oriented magnetic field, it was expected to reveal interesting quantum transport properties, complementary to the conventional quantum Hall effect (QHE) discovered in semiconductor 2D electron systems [5, 6]. The experimental and theoretical studies of this system since carried out have resulted in remarkable discoveries that have more than fulfilled these early expectations.

The first observation of the Wigner solid reported by Grimes and Adams [7] for SEs on superfluid helium unambiguously revealed the hexagonal lattice structure of the electron layer when the plasma parameter $\Gamma^{(\mathrm{pl})} = e^2\sqrt{\pi n_{\mathrm{s}}}/T$ exceeded the critical value $\Gamma_{\mathrm{c}}^{(\mathrm{pl})} \simeq 137$ (where n_{s} is the areal electron density). This number introduced an important landmark for further investigations of strongly interacting 2D electron systems. Nowadays, the properties of the 2D Wigner solid are so well established that it is even used as a practical probe for superfluid properties of Fermi liquid ^{3}He at ultra-low temperatures.

The above-mentioned landmark value of $\Gamma_c^{(pl)}$ indicates that at $\Gamma^{(pl)} < \Gamma_c^{(pl)}$ 2D electrons represent a unique Coulomb liquid in which the interaction energy per electron can be more than a hundred times larger than the mean kinetic energy. In typical metals, these two characteristic energies are of the same order of magnitude. Therefore, in the liquid state, the system of SEs is usually under the extremely strong coupling regime with regard to internal forces. At the same time, in the presence of a strong magnetic field, this 2D electron system represents a nondegenerate quantum liquid with unusual properties. The Coulomb liquid has no weakly interacting quasiparticles or excitations. Therefore the description of quantum transport in it requires the development of new methods stretching far beyond the conventional Fermi-liquid concept. Such methods were introduced in parallel with experimental studies and tested for the description of the unconventional Hall effect and quantum cyclotron resonance (CR) of SEs on superfluid helium. As we shall see in Chap. 3, these methods can be conveniently represented as a universal framework for practical evaluations applicable to highly correlated electrons.

The first important advantage of the use of liquid helium as a cryogenic substrate for the 2D electron layer is accounted for the extremely weak polarizability of helium atoms. The dielectric constant ϵ of liquid helium is very close to unity ($\epsilon_4 - 1 \simeq 0.056$ for ^{4}He and $\epsilon_3 - 1 \simeq 0.042$ for ^{3}He). According to quantum mechanical principles, this allows the surface electrons to hover above the free surface with average height $\langle z \rangle \simeq 10^{-6}$ cm, which is two orders of magnitude larger than the typical atomic scale. Additionally, at $T = 0$ the free surface of superfluid helium has no static irregularities or defects, forming a perfectly flat substrate for the 2D electron system. The only scatterers available in a typical experiment with SEs are helium vapor atoms and surface excitations (ripplons) whose number density decreases with cooling.

With a steady increase in the accumulated charge, the surface of the liquid dielectric eventually becomes unstable [8]. This restricts the electron density of SEs on liquid helium to the range $n_s < n_s^{(c)} \approx 2 \times 10^9\ \mathrm{cm}^{-2}$ [9], where they generally obey nondegenerate statistics. This range is of particular interest for studying the Wigner solid (WS) transition in the dislocation melting regime. The quantum melting regime of the WS requires substantially higher electron densities.

The use of other cryogenic substrates, such as solid hydrogen or neon, allows one to reach higher electron densities at the expense of the cleanness of the system. With a superfluid helium film covering a dielectric substrate, the electron binding to the interface can be changed smoothly by varying the film thickness. In this instance, the van der Waals forces forming the helium film increase the stability region of the charged surface up to electron densities that are typical for semiconductor 2D electron systems.

In this chapter we discuss basic single-electron properties of interface electrons formed on cryogenic substrates, disregarding their mutual interactions. We focus on SEs formed on superfluid ^{4}He, presenting the theory of electron interaction with available scatterers (vapor atoms and an uneven interface), and analysing different descriptions of the collision broadening of Landau levels for 2D electrons exposed to a magnetic field applied normally to the layer.

1.2 Hovering Electrons Above Superfluid Helium ^{4}He

The experimental cell usually used to study SEs on superfluid helium is remarkably reminiscent of the famous Si-MOSFET device (metal oxide semiconductor field effect transistor) [10, 11]. The similarity of these systems is illustrated in Fig. 1.1. The MOSFET device represents a parallel plate condenser in which one of the metal plates is replaced by the Si semiconductor. The 2D system of band electrons at the Si–SiO_2 interface is induced by applying a strong enough positive voltage V_g to the metal electrode (gate), as shown in Fig 1.1a, which creates an electric field. This field bends the conduction band in such a way that a part of it, near the interface, becomes lower than the Fermi level (see Fig. 1.2a). The areal electron density n_s appearing near the interface is proportional to $V_g - V_t$, where V_t is a threshold voltage, and is easily varied by changing V_g. Under typical conditions, $n_s \simeq 10^{11}$–$10^{12}\,\mathrm{cm}^{-2}$.

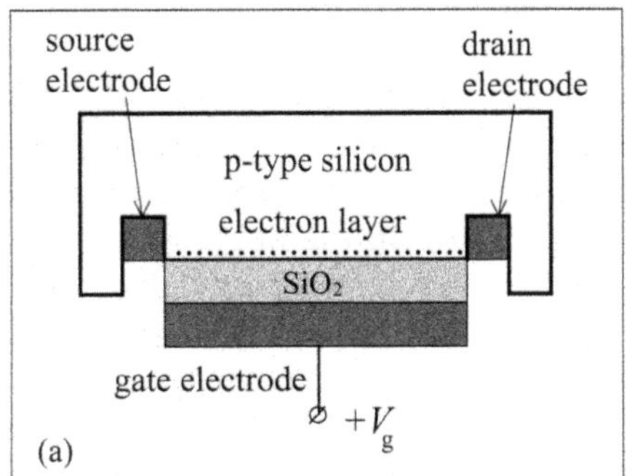

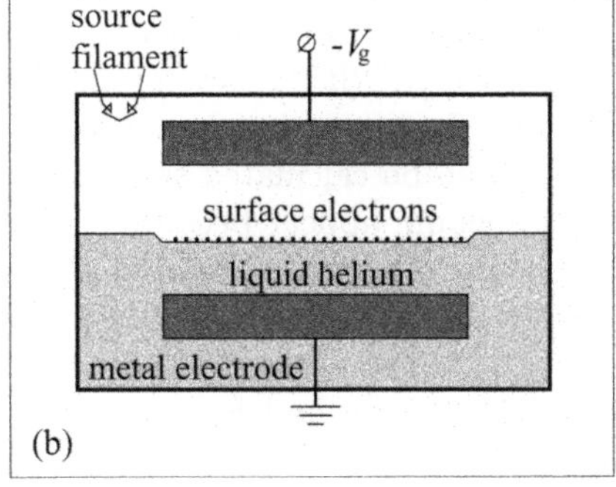

Fig. 1.1. Schematic view of the semiconductor Si-MOSFET device (**a**) and the surface electron system on liquid helium (**b**)

The 2D electron layer on a liquid helium surface is arranged in a similar way, as shown in Fig. 1.1b. Because it is impossible to attach leads to SEs, the liquid helium surface is placed inside a parallel plate condenser. The heated filament acts as a source of free electrons. The electrons proceed to the liquid helium through the hole in the upper electrode and accumulate at its surface until the total electric field above the electron layer $E_\perp^\wedge = E_\perp - 2\pi e n_s$ becomes zero ($-2\pi e n_s$ is the z-component of the field of the electron layer). In equilibrium, the electron density is proportional to the external electric field or the gate voltage: $n_s = E_\perp / 2\pi e$. Owing to the condition $E_\perp^\wedge = 0$, the

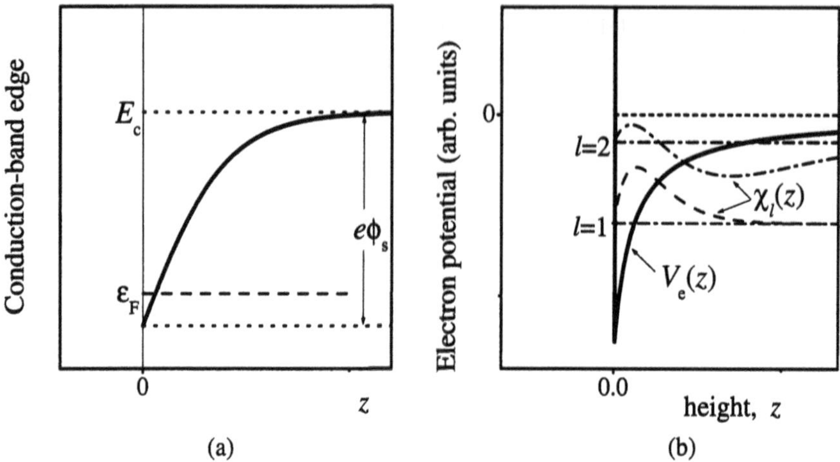

Fig. 1.2. The energy-band bending diagram near the semiconductor–dielectric interface (**a**), and a schematic view of the single-electron potential energy above the liquid helium surface (**b**)

electron layer at the helium surface plays the role of the upper electrode of the parallel plate condenser. Although the SE states above a liquid helium surface exist even in the absence of the holding electric field, one has to apply $E_{\perp} \geq 2\pi e n_s$ in order to create a substantial surface charge.

For the MOSFET device, the shape of the one-dimensional potential well deviates from a triangular shape owing to the charge of the inversion layer, and has to be calculated self-consistently. The surface electron states above liquid helium appear because of competition between the polarization force attracting electrons to helium atoms and the hard-core repulsive force. If a holding electric field $E_{\perp}$ is present, one has to include its potential as well. The attraction is so weak that a single helium atom in three dimensions cannot trap an electron and form a negative ion. This property is called negative electron affinity. The hard-core repulsion is responsible for the appearance of the potential barrier $V_0 \simeq 1\,\mathrm{eV}$ for electron penetration into the bulk of the liquid helium [12].

As the SEs are quite distant from the liquid–vapor interface, the attractive force can be described macroscopically in terms of the image potential. The one-dimensional potential well for an SE can be approximated by the following model, also shown in Fig. 1.2b:

$$V_e(z) = V_0 \theta(-z) - \frac{\Lambda}{z + z_0} \theta(z) + eE_{\perp} z \,, \tag{1.1}$$

where

$$\Lambda = \frac{e^2(\epsilon - 1)}{4(\epsilon + 1)} \,,$$

$\theta(z)$ is the unit step function, $E_\perp$ is the external holding electric field always present in experiments with SEs, and $-e$ is the charge of a free electron. In order to avoid the singularity of the image potential for the sharp surface profile, following [13], we introduce the parameter z_0 of the order of the typical atomic scale, which shifts the singularity of the image potential beyond its action range. This parameter is usually adjusted to fit the surface level positions measured by a spectroscopic technique. The important point is that the adjusted value of z_0 appears to be around 1 Å, which is much smaller than the average distance $\langle z \rangle$ of the SE from the interface.

The structure of the image potential of (1.1) is very convenient for predicting the electron spectrum, as shown by Sanders and Weinreich [14] and by Hipólito, Felicio and Farias [15]. The actual potential of the SE accurately computed by Cheng, Cole and Cohen [16], taking into account the real density profile of liquid helium, deviates from (1.1) in the vicinity of $z = 0$. Still, the overall effect of this deviation on the electron spectrum is rather small. In most cases considered in this book, it is quite sufficient to use an even simpler model, viz., $V_0 \to \infty$ and $z_0 \to 0$, discussed below.

For the flat surface, the symmetry of the system allows us to separate the in-plane motion of SEs from the motion in the z-direction by writing the wave function and energy spectrum as

$$\psi_l(\boldsymbol{r}, z) = \frac{1}{\sqrt{S_\mathrm{A}}} \mathrm{e}^{\mathrm{i}\boldsymbol{k}\cdot\boldsymbol{r}} \chi_l(z) , \tag{1.2}$$

$$\varepsilon_l(k) = \varepsilon_l^{(\perp)} + \frac{\hbar^2 k^2}{2m_\mathrm{e}} , \tag{1.3}$$

where $\boldsymbol{k}$ is the 2D wave vector parallel to the plane, l is the quantum number describing electron motion in the perpendicular direction and S_A is the surface area. As a result, the Schrödinger equation for $\chi_l(z)$ reduces to

$$\frac{\mathrm{d}^2\chi_l}{\mathrm{d}z^2} + \frac{2m_\mathrm{e}}{\hbar^2}\left[\varepsilon_l^{(\perp)} - V(z)\right]\chi_l = 0 . \tag{1.4}$$

It is generally quite a complicated task to solve this equation. A significant simplification comes from the fact that the potential barrier V_0 is very high compared to typical energies of the SEs. Therefore the simplest model with $V_0 \to \infty$ and $z_0 \to 0$ provides us with quite accurate results for the SE spectrum. In this model, the replacement $V(z) \to -\Lambda/z$ (we consider $E_\perp \to 0$) should be accompanied by the boundary condition

$$\chi_l(0) = 0 .$$

As a result, the Schrödinger equation for $\chi_l(z)$ becomes the same as the equation describing the electron spectrum of a hydrogen atom. In the latter case, the boundary condition $\chi(R \to 0) = 0$ appears because of the replacement of the electron wave function $\psi(R) = \chi(R)/R$, where $\boldsymbol{R}$ is the 3D radius vector.

In order for the electron wave function to be finite at $R = 0$, the auxiliary function $\chi(R)$ must vanish at this point.

The spectrum of SE states can thus be found from the hydrogen atom spectrum by replacing e^2 by Λ as defined in (1.1) [2]:

$$\varepsilon_l^{(\perp)} = -\frac{m_e \Lambda^2}{2\hbar^2 l^2}, \qquad l = 1, 2, 3, \ldots . \tag{1.5}$$

In the case of ^{4}He, inserting the right numbers gives $\varepsilon_1^{(\perp)} \simeq -8\,\mathrm{K}$ for the position of the ground surface level, and $(\varepsilon_2^{(\perp)} - \varepsilon_1^{(\perp)})/\hbar \simeq 119.2\,\mathrm{GHz}$ and $(\varepsilon_3^{(\perp)} - \varepsilon_1^{(\perp)})/\hbar \simeq 141.3\,\mathrm{GHz}$ for the transition frequencies. The spectroscopy measurement data $(\varepsilon_2^{(\perp)} - \varepsilon_1^{(\perp)})/\hbar \simeq (125.9 \pm 0.2)\,\mathrm{GHz}$ and $(\varepsilon_3^{(\perp)} - \varepsilon_1^{(\perp)})/\hbar \simeq (148.6 \pm 0.3)\,\mathrm{GHz}$ [13] agree well with these estimates. The exact solution of the Schrödinger equation for the image potential model $V(z) = -\Lambda/(z + z_0)$ was found in [15]. Fitting the data of [13] gives $z_0 = 1.01\,\text{Å}$.

The wave functions of the two lowest surface states can be written as

$$\chi_1(z) = 2\gamma^{3/2} z \mathrm{e}^{-\gamma z}, \tag{1.6}$$

$$\chi_2(z) = \frac{\gamma^{3/2}}{\sqrt{2}} z(1 - \gamma z/2)\mathrm{e}^{-\gamma z/2}, \tag{1.7}$$

where

$$\gamma = \frac{m_e \Lambda}{\hbar^2}.$$

In the case of liquid ^{4}He, the average distances $\langle 1|z|1\rangle = 3/(2\gamma) \simeq 114\,\text{Å}$ and $\langle 2|z|2\rangle \simeq 456\,\text{Å}$ of the hovering electrons from the interface considerably exceed the atomic scale, which justifies the approximations made for the electron potential above the liquid helium surface. For the two lowest surface levels, the shape of the wave function is shown by dashed lines in Fig. 1.2b.

In order to estimate the accuracy of the hydrogenic model ($V_0 = \infty$), it is instructive to find the characteristic length of electron penetration into liquid helium. According to quantum mechanics, in the region $z < 0$, where $V_e(z) \simeq V_0$, the electron wave function decreases exponentially: $\chi_1(z) = \chi_1(0)\exp(\kappa_0 z)$. The penetration length κ_0^{-1} is determined by the value of the potential step V_0 as

$$\kappa_0 = \frac{1}{\hbar}\sqrt{2m_e(V_0 - \varepsilon_l^{(\perp)})} \simeq \frac{1}{\hbar}\sqrt{2m_e V_0}. \tag{1.8}$$

It should be noted that $\kappa_0 \simeq 5.1 \times 10^7\,\mathrm{cm}^{-1}$ is nearly two orders of magnitude larger than γ, which means that the electron penetration into liquid helium is very small. Employing the continuity condition for the first derivative of the electron wave function, one finds that

$$\chi_1(0) = \chi_1'(0)/\kappa_0 \simeq 2\gamma^{3/2}/\kappa_0. \tag{1.9}$$

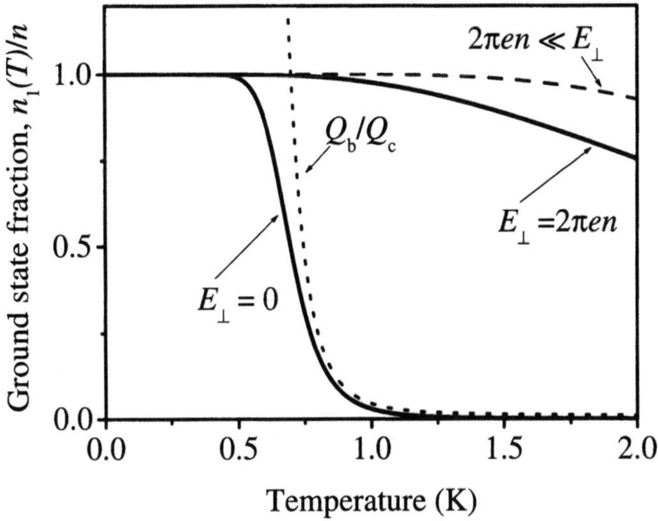

Fig. 1.3. The population of the ground surface level under different conditions concerning the holding electric field $E_\perp$ (*solid* and *dashed curves*), as labelled in the graph. The ratio Q_b/Q_c is shown by the *dotted curve*

Here we have used the fact that $\chi_1'(0)$ is finite in the hydrogenic limit ($V_0 \to \infty$) and can be approximated by the derivative of (1.6). A small change in the normalization constant due to electron penetration into the liquid has been disregarded. From (1.9), we can show that $\chi_1(0)$ is smaller than the maximum value of $\chi_1(z)$ by the factor $\gamma/\kappa_0 \sim 0.026$.

According to (1.5), the energy levels of SEs, evaluated under the condition $E_\perp = 0$, become closely spaced for large numbers l as $\varepsilon_l^{(\perp)}$ approaches the continuum edge. For a spectrum of this kind, the partition function of the bound states has a well-known divergence:

$$Q_b = \sum_{l=1}^{\infty} e^{-\varepsilon_l^{(\perp)}/T} .$$

(For the hydrogen atom, this problem was discussed by Fermi in 1924 [17].) The cutoff l_{max} is found from the condition that the average 'radius' $\langle l|z|l\rangle$, proportional to l^2, does not exceed the container length L_z:

$$l_{max} \approx \sqrt{\frac{L_z}{\langle 1|z|1\rangle}} .$$

For $L_z = 1\,\text{cm}$, we estimate $l_{max} \approx 935$.

Even more importantly, the number of states for the continuous spectrum is very large, and the partition function Q_c for these states is proportional to the container length L_z:

$$Q_{\mathrm{c}} = \sqrt{\frac{m_{\mathrm{e}}T}{8\pi\hbar^2}}L_z \ .$$

At $T > 0.8\,\mathrm{K}$, the ratio $Q_{\mathrm{b}}/Q_{\mathrm{c}}$, shown by the dotted curve in Fig. 1.3, is small. Generally, these two large factors, Q_{b} and Q_{c}, cause the ground surface level to be poorly populated at $T > 0.5\,\mathrm{K}$ [18] in spite of the condition $\left|\varepsilon_1^{(\perp)}\right| > T$, as shown by the solid curve marked with $E_\perp = 0$ in Fig. 1.3. The inclusion of the external holding field potential, even a very weak one, ensures a high population in the ground surface level which drastically changes this situation [19, 20]. We shall discuss this problem after a short description of the SE states in the presence of the holding electric field.

Let as first consider the influence of a weak holding electric field on the electron spectrum and wave functions. Small corrections to level positions can be found by means of the usual perturbation theory:

$$\delta\varepsilon_l^{(\perp)} \simeq eE_\perp \langle l|z|l\rangle \ . \tag{1.10}$$

This can be considered as a linear Stark effect for SE states. The applicability of this approach is limited by the condition $\delta\varepsilon_l^{(\perp)} \ll \varepsilon_{l+1}^{(\perp)} - \varepsilon_l^{(\perp)}$. This Stark effect was employed by Grimes et al. [13] to tune the splitting between electronic surface states on liquid helium to resonance with the applied radiation, as shown in Fig. 1.4.

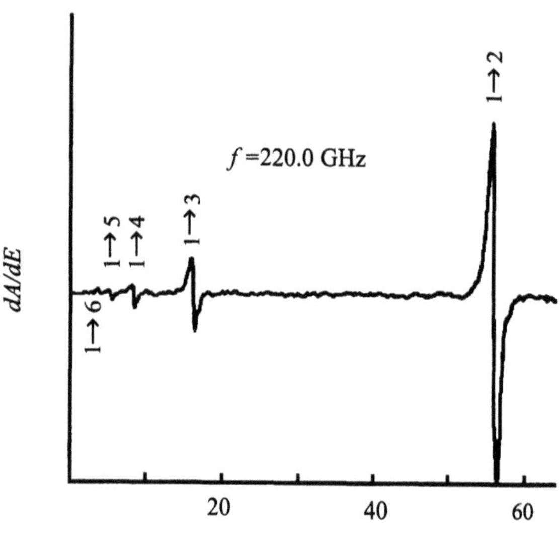

Fig. 1.4. Experimental recording of mm wave absorption derivative vs. applied voltage V_{g} taken at a frequency of 220 GHz and $T = 1.2\,\mathrm{K}$ [13]. The $1 \to 2$, $1 \to 3$, etc., transitions are analogous to the Lyman α, β, etc., transitions of the hydrogen atom

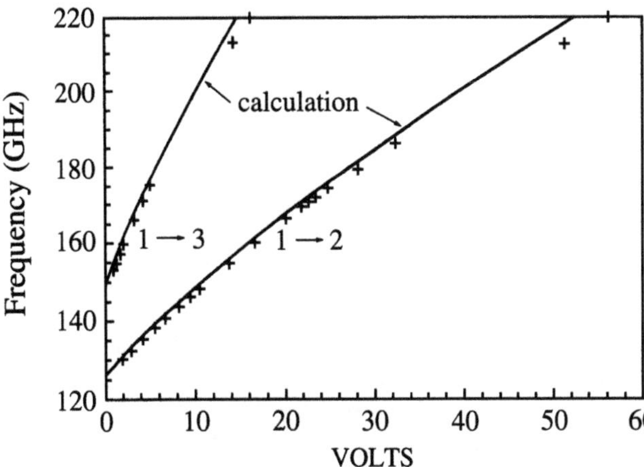

Fig. 1.5. Plot of transition frequencies vs. voltage across the experimental cell [13]. *Crosses* are measured data points while the *solid curves* are the result of the variational calculation

At stronger holding fields, the perturbation treatment breaks down and the corrections to the hydrogen-like spectrum become nonlinear functions of $E_{\perp}$. In this regime, a simple variational calculation of transition frequencies fits experimental data quite satisfactorily, as shown in Fig. 1.5.

In the limit of strong holding fields or for large quantum numbers l, one can disregard the image potential. We then have a triangular-shaped potential and the boundary condition $\chi_l(0) = 0$. The solution for this 1D potential well model is known to be

$$\varepsilon_l^{(\perp)} = \zeta_l e E_{\perp}/\gamma_{\mathrm{F}}, \qquad \gamma_{\mathrm{F}} = \left(\frac{2m_{\mathrm{e}} e E_{\perp}}{\hbar^2}\right)^{1/3}, \tag{1.11}$$

$$\chi_l(z) = \text{const.} \times \mathrm{Ai}\left[\left(z - \varepsilon_l^{(\perp)}/eE_{\perp}\right)\gamma_{\mathrm{F}}\right]. \tag{1.12}$$

The discrete numbers ζ_l are determined by the boundary condition $\mathrm{Ai}(-\zeta_l) = 0$, where $\mathrm{Ai}(z)$ is the Airy function, $\zeta_1 \simeq 2.238$, $\zeta_2 \simeq 4.088$, $\zeta_3 \simeq 5.521, \ldots$. For high levels, it is possible to use the quasi-classical asymptote $\zeta_l \simeq (3\pi l/2)^{2/3}$.

In the transport theory, the cross-section of an electron bound to the interface depends on the wave function $\chi_1(z)$, which is generally affected by the holding electric field in a complicated way. A convenient approximation for the ground state wave function, valid at any value of $E_{\perp}$, can be found by a variational method using $\chi_1(z)$ from (1.6) as a trial function. This is possible because both wave functions describing extreme limiting cases $E_{\perp} = 0$ and $E_{\perp} \to \infty$ have similar shapes. The parameter γ, which determines $\chi_1(z)$, depends on the holding field [21]:

$$\frac{\gamma(E_{\perp})}{\gamma_0} = \frac{1}{3}\left[1 + \left(\beta + \sqrt{\beta^2 - 1}\right)^{1/3} + \left(\beta - \sqrt{\beta^2 - 1}\right)^{1/3}\right], \tag{1.13}$$

where $\gamma_0 = \gamma(0)$ and

$$\beta = 1 + \frac{81}{8}\lambda^2 , \qquad \lambda = \sqrt{\frac{2m_e e E_\perp}{\hbar^2 \gamma_0^3}} . \tag{1.14}$$

Another equivalent representation for the function $\gamma(E_\perp)$ was introduced in [22]:

$$\frac{\gamma(E_\perp)}{\gamma_0} = \frac{3\lambda}{4}\left\{\sinh\left[\frac{1}{3}\operatorname{arcsinh}\left(\frac{9\lambda}{4}\right)\right]\right\}^{-1} . \tag{1.15}$$

The trial form of the variational wave function discussed here significantly simplifies evaluation of all matrix elements entering electron collision rates.

It is instructive to discuss here the dependence of the electron wave function at the helium surface $\chi_1(0)$ on the applied holding field $E_\perp$. We cannot rely on (1.9) with $\gamma(E_\perp)$ defined by the variational approach because the shape of the variational function used to find it is approximate. In contrast, matrix elements evaluated by means of the variational wave function are usually quite accurate, and these can be used. In order to find $\chi_1(0)$, we can use the important property of a bound state

$$\langle 1 | \partial V_e(z)/\partial z | 1 \rangle = 0 , \tag{1.16}$$

which means that the average force, acting on the bound electron in the z-direction, is zero. Let us consider the following model describing the electron potential near the flat interface:

$$V_e(z) = (V_0 + U_m)\theta(-z) + V_\epsilon(z) + eE_\perp z , \tag{1.17}$$

where $V_\epsilon(z)$ is a continuous function which coincides with the image potential at macroscopic values of z, approaches smoothly (within the atomic scale) a constant negative value $-U_m$ ($U_m \ll V_0$) as $z \to 0$, and remains at this value when z is negative. This model is closer to the real potential behavior discussed in [16] than (1.1). Substituting the form of (1.17) for V_e into (1.16), we arrive at the following equation:

$$(V_0 + U_m)\chi_1^2(0) = eE_\perp + \int_{-\infty}^{\infty} \chi_1^2(z) \frac{\partial V_\epsilon}{\partial z} \mathrm{d}z . \tag{1.18}$$

According to the model considered here, the integrand of the second term vanishes at $z \leq 0$. The important point is that the second term is finite in the limit $V_0 \to \infty$. For the trial wave function, which has the same shape as the ground surface level wave function, (1.18) yields

$$\chi_1(0) = 2\gamma_0^{1/2} \frac{\gamma}{\kappa_0} \sqrt{1 + \frac{eE_\perp}{2\Lambda\gamma^2}} . \tag{1.19}$$

Firstly, we note that in the limit $E_\perp \to 0$, this equation transforms into the previously found (1.9). The direct dependence on $E_\perp$ becomes important only

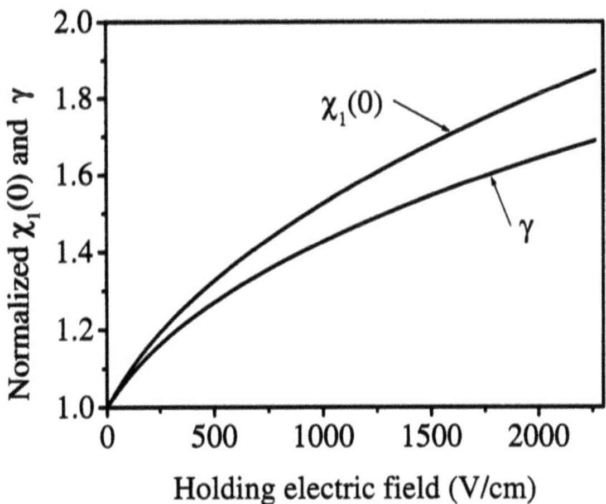

Fig. 1.6. Normalized parameters $\chi_1(0)$ and γ of the ground state wave function vs. the holding electric field $E_\perp$

for high holding fields, when $E_\perp \sim 2\Lambda\gamma_0^2/e \sim 3.46 \times 10^3$ V/cm. At the same time, the indirect dependence on the holding electric field, through $\gamma(E_\perp)$, is even more important. Figure 1.6 compares the holding field dependencies of $\chi_1(0)$ and γ which follow from (1.15) and (1.19). Generally, it should be noted that these dependencies are rather weak and, in the broad range considered, the normalized values of γ and $\chi_1(0)$ do not exceed 2.

Returning to the problem of the population of surface levels, let us assume that the electron density on the ground level is a bit less than the saturated value $n_{\max} = E_\perp/2\pi e$, which means that a small fraction of electrons are somewhere above the average distance $\langle 1|z|1\rangle$. In this case, the ground state surface electrons are affected by the holding field, which is close to the external field $E_\perp$, while the outer electrons are affected by a smaller field $E_\perp^\wedge = E_\perp - 2\pi e n_1$, where n_1 is the areal density of the ground state electrons. For simplicity we consider the model which consists of one surface level, positioned at $\varepsilon_1^{(\perp)} < 0$, and 'outer' electron states which are described by (1.11) with $E_\perp = E_\perp^\wedge$, and by the quasi-classical asymptote for $\zeta_l \simeq (3\pi l/2)^{2/3}$. The fraction of SEs on the ground level, viz.,

$$\frac{n_1}{n} = \frac{\exp\left(-\varepsilon_1^{(\perp)}/T\right)}{\sum_l \exp\left(-\varepsilon_l^{(\perp)}/T\right)}, \tag{1.20}$$

(where $n = N_e/S_A$ is the total electron density) can be represented in the form

$$\frac{n_1}{n} = \frac{1}{1 + \lambda(E_\perp^\wedge)\exp\left(\varepsilon_1^{(\perp)}/T\right)}, \tag{1.21}$$

where the function

$$\lambda(E_{\perp}^{\wedge}) \simeq \frac{T^{3/2}\sqrt{m}}{\sqrt{2}\pi\hbar e E_{\perp}^{\wedge}} \, . \tag{1.22}$$

is found by means of the Poisson summation formula. Since $E_{\perp}^{\wedge} = E_{\perp} - 2\pi e n_1$ depends on n_1, (1.21) is actually a self-consistent equation for the ratio n_1/n.

The solution of (1.21) can be found in the form [20]

$$\frac{n_1}{n} = \frac{1}{2}\left[1 + \frac{1+\delta}{\nu} - \sqrt{\left(1 + \frac{1+\delta}{\nu}\right)^2 - \frac{4}{\nu}}\right] , \tag{1.23}$$

where

$$\delta(E_{\perp}) = \frac{T^{3/2}\sqrt{m}}{\sqrt{2}\pi\hbar e E_{\perp}} \exp\left(\varepsilon_1^{(\perp)}/T\right) , \tag{1.24}$$

and $\nu = n/n_{\max}$. If the number of electrons is much less than the saturated value ($n \ll n_{\max}$), then (1.23) transforms into the simpler formula [19]

$$\frac{n_1}{n_{\mathrm{s}}} \simeq \frac{1}{1 + \delta(E_{\perp})} , \tag{1.25}$$

which shows that the relative number of outer electrons, $n_{\mathrm{out}}/n \simeq \delta(E_{\perp})$, becomes small if $\delta \ll 1$. At $T = 1\,\mathrm{K}$, this condition is fulfilled even for very weak fields $E_{\perp} \gg 4 \times 10^{-3}\,\mathrm{V/cm}$. At $E_{\perp} = 9\,\mathrm{V/cm}$, the ratio in (1.25) is shown in Fig. 1.3 by the dashed curve.

Under the saturation condition ($n = n_{\max}$, $\nu = 1$), according to (1.23), we have another type of asymptotic behavior $n_{\mathrm{out}}/n \simeq \sqrt{\delta(E_{\perp})}$ if $\delta \ll 1$. In this case, the ratio n_1/n vs. temperature is shown by the solid curve marked $E_{\perp} = 2\pi e n$. The holding field $E_{\perp} = 9\,\mathrm{V/cm}$ employed in this plot corresponds to a rather low electron density $n = 10^7\,\mathrm{cm}^{-2}$. It should be noted that for the typical value $n_{\max} \sim 10^8\,\mathrm{cm}^{-2}$, the holding electric field is much stronger, viz., $E_{\perp} \sim 100\,\mathrm{V/cm}$, substantially larger than the critical field $E_{\perp}^{*}$ determined by the condition $\delta(E_{\perp}^{*}) = 1$.

In the next chapter we shall see that correlation effects substantially increase the binding energy of the ground state electrons. Therefore, at temperatures and holding electric fields typically used in experiments with SEs, the population of the ground surface level is close to 100%. Interesting distributions of outer electrons are found in [23] for the opposite limiting case corresponding to the condition $\delta \gg 1$, which can be realized for electrons heated by a low-frequency driving electric field [24–26] or under the nonlinear CR condition.

1.3 Electrons Bound to a Helium Film

Owing to the van der Waals forces acting between a helium atom and the solid substrate, the superfluid helium film covers the walls of the experimental cell. The thickness of the helium film d decreases with height z according to the equipotential condition

$$-\frac{\alpha_{\rm He}}{d^3(z)} = m_{\rm He} G z \,, \tag{1.26}$$

where $\alpha_{\rm He}$ is the van der Waals constant and G is the acceleration due to gravity. The typical film thickness is about 3×10^{-6} cm. Generally, this film collects surface electrons as well.

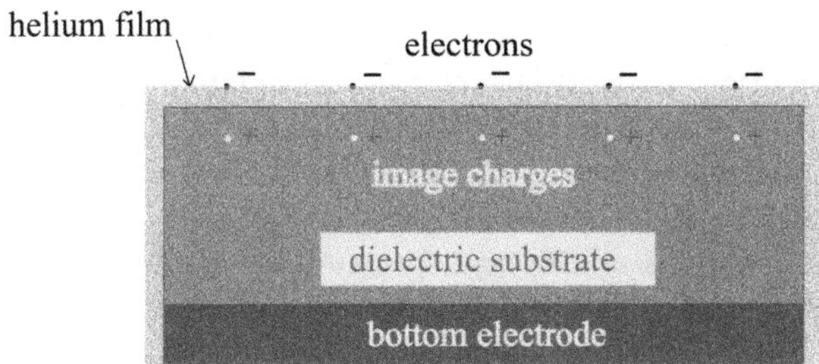

Fig. 1.7. Schematic view of SEs on a helium film and major image charges

The helium film can be placed inside a parallel plate capacitor, as shown in Fig. 1.7. In this case, the film thickness d is constant and the surface electrons collected on this film are mostly under uniform conditions. The potential energy of an electron above a liquid helium film ($z > 0$) can be found as a series of contributions from the infinite number of fictitious charges [27, 28]:

$$V_{\rm e}(z) = -\frac{\Lambda}{z} - \Lambda_{\rm s} \sum_{n=1}^{\infty} \frac{(-\lambda)^{n-1}}{z + nd} \,, \tag{1.27}$$

where

$$\Lambda_{\rm s} = \frac{e^2 \epsilon (\epsilon_{\rm s} - \epsilon)}{(1+\epsilon)^2 (\epsilon_{\rm s} + \epsilon)} \,, \qquad \lambda = \frac{(\epsilon - 1)(\epsilon_{\rm s} - \epsilon)}{(\epsilon + 1)(\epsilon_{\rm s} + \epsilon)} \,, \tag{1.28}$$

and $\epsilon_{\rm s}$ is the dielectric constant of the substrate. Because the dielectric constant of liquid helium is very close to unity, $(\epsilon - 1) \ll 1$, the parameter λ proportional to $(\epsilon - 1)$ is very small and one can disregard terms with $n \geq 2$. The major image charges are shown by hollow dots in Fig. 1.7.

For thick helium films with $d > 10^{-6}$ cm, the average distance $\langle 1|z|1 \rangle$ is much smaller than d, owing to the attraction force induced by the solid substrate. One can therefore expand the potential $V_{\rm e}(z)$ in z/d, which gives

$$V_e(z) \simeq -\frac{\Lambda_s}{d} - \frac{\Lambda}{z} + F_d z \,, \tag{1.29}$$

where $F_d = \Lambda_s/d^2$. The first term represents the main correction to the SE binding energy. For typical $d \simeq 3 \times 10^{-6}$ cm, it is very high ($\Lambda_s/d \approx 140$ K), if ϵ_s is substantially larger than unity. Therefore, at zero holding fields, the surface electrons of the massive liquid helium all leave its surface, charging the film which covers the walls of the experimental cell. If the film is placed inside the cell capacitor parallel to the metal electrodes, this correction guarantees a high population of the ground surface state.

The second term in (1.29) is the usual polarization attraction to liquid helium, while the third term is the potential of the image force acting on electrons because of the solid substrate. The effective holding field of the substrate $E_d = F_d/e$ depends strongly on the film thickness $E_d \propto d^{-2}$. As d varies from 10^{-5} to 10^{-6} cm, the effective holding field ranges from 300 to 3×10^4 V/cm. As the holding force eE_d is also the coupling parameter for electron interactions with surface excitations of liquid helium, a helium film can be used to study 2D electrons in the regime of intermediate and strong coupling with medium vibrations, which represents an interesting case for the polaron problem.

The wave functions and energy spectrum of an electron above a helium film can be found from the equations established for SEs in the presence of the holding electric field. In this case, one should substitute $F_d + eE_\perp$ for $eE_\perp$ and include the correction $-\Lambda_s/d$ to the binding energy. For example, in the limit of strong holding fields E_d, the energy spectrum of SEs can be written as

$$\varepsilon_l^{(\perp)} \simeq -\frac{\Lambda_s}{d} + \zeta_l \left(\frac{\hbar}{2m}\right)^{1/3} (F_d + eE_\perp)^{2/3} \,. \tag{1.30}$$

Since the first term has a very large absolute value, small variations in the film thickness $\delta d(\boldsymbol{r}) = d(\boldsymbol{r}) - d_0$ can cause localization of the SEs in a parallel plane. An accurate description of the effect of substrate irregularities requires knowledge of the interaction of bound electrons with an uneven interface, and this is not trivial. We shall discuss this problem in Sect. 1.8.

A solid substrate with a large dielectric constant ϵ_s reduces the strength of the electron–electron interaction due to the screening effect. A helium film covering a metal substrate ($\epsilon_s \to \infty$) represents an extreme example of such screening. In this case, just above the film surface the electric field produced by an electron is a mixture of the simple Coulomb fields of the electron and its image charge placed at $z = -2d$. As a result, electrons interact via the dipole potential [29, 30]

$$V(r) = e^2 \left[\frac{1}{r} - \frac{1}{\sqrt{r^2 + (2d)^2}}\right] \to \frac{2e^2 d^2}{r^3} \,, \tag{1.31}$$

if $d \ll r$. This strong reduction in the interaction potential affects the spectrum of collective excitations of the interface electron liquid and solid, and leads to an important conclusion with regard to the phase diagram of such systems. As pointed out by Peeters and Platzman [31, 32], in the low density range, such systems represent a 2D quantum liquid which remains in the liquid state even at zero temperature. This conclusion is due to the fact that the average potential energy per electron which follows from (1.31) decreases with the mean distance between electrons faster than the kinetic energy. This behavior is opposite to the behavior of the Coulomb liquid, which undergoes the Wigner solid transition at low densities if the temperature T is low enough.

Electrons collected on the helium film exert a pressure $P_{\rm el} = (F_d + eE_\perp) n_{\rm s}$ on its surface and this can reduce significantly the film thickness $d_0 \to d$ according to Etz et al. [33]. (Here d_0 is the initial thickness in the absence of charges.) The density dependence of the film thickness can be found by equating $P_{\rm el}$ with the pressure difference between levels $z = d$ and $z = d_0$ induced by the nonuniform van der Waals force in the part of the helium film which is not covered with electrons:

$$\frac{d}{d_0} = \frac{1}{(1 + eE_\perp n_{\rm s}/\rho G H)^{1/3}} ,$$

where H is the height above the bath surface level and the polarization part F_d is disregarded. The narrowing of the liquid helium film induced by SEs is very important for initially thick films ($d_0 > 300\,\text{Å}$), as shown in Fig. 1.8. In contrast, thin films with $d_0 < 200\,\text{Å}$ are much less affected by the surface charge because the electron density correction in the denominator of the formula given above is proportional to d_0^3, according to the relation $H = \alpha_{\rm He}/\rho G d_0^3$.

The additional liquid–solid interface allows one to create charged 'impurities' and even artificial atoms with large spatial separation of charges called diplons [34]. Free charges can penetrate liquid helium in the form of negative or positive ions of a complicated structure. In a simplified way, a negative ion in liquid helium can be described as a bubble of quite large radius $R_- \simeq 17\,\text{Å}$, while a positive ion can be imagined as a snowball of smaller radius $R_+ \simeq 7\,\text{Å}$. Such a big difference in the structures of negative and positive ions is caused by the high potential barrier for electron penetration into liquid helium and by the small value of the free electron mass. Therefore, for an electron, it is energetically favorable to be localized in a hollow sphere, reducing its interaction energy with a continuous medium. A positive charge (He^+) has no such barrier and, entering liquid helium, it attracts atoms, thereby increasing the liquid pressure in its vicinity according to the law $P \propto 1/R^4$. At $R < 7\,\text{Å}$ this pressure exceeds the critical pressure necessary for the solidification of liquid helium.

The liquid–solid interface gathers ions. Negative ions interfere with SEs on the helium film creating an interesting bilayer system which undergoes structural phase transitions of various kinds [35]. We shall discuss these effects in Sect. 7.7 in relation with the Wigner solid. A positive ion placed at the

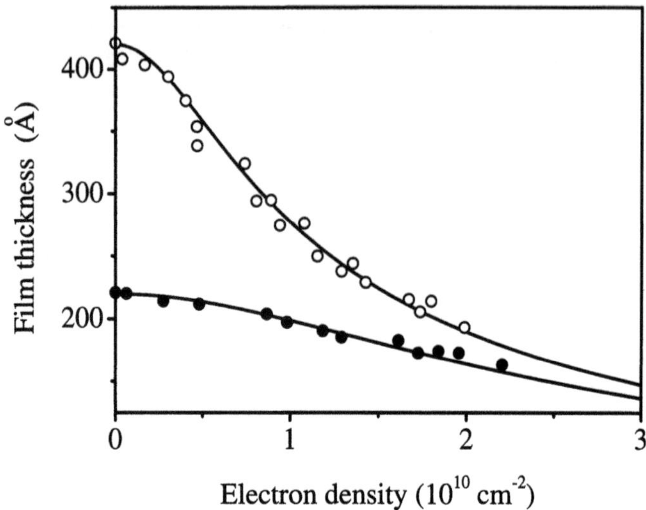

Fig. 1.8. Density dependence of a charged ^{4}He film wetting a glass substrate at $T = 1.6$ K [33]. The thickness d_0 of the uncharged films was 220 and 420 Å, respectively

bottom of the helium film creates a strong attractive potential for an SE moving along the film surface:

$$V(z,r) = -\frac{2}{\epsilon_s + 1}\frac{e^2}{\sqrt{(d+z)^2 + r^2}} \, . \tag{1.32}$$

This situation is shown in Fig. 1.9. Expanding this potential in powers of r/d and z/d, we have in the simplest approximation

$$V(r,z) \simeq -\frac{2e^2}{(\epsilon_s + 1)d} + eE_i z + \frac{1}{2} m\omega_d^2 r^2 \, , \tag{1.33}$$

where

$$E_i = \frac{2e}{(\epsilon_s + 1)d^2} \, , \qquad \omega_d^2 = \frac{2e^2}{(\epsilon_s + 1)m_e d^3} \, . \tag{1.34}$$

The first and second terms of (1.33) increase electron binding to the vapor–liquid interface, while the third term, which has an oscillatory form, causes electron localization in the plane of the surface with the localization radius $L_e = \sqrt{\hbar/m_e\omega_d}$. For typical parameters of the system, $\epsilon_s = 3$ and $d = 3.35 \times 10^{-6}$ cm (glass substrate), the level spacing $\hbar\omega_d \simeq 13.3\,\mathrm{K} \gg T$ and $L_e \simeq 0.77 \times 10^{-6}$ cm. Employing simple estimates, one finds that the electron and positive ion of the diplon cannot collapse, if the film thickness is rather thick. The condition $d > 0.7 \times 10^{-6}$ cm usually guarantees the stability of this dipole complex.

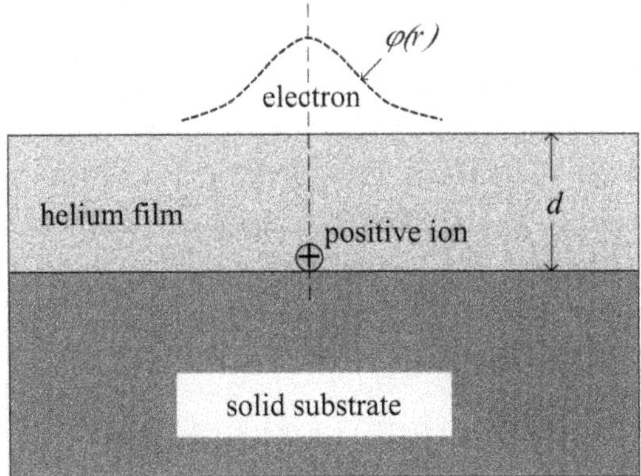

Fig. 1.9. Schematic view of the diplon structure

The system of positive (or even negative) ions at the liquid–solid interface and SEs on the free surface of the helium film opens the way to interesting physics if the solid substrate is smooth enough. For example, a Wigner solid of positive ions induces a periodic potential for an SE moving along the liquid helium surface:

$$V(\boldsymbol{r}) = \sum_{\boldsymbol{g}} V_g \exp\left(\mathrm{i}\boldsymbol{g} \cdot \boldsymbol{r}\right) , \qquad V_g = \frac{4\pi e^2 n_+}{(\epsilon_s + 1) g} \exp(-gd) . \tag{1.35}$$

This causes a mini-band structure in the SE spectrum above the helium film and creates an artificial metal. (Here $\boldsymbol{g}$ is the reciprocal lattice vector and n_+ the areal density of ions.) By changing the film thickness, one can vary the strength of this potential over a wide range. In the opposite limiting case, when SEs form a Wigner solid, a positive ion at the bottom of the helium film couples with an electron and creates a heavy impurity in the Wigner solid. Analogous bilayer systems of 2D charges can be created in semiconductor heterostructures.

1.4 Scattering by Vapor Atoms

For SEs on superfluid helium, the number of available scatterers (vapor atoms and ripplons) decreases with cooling. In this and the following sections, we introduce basic equations describing electron interaction with these two kinds of scatterer, which will be used frequently throughout this book.

Vapor atoms represent a nearly ideal example of short-range scatterers, which are very convenient for testing theoretical approaches to quantum magnetotransport in two dimensions. The 3D density of vapor atoms above the liquid helium surface varies exponentially with temperature:

$$n^{(\mathrm{a})}(T) = \left(\frac{M_{\mathrm{a}}T}{2\pi\hbar^2}\right)^{3/2} \exp\left(-\frac{Q}{T}\right) , \qquad (1.36)$$

where Q is the evaporation constant (for the ^{4}He isotope, we have $Q_4 \simeq 7.17\,\mathrm{K}$) and M_{a} is the mass of a helium atom. The number of surface excitations (ripplons) is described by the Bose distribution function $N_q^{(\mathrm{r})}$ which also changes fast with temperature. The interface excitations involved in the momentum relaxation of SEs ($q \leq 2k^{(\mathrm{el})}$) nevertheless have low energies, and their average number decreases linearly with temperature: $N_q^{(\mathrm{r})} \simeq T/\hbar\omega_q \gg 1$. For liquid ^{4}He, the characteristic temperature $T \sim 1\,\mathrm{K}$ separates two regimes where electron scattering is dominated solely by vapor atoms ($T > 1.2\,\mathrm{K}$) and surface excitations ($T < 0.7\,\mathrm{K}$).

The interaction potential of an electron and a helium atom has a hard repulsion core and a very weak polarization attraction. The latter can be disregarded when describing scattering events. Because all typical scales of the SE wave function are much larger than the hard core radius of the interaction potential and the electron–atom scattering length s_0, it is conventional to adopt the effective potential approximation

$$V(\boldsymbol{R} - \boldsymbol{R}_a) = V^{(\mathrm{a})}\delta(\boldsymbol{R} - \boldsymbol{R}_a) , \qquad V^{(\mathrm{a})} = 2\pi\hbar^2 s_0/m_{\mathrm{e}} . \qquad (1.37)$$

The many-body interaction Hamiltonian is usually written as

$$H_{\mathrm{int}}^{(\mathrm{e-a})} = \sum_e \sum_a V^{(\mathrm{a})}\delta(\boldsymbol{R}_e - \boldsymbol{R}_a)$$
$$= \frac{V^{(\mathrm{a})}}{\Omega_{\mathrm{v}}} \sum_e \sum_{\boldsymbol{K},\boldsymbol{K}'} \mathrm{e}^{-\mathrm{i}(\boldsymbol{q}-\boldsymbol{q}')\boldsymbol{r}_e - \mathrm{i}(k-k')z_e} a_{\boldsymbol{K}}^{\dagger} a_{\boldsymbol{K}'} , \qquad (1.38)$$

where $a_{\boldsymbol{K}}^{\dagger}$ and $a_{\boldsymbol{K}}$ are the creation and destruction operators of vapor atoms, $\boldsymbol{K} = \{\boldsymbol{q}, k\}$ is the 3D wave vector of atoms, with $\boldsymbol{q}$ chosen to be its in-plane component, and Ω_{v} is the volume of the helium vapor phase. The second line of (1.38) is written assuming that the vapor atoms are described by 3D plane waves, which results in a simple exponential form for the single-particle matrix elements of the interaction potential. For reasons of simplicity, we consider the vapor atoms to be ^{4}He atoms. However, the final result does not depend on the particular quantum statistics, because the vapor has low density.

Let us assume that all electrons are in the ground surface level. Then, foreseeing applications in transport theory, it is convenient to average the interaction Hamiltonian over the ground surface state $|1\rangle$ and rename the summation vector $\boldsymbol{K}' - \boldsymbol{K} \to \boldsymbol{K}$ [36]:

$$\langle 1|H_{\mathrm{int}}^{(\mathrm{e-a})}|1\rangle = \frac{V^{(\mathrm{a})}}{\Omega_{\mathrm{v}}} \sum_{\boldsymbol{q}} n_{-\boldsymbol{q}} A_{\mathrm{a},\boldsymbol{q}} , \qquad A_{\mathrm{a},\boldsymbol{q}} = \sum_k \langle 1|\mathrm{e}^{\mathrm{i}kz}|1\rangle \rho_{\boldsymbol{K}} , \qquad (1.39)$$

where

$$n_{\boldsymbol{q}} = \sum_{e} \mathrm{e}^{-\mathrm{i}\boldsymbol{q}\boldsymbol{r}_e} , \qquad \rho_{\boldsymbol{K}} = \sum_{\boldsymbol{K}'} a^{\dagger}_{\boldsymbol{K}'-\boldsymbol{K}} a_{\boldsymbol{K}'} , \tag{1.40}$$

are the density fluctuation operators of electrons and vapor atoms, respectively, written in different representations. In this notation, $\boldsymbol{q}$ is usually equal to the in-plane momentum exchange between an electron and the scatterer. The many-body operator of vapor atoms $A_{\mathrm{a},\boldsymbol{q}}$ represents a sort of projection of the 3D vapor atom system onto the plane of the 2D electron system. It plays the same role as the usual phonon operator $b_{\boldsymbol{q}} + b^{\dagger}_{-\boldsymbol{q}}$ for the electron–phonon interaction.

In the zero temperature limit, many-body calculations are often carried out by expanding the quantum-mechanical time evolution operator in a perturbation series and performing partial summations of the most important infinite sub-series for a certain averaged quantity. The evolution operator is generally represented in the so called time-ordered exponential form, containing an integral of the interaction operator $V(t)$ whose time dependence is governed by the unperturbed Hamiltonian H_0. An example of such a treatment will be given in Sect. 1.9.2 when discussing the collision broadening of Landau levels. Such a perturbation series for an averaged quantity contains smaller mathematical objects which represent the ground state average of the time-ordered product of the interaction operators: $\hat{T}[V(t_1)V(t_2)]$, where $\hat{T}$ is the conventional time-ordering operator. Thus the many-body perturbation theory contains small 'bricks' usually called Green's functions of scatterers:

$$\begin{aligned} D^{(0)}_{\mathrm{s}}(\boldsymbol{q}, t-t') &= -\mathrm{i}\left\langle \hat{T} A_{\mathrm{s},\boldsymbol{q}}(t) A_{\mathrm{s},-\boldsymbol{q}}(t') \right\rangle \\ &= \sum_{k} |\eta_k|^2 (-\mathrm{i}) \left\langle \hat{T} \rho_{\boldsymbol{K}}(t) \rho^{\dagger}_{\boldsymbol{K}}(t') \right\rangle , \end{aligned} \tag{1.41}$$

where $\eta_k = \langle 1|\mathrm{e}^{\mathrm{i}kz}|1\rangle$, and the subscript s labels the scatterers (s = a for vapor atoms and s = r for ripplons). The second line of (1.41) is written for s = a. In the perturbation series, use of the Green's function for operators $A_{\mathrm{a},\boldsymbol{q}}$ is similar to use of the usual phonon Green's function for operators $A_{\mathrm{r},\boldsymbol{q}} = b_{\boldsymbol{q}} + b^{\dagger}_{\boldsymbol{q}}$.

The operator $A_{\mathrm{a},\boldsymbol{q}}(t)$ contains a time-dependent factor

$$\exp\left(-\mathrm{i}\Delta_{\boldsymbol{K}',\boldsymbol{K}'-\boldsymbol{K}} t\right) ,$$

where

$$\hbar \Delta_{\boldsymbol{K},\boldsymbol{M}} = \varepsilon^{(\mathrm{a})}_{K} - \varepsilon^{(\mathrm{a})}_{M} \tag{1.42}$$

is the energy exchange or the excitation spectrum of the vapor system. Straightforward evaluation of (1.41) for the ideal gas of vapor atoms yields

$$\begin{aligned} D^{(0)}_{\mathrm{a}}(\boldsymbol{q}, t) &= -\mathrm{i} \sum_{k} |\eta_k|^2 \sum_{\boldsymbol{K}'} \exp\left(-\mathrm{i}\Delta_{\boldsymbol{K}',\boldsymbol{K}'-\boldsymbol{K}} t\right) \\ &\times \left[\theta(t) N^{(\mathrm{a})}_{\boldsymbol{K}'} \left(N^{(\mathrm{a})}_{\boldsymbol{K}'-\boldsymbol{K}} + 1\right) + \theta(-t) N^{(\mathrm{a})}_{\boldsymbol{K}'-\boldsymbol{K}} \left(N^{(\mathrm{a})}_{\boldsymbol{K}'} + 1\right)\right] , \end{aligned} \tag{1.43}$$

where $N_{\boldsymbol{K}}^{(\mathrm{a})}$ is the distribution function of the vapor atoms. Although the Green's function $D_{\mathrm{a}}^{(0)}(q,t)$ is determined by two-particle processes, it is reminiscent of the phonon Green's function $D_{\mathrm{r}}^{(0)}(q,t)$, which will be discussed in Sect. 1.5.1. Equation (1.43) can be significantly simplified if one takes into account the fact that $N_{\boldsymbol{K}}^{(\mathrm{a})} \ll 1$ (dilute gas) and the momentum exchange K is much smaller than thermal values of vapor atom wave numbers K'. Then renaming the summation vector in the second term, one arrives at

$$D_{\mathrm{a}}^{(0)}(\boldsymbol{q},t) \simeq -\mathrm{i}\sum_{k}|\eta_k|^2\sum_{\boldsymbol{K}'}N_{\boldsymbol{K}'}^{(\mathrm{a})}\exp(-\mathrm{i}\Delta_{\boldsymbol{K}',\boldsymbol{K}'-\boldsymbol{K}}t)\,. \tag{1.44}$$

In the perturbation series and in the electron self-energy equation, we shall deal with the Fourier-transformed function

$$D_{\mathrm{a}}^{(0)}(\boldsymbol{q},\omega) = -\mathrm{i}2\pi\sum_{k}|\eta_k|^2\sum_{\boldsymbol{K}'}N_{\boldsymbol{K}'}^{(\mathrm{a})}\delta(\omega-\Delta_{\boldsymbol{K}',\boldsymbol{K}'-\boldsymbol{K}})\,. \tag{1.45}$$

This equation will be used frequently later to describe the collision broadening of Landau levels induced by vapor atoms, and to analyse more complicated phenomena caused by the strong internal forces in the Coulomb liquid.

The Green's function of scatterers $D_{\mathrm{a}}^{(0)}(q,\omega)$ has singularities at frequencies equal to the excitation spectrum of the vapor atom system $\omega = \Delta_{\boldsymbol{K}',\boldsymbol{K}'-\boldsymbol{K}}$. Later we will see that this is a quite general property of the Fourier transforms of the correlation functions which involve density operators taken at different times, as indicated in (1.41), because they relate to the dynamical structure factors of certain systems.

In the laboratory reference frame, $D_{\mathrm{s}}^{(0)}(q,\omega)$ for an isotropic system depends only on the absolute value of the momentum exchange $\boldsymbol{q}$. When studying nonlinear phenomena or the effect of strong internal forces, we have to consider electrons with orbit centers (under a magnetic field) moving fast with respect to scatterers. In this case, we need to know the form of the Green's function of scatterers in a moving reference frame, where it depends on the direction of $\boldsymbol{q}$. Consider the frame F' relative to which the center-of-mass frame of scatterers F is moving as a whole with a constant velocity $\boldsymbol{V}$. The vapor atom radius vectors defined in different reference frames are related by $\boldsymbol{R}_{\mathrm{a}}' = \boldsymbol{R}_{\mathrm{a}} + \boldsymbol{V}t$, and the density fluctuation operator is $\rho_{\boldsymbol{K}}' = \exp(-\mathrm{i}\boldsymbol{K}\cdot\boldsymbol{V}t)\rho_{\boldsymbol{K}}$. Therefore when we change the reference frame $F \to F'$, we have the following transcription rule:

$$D_{\mathrm{s}}'(\boldsymbol{q},\omega) = D_{\mathrm{s}}^{(0)}(q,\omega-\boldsymbol{q}\cdot\boldsymbol{V})\,. \tag{1.46}$$

Physically, this means that the frequency argument of the proper correlation function acquires a Doppler shift. It is also consistent with the conventional rule which states that the excitation spectrum of the system or the energy exchange acquires the Doppler shift $\hbar\boldsymbol{q}\cdot\boldsymbol{V}$ when we change reference frame.

This affects the position of the singularities (or maxima) of the corresponding Green's function in the right way.

The presentation of the interaction Hamiltonian in the form (1.39) contains the electron density operator $n_{\boldsymbol{q}}$, which is very useful for the description of Coulombic effects in the electron transport theory. As we shall see in Chap. 2, this property of the interaction Hamiltonian allows one to express the effective collision frequency of the conductivity tensor in terms of the equilibrium density–density correlation function of the electron system or the dynamical structure factor $S(q, \omega)$.

Transport properties of electrons interacting with motionless scattering centers of different nature can be described in a similar way by introducing fictitious heavy atoms with $M_{\mathrm{a}} \to \infty$, and by using an appropriate electron–atom interaction potential $V(R)$. In this case, the constant $V^{(\mathrm{a})}$ in the interaction Hamiltonian of (1.38) should be replaced by the corresponding momentum exchange function $V_K^{(\mathrm{a})}$, where $K = \sqrt{q^2 + k^2}$. The necessary transport properties can be found from the final conductivity equations, using the limiting condition $M_{\mathrm{a}} \to \infty$.

1.5 Electron Scattering at an Uneven Interface

1.5.1 Capillary Wave Quanta (Ripplons)

At low temperatures $T < 0.7\,\mathrm{K}$, the vapor density becomes extremely low and the mobility of SEs is limited by their interaction with ripplons or, if SEs are formed on the helium film, with substrate irregularities. Ripplons represent a sort of 2D phonon with an unusual spectrum (ω_q) which is well established in the long-wavelength range $(q \ll 10^8\,\mathrm{cm}^{-2})$:

$$\omega_q^2 = \frac{\alpha}{\rho}(q^2 + \kappa^2) q \tanh(qd) \,, \tag{1.47}$$

where

$$\kappa^2 = \frac{\rho}{\alpha}(G + G_d) \,, \qquad G_d = \frac{3\alpha_{\mathrm{He}}}{\rho d^4} \,,$$

α is the surface tension, ρ is the mass density of liquid helium, d is the thickness of the liquid helium film, and G_d is the additional acceleration induced by van der Waals forces. For thick liquid helium films and $q \gg \kappa$, medium vibrations are described by the capillary wave dispersion $\omega_q = \sqrt{\alpha/\rho}q^{3/2}$. The wave number q of ripplons taking part in electron scattering is usually of the order of the thermal electron wave number, $q \sim 2k_T = \sqrt{8m_{\mathrm{e}}T}/\hbar$, which is much smaller than the atomic wave numbers of liquid helium, and smaller even than the thermal ripplon wave number q_{T}. In the presence of a high magnetic field, the wave number of ripplons involved in scattering is restricted by another condition, $q \sim \sqrt{2}/l_B$, where $l_B \propto 1/\sqrt{B}$

is the electron magnetic length. In any case, ripplons involved in electron scattering belong to the long-wavelength part of the capillary spectrum which is known to high accuracy.

In order to describe electron scattering by these 2D phonons, one has to represent the helium surface displacement operator $\xi(\boldsymbol{r})$ or its Fourier transformation $\xi_{\boldsymbol{q}}$ in terms of the many-body Bose creation and destruction operators, $b_{\boldsymbol{q}}^{\dagger}$ and $b_{\boldsymbol{q}}$. In other words, we need to quantize the capillary waves. In the harmonic approximation, the ripplon Hamiltonian can be written as [27,37]

$$H_{\mathrm{r}} = \frac{\rho}{2} \int \boldsymbol{v}^2(\boldsymbol{R}) \mathrm{d}^2\boldsymbol{r} \mathrm{d}z + \int \mathcal{F}(|\boldsymbol{r}' - \boldsymbol{r}|) \xi(\boldsymbol{r}) \xi(\boldsymbol{r}') \mathrm{d}^2\boldsymbol{r}' \mathrm{d}^2\boldsymbol{r} , \tag{1.48}$$

where $\boldsymbol{v}(\boldsymbol{R}) = \nabla\Phi$ is the velocity field of the liquid helium induced by surface displacements and $\mathcal{F}(|\boldsymbol{r}' - \boldsymbol{r}|)$ is an unknown structure factor. The hydrodynamic potential Φ is found from the Laplace equation

$$\nabla^2 \Phi = 0 , \tag{1.49}$$

with boundary conditions

$$\left(\frac{\partial \Phi}{\partial z}\right)_{z=0} = \dot{\xi} , \qquad \left(\frac{\partial \Phi}{\partial z}\right)_{z=-d} = 0 . \tag{1.50}$$

Introducing the Fourier-transformed function $\Phi_{\boldsymbol{q}}(z)$, the required solution can be found as

$$\Phi_{\boldsymbol{q}}(z) = \dot{\xi}_{\boldsymbol{q}} \frac{\cosh[q(z+d)]}{q \sinh(qd)} . \tag{1.51}$$

Substituting this solution into (1.48), the ripplon Hamiltonian can be represented in the canonical form

$$H_{\mathrm{r}}^{(0)} = \frac{1}{2S_{\mathrm{A}}} \sum_{\boldsymbol{q}} \left(\frac{1}{\mu_q} |\pi_{\boldsymbol{q}}|^2 + \mu_q \omega_q^2 |\xi_{\boldsymbol{q}}|^2 \right) , \tag{1.52}$$

where

$$\mu_q = \frac{\rho}{q \tanh(qd)} , \qquad \pi_{\boldsymbol{q}} = \mu_q \dot{\xi}_{\boldsymbol{q}} , \tag{1.53}$$

and, according to quantum mechanics, the commutator $[\pi_{\boldsymbol{q}}, \xi_{-\boldsymbol{q}'}] = -\mathrm{i}\hbar\delta_{\boldsymbol{q},\boldsymbol{q}'}$. The spectrum of ripplons ω_q and the Fourier transformation $\mathcal{F}_q$ of the structure factor $\mathcal{F}(|\boldsymbol{r}' - \boldsymbol{r}|)$ are related by

$$\omega_q^2 = 2\mathcal{F}_q / \mu_q . \tag{1.54}$$

The standard procedure for introducing Bose operators $b_{\boldsymbol{q}}^{\dagger}$ and $b_{\boldsymbol{q}}$ allows one to represent the displacement operator ξ in the form [27]

$$\xi(\boldsymbol{r}) = \frac{1}{\sqrt{S_{\mathrm{A}}}} \sum_{\boldsymbol{q}} A_{\mathrm{r},\boldsymbol{q}} Q_q \exp(\mathrm{i}\boldsymbol{q} \cdot \boldsymbol{r}) , \tag{1.55}$$

where

$$A_{\mathrm{r},q} = b_{\boldsymbol{q}} + b^{\dagger}_{-\boldsymbol{q}} \,, \qquad Q_q = \sqrt{\frac{\hbar}{2\mu_q \omega_q}} = \sqrt{\frac{\hbar q \tanh(qd)}{2\rho\omega_q}} \,. \tag{1.56}$$

The factor $\tanh(qd)$ entering the definition of Q_q is very important for the description of SE transport above a liquid helium film. In most cases, it compensates the same factor entering the ripplon dispersion [see (1.47)], when these functions appear in the combination $Q_q^2 N_q^{(\mathrm{r})} \simeq Q_q^2 T/(\hbar\omega_q)$. (Here $N_q^{(\mathrm{r})}$ is the ripplon distribution function.) Disregarding this factor in the displacement operator, one can arrive at an incorrect mobility equation.

The ripplon Green's function, which enters the perturbation series for the properties of electrons interacting with medium vibrations, is conventionally defined as

$$D_{\mathrm{r}}^{(0)}(q, t - t') = -\mathrm{i} \left\langle \hat{T} A_{\mathrm{r},q}(t) A_{\mathrm{r},q}(t') \right\rangle \,. \tag{1.57}$$

This function has the usual form of the phonon Green's function [38]:

$$D_{\mathrm{r}}^{(0)}(q, t - t') = -\mathrm{i} \left[(N_q + 1)\, \mathrm{e}^{-\mathrm{i}\omega_q |t-t'|} + N_q \mathrm{e}^{\mathrm{i}\omega_q |t-t'|} \right] \,. \tag{1.58}$$

The Fourier-transformed function $D_{\mathrm{r}}^{(0)}(q, \omega)$ can be found by recalling that the Fourier transformation of the unit step function $\theta(t)$ is

$$\theta_\omega = \frac{\mathrm{i}}{\omega + \mathrm{i}0} \,. \tag{1.59}$$

We shall frequently employ the relation

$$\mathrm{Im} D_{\mathrm{r}}^{(0)}(q, \omega) = -\pi (2N_q + 1) \left[\delta(\omega - \omega_q) + \delta(\omega + \omega_q) \right] \,. \tag{1.60}$$

For example, in Sect. 1.9 we shall see that it is very convenient to describe collision broadening of Landau levels, if the SEs are subject to a magnetic field.

1.5.2 Bloch Approach for Bound Electrons

In this section we shall describe the interaction Hamiltonian of an electron, bound to the interface, with interface irregularities like surface roughness. The unevenness of the interface can be caused by surface excitations of liquid helium (ripplons), or it can be static surface roughness if SEs are formed on a solid cryogenic substrate. We focus in particular on description of electron–ripplon scattering for SEs above superfluid helium since this cannot be considered as a simple extension of the conventional approaches developed for the electron–phonon system.

At first glance, the description of the electron–ripplon interaction and electron scattering cannot cause any problem. The standard procedure consists in finding the potential energy of an electron above an uneven helium surface as a functional of the surface displacement $\xi(\mathbf{r})$, expanding it in a series if the displacements are small, and replacing $\xi(\mathbf{r})$ by the quantum mechanical operator according to (1.55). Conventionally, the interaction Hamiltonian V_{int} is defined as a correction to the electron potential energy that is linear in ξ. In the Born approximation, the probability of scattering between states $\langle j|$ and $\langle j'|$ is proportional to $|\langle j| V_{\text{int}} |j'\rangle|^2$. However, historically, the first attempts to perform this procedure in the usual way, employing the Bloch approach [39] and (later on) the adiabatic approximation [40], all failed to describe SE transport at low temperatures, resulting in mobility equations which disagree with experiment even on a qualitative level. The electron–ripplon collision rate evaluated in these theories appeared to be more than one order of magnitude larger than what was observed experimentally.

Problems arise when seeking an accurate description of electron–ripplon scattering because small interface displacements are naturally directed along the strong binding between SEs and liquid helium (z-direction). In this direction, the electron potential energy has a complicated sharp structure: in a narrow range in the vicinity of the interface, the potential energy changes from relatively small negative values to the huge positive values typical in atomic physics. Therefore in the conventional procedure, a small inaccuracy in the model potential or in the electron wave function $\chi_l(z)$ can cause a serious error in matrix elements describing electron scattering. The proper perturbation theory for interface electrons should be rearranged in a way that makes the matrix elements describing electron scattering weakly dependent on the actual electron potential near the interface and on the actual behavior of the electron wave function in this range. Such a theory of electron–ripplon scattering was first introduced with the assumption that the electron potential inside the liquid helium $V_0 = \infty$ [27]. The theory was then extended to any finite value of V_0 within the framework of the adiabatic approximation [41]. Nowadays, the predictions and results of the theory of electron–ripplon scattering agree well with the available experimental data, describing electron transport along the interface quite accurately down to ultra-low temperatures.

In order to make the problem clear, let us consider the potential energy of an electron in the presence of surface distortions $\xi(\mathbf{r})$. By analogy with the flat surface case, the potential energy of the electron $V_{\text{e}}(z, \{\xi_{\mathbf{q}}\})$ can be written as a sum of three major terms corresponding to the hard-core repulsion, polarization attraction and holding electric field:

$$V_{\text{e}}(z, \{\xi_{\mathbf{q}}\}) = V_{\text{rep}}(z - \xi) + V_{\text{att}}(z, \{\xi_{\mathbf{q}}\}) + eE_{\perp} z \,. \tag{1.61}$$

This is a quite general form for the interaction potential. The only assumption required to obtain it is that the repulsion term should have a short-range nature and act on the SE as the local potential $V_{\text{rep}}(z, \{\xi_{\mathbf{q}}\}) = V_{\text{rep}}(z - \xi)$ of an uneven potential wall. To avoid perturbations in the boundary conditions, it is supposed that $\theta(z)$, entering (1.1) and the definition of $V_{\rm rep} = V_0\theta(\xi - z)$, is a slightly smoothed unit step function corresponding to the real liquid helium density profile at the surface. The attractive term $V_{\rm att}(z, \{\xi_{\boldsymbol q}\})$ is a functional of surface displacements: the displacement function $\xi(\boldsymbol r)$ enters into an integrand expression which sums up the polarization interaction potentials of an electron with liquid helium atoms. At $\xi = 0$, the potential $V_{\rm e}(z, \{\xi_{\boldsymbol q}\}) \to V_{\rm e}(z, 0)$ transforms into $V_{\rm e}(z)$ of (1.1) or (1.17) introduced for the flat surface.

The first important conclusion that comes out of (1.61) is that the holding electric field term $eE_\perp z$ does not enter the perturbation potential

$$\begin{aligned} V_{\rm int} &= V_{\rm e}(z, \{\xi_{\boldsymbol q}\}) - V_{\rm e}(z, 0) \\ &= V_{\rm rep}(z - \xi) - V_{\rm rep}(z) + V_{\rm att}(z, \{\xi_{\boldsymbol q}\}) - V_{\rm att}(z, 0) \, . \end{aligned} \tag{1.62}$$

Therefore, expanding $V_{\rm int}$ in ξ in the usual way, one cannot arrive at an electron–ripplon coupling proportional to the holding force $eE_\perp$. However, we shall see that in the strict theory the electron–ripplon coupling contains a term proportional to $eE_\perp$ which plays an important role in electron transport along the interface. The crucial point of this theory is that the SE, moving freely along the surface, is firmly bound to the helium surface in the perpendicular direction, and the perturbation theory should take this effect into account in a general way before any approximation for the SE potential $V_{\rm e}(z)$ and the wave function $\chi_l(z)$ is used.

Firstly, we note that all models of the SE potential and wave function should satisfy the basic property of the bound state $\langle 1|F_z|1\rangle = 0$ (F_z is the force acting on an SE in the z-direction), as discussed in Sect. 1.2, where it was written in the form (1.16). This condition implies the absence of an electron current in the vertical direction, i.e., $\langle 1|p_z|1\rangle = 0$. For example, from this condition one finds the very instructive relation

$$\left\langle 1 \left| \frac{\partial V_{\rm rep}(z)}{\partial z} + \frac{\partial V_{\rm att}(z,0)}{\partial z} \right| 1 \right\rangle = -eE_\perp \, , \tag{1.63}$$

which was used in Sect. 1.2 to determine the holding field dependence of $\chi_1(0)$. This equation shows how the matrix element of the interaction potential given in (1.62), which does not depend directly on $E_\perp$, can depend linearly on the holding electric field.

In order to bring the above-mentioned property of the SE states into the perturbation treatment of the interaction potential $V_{\rm int}$, we add and subtract an artificial potential $V_{\rm e}(z - \xi, 0)$. The reason is that the linear expansion term $-\xi(\boldsymbol r)\partial V_{\rm e}(z,0)/\partial z$ of this artificial potential has zero matrix elements according to the condition $\langle 1|F_z|1\rangle = 0$. Then a simple rearrangement of different terms in the interaction potential yields

$$V_{\rm int} = eE_\perp\xi + V_{\rm att}(z, \{\xi_{\boldsymbol q}\}) - V_{\rm att}(z - \xi, 0) + \left[V_{\rm e}(z - \xi, 0) - V_{\rm e}(z, 0)\right] \, . \tag{1.64}$$

It should be noted that the last two terms in square brackets cannot contribute to the linear interaction Hamiltonian owing to the basic property of bound states mentioned above. Thus, the linear interaction potential is reduced to the form

$$V_{\mathrm{int}} = eE_{\perp}\xi + V_{\mathrm{att}}(z,\{\xi_{\boldsymbol{q}}\}) - V_{\mathrm{att}}(z-\xi,0)\,, \tag{1.65}$$

which should be used as a starting point for a practical evaluation of bound electron scattering at an uneven interface. The treatment discussed here eliminates the repulsion interaction potential because of its local nature, and strongly reduces the polarization interaction with long-wavelength displacements because, in the long-wavelength limit $q \ll \gamma$, the attractive potential asymptotically attains the quasi-local form $V_{\mathrm{att}}(z,\xi) \approx V_{\mathrm{att}}(z-\xi,0)$. At the same time, this treatment yields a coupling term proportional to the holding electric field $eE_{\perp}\xi$. This term means physically that a strongly bound SE follows surface distortions $[z = \xi(\boldsymbol{r})]$ when moving along the interface, and this changes its potential energy in the holding electric field accordingly: $V(\boldsymbol{r}) = eE_{\perp}\xi(\boldsymbol{r})$.

Depending on the interface system considered, one has to be more specific about the attractive potential $V_{\mathrm{att}}(z,\{\xi_{\boldsymbol{q}}\})$ in order to proceed with further calculations. For the SE system above liquid helium (or above a solid cryogenic substrate with weak atomic polarizability), the polarization attraction term can be written as an integral contribution from all liquid helium (cryogenic substrate) atoms:

$$V_{\mathrm{att}}(z,\{\xi_{\boldsymbol{q}}\}) = -\frac{\Lambda}{\pi}\int \mathrm{d}^3\boldsymbol{R}'\,\frac{\theta(\xi'-z')}{|\boldsymbol{R}'-\boldsymbol{R}|^4}\,, \tag{1.66}$$

where $\xi' = \xi(\boldsymbol{r}')$ and, as stated above, $\theta(z)$ is the slightly smoothed unit step function. If $\xi \to 0$, (1.66) transforms into the usual image potential $-\Lambda/z$. The perturbation of the polarization potential employed by Cole [39], i.e.,

$$V_{\mathrm{att}}(z,\{\xi_{\boldsymbol{q}}\}) - V_{\mathrm{att}}(z,0) = -\frac{\Lambda}{\pi}\int \mathrm{d}^2\boldsymbol{r}'\int_0^{\xi(\boldsymbol{r}')}\frac{\mathrm{d}z'}{[(\boldsymbol{r}'-\boldsymbol{r})^2+(z'-z)^2]^2}\,, \tag{1.67}$$

is very strong. At the same time, according to the perturbation theory established for bound electrons, the interaction potential of (1.65) yields a substantially different result for the polarization term, i.e.,

$$V_{\mathrm{att}}(z,\{\xi_{\boldsymbol{q}}\}) - V_{\mathrm{att}}(z-\xi,0) = -\frac{\Lambda}{\pi}\int \mathrm{d}^2\boldsymbol{r}'\int_{\xi(\boldsymbol{r})}^{\xi(\boldsymbol{r}')}\frac{\mathrm{d}z'}{[(\boldsymbol{r}'-\boldsymbol{r})^2+(z'-z)^2]^2}\,. \tag{1.68}$$

This is much smaller than (1.67) and, more importantly, it vanishes in the long-wavelength limit $\xi(\boldsymbol{r}') - \xi(\boldsymbol{r}) \to 0$. The latter means physically that shifting the flat surface up and down as a whole $[\xi(\boldsymbol{r}) = \mathrm{const.}]$ cannot change the energy of bound electrons if $E_{\perp} = 0$.

Disregarding $z' \sim \xi$ as compared to $z \sim \gamma^{-1}$, the linear terms of the electron–ripplon interaction potential can finally be written in the form [27, 41]

$$\langle 1|\, V_{\rm int}\, |1\rangle = \frac{1}{\sqrt{S_{\rm A}}} \sum_{\boldsymbol{q}} \xi_q V_q \exp(i\boldsymbol{q}\cdot\boldsymbol{r}) \,, \qquad V_q = e(E_\perp + E_q) \,, \tag{1.69}$$

where

$$eE_q = \left\langle 1 \left| \frac{\Lambda q}{z} \left[\frac{1}{qz} - K_1(qz) \right] \right| 1 \right\rangle \,, \tag{1.70}$$

ξ_q is the linear combination of the Bose creation and destruction operators according to (1.55), and $K_n(x)$ is the modified Bessel function of the second kind. The second term in the brackets of (1.70), containing $K_1(qz)$, corresponds to Cole's interaction potential. The first term of (1.70) originates from the repulsion term due to the basic property of bound states $\langle 1|F_z|1\rangle = 0$. In the limiting case $qz \ll 1$, we have $K_1(qz) \simeq 1/qz$, and the terms in the square brackets compensate each other significantly. The form of (1.69) and (1.70) was also reproduced in [22, 42].

It is very instructive to point out the qualitative changes in the polarization interaction appearing for bound electrons. If electrons were not bound to the interface, as shown in Fig. 1.10a, the polarization interaction would be negative (for $\xi > 0$) and very strong in accordance with Cole's result

$$eE_q^{\rm (free)} = -\left\langle 1 \left| \frac{\Lambda q}{z} K_1(qz) \right| 1 \right\rangle \simeq -\Lambda \left\langle 1/z^2 \right\rangle = -2\Lambda\gamma^2 \,, \tag{1.71}$$

where we have assumed that $qz \ll 1$. For $\xi > 0$, the negative sign of the polarization interaction is caused by the appearance of additional atoms attracting electrons according to Fig. 1.10a. For long-wavelength displacements, the polarization interaction does not depend on q and becomes equivalent to a fictitious holding field term $V_{\rm int} \simeq -2\Lambda\gamma^2\xi(\boldsymbol{r})$. Figure 1.10 is a qualitative graph: for thermal vibrations of the liquid helium surface, the amplitude of $\xi(\boldsymbol{r})$ is much less than γ^{-1}. If electrons are bound to the interface, as shown in Fig. 1.10b, the polarization interaction changes sign (!) and becomes positive because, for $\xi > 0$, there are fewer liquid helium atoms near an electron than on the flat surface. The deficiency of attractive helium atoms depends rather on the curvature of the surface displacement than on its amplitude, which introduces the strong q-dependence of the polarization interaction according to (1.70). The q-dependence of eE_q is crucial for the electron–ripplon coupling because it determines the temperature dependence of the electron mobility. Later we shall see that the polarization interaction given by (1.67) and (1.71) yields temperature-independent mobility, while in the theory established for bound electrons the polarization term results in an electron mobility which increases with cooling, approximately as $\mu(T) \propto 1/T$.

For typical wave numbers of surface excitations involved in scattering of electrons $q \sim k_T$, the polarization term eE_q of (1.70) cannot be completely disregarded. It gives a substantial contribution even at rather high electron densities. To find the matrix element $\langle 1| \cdots |1\rangle$, we can use the electron wave function of (1.6) with $\gamma(E_\perp)$ defined by the variational method according to (1.13). Then direct evaluation of the integral over z yields [43]

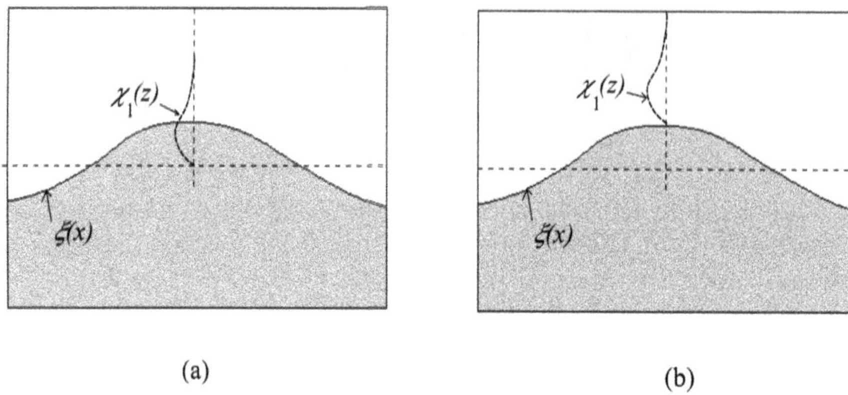

Fig. 1.10. SE on an uneven interface. The diagram shows the difference in the polarization interaction for electrons which do not follow the interface displacement (**a**) and for bound electrons (**b**)

$$eE_q = \frac{\Lambda q^2}{2} w_\mathrm{C}\left(\frac{q^2}{4\gamma^2}\right) , \tag{1.72}$$

where

$$w_\mathrm{C}(x) = -\frac{1}{1-x} + \frac{1}{(1-x)^{3/2}} \ln\left(\frac{1+\sqrt{1-x}}{\sqrt{x}}\right) . \tag{1.73}$$

At $T = 0.5\,\mathrm{K}$, the polarization term is approximately equivalent to the effective holding field of strength 230 V/cm, which is much less than the average polarization field $2\Lambda\gamma^2/e \sim 3.5 \times 10^3\,\mathrm{V/cm}$ acting on the SEs. In the long-wavelength limit, $w_\mathrm{C}(x) \simeq w_\mathrm{LT}(x) = 0.5 \ln(4/x) - 1$, which is valid only at ultra-low temperatures ($T < 0.1\,\mathrm{K}$ or $x < 0.01$). At medium temperatures $T > 0.5\,\mathrm{K}$, one can use the simple interpolation formula $w_\mathrm{C}(x) \simeq 1/(3\sqrt{x})$, which is remarkably close to the exact function over a wide range of values of x. The validity ranges of these two approximations are shown in Fig. 1.11. Thus, at low temperatures, when $\sqrt{x} = k_T/\gamma < 0.15$, the polarization interaction term eE_q is approximately proportional to q^2. At higher temperatures $x = k_T/\gamma > 0.3$ (for liquid helium this occurs at $T > 0.5\,\mathrm{K}$), we have a weaker dependence $eE_q \propto q$.

It should be noted that, although numerical evaluations of the electron conductivity with the exact form (1.73) of $w_\mathrm{C}(x)$ are quite simple nowadays, the asympotic and interpolative forms given above are very useful for obtaining instructive analytic results. The advantage of presenting the electron–ripplon interaction in the form of (1.69) and (1.70) is that V_q does not depend much on the detailed behavior of the electron wave function at the surface $z = 0$, and therefore the approximation of an infinite potential barrier ($V_0 \to \infty$, $z_0 \to 0$) at the helium surface gives accurate results.

The electron–ripplon interaction Hamiltonian can be rewritten similarly to the electron–vapor atom interaction Hamiltonian [(1.39)], so that the scat-

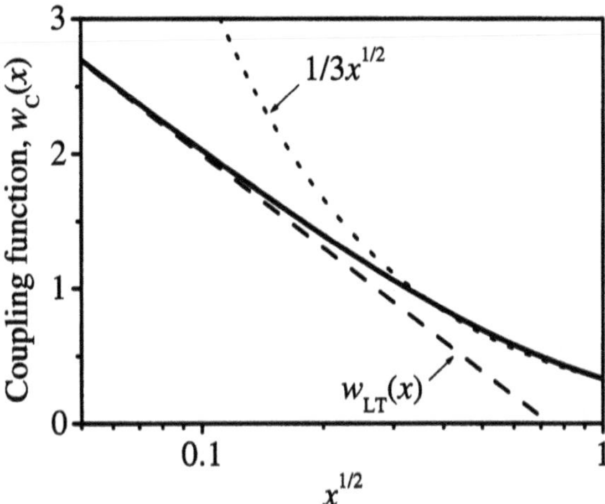

Fig. 1.11. Electron–ripplon coupling function $w_C(x)$ and its approximations $w_{LT}(x)$ (*dashed line*) and $1/3\sqrt{x}$ (*dotted line*) vs. $\sqrt{x}$

tering of electrons on the ground surface level can be described in a universal form by

$$H_{\text{int}} = \sum_{\text{s}=\text{a,r}} \sum_{q} U_{\text{s}} n_{-q} A_{\text{s},q} \,. \tag{1.74}$$

For ripplons (s = r), we use the following notation: $U_{\text{r}} = V_q Q_q / \sqrt{S_{\text{A}}}$ is the electron–ripplon coupling and $A_{\text{r},q} = b_q + b^{\dagger}_{-q}$. For vapor atoms (s = a), $U_{\text{a}} = V^{(\text{a})}/\Omega_{\text{v}}$, and $A_{\text{a},q}$ is as defined previously in (1.39). Thus, owing to (1.74), for both kinds of available scatterer, electron transport over the liquid helium surface can be treated in similar ways.

When discussing the matrix elements of the interaction Hamiltonian, we have confined ourselves to the linear terms which determine the probability of one-ripplon scattering processes: an electron emits or absorbs one quantum of surface waves. These terms are quite sufficient to describe linear electron transport along the interface or find the electron momentum relaxation rate. As we shall see in Chap. 3, to describe the energy relaxation rate we need to know the interaction terms to second order in surface displacements ξ_q. In this case, we should not disregard the last terms of (1.64) collected in square brackets:

$$V_{\text{e}}(z - \xi, 0) - V_{\text{e}}(z, 0) \,. \tag{1.75}$$

Expanding (1.75) up to the second order term in ξ yields the following correction to the interaction Hamiltonian:

$$V^{(2)}_{\text{int}} = \frac{1}{2} \left\langle 1 \left| \frac{\partial^2 V_{\text{e}}(z,0)}{\partial z^2} \right| 1 \right\rangle \xi^2(\boldsymbol{r}) \,. \tag{1.76}$$

Because $V_{\text{e}}(z, 0)$ is the electron potential above the flat surface, when evaluating the matrix element in (1.76), we can use the model potential of (1.17).

Then integrating by parts over z yields

$$\left\langle 1 \left| \frac{\partial^2 V_e(z,0)}{\partial z^2} \right| 1 \right\rangle = 2V_0\chi_1(0)\chi_1'(0) - 2\int_0^\infty \chi_1(z)\chi_1'(z)\frac{\partial V_\epsilon}{\partial z}\mathrm{d}z\,, \qquad (1.77)$$

where $\chi_1'(z)$ is the derivative of $\chi_1(z)$ and $V_\epsilon(z)$ is the polarization term of the electron potential, as discussed in Sect. 1.2. If one substitutes here the hydrogenic function $\chi_1(z)$ and the approximation $V_\epsilon(z) = -\Lambda/z$, the second term will have a logarithmic divergence at the lower limit of the integral. For a smooth potential $V_\epsilon(z)$, this integral is proportional to $[\ln(1/\gamma b) - 1.08]$, where the parameter b characterizes the distance from the helium surface where the polarization potential transforms from the image potential asymptote into the constant $-U_m$ (estimated to be of the order of a few Å).

A simple comparison between the two terms of (1.77) indicates that the second term is much smaller than the first and can be disregarded because $\kappa_0 \gg \gamma \ln(\gamma b)$. Therefore, the second order correction to the interaction Hamiltonian can be rewritten in the form [44]

$$V_{\mathrm{int}}^{(2)} = \kappa_0 V_0 \chi_l^2(0)\xi^2\,. \qquad (1.78)$$

According to Fig. 1.11, $\chi_1^2(0)$ and, consequently, $V_{\mathrm{int}}^{(2)}$ depend rather weakly on the holding electric field [mostly due to the dependence $\gamma(E_\perp)$]. In contrast to the linear interaction term, $V_{\mathrm{int}}^{(2)}$ depends strongly on the actual value of the potential barrier: $V_{\mathrm{int}}^{(2)} \propto \sqrt{V_0}$.

For higher surface levels $l > 1$, the value of the electron wave function $\chi_l(0)$ should be smaller than the one found for the ground level. Therefore, a kind of Lamb shift of the order of

$$-\frac{\hbar^2\gamma^2}{2m}4\kappa_0\gamma\left\langle \xi^2 \right\rangle \qquad (1.79)$$

can be expected for the SE transition frequencies $\omega_{1\to l}$. Here $\chi_1(0)$ was estimated according to (1.19). In the following section, we shall see that the long-wavelength ripplons with $q < \gamma$ should be cut off from the averaging $\langle \xi^2 \rangle$ present in (1.79). It should be noted that the accurate evaluation of ripplon-induced shifts in surface level positions requires inclusion of the linear expansion term of (1.75) in the second order perturbation term. This is not zero because the matrix elements entering it involve different surface states $l' \neq l$.

1.5.3 Adiabatic Approximation

In this section, we discuss another possible treatment of electron scattering induced by medium vibrations, which is conventionally called the Born–Oppenheimer adiabatic approximation. Our intention is to emphasize the importance of the rearrangement of the conventional adiabatic perturbation treatment so that it applies to electrons strongly bound to the interface. Following [41], we shall see that the adiabatic approach which does not take into account the basic property of bound states $\langle 1|F_z|1\rangle = 0$ (such as reported in [40]) can cause serious errors in matrix elements describing electron scattering. In the conventional perturbation treatment of 3D electron–phonon systems, both these approaches are known to yield the same scattering probabilities. The most important practical conclusion which follows from the analysis presented below is that, for long-wavelength ripplons $q < \gamma$, the adiabatic approach significantly reduces electron scattering in the z-direction, along which they are strongly bound to the interface. At the same time probabilities of electron scattering along the surface $l' = l$ are unchanged as compared to the result given by the modified Bloch method. Those readers who are not interested in such theoretical details are advised to skip this section and proceed directly to the following section which deals with electron mobility along the interface.

In comparison with electrons, ripplons can be considered as a slow subsystem. This means that the interaction adjusts the electron wave function adiabatically to the slow motion of surface displacements $\xi(\boldsymbol{r}, t)$, so that it becomes a function of phonon coordinates $\{\xi_{\boldsymbol{q}}\}$. In this picture, called the Born–Oppenheimer approach (BOA), the interaction Hamiltonian itself does not cause electron scattering because its influence is included in the definition of the electron wave function $\psi_{\mathrm{s}}(\boldsymbol{R}, \{\xi_{\boldsymbol{q}}\})$:

$$\left[H_{\mathrm{e}}^{(0)} + V_{\mathrm{e}}(\boldsymbol{R}, \{\xi_{\boldsymbol{q}}\}) - \varepsilon_s(\{\xi_{\boldsymbol{q}}\})\right] \psi_{\mathrm{s}}(\boldsymbol{R}, \{\xi_{\boldsymbol{q}}\}) = 0 \,. \tag{1.80}$$

Electron scattering appears only because the wave function of the whole system, chosen to be the product of the electron and ripplon (phonon) functions

$$\Psi_{\boldsymbol{n},s} = \psi_{\mathrm{s}}(\boldsymbol{R}, \{\xi_{\boldsymbol{q}}\}) \Phi_{\boldsymbol{n},s}(\{\xi_{\boldsymbol{q}}\}) \,, \tag{1.81}$$

does not completely uncouple the Schrödinger equation into two independent parts: (1.80) for electrons and the equation

$$\left[H_{\mathrm{r}}^{(0)} + \varepsilon_s(\{\xi_{\boldsymbol{q}}\}) - \mathcal{E}_{\boldsymbol{n},s}\right] \Phi_{\boldsymbol{n},s} = 0 \tag{1.82}$$

for ripplons. The coupling (electron scattering) appears because the ripplon Hamiltonian $H_{\mathrm{r}}^{(0)}$ no longer commutes with the electron wave function $\psi_{\mathrm{s}}(\boldsymbol{R}, \{\xi_{\boldsymbol{q}}\})$. The nonadiabatic terms are usually written as

$$H_{\mathrm{NA}} \Psi_{\boldsymbol{n},s} = \left[H_{\mathrm{r}}^{(0)}, \psi_{\mathrm{s}}\right] \Phi_{\boldsymbol{n},s} \,. \tag{1.83}$$

The matrix elements of H_{NA} are assumed to determine the probability of electron scattering in the adiabatic approach.

In the 3D case of nearly free electrons, the corrections $\delta\psi_{\mathrm{s}}$ to the electron wave function, dependent on phonon coordinates, are usually found using the perturbation treatment. As a result, to lowest order, the BOA gives the same scattering probability as the conventional Bloch approach [45]. In the case of 2D electrons strongly bound to the interface, one cannot rely on perturbation theory. Luckily, the effect of strong electron binding to the interface can be taken into account in a quite simple way. Consider the states

$$\left|\Psi_{\boldsymbol{n},s}^{(0)}\right\rangle = \exp\left[-\mathrm{i}k_z\xi(\boldsymbol{r})\right]|l,\boldsymbol{k}\rangle\,|\boldsymbol{n}\rangle \ , \tag{1.84}$$

which are constructed from the free electron–ripplon states by means of the unitary transformation. (Here k_z is the operator $p_z/\hbar$.) In the usual Schrödinger picture, the corresponding electron wave function is written as

$$\psi_{\mathrm{s}}^{(0)}(\boldsymbol{R},\{\xi_{\boldsymbol{q}}\}) = f_l(z-\xi)\frac{1}{\sqrt{S_{\mathrm{A}}}}\exp(\mathrm{i}\boldsymbol{k}\cdot\boldsymbol{r}) \ , \tag{1.85}$$

which reveals a quite clear physical meaning: the electron wave function describing electron motion in the z-direction is adjusted and follows the smooth and slow interface distortions $\xi(\boldsymbol{r})$.

The wave functions $\psi_{\mathrm{s}}^{(0)}(\boldsymbol{R},\{\xi_{\boldsymbol{q}}\})$ introduced above are eigenfunctions of a certain Hamiltonian which we denote as $H_{\mathrm{A}}^{(0)}$:

$$H_{\mathrm{A}}^{(0)}\mathrm{e}^{-\mathrm{i}k_z\xi}\,|l,\boldsymbol{k}\rangle = \varepsilon_{l,\boldsymbol{k}}\mathrm{e}^{-\mathrm{i}k_z\xi}\,|l,\boldsymbol{k}\rangle \ . \tag{1.86}$$

From this equation one can easily show that

$$H_{\mathrm{A}}^{(0)} = \mathrm{e}^{-\mathrm{i}k_z\xi}H_{\mathrm{e}}^{(0)}\mathrm{e}^{\mathrm{i}k_z\xi} = \mathrm{e}^{-\mathrm{i}k_z\xi}K_{\mathrm{e}}\mathrm{e}^{\mathrm{i}k_z\xi} + V_{\mathrm{e}}(z-\xi,0) \ , \tag{1.87}$$

where $H_{\mathrm{e}}^{(0)} = K_{\mathrm{e}} + V_{\mathrm{e}}(z,0)$ is the electron Hamiltonian above the flat surface, K_{e} is the kinetic energy operator, and we have used the identity $\mathrm{e}^{-\mathrm{i}k_z\xi}V(z)\mathrm{e}^{\mathrm{i}k_z\xi} = V(z-\xi)$. It is clear that $H_{\mathrm{A}}^{(0)}$ includes the whole perturbation of the repulsion potential $V_{\mathrm{rep}}(z-\xi)$ and a substantial part of the polarization interaction. Owing to the kinetic energy term, it also contains new perturbations which are definitely small for long-wavelength ripplons. The important point is that the interaction terms included in the new, exactly solvable Hamiltonian $H_{\mathrm{A}}^{(0)}$ do not change the SE spectrum: it is the same as that found for the flat surface $\varepsilon_{l,\mathrm{k}} = \varepsilon_{l,\mathrm{k}}^{(0)}$.

The rest of the electron Hamiltonian $\delta H_{\mathrm{A}} = H_{\mathrm{e}} - H_{\mathrm{A}}^{(0)}$ can be written as

$$\delta H_{\mathrm{A}} = \delta K_{\mathrm{e}} + V_{\mathrm{e}}(z,\{\xi_q\}) - V_{\mathrm{e}}(z-\xi,0) \ , \tag{1.88}$$

where

$$\delta K_{\mathrm{e}} = K_{\mathrm{e}} - \mathrm{e}^{-\mathrm{i}k_z\xi}K_{\mathrm{e}}\mathrm{e}^{\mathrm{i}k_z\xi} = -\frac{p_z}{2m_{\mathrm{e}}}\left(\boldsymbol{p}\cdot\boldsymbol{\nabla}\xi + \boldsymbol{\nabla}\xi\cdot\boldsymbol{p}\right) - \frac{p_z^2}{2m_{\mathrm{e}}}(\nabla\xi)^2 \ , \tag{1.89}$$

and $\boldsymbol{p}$ is the in-plane momentum operator. Now we can really see that, in the long-wavelength limit, the perturbations of the kinetic energy are small because $\nabla\xi \ll 1$. The difference between the potential energies entering (1.88) is also small:

$$V_{\rm e}(z,\{\xi_{\boldsymbol{q}}\}) - V_{\rm e}(z-\xi,0) = eE_{\perp}\xi + V_{\rm att}(z,\xi) - V_{\rm att}(z-\xi,0)\,. \qquad (1.90)$$

It is interesting to note that $\delta H_{\rm A}$ of (1.88) is similar to $V_{\rm int}$ found by means of the Bloch approach for bound electrons [see (1.64)]. The important difference is that the two terms collected in the square brackets $[V_{\rm e}(z-\xi,0) - V_{\rm e}(z,0)]$ are replaced by the kinetic energy perturbation term $\delta K_{\rm e}$. If the surface level number l is preserved, the linear part of this term has zero matrix elements as well, owing to the property of bound states: $\langle l|p_z|l\rangle = 0$.

The approximation for the electron wave function given in (1.84) and (1.85) is not sufficient to describe electron scattering within the ground surface level ($l' = l = 1$) because the corresponding matrix elements of the nonadiabatic terms in (1.83) are still zero, as we shall see later. Nevertheless, the use of these functions as the main approximation for $\psi_{\rm s}(\boldsymbol{R},\xi_{\boldsymbol{q}})$ allows us to rearrange the perturbation theory in such a way that the residual interaction $\delta H_{\rm A}$ becomes small. In order to find the corrections to the wave function $\psi_{\rm s}^{(0)}(\boldsymbol{R},\{\xi_{\boldsymbol{q}}\})$ or energy spectrum $\varepsilon_{l,\boldsymbol{k}}^{(0)}$, one can use conventional perturbation theory with respect to $\delta H_{\rm A}$, which is much smaller than the initial interaction Hamiltonian $V_{\rm e}(z,\{\xi_{\boldsymbol{q}}\}) - V_{\rm e}(z,0)$. Recalling, the main result of the adiabatic approach for the description of electron–phonon scattering in typical solids, one can already foresee at this point that in the end the matrix elements of the nonadiabatic terms will be the same as the matrix elements of the residual interaction $\delta H_{\rm A}$, as long as it is small.

In order to prove the above statement and to evaluate the nonadiabatic terms, it is convenient to perform a unitary transformation of electron states $|\psi\rangle$ and operators H:

$$\left|\tilde{\psi}\right\rangle = {\rm e}^{{\rm i}k_z\xi}\,|\psi\rangle\,, \qquad \tilde{H} = {\rm e}^{{\rm i}k_z\xi} H {\rm e}^{-{\rm i}k_z\xi}\,, \qquad (1.91)$$

which does not change the matrix elements. The reason for using such a transformation is that $\left|\tilde{\psi}_{\rm s}^{(0)}\right\rangle = |l,\boldsymbol{k}\rangle$ and $\tilde{H}_{\rm A}^{(0)} = H_{\rm e}^{(0)}$ due to (1.84) and (1.87), which simplifies evaluations. In the new representation, the operator $\tilde{H}_{\rm NA}$ can be rewritten as

$$\tilde{H}_{\rm NA} = \sum_{\boldsymbol{q}} \pi_{\boldsymbol{q}}^{(\rm e)}\dot{\xi}_{-\boldsymbol{q}} + \sum_{\boldsymbol{q}} \frac{1}{2\mu_{\boldsymbol{q}}}\pi_{\boldsymbol{q}}^{(\rm e)}\pi_{-\boldsymbol{q}}^{(\rm e)} - p_z\dot{\xi} + \frac{p_z^2}{2M^*}\,, \qquad (1.92)$$

where $\pi_{\boldsymbol{q}}^{(\rm e)}$ acts only on the electron wave function

$$\pi_{\boldsymbol{q}}^{(\rm e)}\psi_{\rm s}(\boldsymbol{R},\{\xi_{\boldsymbol{q}}\}) = -{\rm i}\hbar\frac{\partial\psi_{\rm s}}{\partial\xi_{-\boldsymbol{q}}}\,, \qquad (1.93)$$

$p_z = \hbar k_z$, and

$$\frac{1}{M^*} = \sum_{q} \frac{1}{\mu_q} . \tag{1.94}$$

The last term of (1.92) causes no scattering and can be disregarded. The second term is zero for the linear corrections to $\left|\tilde{\psi}_s^{(0)}\right\rangle = |l, \boldsymbol{k}\rangle$ and can only contribute to the scattering probability in higher orders of the perturbation treatment. The third term is proportional to two small factors $\gamma\xi \ll 1$ and $\hbar\omega_q$. The latter is very small compared with typical electron energies if $q \leq 2k$.

We therefore consider only the first term of (1.92) and rewrite its matrix elements as

$$\langle \boldsymbol{n}', \tilde{s}' | \tilde{H}_{\mathrm{NA}} | \boldsymbol{n}, \tilde{s} \rangle \simeq \frac{\mathrm{i}}{\hbar} \sum_{q} \langle \boldsymbol{n}' | \langle \tilde{s}' | \pi_{q}^{(\mathrm{e})} | \tilde{s} \rangle [H_{\mathrm{r}}^{(0)}, \xi_{-q}] | \boldsymbol{n} \rangle , \tag{1.95}$$

where we have employed the quantum-mechanical definition of the time derivative $\dot{\xi}$, and set $|\tilde{s}\rangle = \left|\tilde{\psi}_s\right\rangle$. When evaluating $\langle \tilde{s}' | \pi_q^{(\mathrm{e})} | \tilde{s} \rangle$, it is not sufficient just to replace $|\tilde{s}\rangle \rightarrow |l, \boldsymbol{k}\rangle$, because this will give zero. We use the strict relations

$$\langle \tilde{s}' | \pi_q^{(\mathrm{e})} | \tilde{s} \rangle = \frac{1}{\varepsilon_{s'} - \varepsilon_s} \langle \tilde{s}' | [\tilde{H}_{\mathrm{e}}, \pi_q^{(\mathrm{e})}] | \tilde{s} \rangle = \frac{1}{\varepsilon_{s'} - \varepsilon_s} \langle \tilde{s}' | [\delta\tilde{H}_{\mathrm{A}}, \pi_q^{(\mathrm{e})}] | \tilde{s} \rangle . \tag{1.96}$$

Now, in the last expression, one can replace $|\tilde{s}\rangle \rightarrow |l, \boldsymbol{k}\rangle$.

In the linear approximation, substituting the above matrix elements into (1.95) yields

$$\langle \boldsymbol{n}', \tilde{s}' | \tilde{H}_{\mathrm{NA}} | \boldsymbol{n}, \tilde{s} \rangle = \left\langle \boldsymbol{n}', \tilde{s}' \left| \frac{\mathcal{E}_{\boldsymbol{n}'}^{(\mathrm{r})} - \mathcal{E}_{\boldsymbol{n}}^{(\mathrm{r})}}{\varepsilon_s - \varepsilon_{s'}} \delta\tilde{H}_{\mathrm{A}} \right| \boldsymbol{n}, \tilde{s} \right\rangle . \tag{1.97}$$

It is clear that for scattering events which preserve the energy of the system ($\mathcal{E}_{\boldsymbol{n}'}^{(\mathrm{r})} - \mathcal{E}_{\boldsymbol{n}}^{(\mathrm{r})} = \varepsilon_s - \varepsilon_{s'}$), the matrix elements of the nonadiabatic terms are the same as the matrix elements of the operator $\delta\tilde{H}_{\mathrm{A}}$.

Applying the unitary transformation of (1.91) to the residual interaction Hamiltonian δH_{A} given in (1.88) yields

$$\delta\tilde{H}_{\mathrm{A}} = eE_{\perp}\xi + V_{\mathrm{att}}(z + \xi, \{\xi_q\}) - V_{\mathrm{att}}(z, 0) + \delta\tilde{K}_{\mathrm{e}}, \tag{1.98}$$

$$\delta\tilde{K}_{\mathrm{e}} = \frac{1}{2m_{\mathrm{e}}} \left[-p_z (\boldsymbol{p} \cdot \nabla\xi + \nabla\xi \cdot \boldsymbol{p}) + p_z^2 (\nabla\xi)^2 \right] . \tag{1.99}$$

Here we have used the following relations: $\delta\tilde{K}_{\mathrm{e}}(\xi) = -\delta K_{\mathrm{e}}(-\xi)$ and $\delta K_{\mathrm{e}}(\xi) = K_{\mathrm{e}} - \tilde{K}_{\mathrm{e}}(-\xi)$. The polarization term of (1.98) can be written as

$$-\frac{\Lambda}{\pi} \int \mathrm{d}^2 \boldsymbol{r}' \int_0^{\xi(\boldsymbol{r}') - \xi(\boldsymbol{r})} \frac{\mathrm{d}z'}{[(\boldsymbol{r}' - \boldsymbol{r})^2 + (z' - z)^2]^2} . \tag{1.100}$$

In the linear approximation, it coincides with (1.68) found previously in the framework of the Bloch approach for bound electrons.

As anticipated, the only important difference between $\delta\tilde{H}_{\rm A}$ and $V_{\rm int}$ of (1.64) is that the term $[V_{\rm e}(z-\xi,0)-V_{\rm e}(z,0)]$ of the Bloch approach is replaced by $\delta\tilde{K}_{\rm e}$ in the adiabatic theory. Neither of these terms lead to electron scattering within the surface level ($l'=l$) due to the property of the bound states $\langle l|p_z|l\rangle=0$. In contrast, if an electron is scattered in the vertical direction along the strong binding ($l'\neq l$), the probability of electron scattering is different for these two approaches. In the long-wavelength limit, the term $\langle l'|\,\partial V_{\rm e}/\partial z\,|l\rangle\,\xi(\boldsymbol{r})$ of the Bloch approach is finite and yields strong scattering. At the same time the BOA gives smaller matrix elements for out-of-layer scattering because $\delta\tilde{K}_{\rm e}\to 0$ when $\boldsymbol{\nabla}\xi\to 0$. This is in accordance with the physical meaning of the adiabatic adjustment of electron states to the slow interface motion.

It should also be emphasized that the second order term in the expression for $\delta\tilde{K}_{\rm e}$ [(1.99)] is proportional to $(\nabla\xi)^2$, and therefore has the strong dependence on the wave numbers of medium vibrations involved in the scattering events, contrary to the result from the Bloch approach [(1.78)]. For long-wavelength ripplons, it is very small, but for short-wavelength excitations, it becomes even larger than (1.78), and disallows the approximation (1.85) employed for the adiabatic SE wave function. This approximation fails because, for short-wavelength excitations, it causes strong disturbances in the electron kinetic energy $\delta\tilde{K}_{\rm e}$. In this case, one has to use the function $\langle\boldsymbol{r}|l,\boldsymbol{k}\rangle$ as a starting point for perturbation theory. Then, as in the 3D case, the adiabatic approximation would give the same matrix elements for electron scattering as those found in the Bloch approach. Additionally, we note that the structure of the residual interaction $\delta\tilde{H}_{\rm A}$ in the BOA shows that the long-wavelength ripplons with $q<\gamma$ should be cut off from the average $\langle\xi^2\rangle$ which enters the SE level shifts found previously using the Bloch approach [(1.79)].

1.6 Mobility Along the Helium Surface

The mobility measurement is the most common way of studying a 2D electron system. It tests the quality of the interface or the purity of the system, and the applicability of theoretical models for major interactions. It also shows the influence of electron–electron interaction on electronic transport. A detailed discussion of quantum transport phenomena in the strongly correlated electron liquid will be given in Chaps. 3, 4 and 5. In this section we restrict ourselves to the single-electron transport properties of SEs and introduce the main experimental achievements in this field.

Early mobility measurements for SEs on liquid helium were confined to a rather high temperature range $T>0.8\,{\rm K}$, where electron mobility is limited by vapor atom scattering and increases with cooling at an exponential rate. For example, the famous experiment of Sommer and Tanner [46] was meant to observe the ripplon limited mobility of SEs predicted by Cole [39]. According to theoretical estimates, at approximately $T \approx 1.1\,\mathrm{K}$, the exponential temperature dependence of the electron mobility was expected to change to a temperature-independent plateau owing to the interaction with capillary waves. However, the mobility data of Sommer and Tanner showed the same mobility increase as the temperature was reduced to approximately $0.9\,\mathrm{K}$, similar to the decrease in the mobility of free 3D electrons in a dilute helium gas.

The explanation of the mobility data of Sommer and Tanner follows straightforwardly from the interaction potential discussed in the last section. Indeed, (1.70) found by means of the Bloch approach for bound electrons shows the strong compensation of the polarization interaction potential for long-wavelength interface excitations. It should be noted that the wave numbers of ripplons taking part in one-ripplon emission and absorption processes are restricted by the electron wave number: $q \leq 2k$. Therefore, at typical helium temperatures, the argument of the function $K_1(qz)$ which enters the electron–ripplon coupling is small, $q\langle z\rangle \ll 1$, and the polarization interaction is reduced. Even at $q\langle z\rangle \sim 1$, the reduction of the electron–ripplon coupling is strong.

In the kinetic equation method, the collision frequency of SEs, the coupling function V_q, and the parameters of the 2D phonon (ripplon) system are related by [27]

$$\begin{aligned} \nu(\varepsilon_k) &= \frac{m_\mathrm{e}}{2\pi\hbar^3}\int\limits_0^{2\pi}\mathrm{d}\varphi(1-\cos\varphi)\frac{\hbar V_q^2}{2\mu_q\omega_q}\left(2N_q+1\right) \\ &\simeq \frac{T}{8\pi\alpha\hbar\varepsilon_k}\int_0^{2\pi} V_q^2\mathrm{d}\varphi \equiv \frac{T}{4\alpha\hbar\varepsilon_k}\left\langle V_q^2\right\rangle_\varphi , \end{aligned} \tag{1.101}$$

where $q = 2k\sin(\varphi/2)$ is the momentum exchange in a collision and ε_k is the electron energy in the ground surface level. We have used the approximation $N_q \simeq T/\hbar\omega_q$ which is valid for ripplons involved in electron momentum relaxation ($q \leq 2k$) down to ultra-low temperatures. The definitions of ripplon parameters such as μ_q and ω_q are taken from (1.47) and (1.53).

Owing to the unusual ripplon dispersion $\omega_q \propto q^{3/2}$, the velocity of the capillary wave $c_q \propto \sqrt{q} \to 0$ if $q \to 0$. Therefore, both emission and absorption processes contribute to the electron momentum relaxation rate, which is the origin of the factor $2N_q + 1$ in the integrand. The second line of (1.101) establishes a quite simple relation between the collision frequency and the electron–ripplon coupling function V_q averaged over the angle of electron scattering φ. At low temperatures the compensation of the polarization interaction discussed above reduces the coupling function V_q and changes its q-dependence. The latter affects the ε_k-dependence of the collision frequency.

In order to obtain the electron mobility, we have to average the momentum relaxation time $1/\nu(\varepsilon)$ over ε_k according to the kinetic equation method [27]:

$$\mu = \frac{e}{m_\mathrm{e}} \int_0^\infty \frac{x \exp(-x)}{\nu(Tx)} \mathrm{d}x \,. \tag{1.102}$$

(for details see also Sect. 3.5). Evaluation of this integral is very simple in the limiting case of strong holding fields, when $V_q \simeq eE_\perp$. In this case, the average over φ is trivial $\nu(\varepsilon_k) \propto 1/\varepsilon_k$, and the integral over x gives the additional numerical factor 2 in the mobility equation:

$$\mu = \frac{e}{m_\mathrm{e}} \frac{8\alpha\hbar}{(eE_\perp)^2} \,. \tag{1.103}$$

The electron mobility does not depend on temperature because the linear decrease in the average ripplon number $N_q \propto T$ is compensated by the increase in the electron momentum relaxation rate $\nu \propto 1/\varepsilon_k$ with cooling.

In the opposite limit of weak holding fields $E_\perp \to 0$, the polarization term eE_q entering the electron–ripplon coupling V_q dominates. In this case, the coupling function has a complicated q-dependence and we have to evaluate integrals of (1.101) and (1.102) numerically. The result of the numerical evaluation is shown in Fig. 1.12 by the dashed curve marked μ_R.

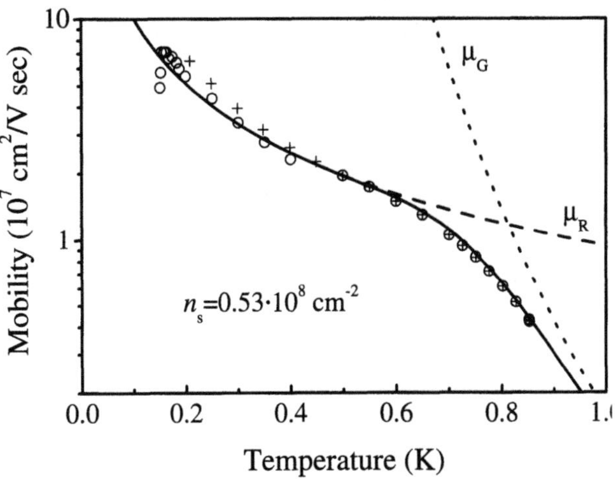

Fig. 1.12. Mobility of SEs along the liquid ^{4}He surface. Theory according to (1.101–1.104) (*curves*) and the data of Mehrotra et al. [50] (*symbols*). *Cross symbols* are data from the second experimental run

In the high temperature range, we have to include the collision frequency due to electron scattering at vapor atoms [22]:

$$\nu_\mathrm{G} = \frac{3[V^{(\mathrm{a})}]^2 n^{(\mathrm{a})} m_\mathrm{e} \gamma}{8\hbar^3} \,. \tag{1.104}$$

The electron mobility limited by electron–atom scattering alone is shown in Fig. 1.12 by the dotted curve marked μ_{G}. The total mobility curve defined by $\nu(\varepsilon) = \nu_{\mathrm{G}} + \nu_{\mathrm{R}}(\varepsilon)$ is shown by the solid curve. In the low temperature regime, the strong temperature dependence of the electron mobility is due to the discussed q-dependence of the electron–ripplon coupling. For intermediate helium temperatures $T \sim 0.5\,\mathrm{K}$, the approximation $w_{\mathrm{C}}(x) \simeq 1/3\sqrt{x}$ in the expression for E_q from (1.72) yields $(eE_q)^2 \propto q^2 \propto \varepsilon_k$. As a result, the electron collision frequency ν is approximately independent of the electron energy ε_k and is proportional to the average number of long-wavelength ripplons available at the given temperature $N_q \propto T$.

The first observation of the ripplon-limited mobility of SEs was independently reported by Grimes and Adams [47], and by Rybalko, Kovdrya and Eselson [48]. The mobility data of the first experiment were obtained from the plasmon resonance linewidth. In the low temperature regime, the mobility was found to be independent of T even for low electron densities. The absolute value of the electron mobility was shown to be a decreasing function of the holding electric field $E_{\perp}$.

In another experiment [48], electron mobility was obtained from the phase shift of the experimental signal of a rather low frequency $\omega \ll \nu$. These data showed a strong temperature dependence in the ripplon-limited mobility of SEs, but its absolute value was substantially lower than what was found in the plasmon resonance experiment. The measurements of Iye [49] revealed the same temperature dependence in the electron mobility as was found in [48], but the absolute values were shown to be approximately one order of magnitude higher.

The most accurate mobility measurements were performed by Mehrotra et al. [50]. Their low density data are shown Fig. 1.12. One can see that the mobility data are in good quantitative agreement with theory, showing the transition from the vapor-atom-dominated regime to ripplon-limited mobility. The sharp drop in electron mobility at $T < 0.16\,\mathrm{K}$ is due to the Wigner solid transition.

At high electron densities, the mobility data of [50] showed a strong decrease in μ with n_{s}, in addition to the decrease caused by the electron–ripplon coupling $V_q = eE_{\perp} + eE_q$ under the saturation condition $E_{\perp} = 2\pi e n_{\mathrm{s}}$. We shall discuss these data in Chap. 3, which deals with the quantum transport properties of highly correlated electrons.

In conclusion, the mobility data currently available indicate that electron interactions with uneven interfaces are now well understood. Experimental data are in good (even quantitative) agreement with theory, which allows us to apply it to more advanced studies of electron transport in strongly correlated Coulomb liquids.

1.7 Other Cryogenic Interfaces

1.7.1 Electrons on the Surface of Fermi Liquid ^{3}He

Liquid ^{3}He is another example of a cryogenic substrate which provides external SE states similar to those described for liquid ^{4}He. The potential barrier for electron penetration into this liquid is expected to be approximately the same as it is for liquid ^{4}He. (Theory estimates $V_0 \simeq 0.9\,\text{eV}$.) On the other hand, the dielectric constant of ^{3}He is even closer to unity, $\epsilon_3 - 1 \simeq 0.0423$, which results in weaker binding to the substrate: the ground state binding energy [(1.5)] is estimated to be $\left|\varepsilon_1^{(\perp)}\right| \simeq 4.2\,\text{K}$, and the average electron height above the surface $\langle 1|z|1\rangle \simeq 153.3\,\text{Å}$.

Surface electrons on liquid ^{3}He were firstly observed by Edel'man [51] in a CR absorption experiment. The spectrum of surface states was studied in [52] by measuring the transition frequencies $f_{1\to2}$ and $f_{1\to3}$ as functions of the holding electric field $E_\perp$. A linear extrapolation to $E_\perp = 0$ gave $f_{1\to2} = 69.8 \pm 0.15\,\text{GHz}$ and $f_{1\to3} = 83.15 \pm 0.25\,\text{GHz}$, which is very close to the theoretical values given by the hydrogenic model: 67.6 and 80.1 GHz.

Regarding the natural scatterers for SEs above liquid ^{3}He, it should be noted that in this case the vapor atom density is substantially higher than above liquid ^{4}He, because the evaporation constant Q_3 entering (1.36) is substantially smaller: $Q_3 \simeq 2.5\,\text{K}$. Therefore, the transition from the vapor-atom scattering regime to ripplon-limited mobility occurs at substantially lower temperatures $T \approx 0.4\,\text{K}$. If SEs are in the liquid or gas state, then at low temperatures they interact mainly with capillary wave quanta ripplons. In liquid ^{3}He, the spectrum of ripplons is softer than in liquid ^{4}He because of the lower surface tension: $\alpha_3/\alpha_4 \simeq 0.43$.

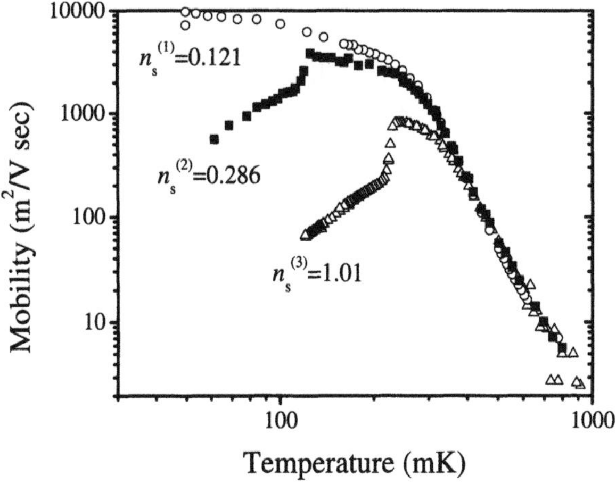

Fig. 1.13. Mobility of SEs above liquid ^{3}He vs. temperature [53]

Generally, the transport properties of free SEs above liquid ^{3}He should be qualitatively the same as were obtained for liquid ^{4}He. Typical mobility data from Shirahama et al. [53] are shown in Fig. 1.13 for three electron densities. The temperature dependence of the experimental data is in accordance with the theory, although the data are lower than the theoretical values by a factor ~ 2 in the whole temperature range, for both scattering regimes. The sharp change in the temperature dependence of the SE mobility occurs when the system undergoes the WS transition. At this point, the WS transport becomes coupled to transport of the lattice of surface dimples appearing under each electron, because of the electron–ripplon interaction. At intermediate temperatures, the dimple mobility is determined by the large viscosity of the Fermi liquid, increasing quickly with cooling: $\eta(T) \propto T^{-2}$. At ultra-low temperatures, the mobility of surface dimples is determined by reflection of bulk excitations from the uneven helium surface [54]. The bulk excitation spectrum of Fermi liquid ^{3}He differs substantially from the excitation spectrum of Bose liquid ^{4}He. As a result, the mobility of the WS over the Fermi liquid substrate decreases by several orders of magnitude. We shall discuss these effects in Chap. 8 which deals with WS transport.

A solution of ^{3}He in ^{4}He undergoes the stratification transition when the concentration of the light isotope exceeds 6.6%. Owing to the existence of the Andreev surface states of ^{3}He atoms [55], the stratification transition begins at the free surface, and under certain conditions it is possible to create a thin film of nearly pure ^{3}He covering the 6.6% solution of massive liquid [56], as shown in Fig. 1.14. This picture is called micro-stratification. Such a double-interface liquid substrate provides additional possibilities for studying 2D electron systems. The film of light isotope can be used for spatial separation of SEs above the free surface and positive ions below the ^{3}He–^{4}He interface, creating bilayer systems [57] similar to the one discussed in Sect. 1.3, which dealt with the SE state above a liquid helium film.

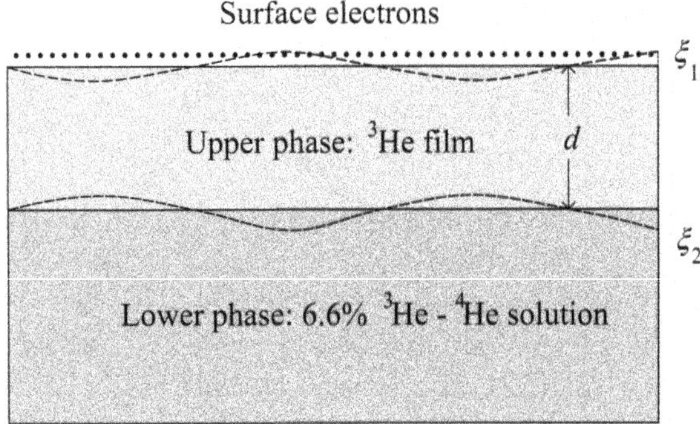

Fig. 1.14. Schematic view of SEs above the double-interface substrate of the ^{3}He–^{4}He solution with micro-stratification

In addition to the usual ripplons, SEs above this film are scattered by excitations representing the thickness oscillations of the thin ^{3}He film. The spectrum of these oscillations has acoustic behavior because of the van der Waals forces acting between the light film and the heavier solution substrate [58]:

$$\omega_s^2 \simeq \sigma q \tanh(qd) , \qquad \sigma = \frac{\pi b n_1^{(h)} \left[n_2^{(h)} - n_1^{(h)} \right]}{2\rho_1 d^4} , \tag{1.105}$$

where $n_1^{(h)}$ and $n_2^{(h)}$ are helium atom densities in the upper and lower liquid phases, respectively, b is the parameter for the interaction potential of two helium atoms, $V(R) \simeq -b/R^6$, and ρ_1 is the corresponding mass density. Numerically, σ exceeds the gravitational acceleration G if $d < 4 \times 10^{-5}$ cm. The pure sound-like dispersion and the ripplon spectrum have a crossing point at $q = q_0 = \rho\sigma d/\alpha$. Therefore the real spectrum of surface excitations of the double-interface substrate deviates from these asymptotes in the vicinity of q_0 due to the mutual 'repulsion' of these modes. Near this point, the lower surface mode transforms from the ripplon spectrum at $q < q_0$ into the sound mode at $q > q_0$, while the higher mode transforms from the sound mode at $q < q_0$ into the ripplon mode at $q > q_0$.

In the case of the thin ^{3}He film covering the liquid substrate, the electron interaction with the sound surface mode can be even stronger than the interaction with ripplons. This happens because the surface displacements $\xi_1(\boldsymbol{r})$ and $\xi_2(\boldsymbol{r})$ oscillate out of phase and the polarization interaction with the lower phase is not compensated as it is for the electron–ripplon interaction $[1/qz - K_1(qz) \to 0]$. The electron interaction with excitations of such a double-interface substrate was analyzed in [58]. For $q(z+d) < 1$, in the most interesting cases, the interaction with the sound mode is proportional to the difference between the dielectric constants ϵ_1 and ϵ_2 of the liquid phases:

$$V_{(s)}(d) \simeq \left\langle 1 \left| \frac{\Lambda_s}{(z+d)^2} \right| 1 \right\rangle , \qquad \Lambda_s = \frac{e^2 \epsilon_1 (\epsilon_2 - \epsilon_1)}{(\epsilon_1 + 1)^2 (\epsilon_2 + \epsilon_1)} . \tag{1.106}$$

In the interaction Hamiltonian, $V_{(s)}$ plays the same role as the electron–ripplon coupling V_q [see (1.69)]. At $d \ll \gamma^{-1}$, the coupling function $V_{(s)}(d) \simeq 2\Lambda_s \gamma^2$. The electron collision frequency induced by the thickness oscillations of the light film is independent of the electron energy ε_k for $\varepsilon_k \ll \hbar^2 \sigma \rho_1 / 8 m_e \alpha_{12}$ [58]:

$$\nu_s = \frac{m_e T V_{(s)}^2(d)}{\hbar^3 \sigma \rho_1} . \tag{1.107}$$

Thus the temperature dependence of the single-electron mobility has the asymptote $\mu \propto 1/T$ at ultra-low temperatures.

It should be noted that charging the free surface of the micro-stratified solutions with SEs is also a very promising way of studying superfluid properties of ^{3}He films whose thickness is comparable with the coherence length.

In this case, a unique flatness of the substrate can be achieved for the ^{3}He film. An additional advantage is that such a film is weakly bound to the substrate. Surface charges can be used as a tool for studying this system under different regimes, and even for monitoring the micro-stratification transition itself. (Previously, this was done by studying the thermodynamic properties of the helium solutions [56, 59].)

1.7.2 Solid Interfaces (H_2, Ne)

The instability of the charged surface of the liquid dielectric which is known to occur at $q \sim \kappa$ prevents one from increasing the SE density beyond a certain critical value $n_s^{(c)}$. For liquid helium ^{4}He, this value $n_s^{(c)} \approx 2 \times 10^9\,\mathrm{cm}^{-2}$. It would be a great advantage to use a solid dielectric with the dielectric constant close to unity in order to increase the density of the interface electron system. Unfortunately, the condensed helium remains liquid even at zero temperature because of the strong quantum effects, if no pressure is applied. The solid phase of the helium isotopes does not have a free surface. Therefore, the best candidates to be used as the solid cryogenic substrate for SEs are solid neon Ne and hydrogen H_2.

In the first experiments with solid hydrogen, the substrate was usually crystallized by cooling liquid hydrogen down to its freezing point in a glass ampoule [60]. Electrons were emitted from a filament turned on for a short time. Then the crystal was slowly cooled down to temperatures $T \sim 4\,\mathrm{K}$. The strong binding of SEs to the solid surface (the electron binding energy on solid H_2 is approximately 21 times larger than above liquid ^{4}He) allows one to use rather high temperatures for transport measurements. The average electron height is about 25 Å, which still belongs to macroscopic scales. The SEs on solid hydrogen interact strongly with the static surface roughness $\xi(\boldsymbol{r})$. To describe this interaction one can use the results obtained for the electron–ripplon interaction. We just need to replace the dielectric constant of the medium $\epsilon \to \epsilon_{H_2}$ in (1.69) and (1.70) and to make certain assumptions with regard to the roughness distribution ξ_q.

The low frequency mobility study by [60] was based on measurements of the energy loss induced in an experimental circuit by the electron layer. Exploiting the good agreement between theory and experiment for SEs on liquid helium, the experimental device was calibrated beforehand with liquid helium condensed in the ampoule. According to the data of Troyanovskii and Khaikin [60], the SE mobility above solid hydrogen increases with cooling approximately as $\mu \propto 1/T$, reaching values of about $8 \times 10^4\,\mathrm{cm^2/V\,s}$ at $T \approx 4\,\mathrm{K}$. It was proven that the molecular gas and Rayleigh waves cannot limit the electron mobility down to these values, and that the mobility is determined by electron scattering at static surface defects.

Edel'man and Faley [61] studied the CR and magnetoresistance of SEs above a solid hydrogen surface at a frequency of 20 GHz and used the helium film to test models describing the solid roughness. They found that SEs on solid hydrogen have effective mass close to the free electron mass and mobility agreeing with the low-frequency data of [60]. In order to explain the data, they assumed that the hydrogen surface has a terrace structure with flat sections about 10^{-5} cm in size. The height of the steps separating the terraces was estimated to be $\approx 10^{-6}$ cm.

A detailed theoretical analysis of the mobility of SEs above the surface of solid hydrogen was given in [62]. According to [60, 62], the electron collision frequency can be written in a form similar to (1.101) given above for the electron–ripplon interaction:

$$\nu(\varepsilon_k) = \frac{m_e}{2\pi\hbar^3} \int_0^{2\pi} (1 - \cos\varphi) V_q^2 \left|\xi_{\boldsymbol{q}}\right|^2 d\varphi \,, \tag{1.108}$$

where $\xi_{\boldsymbol{q}}$ is the Fourier-transformed function of the static displacements of the solid surface from the flat shape, and $V_q = eE_\perp + eE_q^{(\mathrm{hyd})}$ is the electron–roughness coupling function obtained from (1.70) by means of the replacement $\epsilon \to \epsilon_{\mathrm{H}_2} \simeq 1.29$. It is reasonable to assume that the roughness correlation has the Gaussian form

$$\langle \xi(\boldsymbol{r}')\xi(\boldsymbol{r})\rangle = \xi_0^2 \exp\left(-\frac{\left|\boldsymbol{r}' - \boldsymbol{r}\right|^2}{L^2}\right) , \tag{1.109}$$

which yields

$$\left\langle \left|\xi_{\boldsymbol{q}}\right|^2 \right\rangle = \zeta^2(q) \equiv \pi\xi_0^2 L^2 \exp\left(-\frac{q^2L^2}{4}\right) , \tag{1.110}$$

where ξ_0 is the average displacement of the interface and L is of the order of the range of the spatial variation of the surface profile in the plane of the surface. Such approximations are frequently used in semiconductor 2D electron systems [10]. With this assumption, the authors of [62] evaluated the electron mobility numerically as a function of temperature for different values of the parameters ξ_0 and L. In the model considered, the factor $\zeta^2(q)$, affecting the electron scattering at the surface roughness, has a maximum at $L \sim 2/q \sim k_T$.

As the quantity $\left\langle \left|\xi_{\boldsymbol{q}}\right|^2 \right\rangle = \zeta^2(q)$ only depends on the absolute value of the vector $\boldsymbol{q}$, (1.108) can be represented in the form

$$\nu(\varepsilon_k) = \frac{4m_e}{\pi\hbar^3} \int_0^1 \frac{x^2}{\sqrt{1 - x^2}} V_{2kx}^2 \zeta^2(2kx) dx \,, \tag{1.111}$$

convenient for analysing the temperature dependence of the electron mobility. This equation indicates that the energy dependence of the collision frequency is completely determined by the q-dependence of the product $V_q^2 \zeta^2(q)$. Knowing V_q^2 and the experimental mobility data, one can make certain conclusions about the surface roughness distribution $\zeta^2(q)$.

Let us find out why the SE mobility increases with cooling for a static roughness potential. Consider qualitatively the high temperature approximation for the coupling function $w_C(x) \simeq 1/3\sqrt{x}$ which enters (1.72) for E_q. In this case, $eE_q^{(\mathrm{hyd})} \propto q$, which leads to $V_q^2 \propto \varepsilon_k \sim T$. According to (1.110), for small $L \ll k_T^{-1}$, the factor $\zeta^2(q) \simeq \pi\xi_0^2 L^2 = \mathrm{const.}$ and therefore $\nu \propto \varepsilon_k$. The collision frequency thus decreases with cooling and the electron mobility increases as $\mu(T) \propto 1/T$. In the most important temperature range $(5 < T < 10\,\mathrm{K})$, the parameter k_T/γ varies between 0.17 and 0.25 which does not allow one to simplify the coupling function $w_C(x)$ by the approximations shown in Fig. 1.11. One must therefore conduct numerical evaluations. In [62] it was shown that, in the temperature range $5 \leq T \leq 10\,\mathrm{K}$, the best agreement with the data of Edel'man and Faley, viz.,

$$\mu \simeq \frac{8 \times 10^4}{T} \left[\mathrm{cm}^2/\mathrm{V\,s}\right] , \qquad (1.112)$$

can be found for $L = 5 \times 10^{-7}\,\mathrm{cm}$ and $\xi_0 = 2.58 \times 10^{-8}\,\mathrm{cm}$.

In other series of experiments concerned with quantum localization effects, solid hydrogen was grown on a sapphire substrate at the triple point ($\approx 14\,\mathrm{K}$) [63]. Electrons were introduced via UV photoemission or a filament. The crystals were then cooled down to liquid-helium temperatures. In the first electron transport experiments with such samples [63], the electron conductance fell gradually with time owing to surface degradation. The election mobility was reduced by the factor 10^{-3}–10^{-4} over a period of a few days. Later it was found that the ambient light passing through the view slit is the origin of this hydrogen surface degradation. Putting the experimental cell in the dark switches off the decrease in electron conductivity with time.

The steady degradation of the solid hydrogen surface was even exploited by Adams and Paalanen [63] for observation of localization effects in the conductivity of the 2D nondegenerate electron gas. The SEs on a fresh hydrogen surface displayed the classical positive magnetic field dependence: $\sigma_{xx}^{-1} \propto 1 + (\mu B)^2$. The mobility value was estimated to be about $2.4\,\mathrm{m}^2/\mathrm{V\,s}$. After one or two days the electron mobility had fallen by approximately one order of magnitude and the magnetoconductivity data had started to show negative magnetoresistance at low fields typical for weak localization effects. These measurements were conducted with a minimal amount of helium in the cell and electron scattering was dominated by surface irregularities. Electron mobilities of the same order of magnitude can be attained at a clean hydrogen surface by introducing helium gas. In this case, the negative magnetoresistance is substantially smaller because of thermal motion of vapor atoms and recoil, both of which reduce localization effects [64, 65].

Another kind of solid hydrogen substrate for SEs used by Kono, Albrecht and Leiderer [66] is the quench-condensed hydrogen film. When preparing

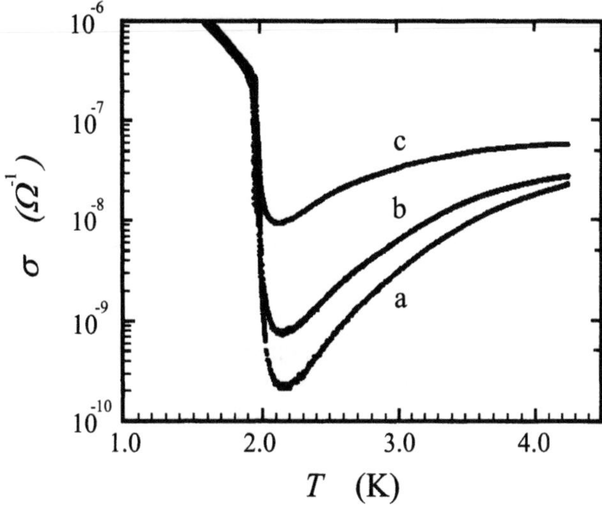

Fig. 1.15. Conductivity above quench-condensed hydrogen films corresponding to different annealing stages [66]

such a film, hydrogen gas is fed into the cell at a constant rate. Atoms release their energy after reaching the cold surface (a glass plate of thickness 0.2 cm) placed on the copper electrode, the lowest part of which is immersed in liquid helium. After preparing a hydrogen film of thickness $d_{\mathrm{H}} \sim 2\,\mu\mathrm{m}$, a small amount of helium gas was introduced into the cell in order to improve charging of the hydrogen surface.

In contrast to solid hydrogen prepared from the liquid state, the freshly prepared quench-condensed hydrogen film has a very bad surface and electron mobility is extremely low (sometimes even below the sensitivity threshold). Repeatedly pursuing the annealing process improves the quality of the surface. Typical electron conductivity data are shown in Fig. 1.15. At low temperatures the helium gas introduced into the cell covers the hydrogen surface, forming a thick helium film. Moving along this film, electrons have high mobility. With increasing temperature, the helium film sharply evaporates at $T > T_{\mathrm{c}}$ (the characteristic temperature T_{c} depends on the amount of helium gas in the cell) and electrons move along the bare hydrogen surface, showing very low mobility. With further heating, the mobility increases because of the thermal improvement of the hydrogen surface and thermally-activated hopping. The smallest activation energy achieved for electron transport above a hydrogen film is about 10 K, which is much smaller than the binding energy of SEs in the perpendicular direction.

The introduction of a fixed amount of helium gas into the experimental cell with the solid hydrogen surface appeared to be essential in order to observe the helium film layering oscillations of the SE mobility on the best surfaces [67, 68] and the helium-film-induced structural transition in a disordered system of localized 2D electrons [69]. In the first instance, the SEs were used as an ultra-sensitive probe for studying thin He films. This indicated the layerwise adsorption for films with up to 9 atomic layers. The helium-film-induced retrapping transition of the SEs will be discussed in the following section.

The 2D interface electron system has also been realized on solid neon [70]. The potential barrier for electron penetration into neon is approximately 2 times smaller than in the case of liquid helium ($V_0 \approx 0.61\,\mathrm{eV}$). Therefore, the binding energy and $\langle z \rangle$ differ substantially from that of the hydrogenic model ($V_0 \to \infty$, $z_0 \to 0$). For example, use of (1.5) gives $\varepsilon_1^{(\perp)} \simeq -116\,\mathrm{K}$, while the more accurate calculation results in stronger binding $\varepsilon_1^{(\perp)} \simeq -203\,\mathrm{K}$ [39]. A detailed study of electronic properties of the charged neon surface has been made by Kajita [71]. It was shown that surface charges are not localized, even at $n_s < 10^8\,\mathrm{cm}^{-2}$. The highest electron density achieved in these studies is about $3 \times 10^{10}\,\mathrm{cm}^{-2}$, which is much higher than on liquid helium. The important advantage of interface electron systems formed on solid substrates such as neon is the possibility of changing the number of scatterers (He gas atoms) with other conditions fixed. It was established that the resistivity of SEs is proportional to the number of scatterers only if it is rather low: $n^{(a)} \leq 2 \times 10^{20}\,\mathrm{cm}^{-3}$. At higher densities of the gas atoms, quantum localization effects lead to a strong increase in the electron resistivity [71].

A strong density dependence of the electron mobility was observed when studying electrons on the solid neon substrate. According to Kajita's results, at $n_s > 10^9\,\mathrm{cm}^{-2}$, the conductivity of SEs is a nonlinear function of electron density n_s which bends to the conductivity saturation. A comparison of this result with the data of Mehrortra et al. [50] indicates that it is an effect introduced by the electron density itself rather than an effect determined by the plasma coupling parameter $\Gamma_{\mathrm{pl}} = e^2\sqrt{\pi n_s}/T$, because under conditions $n_s > 10^9\,\mathrm{cm}^{-2}$ and $T = 4.2\,\mathrm{K}$, the plasma parameter $\Gamma_{\mathrm{pl}} \approx 10$. At the same time we know that in the low electron density regime, the electron mobility does not depend on the electron density up to $\Gamma \approx 100$. We shall discuss these many-electron effects in Chap. 3.

Another interesting many-electron phenomenon observed on solid neon is the influence of electron correlations on the temperature dependence of electron conductivity [71, 72]. At low temperatures (about 1–2 K), the conductivity becomes independent of temperature if n_s is small. At high densities, which correspond to $\Gamma_{\mathrm{pl}} \approx 135$, the electron conductivity drops sharply with cooling. This effect is reminiscent of the strong decrease in the Wigner solid mobility above liquid ^{3}He which we discussed above. It is assumed to be caused by the pinning of the Wigner crystal by surface roughness.

1.8 Retrapping Transition

As discussed in Sect. 1.5.2, the polarization interaction potential of an electron with an uneven interface changes its sign because of the adiabatic adjustment of the electron wave function to surface displacements: eE_q is negative for free electron states, but it is positive for bound electrons according to (1.70). This change of sign of the polarization interaction potential is the reason for an interesting anomaly in the electron conductivity above thin helium films on quench-condensed solid hydrogen. We shall now discuss this feature.

The conductivity anomaly observed by Kono, Albrecht and Leiderer [73] appears as a pronounced successive maximum and minimum in the electron conductivity with a gradual increase in the helium film thickness, as shown in Fig. 1.16. For a long while, the reported anomaly in the electron conductivity was an unexplained phenomenon. Later, in [69], it was shown that this conductivity anomaly represents an unusual structural transition in the system of electrons localized at substrate irregularities. The origin of this transition relates directly to the change in sign of the electron–roughness interaction potential mentioned above.

The SEs on a thin helium film mainly interact with the roughness of the solid hydrogen substrate via the polarization attraction potential. The solid substrate roughness $\xi^{(\mathrm{S})}(\boldsymbol{r})$ is a static, long-wavelength distortion of the hydrogen surface. If there is no helium film, the interaction of the bound electron with the interface roughness is reduced because of the adjustment of the electron wave function, as discussed in Sect. 1.5. In the presence of a liquid helium film, the repulsion barrier V_0 for SEs is created by the helium surface. Therefore, the adjustment of the electron wave function to the solid roughness is governed by the static displacement of the free surface of liquid helium $\xi^{(\mathrm{L})}(\boldsymbol{r})$ induced by van der Waals forces. The extent of this adjustment depends strongly on the film thickness d. For thick films, the helium surface is nearly flat and, with regard to the solid interface, SEs behave like unbound electrons. Their wave function $f_1(z)$ does not follow the displacements of the solid surface. In this instance, the flatness of the helium surface remarkably restores Cole's interaction potential given in (1.67) with $eE_q < 0$. (There, one has to change Λ and z in accordance with the properties of the solid substrate.) When d is decreased, the helium surface begins to follow the solid roughness and, finally, at $d \to 0$, we arrive at the result found for bound electrons $eE_q > 0$. This means that at a certain film thickness $eE_q = 0$ for a given roughness wave number.

Experimentally, the thickness of the helium film was varied by changing the temperature in a closed experimental cell. For thin films, the increase in the helium film thickness with cooling can be described by the simple van der Waals relation

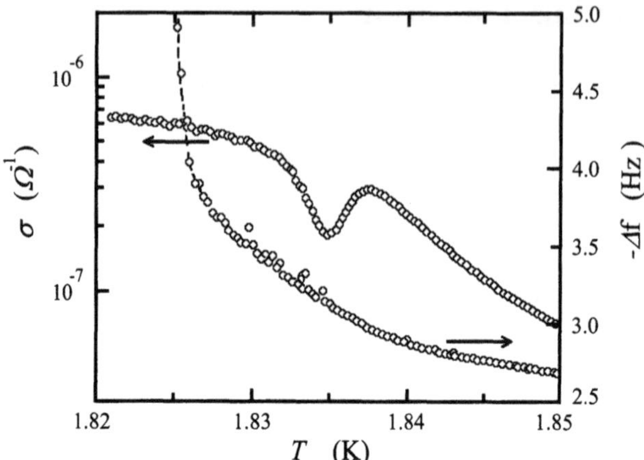

Fig. 1.16. The conductivity anomaly observed with growth of a helium film above an uneven hydrogen surface, and the frequency change of the quartz oscillator used to monitor the helium thickness [73]

$$d^{-3}(T) = \frac{Tn_{\text{He}}^{(\text{L})}}{\alpha_{\text{He}}}\left[\frac{3}{2}\ln\left(\frac{T}{T_{\text{c}}}\right) + \frac{T - T_{\text{c}}}{T_{\text{c}}T}Q\right] , \tag{1.113}$$

where α_{He} is the van der Waals constant, $n_{\text{He}}^{(\text{L})}$ is the liquid helium density, and $Q \simeq 7.17\,\text{K}$. This relation is based on the temperature dependence of the saturated vapor atom density given in (1.36). Under this approximation, the film thickness $d \to \infty$ when $T \to T_{\text{c}}$, if the vapor atom density $n^{(\text{a})}$ is fixed. The critical temperature T_{c} is determined by the total amount of helium in the cell. In the actual experiment, the amount of helium vapor atoms depends self-consistently on d, which makes $d(T)$ finite at low temperatures $T < T_{\text{c}}$. Thus, in the temperature range of Fig. 1.16, the helium film is a decreasing function of T.

Consider now an SE above a helium film covering an uneven substrate of solid hydrogen. The displacements $\xi^{(\text{S})}(\boldsymbol{r})$ of the solid interface are assumed to be fixed. The substrate roughness causes a static distortion $\xi^{(\text{L})}(\boldsymbol{r})$ of the free surface of the liquid helium whose amplitude depends on the film thickness according to the relation

$$\frac{\xi_q^{(\text{L})}}{\xi_q^{(\text{S})}} = \frac{0.5(qd)^2 K_2(qd)}{1 + (d/d_q)^4} , \tag{1.114}$$

where

$$d_q^4 = \frac{2\alpha_{\text{He}}\rho}{\alpha n_{\text{He}}^{(\text{L})} M_{\text{He}} q^2} , \tag{1.115}$$

and the surface tension α should be distinguished from the van der Waals constant α_{He}. For thin films $d \ll d_q$, the ratio in (1.114) is very close to unity, or $\xi^{(\mathrm{L})}(\boldsymbol{r}) \to \xi^{(\mathrm{S})}(\boldsymbol{r})$. In the opposite limit $d \gg d_q$, the free surface of the liquid helium film is nearly flat $\xi^{(\mathrm{L})} \ll \xi^{(\mathrm{S})}$.

The electron wave function adjusts itself to the displacements of the surface of the liquid helium film $\xi^{(\mathrm{L})}$, which are smaller than the solid surface roughness $\xi^{(\mathrm{S})}$. Therefore, the long-wavelength compensation of the electron polarization interaction with the solid roughness is not as complete as it is for the electron–ripplon interaction. When d is decreased, the ratio $\xi_q^{(\mathrm{L})}/\xi_q^{(\mathrm{S})}$ varies from zero to unity, as noted above, and the electron interaction with the solid roughness transforms gradually from Cole's interaction form with $eE_q < 0$ to the adiabatic form with $eE_q > 0$, changing its sign at a certain value of d. In order to see this clearly, we simplify the electron potential as follows:

$$V_{\mathrm{e}}(z, \{\xi^{(\mathrm{S})}\}) = V_{\mathrm{rep}}(z - \xi^{(\mathrm{L})}) + V_{\mathrm{att}}(z, \{\xi^{(\mathrm{S})}\}) . \qquad (1.116)$$

Here we consider only the electron attraction to the solid substrate (the second term) and the strong repulsion from liquid helium (the first term). The repulsion term depends on the liquid helium displacements in the usual way, while the attractive potential is due to the electron interaction with the substrate atoms:

$$V_{\mathrm{att}}(z, \{\xi^{(\mathrm{S})}\}) = -\frac{\Lambda_{\mathrm{S}}}{\pi} \int \mathrm{d}^2\boldsymbol{r}' \int_{-\infty}^{-d+\xi^{(\mathrm{S})}(\boldsymbol{r}')} \frac{\mathrm{d}z'}{[(\boldsymbol{r}' - \boldsymbol{r})^2 + (z' - z)^2]^2} . \qquad (1.117)$$

Once again, adding and subtracting the artificial potential $V_{\mathrm{e}}(z - \xi^{(\mathrm{L})}, 0)$ the polarization interaction is found to be

$$V_{\mathrm{att}}(z, \{\xi^{(\mathrm{S})}\}) - V_{\mathrm{att}}(z - \xi^{(\mathrm{L})}, 0) \simeq -\frac{\Lambda_{\mathrm{S}}}{\pi} \int \mathrm{d}^2\boldsymbol{r}' \frac{\xi^{(\mathrm{S})}(\boldsymbol{r}') - \xi^{(\mathrm{L})}(\boldsymbol{r})}{[(\boldsymbol{r}' - \boldsymbol{r})^2 + (z + d)^2]^2} . \qquad (1.118)$$

Finally, the interaction with the solid roughness can be rewritten in a form similar to (1.69):

$$\langle 1 | V_{\mathrm{int}} | 1 \rangle = \sum_{\boldsymbol{q}} \xi_{\boldsymbol{q}}^{(\mathrm{S})} V_q \exp(\mathrm{i}\boldsymbol{q} \cdot \boldsymbol{r}) , \qquad V_q = eE_{\perp} \xi_{\boldsymbol{q}}^{(\mathrm{L})}/\xi_{\boldsymbol{q}}^{(\mathrm{S})} + eE_q , \quad (1.119)$$

where

$$eE_q = \left\langle 1 \left| \frac{\Lambda_{\mathrm{S}} q}{z + d} \left[\frac{1}{q(z + d)} \frac{\xi_{\boldsymbol{q}}^{(\mathrm{L})}}{\xi_{\boldsymbol{q}}^{(\mathrm{S})}} - K_1 [q(z + d)] \right] \right| 1 \right\rangle . \qquad (1.120)$$

As expected, the holding field term is proportional to the liquid helium surface displacements $\xi_{\boldsymbol{q}}^{(\mathrm{L})}$. For thick films, $\xi_{\boldsymbol{q}}^{(\mathrm{L})} \to 0$ and eE_q coincides with Cole's polarization interaction potential ($eE_q < 0$), while the holding field term tends to zero. In the opposite limit of thin films, $\xi_{\boldsymbol{q}}^{(\mathrm{L})} \to \xi_{\boldsymbol{q}}^{(\mathrm{S})}$ and at $q(z+d) \ll 1$, the main asymptote of $K_1(x)$ is compensated by the first term of (1.120) and we have $eE_q > 0$. This means that, at an intermediate film thickness d^*,

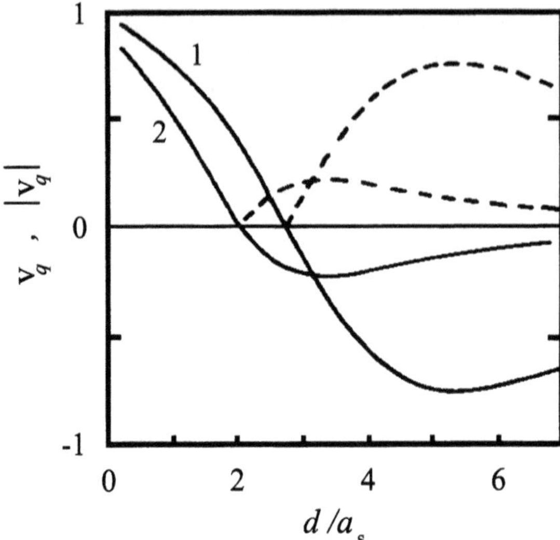

Fig. 1.17. Electron–roughness interaction potential $v_q(d)$ (*solid curves*) and its absolute value (*dashed curves*) as functions of the normalized film thickness for $n_s = 3.1 \times 10^8\,\mathrm{cm}^{-2}$ and two roughness wave numbers: (1) $q = \pi \times 10^{-5}\mathrm{cm}^{-1}$ and (2) $q = 3\pi \times 10^{-5}\mathrm{cm}^{-1}$ [69]

the interaction potential $eE_q(d^*) = 0$ and electrons do not feel the substrate roughness.

The dimensionless interaction potential $v_q(d) = E_q(d)/E_q(0)$ and its absolute value are shown in Fig. 1.17 for two typical values of q. Here we use the notation $a_s \simeq 15.7\,\text{Å}$ for the effective Bohr radius of the SEs above bare hydrogen. In the general case, the effective Bohr radius of the SEs depends on the helium film thickness d.

The physical picture behind the change of sign of the interaction potential is sketched schematically in Fig. 1.18. In the regime of strong coupling with the substrate roughness, electrons are mostly trapped at surface irregularities and their mobility is dominated by thermally-activated hopping. The important point is that the electrons are trapped differently for the following two extreme cases:

- the liquid helium surface is nearly flat ($\xi_q^{(L)} \simeq 0$),
- the helium surface follows the solid substrate roughness ($\xi_q^{(L)} = \xi_q^{(S)}$).

In the latter case, electrons are trapped in the valleys of the surface roughness, as shown in Fig. 1.18a. In the opposite extreme case (flat film surface), SEs are localized just above the maxima of the solid interface displacements (Fig. 1.18b). These two states of the system of localized electrons are topologically different and therefore, when varying d, there should be a retrapping

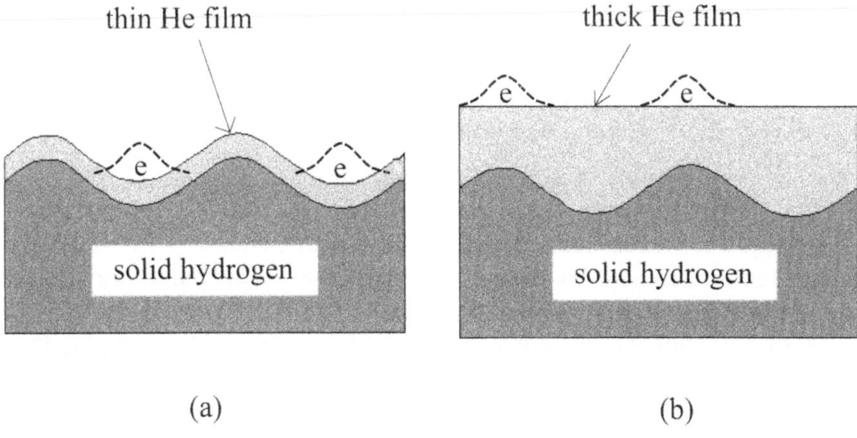

Fig. 1.18. Schematic view of the two states involved in the structural transition of the SEs above a rough solid substrate covered with a helium film: SEs are localized in the valleys of the surface roughness for thin films (**a**), and SEs are localized opposite the tops of the solid surface profile for thick films (**b**)

transition regime from one state into the other. In this regime, the electrons are nearly free and have high mobility.

The retrapping transition takes place at one point $d = d^*$, if the solid roughness is composed of the harmonics $\xi_{\boldsymbol{q}}^{(\mathrm{S})}$ of the same $|\boldsymbol{q}|$. In a real system, the transition should be broadened due to the distribution of q. Luckily, the calculations show that, for large enough q, the critical value d^* is nearly independent of q, and a change in q mainly affects the magnitude of the interaction potential. Under these circumstances, we can greatly simplify the problem by using only one typical absolute value of the wave vector $\boldsymbol{q}$ when describing the surface roughness of the solid substrate. With this model, we expect the evaluated anomalies of the SE conductivity as a function of d to be slightly sharper than the experimentally observed anomalies.

For a quantitative test of the discussed behavior, we can use the results of experimental studies on bare hydrogen surfaces [73], which indicate a thermally-activated behavior of the electron conductivity:

$$\sigma^{(\mathrm{S})} = \sigma_0 \exp\left(-\frac{\Delta\varepsilon^{(\mathrm{S})}}{T}\right) , \tag{1.121}$$

where the activation energy above bare hydrogen $\Delta\varepsilon^{(\mathrm{S})} \geq 10\,\mathrm{K}$. For the simplified model with only a single $|\boldsymbol{q}|$ dominating the electron–roughness interaction, the activation energy above a helium film $\Delta\varepsilon^{(\mathrm{F})}(d)$ can be found in the following way. The electron–roughness potential creates randomly distributed potential wells. For a trapped electron, the position of the ground level with regard to the continuous spectrum, which determines the activation energy, is proportional to the amplitude of the electron–roughness interaction potential. Quantum tunnelling along the surface is negligibly small. Because $\Delta\varepsilon^{(\mathrm{F})}(d)$ should satisfy the limiting condition $\Delta\varepsilon^{(\mathrm{F})}(0) = \Delta\varepsilon^{(\mathrm{S})}$, the activation energy $\Delta\varepsilon^{(\mathrm{F})}(d)$ can be written with the proportionality factor as

$$\Delta\varepsilon^{(\mathrm{F})}(d) = \Delta\varepsilon^{(\mathrm{S})}|v_q(d)| \,, \tag{1.122}$$

where $v_q(d)$ is the normalized interaction potential shown in Fig. 1.17. The detailed information about the substrate roughness is hidden in the quantity $\Delta\varepsilon^{(\mathrm{S})}$ which has been measured experimentally.

Assuming that the SEs are localized at both sides of the retrapping transition ($\Delta\varepsilon^{(\mathrm{F})} > T$), with the exception of the near vicinity of it where $\Delta\varepsilon^{(\mathrm{F})}$ is smaller than T, the thermally-activated conductivity

$$\sigma^{(\mathrm{F})} = \sigma_0 \exp\left(-|v_q(d)|\frac{\Delta\varepsilon^{(\mathrm{S})}}{T}\right) \tag{1.123}$$

represents a good approximation. According to this, the conductivity of SEs exhibits two anomalies. The peak anomaly occurs at $d \to d^*$ where the absolute value of the interaction potential attains the minimum $|v_q(d)| \to 0$. The dip anomaly occurs when $|v_q(d)|$ attains a local maximum at a certain $d = d_{\mathrm{m}} > d^*$, as shown in Fig. 1.17. In this case, electrons are trapped opposite peaks of solid roughness.

In order to guess the most relevant value of q, which determines the SE interaction with the quench-condensed hydrogen film, we compare the experimental data with the conductivity curves evaluated by means of (1.123) for different values of the roughness wave number. The results of the numerical evaluations, taking into account electron interaction with solid and liquid roughness, are shown in Fig. 1.19a. Here the activation energy $\Delta\varepsilon^{(\mathrm{S})} = 18\,\mathrm{K}$ and the electron density $n_s \simeq 3.1 \times 10^8\,\mathrm{cm}^{-2}$ correspond to the experimental conditions. Comparing these theoretical curves with the conductivity data vs. frequency of the quartz balance measuring the helium film thickness (Fig. 1.16), the dominant surface roughness wave number is estimated to be typically around or above $3\pi \times 10^5\,\mathrm{cm}^{-1}$. Figure 1.19a shows that, in the short wavelength limit, the position of the conductivity maximum is nearly independent of q, unlike the position of the minimum, justifying the single $|\boldsymbol{q}|$ model of solid roughness employed above.

The black circles in Fig. 1.19b show the experimental dependence $\sigma^{(\mathrm{F})}(T)$ due to [69, 73] for a freshly prepared quench-condensed hydrogen film which is extremely rough. The conductivity maximum appears as a pronounced feature. The sharp tip of the peak indicates, according to Fig. 1.19a, that the electrons tend to occupy rather long-wavelength potential wells induced by the solid substrate. Data taken after annealing, also shown in this figure by white circles, exhibit a much smoother maximum which is only slightly shifted into the range of smaller thickness values (i.e., higher T), in accordance

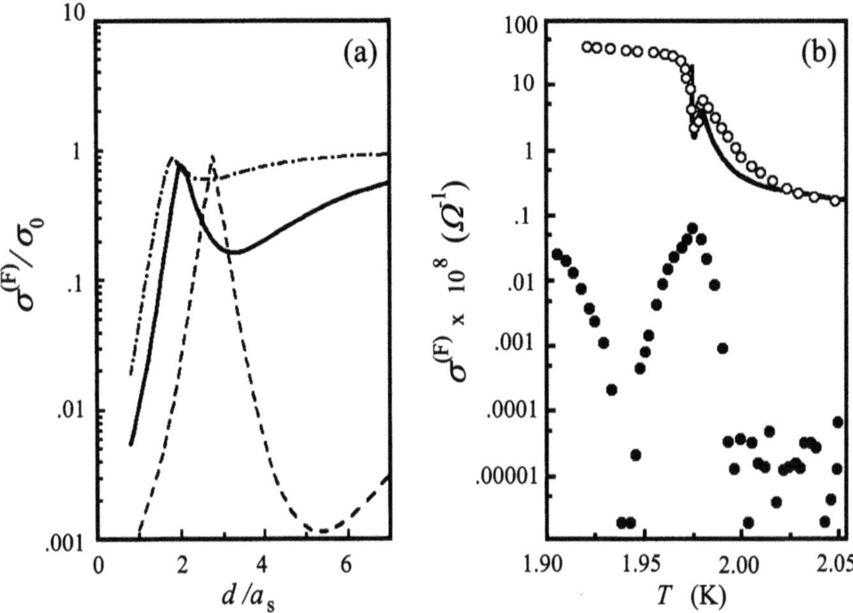

Fig. 1.19. Conductivity anomalies of SEs above a helium film. (**a**) Theoretical evaluations of $\sigma(d)$ for three values of q: $3.2 \times 10^5\,\mathrm{cm}^{-1}$ (*dashed*), $10^6\,\mathrm{cm}^{-1}$ (*continuous*), and $2 \times 10^6\,\mathrm{cm}^{-1}$ (*dash-dotted*). (**b**) Experimental $\sigma(T)$ data for a freshly prepared hydrogen substrate (*black circles*), after annealing (*white circles*), and in theory (*solid curve*) fitted by $\Delta\varepsilon^{(S)} = 12\,\mathrm{K}$ and $q = 3\pi \times 10^5\,\mathrm{cm}^{-1}$ [69]

with the theoretical concept presented in Fig. 1.19a. At the same time, the conductivity minimum is much more strongly shifted than the conductivity maximum towards higher T (thinner films), which agrees remarkably well with the theoretical evaluations shown in Fig. 1.19a. One can conclude that the annealing process reduces the long-wavelength roughness – the observed behavior is then in perfect accordance with the theoretical concept of the structural retrapping transition discussed above.

1.9 Cyclotron Motion: Quantization and Collision Broadening

A 2D electron gas, subject to the magnetic field $\boldsymbol{B}$ oriented normally to the electron layer, represents a unique system of free particles with a purely discrete energy spectrum. The symmetry of the system uncouples the electron Hamiltonian into two independent parts corresponding to electron motion in the plane of the system and in the perpendicular direction $H_\mathrm{e} = H_\mathrm{e}^{(\parallel)} + H_\mathrm{e}^{(\perp)}$. Here we disregard $H_\mathrm{e}^{(\perp)}$, which was discussed in Sect. 1.2. The in-plane part of the electron Hamiltonian has the standard form

$$H_{\rm e}^{(\parallel)} = \frac{1}{2m_{\rm e}} \left(\boldsymbol{p} + \frac{e}{c} \boldsymbol{A} \right)^2 . \tag{1.124}$$

We use the Landau gauge for the vector potential, i.e., $\boldsymbol{A} = (0, Bx, 0)$ and the Landau basis wave functions $\langle \boldsymbol{r} | N, X \rangle$, with $N = 0, 1, 2, \ldots$ and X the center coordinate of the cyclotron motion:

$$\langle \boldsymbol{r} | N, X \rangle = \frac{1}{S_{\rm A}^{1/4}} \varphi_N(x - X) \exp\left(-{\rm i} X y / l_B^2\right) , \tag{1.125}$$

where $\varphi_N(x)$ is a one-dimensional harmonic oscillator wave function, and $l_B = \sqrt{\hbar c / eB}$ is the magnetic length.

The energy spectrum of the Hamiltonian of (1.124) is well known. It is a set of equally spaced Landau levels

$$\varepsilon_{N,X}^{(\parallel)} = \hbar \omega_{\rm c} (N + 1/2) , \tag{1.126}$$

where $\omega_{\rm c} = eB/m_{\rm e}c$ is the cyclotron frequency.

Each Landau level contains $n_B = 1/2\pi l_B^2$ states per unit area owing to the degeneracy with regard to the position of the orbit center coordinate X. The ratio $n_{\rm s}/n_B$ is the filling factor at $T \to 0$. For $B = 1\,{\rm T}$, the value of n_B is estimated to be approximately $2.4 \times 10^{10}\,{\rm cm}^{-2}$, which is significantly larger than the electron density $n_{\rm s}$ realized on a free surface of liquid helium (although such densities can be realized on helium films). As the filling factor is very small, at $\hbar\omega_{\rm c} \gg T$, SEs predominantly populate the lowest Landau level.

In the presence of the magnetic field, the Fourier transform of the electron density operator $n_{\boldsymbol{q}} = \sum_e \exp(-{\rm i}\boldsymbol{q} \cdot \boldsymbol{r}_e)$ can be conveniently written in terms of the Fermi creation and annihilation operators defined for the Landau states

$$n_{\boldsymbol{q}} = \sum_X \sum_{NN'} J_{N,N'}(\boldsymbol{q}) c_{N,X}^{\dagger} c_{N',X - l_B^2 q_y} , \tag{1.127}$$

where the matrix element is

$$J_{NN'} = \left\langle N, X \left| \exp(-{\rm i}\boldsymbol{q} \cdot \boldsymbol{r}) \right| N', X - q_y l_B^2 \right\rangle .$$

The quantity $|J_{N,N'}|^2$ enters the quantum transport equations for the electron conductivity and restricts the momentum exchange at a collision because it is proportional to $\exp(-q^2 l_B^2/2)$:

$$|J_{N,N+M}|^2 = \frac{N!}{(N+M)!} x^M {\rm e}^{-x} \left[L_N^M(x) \right]^2 , \tag{1.128}$$

where $L_N^M(x)$ are the associated Laguerre polynomials and we have introduced the dimensionless parameter $x = q^2 l_B^2/2$. In the quantum limit, the restriction $\hbar q \sim \sqrt{2}\hbar/l_B$ is stronger than the condition $q \leq 2k$ given in the semi-classical treatment.

Knowledge of the electron spectrum of the 2D electron gas under a magnetic field is not sufficient to describe quantum transport phenomena. The singular nature of this electron system causes quite uncertain conductivity results in the conventional approximations. For elastic scattering, the effective collision frequency in the DC case $\nu_{\text{eff}}(\omega = 0) \to \infty$, because of multiple electron scattering with the same scatterer. On the other hand, inelastic scattering with an energy exchange, smaller than the Landau level spacing $\hbar\omega_c$, makes $\nu_{\text{eff}}(0) = 0$. Although the energy exchange $\Delta\varepsilon_s$ may be extremely small, the fact that the real scattering is inelastic means that the final result for σ_{xx} depends strongly on the relation between $\Delta\varepsilon_s$ and the width of the electron density of states.

Owing to the singular nature of a 2D electron gas in a normal magnetic field, the electron density-of-states function consists of a series of delta functions located at $\varepsilon = \varepsilon_N$ [from now on we shall drop the superscript $(\|)$], with the level degeneracy $n_B = 1/2\pi l_B^2$ as proportionality factor:

$$\Delta_{2D}(\varepsilon) = n_B \sum_N \langle N, X|\, \delta(\varepsilon - H_e)\, |N, X\rangle \ . \tag{1.129}$$

Here H_e is the electron Hamiltonian. We disregard the small complication of including the electron spin, bearing in mind that it does not change the final result. The interaction with scatterers broadens the peaks of the density of states because we have to average (1.129) over the configuration of scatterers. Before discussing the effect of collision broadening, we need to introduce some basic definitions.

In order to include the interaction with scatterers, one should represent the matrix elements of (1.129) as the imaginary part of the single-electron Green's function:

$$\langle N, X|\, \delta(\varepsilon - H_e)\, |N, X\rangle = -\frac{1}{\pi\hbar}\text{Im}G_N(\varepsilon) \ .$$

We use the conventional definition [10]

$$G_N(\varepsilon)\delta_{NN'}\delta_{XX'} = \hbar \left\langle \middle| c_{N,X}(\varepsilon - H_e + \mathrm{i}0)^{-1} c^\dagger_{N',X'} \middle| \right\rangle , \tag{1.130}$$

where $\langle\ |$ is the ground state of H_e, which includes the interactions, and $c_{N,X}$ is the electron destruction operator. In the perturbation series for the Green's function of an electron in an empty energy band, it is sometimes useful to have the Green's functions of scatterers taken at a finite temperature. In this case, following [38], we assume that the thermal average is taken over the occupation numbers of scatterers. In other words, one can assume that the average over the scatterer variables $\langle\ \ \rangle_{\text{sc}}$ is included in the definition of (1.130).

There are other useful representations of the electron Green's function which we also use in this book. Firstly, it should be noted that

$$G_N(\varepsilon) = -\mathrm{i}\left\langle \left| c_{N,X} \int_0^{\infty} \exp\left[\mathrm{i}\left(\varepsilon - H_{\mathrm{e}} + \mathrm{i}0\right) t/\hbar\right] \mathrm{d}t\, c_{N,X}^{\dagger} \right| \right\rangle , \tag{1.131}$$

which shows that the electron Green's function is the Laplace transform of the unitary time-evolution operator averaged in the proper way. Here and below the average $\langle \ \ \rangle_{\mathrm{sc}}$ is assumed but not written explicitly. If $|\ \rangle$ is the vacuum state of H_{e}, one can rewrite (1.131) as the Fourier transform of the time-ordered Green's function:

$$G_N(\varepsilon) = \int_{-\infty}^{\infty} \mathrm{e}^{\mathrm{i}\varepsilon t/\hbar} G_N(t) \mathrm{d}t\,, \qquad G_N(t) = -\mathrm{i}\left\langle \left| \hat{T} c_{N,X}(t) c_{N,X}^{\dagger} \right| \right\rangle , \tag{1.132}$$

where $\hat{T}$ is the conventional time-ordering operator and

$$c_\alpha(t) = \mathrm{e}^{\mathrm{i}H_{\mathrm{e}}t/\hbar} c_\alpha \mathrm{e}^{-\mathrm{i}H_{\mathrm{e}}t/\hbar} . \tag{1.133}$$

The perturbation series for this function is well known from the Feynman diagram technique. In the pole approximation usually used for systems with a continuous spectrum,

$$G_\alpha(t) \simeq -\mathrm{i} \exp\left(-\mathrm{i}\varepsilon_\alpha t/\hbar - \gamma_\alpha t/\hbar\right) , \tag{1.134}$$

which describes the exponential decay of an electron state for long times. We shall see that this conventional quasiparticle behavior fails for systems with discrete energy spectrum, such as the 2D electron gas under a quantizing magnetic field.

In order to arrive at a finite magnetoconductivity for the 2D electron gas interacting with a static impurity potential, one must introduce the collision broadening of Landau levels Γ_N. There are two main approaches to describing the collision broadening of Landau levels. The self-consistent Born approximation (SCBA) was introduced by Ando and Uemura [74] to describe the DC magnetoconductivity of the degenerate 2D electron gas. In the SCBA, the electron density of states is treated in the framework of the self-consistent perturbation theory for the electron Green's function [(1.132)] and it is found to have a semi-elliptical shape with sharp edges. In another treatment proposed by Gerhardts [75], the time evolution operator entering (1.131), averaged over the distribution of impurities, is presented in the form of a cumulant expansion. Restricting the analysis to the second order terms, Gerhardts found that the electron density of states is a Gaussian with smooth tails described by the same broadening parameter Γ_N. In this section, we discuss both these approaches applied to the different kinds of scatterers available for interface electrons.

1.9.1 Self-Consistent Approximation

The great advantage of the conventional perturbation theory for the Green's function is that it is possible to obtain a general form for $G_N(\varepsilon)$ by formally summing the infinite series. The best known example is Dyson's equation (see [38]):

$$G_{\alpha}(\varepsilon) = \frac{\hbar}{\varepsilon - \varepsilon_{\alpha} + \mathrm{i}0 - \Sigma_{\alpha}(\varepsilon)} , \tag{1.135}$$

where the electron self-energy $\Sigma_N(\varepsilon)$ is the sum of the infinite number of self-energy terms. The structure of these terms depends on the interaction Hamiltonian V_{int}. In the case of SEs, V_{int} is generally written as

$$V_{\mathrm{int}} = \sum_{\boldsymbol{q}} U_{\mathrm{s},q} A_{\mathrm{s},\boldsymbol{q}} n_{-\boldsymbol{q}} ,$$

where the particular forms of the coupling function $U_{\mathrm{s},q}$ and the many-body operator of scatterers $A_{\mathrm{s},\boldsymbol{q}}$ are defined in (1.39) and (1.56), and $n_{\boldsymbol{q}}$ is the electron density fluctuation operator.

Using the general form of the interaction Hamiltonian and the presentation for $n_{\boldsymbol{q}}$ under a magnetic field [(1.127)], the simplest contribution to the electron self-energy can be written as [36, 76]

$$\Sigma_N(\varepsilon) = \mathrm{i} \sum_{q} \sum_{N'} |J_{N,N'}|^2 \int \frac{\mathrm{d}\omega}{2\pi\hbar} G_{N'}^{(0)}(\varepsilon - \hbar\omega) U_{\mathrm{s},q}^2 D_{\mathrm{s}}^{(0)}(q,\omega) . \tag{1.136}$$

This equation is valid for both kinds of scatterer available in the system of SEs, which are distinguished here by subscripts $\mathrm{s} = \mathrm{a}, \mathrm{r}$. Such a self-energy contribution is typical for electron–phonon systems. The Green's functions of scatterers, i.e.,

$$D_{\mathrm{s}}^{(0)}(q,\omega) = -\mathrm{i} \left\langle \hat{T} A_{\mathrm{s},\boldsymbol{q}}(t) A_{\mathrm{s},\boldsymbol{q}}^{\dagger}(t') \right\rangle_{\omega} ,$$

were discussed in Sects. 1.4 and 1.5. The main approximation of the SCBA theory is that the unperturbed electron Green's function $G_{N'}^{(0)}$ entering the right-hand side of (1.136) is replaced by the exact Green's function $G_{N'}$. In this approximation, (1.135) and (1.136) become a set of two self-consistent equations for $G_N(\varepsilon)$. This procedure is assumed to be a substitution for the partial summation of the self-energy diagrams.

Quite generally, $D_{\mathrm{s}}^{(0)}(q,\omega)$ (or its imaginary part) is proportional to $\delta(\omega - \Delta\omega)$, where $\hbar\Delta\omega$ is the energy exchange at collision. For vapor atoms, $\Delta\omega = \Delta_{\boldsymbol{K}',\boldsymbol{K}'-\boldsymbol{K}}$, according to (1.45), while for ripplons $\Delta\omega = \pm\omega_q$. If the energy exchange at collisions is small compared to the broadening of Landau levels $\hbar\Delta\omega \ll \Gamma_N$ (quasi-elastic scattering), the frequency shift in the argument of the electron Green's function $G_{N'}(\varepsilon - \hbar\omega)$ can be disregarded. Then, taking into account the fact that $D_{\mathrm{s}}(q, t \to 0) = \int D_{\mathrm{s}}^{(0)}(q,\omega)\mathrm{d}\omega/2\pi$ is proportional to $-\mathrm{i}$ and disregarding mixing between different Landau levels, we find

$$\Sigma_N(\varepsilon) = \frac{1}{4\hbar} \Gamma_N^2 G_N(\varepsilon) , \tag{1.137}$$

where

$$\Gamma_N^2 = 4 \sum_q U_{s,q}^2 \left| J_{N,N} \right|^2 \left| D_{\mathrm{s}}(q, t \to 0) \right| . \tag{1.138}$$

For vapor atoms, according to (1.127),

$$D_{\mathrm{a}}(q, t \to 0) = -\mathrm{i} N^{(\mathrm{a})} \sum_k \left| \eta_k \right|^2 , \tag{1.139}$$

while for 2D phonons (ripplons),

$$D_{\mathrm{s}}(q, t \to 0) = -\mathrm{i}(2N_q + 1) . \tag{1.140}$$

[see (1.58)].

Substituting the form of (1.137) for the electron self-energy in Dyson's equation yields a simple quadratic equation:

$$G_N^2 - \frac{4(\varepsilon - \varepsilon_N)\hbar}{\Gamma_N^2} G_N + \frac{4\hbar^2}{\Gamma_N^2} = 0 , \tag{1.141}$$

whose solution can be written as

$$G_N(\varepsilon) = \frac{2(\varepsilon - \varepsilon_N)\hbar}{\Gamma_N^2} - \mathrm{i} \frac{2\hbar}{\Gamma_N} \sqrt{1 - \frac{(\varepsilon - \varepsilon_N)^2}{\Gamma_N^2}} . \tag{1.142}$$

Thus the Landau level density defined as $-(n_B/\pi\hbar)\mathrm{Im}G_N(\varepsilon)$ has a semi-elliptical shape with sharp edges at $\varepsilon = \varepsilon_N \pm \Gamma_N$. A more rigorous theory due to Ando [77] shows that the level shape is generally a kind of average of elliptical and Gaussian forms. The lowest Landau level is shaped like a Gaussian. However, the level density gradually approaches the semi-elliptical shape with increasing level number N.

For electron scattering at a short-range 'impurity' potential like the one induced by vapor atoms, use of (1.139) yields

$$\Gamma_N = \Gamma_{\mathrm{se}} \equiv \sqrt{\frac{2}{\pi} \hbar^2 \omega_c \nu_0} , \tag{1.143}$$

where ν_0 coincides with the collision frequency under zero magnetic field. The Landau level broadening does not depend on the level number N and it increases with the magnetic field as $\Gamma_{\mathrm{se}} \propto \sqrt{B}$. For vapor-atom scattering,

$$\nu_0 = \frac{[V^{(\mathrm{a})}]^2 n^{(\mathrm{a})} m_e}{\hbar^3} \frac{1}{L_z} \sum_k \left| \eta_k \right|^2 \equiv \nu_0^{(\mathrm{a})} . \tag{1.144}$$

Thus, Γ_{se} decreases at an exponential rate with cooling owing to the vapor atom density $n^{(\mathrm{a})}(T)$.

Following Ando and Uemura [74], it is instructive to study the range dependence of Γ_N by introducing the scatterers which interact with the SEs by means of an artificial Gaussian potential

$$V(R) = V^{(\mathrm{a})}\delta(z)\frac{1}{\pi d^2}\exp\left(-\frac{r^2}{d^2}\right) . \tag{1.145}$$

This potential approaches the usual δ-function potential if the range parameter $d \to 0$. The new potential introduces an additional factor $\exp(-q^2 d^2/2)$ into the integrand of (1.138). For the lowest Landau level, straightforward evaluation of (1.138) gives

$$\Gamma_0 = \frac{\Gamma_{\mathrm{se}}}{\sqrt{1 + d^2/l_B^2}} . \tag{1.146}$$

For higher Landau levels, the range dependence of Γ_N can be evaluated numerically. According to [74], the ratio Γ_N/Γ_0 has a minimum at $d/l_B \sim 1$ whose position lowers with N.

For the electron–ripplon scattering, using the property of the ripplon Green's function given in (1.140), we find that

$$\Gamma_N^2 = \frac{4}{S_{\mathrm{A}}}\sum_q V_q^2 Q_q^2 \mathrm{e}^{-q^2 l_B^2/2}\left[L_N\left(q^2 l^2/2\right)\right]^2\left(2N_q^{(\mathrm{r})} + 1\right) . \tag{1.147}$$

In this instance, the level broadening depends on the Landau level number N and it has much weaker temperature dependence: $\Gamma_N \propto \sqrt{T}$, owing to the fact that $N_q^{(\mathrm{r})} \approx T/\hbar\omega_q$. Another distinguishing feature of the collision broadening induced by ripplons is that it has a logarithmic divergence for the pure capillary spectrum $\omega_q \propto q^{3/2}$, the same as the divergence in $\langle \xi^2 \rangle$. The inclusion of gravity or van der Waals forces eliminates this divergence, although there might be other reasons for the cutoff at small q.

1.9.2 Cumulant Approach

The cumulant expansion method [75] results in a level shape which has the appropriate smooth behavior at the edges. Another advantage of this method, convenient for analytical evaluations, is that the electron density of states is described by a simple Gaussian function. The cumulant approach starts with (1.131), which establishes the relationship between the electron Green's function and the Laplace transform of the diagonal matrix element of the time evolution operator

$$J_H(t) = \exp\left(-\mathrm{i}Ht/\hbar\right) . \tag{1.148}$$

This operator obeys the equation

$$\frac{\partial J_H}{\partial t} = -\frac{\mathrm{i}}{\hbar} H J_H \tag{1.149}$$

and the condition $J_H(0) = 1$.

In the presence of an interaction with scatterers $H = H_0 + V$, one cannot uncouple the exponent of (1.148) in a trivial way. We must use the algebra of non-commutative operators. The following result is widely used in practical evaluations:

$$J_{H_0+V}(t) = J_{H_0}(t) J_{\tilde{V}}(t) , \tag{1.150}$$

where

$$\tilde{V} = \mathrm{e}^{\mathrm{i}H_0 t/\hbar} V \mathrm{e}^{-\mathrm{i}H_0 t/\hbar} , \tag{1.151}$$

and $J_{\tilde{V}}(t)$ is defined as a solution of the equation

$$\frac{\partial J_{\tilde{V}}}{\partial t} = -\frac{\mathrm{i}}{\hbar} \tilde{V} J_{\tilde{V}} , \tag{1.152}$$

which satisfies the condition $J_{\tilde{V}}(0) = 1$. It should be noted that $\tilde{V}$ is just an operator in the interaction representation, whose time development is governed by the unperturbed Hamiltonian. Therefore in the following we shall also use the notation $\tilde{V} = V(t)$, assuming that the time evolution of V is described by H_0.

From (1.152), we find the useful integral equation

$$J_{\tilde{V}}(t) = 1 + \int_0^t \mathrm{d}\tau \frac{(-\mathrm{i})}{\hbar} V(\tau) J_{\tilde{V}}(\tau) .$$

By iteration, $J_{\tilde{V}}(t)$ is represented as the conventional time-ordered exponent

$$\exp\left(-\mathrm{i}Ht/\hbar\right) = \exp\left(-\mathrm{i}H_0 t/\hbar\right) \hat{T} \exp\left[-(\mathrm{i}/\hbar) \int_0^t \mathrm{d}\tau V(\tau)\right] . \tag{1.153}$$

The abbreviation $\hat{T}\exp(\cdots)$ means that the time ordering operator $\hat{T}$ is applied to all terms in the expansion of the corresponding exponent.

In order to obtain the electron Green's function G_N, we need to average the time-ordered exponent over the scatterer variables. It would be very simple if it were an ordinary exponent. For example, in the case of the electron interaction with phonons (ripplons), the interaction Hamiltonian V is a linear function of the Bose creation and destruction operators, $b^\dagger$ and b, respectively. When averaging such an exponent, one can use the Bloch identity

$$\left\langle \exp \hat{Q} \right\rangle = \exp\left(\frac{1}{2}\left\langle \hat{Q}^2 \right\rangle\right) , \tag{1.154}$$

where Q is a linear combination of b and $b^\dagger$. For the time-ordered exponent, we may expect something similar to this equation. Gerhardts [75] suggested using the cumulant expansion

$$\left\langle T \exp(\lambda \hat{Q}) \right\rangle = \exp\left(\sum_{\nu=1}^{\infty} \frac{\lambda^\nu}{\nu!} C_\nu\right) , \tag{1.155}$$

where the cumulants C_ν are found by equating the proportionality factors of equal power terms with regard to the dimensionless coupling parameter λ. For the electron–phonon interaction, the first order term C_1 is zero. The second order term

$$C_2 = \left(-\frac{\mathrm{i}}{\hbar}\right)^2 \int_0^t \mathrm{d}t_1 \int_0^t \mathrm{d}t_2 \left\langle \hat{T} V(t_1) V(t_2) \right\rangle \tag{1.156}$$

is finite and plays the main role in the cumulant approach.

Using the general form of the interaction Hamiltonian, we find that

$$\begin{aligned} \left[\left\langle \hat{T} V(t_1) V(t_2) \right\rangle\right]_{\alpha,\alpha} &= \frac{1}{S_\mathrm{A}} \sum_{\boldsymbol{q}} U_{\mathrm{s},q}^2 \sum_{N'} |J_{N,N'}|^2 \\ &\times \exp\left[\mathrm{i}\frac{(\varepsilon_N - \varepsilon_{N'})}{\hbar} |t_1 - t_2|\right] \mathrm{i} D_\mathrm{s}^{(0)}(q, t_1 - t_2) \,, \end{aligned} \tag{1.157}$$

where $D_\mathrm{s}^{(0)}(q,t)$ is the Green's function of scatterers. Even at this point one can see that, for quasi-elastic scattering [$\omega_q \to 0$ and $D_\mathrm{s}^{(0)}(q,t) \simeq D_\mathrm{s}^{(0)}(q,0)$] and $N' = N$,

$$\frac{1}{2}\left[C_2(t)\right]_{\alpha,\alpha} \simeq -\frac{\Gamma_N^2 t^2}{8\hbar^2} \,, \tag{1.158}$$

with Γ_N is defined by (1.138).

A more detailed analysis carried out for the electron–phonon interaction yields

$$\begin{aligned} \frac{1}{2}\left[C_2(t)\right]_{\alpha,\alpha} &= -\frac{1}{\hbar^2 S_\mathrm{A}} \sum_{\boldsymbol{q}} V_q^2 Q_q^2 \sum_{N'} |J_{N,N'}|^2 \\ &\times \left[(N_q + 1)\phi(\omega_{N',N} + \omega_q, t) + N_q \phi(\omega_{N',N} - \omega_q, t)\right] \,, \end{aligned} \tag{1.159}$$

where $\omega_{N',N} = (\varepsilon_{N'} - \varepsilon_N)/\hbar$ and

$$\phi(\omega, t) = \frac{t}{\mathrm{i}\omega} + \frac{1}{\omega^2}\left[1 - \exp(-\mathrm{i}\omega t)\right] \,. \tag{1.160}$$

It is instructive to note that nearly the same result could be found disregarding the time ordering and using the Bloch identity of (1.155). In this case, $\phi(\omega,t)$ would have the slightly different form

$$\phi(\omega, t) = P(\omega, t) \equiv \frac{1 - \cos(\omega t)}{\omega^2} \,. \tag{1.161}$$

The function $P(\omega,t)$ appears frequently in the conventional analysis of the probability of electron scattering. In the long-time limit, we have

$$P(\omega, t) \simeq \pi t \delta(\omega) \,. \tag{1.162}$$

Nevertheless, for $N_q \gg 1$, the contribution of the main term of (1.159) with $N' = N$, playing the major role in the cumulant approach, is (remarkably) the same as the one found by means of the Bloch identity:

$$(N_q + 1)\phi(\omega_q, t) + N_q\phi(-\omega_q) \simeq (2N_q + 1)P(\omega_q, t) \, .$$

Therefore the difference between the ordinary exponent and the time-ordered exponent in (1.153) becomes important only when the number of medium vibration quanta is small, i.e., $N_q \sim 1$. For SEs on liquid helium, we usually have the opposite situation, i.e., $N_q^{(\mathrm{r})} \gg 1$, because the ripplon wave number q is restricted by the electron magnetic length: $q \sim 1/l_B$. Thus, one may conclude that the Bloch identity can also be used to describe the broadening of Landau levels, if the medium vibrations involved can be treated quasi-classically.

The important property of the function $\phi(\omega, t)$ is that, for low frequencies $\omega \to 0$, it does not depend on ω and is proportional to t^2:

$$\phi(\omega, t) \simeq t^2/2 \, . \tag{1.163}$$

The same limiting behavior is satisfied by $P(\omega, t)$ from (1.161). This means that, for quasi-elastic scattering, when the energy exchange at collisions is small compared to the Landau level broadening Γ_N, the second order cumulant can be approximated by (1.158).

A similar result can be found for electrons scattered by vapor atoms. In this case, substitution of the proper Green's function of scatterers from (1.58) yields

$$\frac{1}{2}\left[C_2(t)\right]_{\alpha,\alpha} = -\frac{1}{\hbar^2}\left(\frac{V^{(\mathrm{a})}}{\Omega_v}\right)^2 \sum_{\boldsymbol{q}}\sum_{N'} |J_{N,N'}|^2 \sum_k |\eta_k|^2 \times \sum_{\boldsymbol{K}'} N_{\boldsymbol{K}'}^{(\mathrm{a})} \phi(\omega_{N',N} + \Delta_{\boldsymbol{K}'-\boldsymbol{K},\boldsymbol{K}'}, t) \, . \tag{1.164}$$

In the limiting case of elastic scattering $\Delta_{\boldsymbol{K}'-\boldsymbol{K},\boldsymbol{K}'} \to 0$, the Landau level broadening parameter Γ_N which follows from (1.164) appears to be the same as that defined in the framework of the self-consistent approximation [(1.138) and (1.147)].

For electrons with a continuous spectrum, one can use the long-time approximation which yields the usual quasiparticle behavior of the electron propagator $G_\alpha(t)$ [see (1.134)]. If the electron system is characterized by the purely discrete Landau spectrum ($\varepsilon_\alpha = \varepsilon_N$) and the scattering is elastic, the long-time limit does not exist, and the electron propagator has a substantially different time-dependence [75]:

$$G_\alpha(t) \approx (-\mathrm{i}/\hbar) \exp\left(-\mathrm{i}\varepsilon_\alpha t/\hbar - \frac{\Gamma_\alpha^2 t^2}{8\hbar^2}\right) \, . \tag{1.165}$$

In the real case, there is an energy exchange between an electron and a scatterer $\hbar\Delta$ (although it may be extremely small). Therefore, (1.165) is valid only at rather short times $t < 1/\Delta$. At longer times the electron propagator has the usual quasiparticle time dependence given in (1.134). Electron scattering can thus be treated quasi-elastically only if $\hbar\Delta \ll \Gamma_N$.

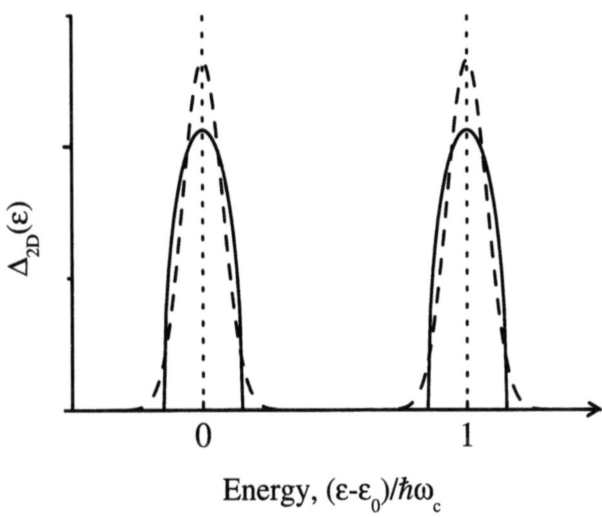

Fig. 1.20. The density-of-state function for two nearest Landau levels: semi-elliptical form (*solid line*) and Gaussian form (*dashed line*)

The shape of the density-of-state function is determined by the imaginary part of the electron Green's function: $\Delta_{2D}(\varepsilon) = -(n_B/\pi\hbar)\mathrm{Im}G_N(\varepsilon)$. By taking the simple time integral, we find

$$-\mathrm{Im}G_N(\varepsilon) = \frac{\sqrt{2\pi}\hbar}{\Gamma_N}\exp\left[-\frac{2(\varepsilon-\varepsilon_N)^2}{\Gamma_N^2}\right] . \tag{1.166}$$

The cumulant approach thus results in the Gaussian shape of Landau levels, which does not have the sharp cutoff of the density of states appearing in the SCBA. Both shapes of the electron density-of-state function are shown in Fig. 1.20.

It is very important to use the density-of-state function broadened by the interaction with scatterers for the electron magnetotransport equations of the two-dimensional electron gas, when the mutual Coulomb interaction can be disregarded. Such a density-of-state function eliminates the uncertainty in the electron conductivity mentioned above. The influence of the strong internal forces acting on an electron of the Coulomb liquid will be discussed in the next chapter.

2 Strongly Correlated Coulomb Liquid

2.1 Fundamental Correlation Functions

In typical liquids and even solids, the average interaction potential energy per particle is approximately of the same order of magnitude as the average kinetic energy. Under these conditions, the Fermi liquid approach, which reduces the many-body system of strongly interacting particles to the system of weakly interacting excitations, has proven to be quite effective. On the other hand, the Coulomb liquid is characterized by extremely strong internal forces acting between electrons. This makes the conventional quasi-particle approach doubtful. As emphasized above, in the 2D Coulomb liquid of interface electrons the average interaction potential energy per electron can be approximately a hundred times larger than the average kinetic energy. One cannot rely on the Fermi liquid approach under such extreme conditions. It is even impossible to introduce a definite excitation spectrum valid for the whole Coulomb liquid if the system is subject to a normal magnetic field.

In order to describe such strongly correlated electron systems, one has to use the general technique of quantum correlation functions. In Chap. 3 we shall see that the conductivity description based on the quasi-particle spectrum and kinetic equations is too detailed: in the quantum transport theory of highly correlated electrons one actually needs to know a more global property of the electron liquid, namely, the density–density correlation function or the dynamical structure factor (DSF). In this chapter we discuss the most general properties of the 2D electron liquid which originate from the strong mutual Coulomb interaction. Remarkably, we shall find that, in spite of the strong internal forces acting between electrons, the single-electron behavior remains for global properties of the Coulomb liquid such as the DSF $S(q,\omega)$, even though the system is not equivalent to a gas-like collection of independent electrons or quasi-particles. For example, it is remarkable that in certain cases the DSF of the Coulomb liquid, and even of the Wigner solid, can be expressed in terms of a simple double integral of the DSF found for the ideal electron gas. Therefore, the use of seemingly more complicated notions and theoretical 'building blocks' such as the quantum correlation functions significantly simplifies the description of strongly interacting electron systems and reveals interesting physics.

2.1.1 General Definitions

In this section, we introduce some basic definitions and relations which will be used frequently throughout this book. Athough it is practically impossible to make this introduction entirely self-contained and a certain general knowledge of condensed matter theory is required of the reader, our intention is to make the following theoretical digressions as easy as possible. We shall deliberately avoid proofs of relations and identities that are quite standard and can be found in practically any contemporary textbook. We shall nevertheless recapitulate the important details needed to understand these standard relations and the results obtained from them, which are the subject matter of this book.

In the last chapter we already discussed certain elements of the many-body theory, such as the Green's functions of scatterers (Sects. 1.4 and 1.5) and the Green's function of an electron in the empty band (Sect. 1.9). The electron Green's function $G_\alpha(t-t')$ is usually introduced as the average of the time-ordered product of the electron destruction c_α and creation $c^\dagger_\alpha$ operators taken at different times. This quantum correlation function is very convenient for the description of the single-electron properties of the system. For example, in Sect. 1.9 we used it to find the single-electron density-of-state function. The Green's function of scatterers $D_{\mathrm{s}}(q,t)$, defined as the average of the time-ordered product of the scatterer operators $A_{\mathrm{s},\boldsymbol{q}}$ and $A_{\mathrm{s},-\boldsymbol{q}}$, appears in the analysis of the perturbation series for the electron Green's function $G_\alpha(t-t')$. Here we are going to discuss briefly the quantum correlation functions of a different kind which appear naturally in the quantum linear response theory, and which are also often called Green's functions. They are very useful for finding the average values of certain physical quantities, such as the electric current and conductivity.

In quantum statistical physics, the average value of a physical quantity described by an operator A is determined by the statistical operator $\rho^{(\mathrm{st})}$:

$$\langle A\rangle = \mathrm{Tr}(\rho^{(\mathrm{st})}A)\ , \tag{2.1}$$

where Tr denotes the trace. The equilibrium statistical operator is usually written in the Gibbs form $\rho_0^{(\mathrm{st})} = \exp[(F-H_0)/T]$, where F is the free energy, and H_0 is the Hamiltonian of the system considered. In the presence of an external perturbation like an electric field, the system Hamiltonian is written as $H = H_0 + V$, where V describes the perturbation. The operator H_0 may not be exactly solvable and, in its turn, it may contain another perturbation term due to mutual interactions. But that is another story, which we discuss later. For the time being, let us just imagine that we can solve the problem of internal interactions, and focus our attention only on the external perturbation.

In the interaction representation, the time evolution of the statistical operator obeys the Liouville equation which, by integrating between $-\infty$ and t, can be written in the very useful integral form

$$\tilde{\rho}^{(\mathrm{st})}(t) = \rho_0^{(\mathrm{st})} - \frac{\mathrm{i}}{\hbar}\int_{-\infty}^{t} \left[\tilde{V}(t_1), \tilde{\rho}^{(\mathrm{st})}(t_1)\right] \mathrm{d}t_1 \,, \tag{2.2}$$

where $[\ ,\ ]$ is the commutator, the tilde indicates the interaction representation, and we assume the condition $\tilde{\rho}^{(\mathrm{st})}(-\infty) = \rho_0^{(\mathrm{st})}$. The linear response approximation is obtained by substituting $\rho_0^{(\mathrm{st})}$ for $\tilde{\rho}^{(\mathrm{st})}(t_1)$ in the integrand of (2.2). Then the average value of A is found as [78]

$$\langle A\rangle - \langle A\rangle_0 = -\frac{\mathrm{i}}{\hbar}\int_{-\infty}^{t} \langle [A(t), V(t')]\rangle_0 \,\mathrm{d}t \,, \tag{2.3}$$

where $\langle\ \rangle_0$ indicates the average over the equilibrium distributions. We have omitted the tilde because, in the interaction representation, the time development of operators (A and V in particular) is governed by the unperturbed Hamiltonian.

Equation (2.3) is the well-known general result from linear response theory. Assuming that the perturbation potential is proportional to an operator B,

$$V(t) = B(t)\mathcal{F}(t) \,, \tag{2.4}$$

a simple rearrangement of (2.3) yields

$$\langle A\rangle - \langle A\rangle_0 = \int_{-\infty}^{\infty} \mathrm{d}\tau G_{AB}^{(\mathrm{R})}(\tau)\mathcal{F}(t-\tau) \,, \tag{2.5}$$

and

$$G_{AB}^{(\mathrm{R})}(t) = -\frac{\mathrm{i}}{\hbar}\theta(t)\langle [A(t), B(0)]\rangle_0 \,, \tag{2.6}$$

where $\mathcal{F}(t)$ is an arbitrary function of time depending on the perturbation. The unit step function $\theta(t)$ entering (2.6) has a clear causal meaning.

The correlation function of (2.6) is called the retarded Green's function for two operators A and B and is sometimes written as $G_{AB}^{(\mathrm{R})}(t) \equiv \langle\langle A(t)B\rangle\rangle$ [79]. In terms of the Fourier transforms, (2.5) can be represented as

$$(\langle A\rangle - \langle A\rangle_0)_\omega = G_{A,B}^{(\mathrm{R})}(\omega)\mathcal{F}(\omega) \,, \tag{2.7}$$

where

$$G_{A,B}^{(\mathrm{R})}(\omega) \equiv \langle\langle A \mid B\rangle\rangle_\omega = -\frac{\mathrm{i}}{\hbar}\int_{0}^{\infty} \mathrm{e}^{\mathrm{i}\omega t}\langle [A(t), B(0)]\rangle_0 \,\mathrm{d}t \,. \tag{2.8}$$

In the following, we shall omit the subscript 0 showing that the average $\langle\ \rangle$ is performed over the equilibrium distributions. For convenience, we use different notations for the temporal Fourier transforms, using a subscript notation, such as $(\cdots)_\omega$, and $\langle\cdots\rangle_\omega$, or just writing the same symbol with a frequency argument like $\mathcal{F}(\omega)$. Care should be taken not to confuse the latter notation with the replacement $t \to \omega$ in the argument of the original function.

Employing the Fourier transform of the unit step function $\theta(t)$ given in (1.59), the important spectral representation can be found for the retarded correlation function:

$$G^{(R)}_{A,B}(\omega) = \int_{-\infty}^{\infty} \frac{d\omega'}{2\pi\hbar} \frac{\left(1 - e^{-\hbar\omega'/T}\right)}{\omega - \omega' + i\delta} \langle A(t)B\rangle_{\omega'} , \tag{2.9}$$

where the subscript ω indicates the Fourier transform of the simple correlation function $\langle A(t)B\rangle$ and $\delta > 0$ is an infinitesimal parameter. It is sometimes convenient to consider the function of (2.9) in the whole complex plane of the frequency variable, where it is a holomorphic function.

Differentiating (2.6) yields the following equation which describes the time evolution of the Green's functions introduced above:

$$-i\hbar \frac{\partial G^{(R)}_{AB}}{\partial t} = -\delta(t)\langle[A,B]\rangle + G^{(R)}_{[H,A]B}(t) , \tag{2.10}$$

where the first term originates from $\dot{\theta}(t) = \delta(t)$. By means of the relation $G^{(R)}_{AB}(t \to 0) = (-i/2\hbar)\langle[A,B]\rangle$, the equation of motion for the respective Fourier transforms can be written in two equally useful forms:

$$\begin{aligned} \hbar\omega G^{(R)}_{A,B}(\omega) &= \langle[A,B]\rangle - G^{(R)}_{[H,A],B}(\omega) \\ &= \langle[A,B]\rangle - G^{(R)}_{A,[B,H]}(\omega) . \end{aligned} \tag{2.11}$$

Here the electron Hamiltonian H generally includes the term V_{int} describing the interaction with scatterers. Both forms of (2.11) will be used in Sect. 3.4 when discussing quantum transport phenomena.

At this point we have finished our short introductory reminder of the mathematical objects arising in the many-body theory. In the following, we shall proceed with the physics of strongly correlated interface electrons.

2.1.2 Density–Density Correlation Function

The very important example of the correlation functions introduced above appears when studying the response of the electron system to electric field perturbations. In this case, the perturbation potential operator B defined by (2.4) is just the density fluctuation operator

$$B = n_{-\boldsymbol{q}} = \sum_{e} e^{i\boldsymbol{q}\cdot\boldsymbol{r}_e} ,$$

and $\mathcal{F}(\omega) = e\varphi_{\boldsymbol{q}}(\omega)$, where $\varphi_{\boldsymbol{q}}(\omega)$ is the Fourier transform of the scalar electric potential. Then the average density response can be found as

$$\langle n_{\boldsymbol{q}} \rangle = G^{(\mathrm{R})}_{n_{\boldsymbol{q}},n^{\dagger}_{\boldsymbol{q}}}(\omega) e\varphi_{\boldsymbol{q}}(\omega) \,. \tag{2.12}$$

The particular retarded Green's function $G^{(\mathrm{R})}_{n_{\boldsymbol{q}},n^{\dagger}_{\boldsymbol{q}}}(\omega) \equiv \langle\langle n_{\boldsymbol{q}} \mid n_{-\boldsymbol{q}} \rangle\rangle_{\omega}$ is conventionally called the density–density response (polarization) function, and sometimes denoted by $\chi(q,\omega)$. This function determines the dielectric function $\epsilon(q,\omega)$ of the electron system by means of the relation

$$\epsilon(q,\omega) = \frac{1}{1 + V_q^{(\mathrm{C})} \chi(q,\omega)} \,, \tag{2.13}$$

where $V_q^{(\mathrm{C})}$ is the Fourier transform of the Coulomb interaction. In the 2D case, $V_q^{(\mathrm{C})} = 2\pi e^2/q$.

The dynamical structure factor (DSF) $S(\boldsymbol{q},\omega)$ is introduced as the space and time Fourier transform of the simple density–density correlation function $N_{\mathrm{e}}^{-1} \langle n(\boldsymbol{r},t) n(\boldsymbol{r}',0) \rangle$:

$$S(\boldsymbol{q},\omega) = \frac{1}{N_{\mathrm{e}}} \int_{-\infty}^{\infty} \mathrm{e}^{\mathrm{i}\omega t} \langle n_{\boldsymbol{q}}(t) n_{-\boldsymbol{q}}(0) \rangle \, \mathrm{d}t \,. \tag{2.14}$$

This correlation function determines the particle cross-section in the theory of thermal neutron scattering by solids, developed intensively around 1950 (for details see [80]). As we shall see in Chap. 3, the effective collision frequency and the CR linewidth of the Coulomb liquid and Wigner solid can be found from an integral form of the electron DSF. The DSF of the 2D Coulomb liquid thus represents a very important property to be discussed in more detail. Usually, with the exception of the case of noninteracting particles, the evaluation of the DSF is a very difficult problem because it requires knowledge of true many-body eigenstates of the system. In Sect. 2.4 we shall see that, in spite of these difficulties, a good approximation to the DSF of the Coulomb liquid under a magnetic field does appear to be possible.

Employing the spectral representation for the retarded Green's function of (2.9), one can find an interesting and important relation between the DSF and the response function:

$$\mathrm{Im} G^{(\mathrm{R})}_{n_{\boldsymbol{q}},n^{\dagger}_{\boldsymbol{q}}}(\omega) \equiv \mathrm{Im}\chi(q,\omega) = -\frac{1 - \mathrm{e}^{-\hbar\omega/T}}{2\hbar} N_{\mathrm{e}} S(q,\omega) \,. \tag{2.15}$$

This is the so called fluctuation–dissipation theorem. It is usually used for an approximate evaluation of the DSF. In the quantum transport theory of Coulomb liquids, it is sometimes easier to evaluate the DSF directly for a certain model of the electron liquid. Equation (2.15) is then used only to establish the general relation between the conductivity relaxation kernel and the equilibrium DSF.

2.1.3 Dynamical Structure Factor

For an isotropic electron liquid, the equilibrium DSF depends only on the absolute value of the vector $\boldsymbol{q}$ which represents the momentum exchange $\hbar\boldsymbol{q}$ at collisions in the transport theory. The next basic property of the equilibrium DSF concerns its frequency argument:

$$S(q,-\omega) = \mathrm{e}^{-\hbar\omega/T} S(q,\omega) \, . \tag{2.16}$$

This property will be used frequently in this book.

Suppose, we know the true eigenstates $|j\rangle$ and energy spectrum ε_j of the electron system. Then the direct quantum mechanical average of $n_{\boldsymbol{q}}(t)n_{-\boldsymbol{q}}$ gives

$$S(q,\omega) = \frac{2\pi\hbar}{N_{\mathrm{e}}} \left\langle \sum_{j'} |\langle j'|n_{-\boldsymbol{q}}|j\rangle|^2 \delta(\varepsilon_{j'} - \varepsilon_j - \hbar\omega) \right\rangle_{\mathrm{T}} \, , \tag{2.17}$$

where the brackets $\langle \, \rangle_{\mathrm{T}}$ denote the thermal average over all true states $|j\rangle$. In the theory of thermal neutron scattering, this equation establishes the well-known relation between the scattering cross-section of the particle flux and the DSF of the target. We shall see that, in the quantum transport theory, it also allows one to find the relation between the effective collision frequency of the electron system and $S(q,\omega)$.

For single-electron states $|\alpha\rangle$ of the electron system, the density fluctuation operator can be conveniently expressed in terms of the creation c_α and destruction $c_\alpha^\dagger$ operators:

$$n_{\boldsymbol{q}} = \sum_{\alpha,\beta} \left(\mathrm{e}^{-\mathrm{i}\boldsymbol{q}\cdot\boldsymbol{r}_{\mathrm{e}}}\right)_{\alpha,\beta} c_\alpha^\dagger c_\beta \, .$$

Inserting this formula into the definition of the DSF given in (2.14) and disregarding the mutual Coulomb interaction, after some many-body algebraic transformations, the DSF can be represented as a trace in the single-electron basis:

$$\begin{aligned} S_{\mathrm{se}}(q,\omega) = & \frac{2\pi\hbar}{N_{\mathrm{e}}} \int \mathrm{d}\varepsilon \, [1 - f(\varepsilon + \hbar\omega)] \, f(\varepsilon) \\ & \times \left\langle \mathrm{Tr} \left[\delta(\varepsilon - H_{\mathrm{e}}) \mathrm{e}^{-\mathrm{i}\boldsymbol{q}\cdot\boldsymbol{r}} \delta(\varepsilon - H_{\mathrm{e}} + \hbar\omega) \mathrm{e}^{\mathrm{i}\boldsymbol{q}\cdot\boldsymbol{r}} \right] \right\rangle_{\mathrm{s}} \, , \end{aligned} \tag{2.18}$$

where H_{e} is the single-electron Hamiltonian including the interaction with scatterers, $f(\varepsilon)$ is the Fermi distribution function, and $\langle \, \rangle_{\mathrm{s}}$ denotes the average with respect to the scatterer variables. Comparing this form of $S(q,\omega)$ with the famous magnetoconductivity equations of Kubo, Miyake and Hashitsume [81], one can see that there should be a very simple relation between σ_{xx} and the electron DSF. This relation and similar topics are discussed in Chaps. 3 and 4.

If there is no applied magnetic field, the DSF of (2.18) can be rewritten in the form

$$S_{\mathrm{se}}(q,\omega) = \frac{4\pi\hbar}{N_{\mathrm{e}}} \sum_{\boldsymbol{k}} f(\varepsilon_{\boldsymbol{k}}) \left[1 - f(\varepsilon_{\boldsymbol{k}+\boldsymbol{q}})\right] \delta(\varepsilon_{\boldsymbol{k}} - \varepsilon_{\boldsymbol{k}+\boldsymbol{q}} + \hbar\omega) , \tag{2.19}$$

where $\varepsilon_{\boldsymbol{k}} = \hbar^2 k^2/2m$ is the free electron energy spectrum. Using the well-known identity

$$f(\varepsilon_{\boldsymbol{k}}) \left[1 - f(\varepsilon_{\boldsymbol{k}'})\right] = \left[f(\varepsilon_{\boldsymbol{k}}) - f(\varepsilon_{\boldsymbol{k}'})\right] \frac{1}{1 - \mathrm{e}^{(\varepsilon_{\boldsymbol{k}} - \varepsilon_{\boldsymbol{k}'})/T}} \tag{2.20}$$

and the fluctuation–dissipation theorem [(2.15)], one can find the response function of the free electron gas:

$$\chi^{(0)}(q,\omega) = 2 \sum_{\boldsymbol{k}} \frac{f(\varepsilon_{\boldsymbol{k}}) - f(\varepsilon_{\boldsymbol{k}+\boldsymbol{q}})}{\hbar\omega - \varepsilon_{\boldsymbol{k}+\boldsymbol{q}} + \varepsilon_{\boldsymbol{k}} + \mathrm{i}\delta} . \tag{2.21}$$

This equation is conventionally found by other methods [82, 83]. It is instructive to note that the result of the kinetic equation method for $\chi^{(0)}(q,\omega)$ is reproduced here by means of the quantum equation for the DSF in the simplest approximation, disregarding the so called self-energy effects and vertex corrections.

The nondegenerate gas (NDG) represents the simplest case for obtaining the electron DSF in two-dimensions. Substituting the distribution function $f(\varepsilon) \simeq \exp[(\mu - \varepsilon)/T]$ into (2.19), one easily arrives at

$$S_{\mathrm{NDG}}(q,\omega) = \frac{1}{q} \sqrt{\frac{2\pi m_{\mathrm{e}}}{T}} \exp\left[-\frac{(\varepsilon_q - \hbar\omega)^2}{4\varepsilon_q T}\right] . \tag{2.22}$$

This form of the electron DSF has the following distinguishing features: the singularity at the point ($q = 0, \omega = 0$) and the maximum at $\hbar\omega$ is equal to the single-electron excitation spectrum ε_q. As a function of frequency, $S(q,\omega)$ has a Gaussian form strongly broadened for excitations with $\varepsilon_q \sim T$, and a sharp peak form for excitations with $\varepsilon_q \gg T$. Although one cannot expect $S(q,\omega)$ evaluated for noninteracting particles to be a reasonable approximation for the Coulomb liquid, we shall see in Chap. 7 that, at high temperatures $T > \theta_{\mathrm{D}}$ (where θ_{D} is the Debye temperature), the Wigner solid DSF is close to the form (2.22) found for the nondegenerate electron gas.

For a degenerate gas (DG), there are two extreme ranges of the frequency argument ω where the DSF can be found in an analytical form. In conventional metals and even semiconductors, the Fermi energy $\varepsilon_{\mathrm{F}} = \pi\hbar^2 n_{\mathrm{s}}/m_{\mathrm{e}}$ is very high and usually $\hbar\omega \ll \varepsilon_{\mathrm{F}}$. In this case, approximating

$$f(\varepsilon_{\boldsymbol{k}}) \left[1 - f(\varepsilon_{\boldsymbol{k}+\boldsymbol{q}})\right] \simeq \frac{\hbar\omega}{1 - \mathrm{e}^{-\hbar\omega/T}} \delta(\varepsilon_{\boldsymbol{k}} - \varepsilon_{\mathrm{F}}) \tag{2.23}$$

yields

$$S_{\mathrm{DG}}(q,\omega) \simeq \frac{2m_e\omega}{\pi n_s\left(1-\mathrm{e}^{-\hbar\omega/T}\right)\sqrt{4\varepsilon_q\varepsilon_{\mathrm{F}}-(\varepsilon_q-\hbar\omega)^2}} \,. \tag{2.24}$$

Here and below we take the DSF to be zero when the argument of the square-root symbol becomes negative. In the opposite limiting case $\hbar\omega \gg \varepsilon_{\mathrm{F}}$ realized for SEs on liquid helium at ultra-low temperatures, $f(\varepsilon_{\boldsymbol{k}}+\hbar\omega)\approx 0$ and

$$\begin{aligned} S_{\mathrm{DG}}(q,\omega) &\simeq \frac{\hbar}{\varepsilon_{\mathrm{F}}\varepsilon_q}\sqrt{4\varepsilon_q\varepsilon_{\mathrm{F}}-(\varepsilon_q-\hbar\omega)^2} \\ &= \frac{\hbar}{\varepsilon_{\mathrm{F}}q^2}\sqrt{(q_+^2-q^2)(q^2-q_-^2)}\,, \end{aligned} \tag{2.25}$$

where

$$q_{\pm} = \sqrt{k_{\mathrm{F}}^2+q_\omega^2} \pm k_{\mathrm{F}}\,, \qquad q_\omega^2 = \frac{2m_e\omega}{\hbar}\,, \tag{2.26}$$

and $k_{\mathrm{F}} = \sqrt{2\pi n_s}$ is the Fermi wave number of electrons. The new quantities q_+ and q_- represent two extreme values of the electron momentum exchange for the limiting case $\hbar\omega > \varepsilon_{\mathrm{F}}$.

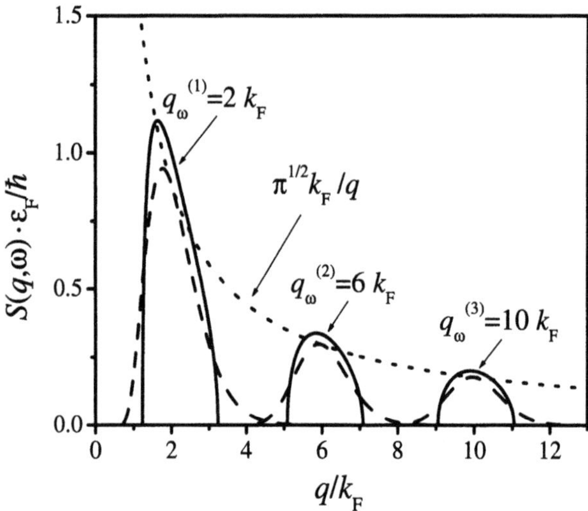

Fig. 2.1. The electron dynamical structure factor vs. the dimensionless wave number $q/\sqrt{2\pi n_s}$ evaluated for degenerate (*continuous curve*) and nondegenerate (*dashed curve*) statistics for three values of the frequency argument, which are marked with the respective values of the parameter $q_\omega = \sqrt{2m_e\omega/\hbar}$

Remarkably, the main features of the DSF of a nondegenerate electron gas remain for degenerate electrons as well. The singularity $S(q,0)\propto 1/q$ at $q\to 0$ is clearly seen in (2.24). At high frequencies $\hbar\omega\gg\varepsilon_{\mathrm{F}}$, the DSF has a maximum at $\varepsilon_q=\hbar\omega$ (or, as a function of q, at $q=q_\omega$), as follows from (2.25).

The DSFs of the degenerate and nondegenerate statistics are compared in Fig. 2.1 for three different values of the frequency argument ω. In order to obtain comparable broadening for different curves, the temperature of the nondegenerate gas is chosen to be equal to ε_F. The curves representing these two extreme approximations for $S(q,\omega)$ have different shapes: the DSF of the degenerate gas has sharp edges while the DSF of nondegenerate electrons varies smoothly. Still, for $T = \varepsilon_F$, the two approximations have nearly the same values at the maxima and nearly the same line widths. Therefore, at any other temperature, one may expect the area under these different curves to be approximately the same. This property of the electron DSF leads to similar high-frequency conductivities for the degenerate and nondegenerate electrons, as we will discuss in Sect. 7.6.

Before considering the DSF of the 2D electron liquid subject to a strong magnetic field, it is very instructive to obtain the self-correlation function of an electron in an isotropic harmonic oscillator potential. The characteristic frequency of this potential is set at ω_0. In this case, anticipating the application to the Wigner solid, it is convenient to use another treatment of the electron DSF by starting with

$$\tilde{S}(q,t) = \prod_{\alpha=x,y} \langle \exp[-\mathrm{i}k_\alpha r_\alpha(t)] \exp[\mathrm{i}k_\alpha r_\alpha] \rangle \ , \tag{2.27}$$

where r_α is the x or y coordinate of the electron, depending on the subscript. When averaging, one has to take into account the non-commuting nature of the operator

$$x(t) = \left(\frac{\hbar}{2m\omega_0}\right)^{1/2} \left[a \exp(-\mathrm{i}\omega_0 t) + a^\dagger \exp(\mathrm{i}\omega_0 t)\right] \tag{2.28}$$

taken at different times.

To proceed further we make use of the identity

$$\exp(A)\exp(B) = \exp(A + B + C) \ ,$$

where

$$C = \frac{1}{2}[A,B] + \frac{1}{12}\Big\{\big[[A,B],B\big] + \big[[B,A]\,A\big]\Big\} + \cdots \, . \tag{2.29}$$

In the case considered here, the commutator $[A, B]$ is a c-number and the series in (2.29) contains only the first term. Then using the Bloch identity given previously in (1.154), we find that

$$\langle \exp(A)\exp(B) \rangle = \exp\left(\frac{1}{2}\left\langle A^2 + B^2 + 2AB \right\rangle\right) \, . \tag{2.30}$$

Thus for the operators entering (2.27), we have

$$\big\langle \exp\big[-\mathrm{i}k_x x(t)\big] \exp\big[\mathrm{i}k_x x(0)\big] \big\rangle = \exp\big[-k_x^2 2w(t)\big] \ , \tag{2.31}$$

where

$$2w(t) = \langle x^2 \rangle - \langle x(t)x(0) \rangle$$
$$= \frac{\hbar}{2m\omega_0}\left[(n_0+1)\left(1-\mathrm{e}^{-\mathrm{i}\omega_0 t}\right) + n_0\left(1-\mathrm{e}^{\mathrm{i}\omega_0 t}\right)\right], \qquad (2.32)$$

and n_0 is the Bose distribution function for the oscillator characterized by the frequency ω_0.

Making use of (2.31), the electron DSF can be rewritten as

$$S(q,\omega) = \frac{1}{N_\mathrm{e}} \int_{-\infty}^{\infty} \mathrm{e}^{\mathrm{i}\omega t - q^2 2w(t)} \mathrm{d}t\,. \qquad (2.33)$$

In the ultra-quantum limit $\hbar\omega_0 \gg T$, one can simplify

$$q^2 2w(t) \simeq x_q\left(1-\mathrm{e}^{-\mathrm{i}\omega_0 t}\right), \qquad (2.34)$$

where

$$x_q = q^2 \left\langle r^2 \right\rangle_0 /2 \equiv \hbar q^2/2m\omega_0\,.$$

The time integral of (2.33) can be evaluated by expanding the exponential in $\exp(-\mathrm{i}\omega_0 t)$, which yields

$$S(q,\omega) = 2\pi\mathrm{e}^{-x_q} \sum_{n=0}^{\infty} \frac{x_q^n}{n!}\delta(\omega - n\omega_0)\,. \qquad (2.35)$$

We observe that the DSF of an electron in the harmonic potential has sharp maxima at $\omega = 0$ and at all excitation frequencies $\omega = n\omega_0$. Later we shall see that the first term, usually called the elastic term, describes electron collisions with zero energy exchange. In this sense, the other terms also have a clear physical meaning. They describe scattering processes in which a scatterer loses its energy by exciting quanta of the electron motion in the harmonic potential. Generally, the DSF of the model considered here also contains terms with $n < 0$ [80], which correspond to the number of units of energy $\hbar\omega_0$ gained by a scatterer. One can find these terms by expanding the exponential of (2.33) in powers of $n_0 \exp(\mathrm{i}\omega_0 t)$.

Comparing the DSF of free noninteracting electrons [(2.19)] with the DSF of electrons localized in the harmonic potential [(2.35)], one can see that both consist of sharp peaks positioned at the electron energy excitation spectrum. This is a quite general property of the DSF. It is only the conversion from the sum $\sum_{\boldsymbol{k}}$ to the integral $\int \mathrm{d}^2\boldsymbol{k}$ which makes the broad Gaussian form given above for nondegenerate electrons, or the semi-elliptic form with sharp edges obtained for the degenerate electron gas. It is reasonable to expect that interactions will broaden these peaks, which is particularly important for electrons with discrete energy spectrum.

In the presence of a strong normal magnetic field, use of the discrete Landau spectrum in (2.18) gives a singular DSF similar to (2.35). In the general conductivity equation, this form of $S(q,\omega)$ yields a divergent result for elastic scattering ($\omega = 0$). To avoid this divergence, following Kubo, Miyake and Hashitsume [81] (damping theoretical approximation) and Ando and Uemura [74] (self-consistent Born approximation), it is commonly accepted to replace the matrix elements $\langle N,X|\,\delta(\varepsilon - H_{\mathrm{e}})\,|N,X\rangle$ entering the trace of (2.18) by the imaginary parts of the proper electron Green's functions [see (1.130)], broadened by the interaction with scatterers. Then, using the Landau basis states, (2.18) can be transformed to

$$S_{\mathrm{se}}(q,\omega) = \frac{m_{\mathrm{e}}\omega_{\mathrm{c}}}{\pi^2 N_{\mathrm{e}}} \int \mathrm{d}\varepsilon\, f(\varepsilon)\,[1 - f(\varepsilon + \hbar\omega)] \times \sum_{N,N'} |J_{N,N'}(x_q)|^2\, \mathrm{Im}G_N(\varepsilon)\mathrm{Im}G_{N'}(\varepsilon + \hbar\omega)\;, \qquad (2.36)$$

where $x_q = q^2 l_B^2/2$. It should be noted that, in the absence of interactions, this equation is just another way of representing the exact form of (2.18).

The above-mentioned treatment of the electron DSF includes self-energy effects but disregards vertex correction terms. The latter are defined as diagrams in which the scattering links different electron Green's functions. From the Feynman diagram formulation of the DC conductivity (for details see [38]), we know that the analogous vertex terms in the conductivity series are important because they provide the $1 - \cos\theta$ factor for the momentum relaxation rate $\Gamma^{(\mathrm{tr})}/\hbar$, making it different from the inverse lifetime $\Gamma/\hbar$ (the average rate of scattering events). As we shall see in Chap. 3, the conductivity equation which we are going to use is organized as a perturbation series for the effective collision frequency, with the $1 - \cos\theta$ factor built in from the very beginning. Therefore, at $B = 0$, the simplest approximation for the DSF gives the conductivity equation which coincides with the result of the kinetic equation method. We may thus assume that inclusion of the vertex terms in $S(q,\omega)$ will not spoil the approximation of (2.36).

Comparison with the strict magnetoconductivity evaluations of Ando and Uemura [74], performed for both the short-range and long-range impurity potentials, agrees well with this assumption. For example, we shall see that the approximation of (2.36) gives the same range dependence of the transport relaxation time $\hbar/\Gamma^{(\mathrm{tr})}(d/l_B)$ or the transverse conductivity $\sigma_{xx}(d/l_B)$ as that found in the conventional conductivity perturbation theory, which includes the vertex part. Physically, the predominance of the self-energy effects for the electron DSF is caused by the singular nature of the 2D electron gas subject to a magnetic field: a small broadening Γ of the Landau levels induces huge changes in $S(q,\omega) \propto 1/\Gamma$.

For the Gaussian form of $\mathrm{Im}G_N(\varepsilon)$ given in (1.166) and nondegenerate statistics ($f \ll 1$), we find

$$S_{se}(q,\omega) = 2\sqrt{\pi}\hbar\left(1 - e^{-\hbar\omega_c/T}\right)\sum_{NN'} e^{-N\hbar\omega_c/T}\frac{|J_{NN'}(x_q)|^2}{\Gamma_{NN'}}$$
$$\times \exp\left\{-\frac{\hbar^2[\omega - (N'-N)\omega_c - \Gamma_N^2/4T\hbar]^2}{\Gamma_{NN'}^2}\right\}, \quad (2.37)$$

where we have introduced the average broadening $\Gamma_{N,N'} = \sqrt{(\Gamma_N^2 + \Gamma_{N'}^2)/2}$ of two Landau levels and used the condition $N_e = \int \Delta_{2D}(\varepsilon) f(\varepsilon) d\varepsilon$ in order to determine the chemical potential entering the distribution function $f(\varepsilon)$.

The position and form of the frequency shift $-\Gamma_N^2/4T\hbar$ in the exponent of (2.37) assumes that Γ_N does not depend (or only very weakly depends) on the level number N, a property used when evaluating the chemical potential. This term is usually very small, with the exception of high magnetic fields and temperatures, where the scattering by vapor atoms dominates. Still, in the latter case, $\Gamma_N \simeq \Gamma_{se}$ is truly independent of N and the form of (2.37) is correct. It is very instructive to note that the presence of the frequency shift $-\Gamma_N^2/4T\hbar$ guarantees the general property of the equilibrium DSF $S(q,-\omega) = e^{-\hbar\omega/T} S(q,\omega)$. In the exponent of the DSF, the parameter $\hbar\omega/2T$ is much smaller than $(\hbar\omega/\Gamma_{NN'})^2$ because of the extremely narrow Landau levels $\Gamma_N \ll T$. When disregarding the parameter $\Gamma_N^2/4T\hbar$ in comparison to $\omega - (N'-N)\omega_c$, we lose the above-mentioned property of the electron DSF in the approximate equation for $S(q,\omega)$. In this case, one should keep in mind that there are small shifts in the positions of maxima of the DSF that provide the necessary condition for the frequency dependence of the DSF.

In the ultra-quantum limit ($\hbar\omega_c \gg T$), the population of high Landau levels is small and the electron DSF approaches the very simple form

$$S_{se}(q,\omega) \simeq 2\sqrt{\pi}\hbar e^{-x_q} \sum_{N=0}^{\infty} \frac{x_q^N}{N!\Gamma_{0,N}} \exp\left[-\frac{\hbar^2(\omega - N\omega_c)^2}{\Gamma_{0,N}^2}\right], \quad (2.38)$$

which we shall use frequently later. In this equation we have disregarded the small frequency shift $-\Gamma_0^2/4T\hbar$ discussed above, because Landau levels of surface electrons on liquid helium are extremely narrow $\Gamma_N \ll T$. For semiconductor 2D electron systems and surface electrons on 'dirty' substrates, one should retain this frequency shift in the exponent of (2.38).

Equation (2.38) is reminiscent of the DSF given above for the oscillator model. Indeed, replacing the δ-functions of (2.35) by proper Gaussians, we reproduce (2.38). For the electron interaction with short-range scatterers, the broadening parameters are independent of the level number $\Gamma_{N,N'} = \Gamma_N = \Gamma_{se}$. Equation (2.38) shows that the electron DSF, as a function of frequency, has pronounced maxima at $\omega = 0$, at the CR position $\omega = \omega_c$ and at all subharmonics $\omega = N\omega_c$. The region $\omega \to 0$ is important for DC magnetotransport, while the properties of the DSF at $\omega \approx \omega_c$ describe the quantum CR.

It has been often emphasised in the literature that one cannot expect the DSF of noninteracting particles to be a good approximation for the DSF of a strongly interacting system. In Sects. 2.4 and 7.5, we shall see that the main structure of the DSF of noninteracting electrons under the magnetic field [(2.38)] is also mysteriously preserved for the Coulomb liquid and even for the Wigner solid. As a function of frequency ω, the Coulomb liquid DSF is still a series of similar Gaussians whose broadening parameter $\Gamma^*_{0,N}$ depends additionally on the electron density and the wave number q.

As for the degenerate electron gas subject to a magnetic field, it is worth mentioning the property of $S(q,0)$ which relates to the DC magnetoconductivity of the semiconductor 2D electron systems. Replacing $-\mathrm{d}f/\mathrm{d}\varepsilon$ by a δ-function, we find that, under a strong magnetic field,

$$S_{\mathrm{DG}}(q,0) = \frac{m_{\mathrm{e}}\omega_{\mathrm{c}}T}{\pi^2 N_{\mathrm{e}}\hbar^2}\sum_N |J_{N,N}(x_q)|^2 \left[\mathrm{Im}G_N(\varepsilon_{\mathrm{F}})\right]^2 . \tag{2.39}$$

Thus as a function of ε_{F}, $S_{\mathrm{DG}}(q,0)$, has sharp maxima at $\varepsilon_{\mathrm{F}} = \varepsilon_N$ due to the properties of the electron density-of-state function discussed above. In the SCBA theory, $\mathrm{Im}G_N(\varepsilon)$ has the semi-elliptic form

$$-\mathrm{Im}G_N(\varepsilon) = \frac{2\hbar}{\Gamma_N}\sqrt{1-\left(\frac{\varepsilon-\varepsilon_N}{\Gamma_N}\right)^2} , \tag{2.40}$$

which yields the parabolic shape of the DSF maxima. In the cumulant approach, the maxima have the Gaussian shape according to (1.166).

The fundamental correlation functions are usually evaluated in the motionless laboratory reference frame, where they depend only on the absolute value of the wave vector if the phase is isotropic. In order to describe nonlinear transport phenomena, and even linear transport properties of the Coulomb liquid, we shall need to know the electron DSF in other reference frames. The proper transcription rule can be found in a similar way to that described in Sect. 1.4 [see (1.46) and discussions there]. For example, let us consider the electron liquid moving as a whole with drift velocity $\boldsymbol{u}$ with respect to the laboratory reference frame. The electron DSF only has the usual equilibrium form, which does not depend on the direction of $\boldsymbol{q}$, in the center-of-mass reference frame. Because of Galilean invariance, the DSF defined in the laboratory frame relates to the DSF in the center-of-mass frame according to the transcription rule [84]

$$S'(\boldsymbol{q},\omega) = S(q,\omega - \boldsymbol{q}\cdot\boldsymbol{u}) . \tag{2.41}$$

The Doppler shift in the frequency argument makes the electron DSF depend on the direction of the wave vector. In this case, the positions of the maxima of the DSF defined in the laboratory frame are in accordance with the well-known correction to the electron excitation spectrum $\varepsilon'_j = \varepsilon_j + \hbar\boldsymbol{q}\cdot\boldsymbol{u}$.

2.2 Fluctuational Electric Field Concept

In equilibrium at $T = 0$, the internal force acting on a chosen classical particle is zero. Thermal and zero-point fluctuations deflect particles from their local potential energy minima so that they become subject to the strong fluctuational field $\boldsymbol{E}_{\mathrm{f}}^{(i)}$ of internal forces. Dykman and Khazan [85] were the first to show the importance of this fluctuational field for quantum magnetotransport in 2D Coulomb liquids. They found that, in the presence of a high magnetic field oriented normally to the layer, this fluctuational electric field can be considered as a quasi-uniform field causing a fast drift motion of the cyclotron orbit center and inducing a continuous correction to the Landau spectrum $eE_{\mathrm{f}}^{(i)}X$, where X is the orbit center coordinate along the fluctuational field.

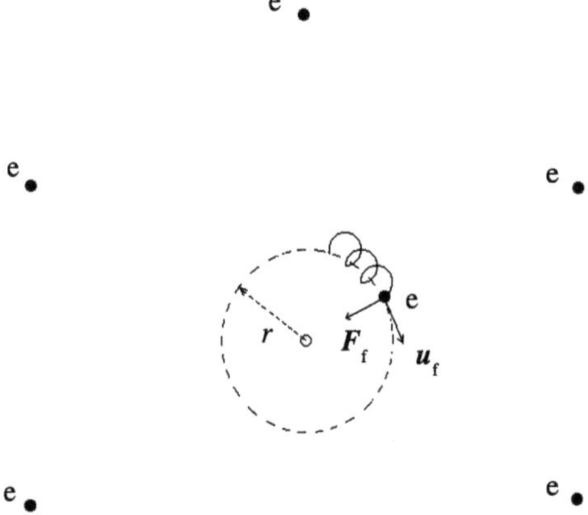

Fig. 2.2. Schematic view of the motion of the electron subject to a magnetic field and an internal force $\boldsymbol{F}_{\mathrm{f}}$ of fluctuational origin

In the presence of the restoring internal force and the magnetic field oriented normally, the electron orbit drifts in the perpendicular direction, as shown in Fig. 2.2. The Coulomb liquid under a magnetic field can thus be considered as an ensemble of localized one-electron currents of average radius r determined by the condition $m_{\mathrm{e}}\omega_{\mathrm{p}}^2 r^2/2 = T$, where ω_{p} is the characteristic frequency of short-wavelength vibrations of the 2D plasma under zero magnetic field:

$$\omega_{\mathrm{p}}^2 = \frac{2\pi e^2 n_{\mathrm{s}}^{3/2}}{m_{\mathrm{e}}} . \tag{2.42}$$

The fluctuational force acting on a chosen electron which moves in the Coulomb potential of other electrons is proportional to the displacement r from equilibrium: $eE_{\mathrm{f}} = m_{\mathrm{e}}\omega_{\mathrm{p}}^2 r$. For the average radius r determined above, this yields $E_{\mathrm{f}} \approx \sqrt{4\pi} n_{\mathrm{s}}^{3/4} T^{1/2}$. The harmonic approximation for the WS gives an additional numerical proportionality factor 0.84 to this estimate [86]. The strength of the fluctuational electric field can thus be varied by changing T and n_s.

For particular evaluations, one needs to know the distribution of the fluctuational electric field. Monte Carlo simulations [87] show that, in the broad range of the electron plasma parameter $20 < \Gamma^{(\mathrm{pl})} < 200$, the distribution of the fluctuational field is close to a Gaussian. In the character of the Gaussian width parameter $\sqrt{\langle E_{\mathrm{f}}^2 \rangle} \equiv E_{\mathrm{f}}^{(0)}$, one can use the following expression $E_{\mathrm{f}}^{(0)} = \mathcal{F}\sqrt{T} n_{\mathrm{s}}^{3/4}$ with the numerical proportionality factor $\mathcal{F} \simeq 2.985 \approx 3$. According to the Monte Carlo study, the variation of the numerical factor $\mathcal{F}$ is very small (less than 10%) for $\Gamma^{(\mathrm{pl})} > 10$.

If the fluctuational electric field can be considered as quasi-uniform, then it is possible to study the Coulomb liquid as an ensemble of 2D electrons with continuous corrections $eE_{\mathrm{f}}^{(i)}X$ to the discrete Landau spectrum, as shown by Dykman and Khazan. The many-electron transport of interface electrons on liquid helium was developed in [88] as a sophisticated theory for 2D electrons with continuous spectrum. Later on, in the series of publications [76, 89, 90], a simpler approach based on the dynamics of electrons with pure discrete spectrum was introduced. The main idea of this approach is that the continuous correction to the Landau spectrum is not the major effect of the fluctuational electric field. Even a single electron, which has the discrete Landau spectrum in the laboratory frame, acquires a continuous correction if it is described in any other frame moving with respect to the laboratory frame with a constant velocity $\boldsymbol{u}$, because of the appearance of the electric field $\boldsymbol{E}' = -(1/c)\boldsymbol{B} \times \boldsymbol{u}$. The most important effect of the strong fluctuational electric field is the ultra-fast relative motion of the electron orbit center with respect to the scatterers, and this effect can be described using either the continuous or the discrete energy spectrum.

At finite temperature, each electron orbit drifts in the crossed magnetic and fluctuational electric fields with velocity $\boldsymbol{u}_{\mathrm{f}}^{(i)}$ relative to the center-of-mass frame of the electron liquid and the system of scatterers ($u_{\mathrm{f}}^{(i)} = cE_{\mathrm{f}}^{(i)}/B$), as shown in Fig. 2.3a. It should be emphasized that the quasi-uniform fluctuational field does not change the singular nature of an electron in two dimensions under a normal magnetic field because this field can be eliminated by proper choice of the reference frame:

$$\boldsymbol{E}_{\mathrm{f}}' = \boldsymbol{E}_{\mathrm{f}} - \frac{1}{c}\boldsymbol{B} \times \boldsymbol{u} \to 0\,. \tag{2.43}$$

This is the frame in which the center of the electron orbit is at rest and the electron spectrum coincides with the discrete Landau spectrum. In this frame, the electron orbit is still while the scatterers are moving as a whole in the opposite direction, as shown in Fig. 2.3b. Therefore the effect of the many-electron fluctuational field on a single electron can be described in terms of the pure discrete Landau spectrum.

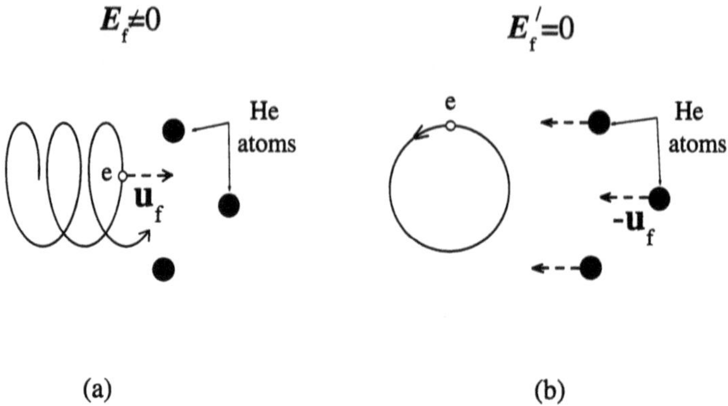

Fig. 2.3. Schematic view of the electron scattering event from the laboratory frame (**a**) and from a frame moving along with the electron orbit center (**b**)

Here it is instructive to note the important differences in the description of the conventional Fermi liquid and the Coulomb liquid. In the first case, all quasi-particles are described by the same energy spectrum $\varepsilon(p)$ influenced by the internal interaction. For the Coulomb liquid in the laboratory frame, one cannot introduce a universal spectrum valid for all quasi-particles, because the correction $eE_{\mathrm{f}}^{(i)}X$ to the Landau spectrum caused by internal forces is different for different electrons. The universal spectrum (the Landau spectrum) can be introduced only in the local reference frames, moving along the electron orbit centers, different for different electrons.

2.3 Coulomb Narrowing of Landau Levels

2.3.1 Electric Field Effect on the Density of States

Before proceeding with the fluctuational electric field effect let us consider a single electron in the crossed magnetic and uniform external electric fields, $\boldsymbol{B}$ and $\boldsymbol{E}$, respectively, in the presence of scatterers. In the laboratory reference frame, its orbit center drifts with the constant velocity $u = cE/B$ in the direction perpendicular to $\boldsymbol{E}$. In the other reference frame, moving along with the electron orbit center, the electric field is

$$\boldsymbol{E}' = \boldsymbol{E} - \frac{1}{c}\boldsymbol{B} \times \boldsymbol{u} = 0 ,$$

and the electron spectrum coincides with the pure discrete Landau spectrum if the interaction with scatterers is disregarded.

When an electron is considered in the moving frame where its spectrum is pure discrete, the broadening of Landau levels can be introduced in the usual way, employing the SCBA or the cumulant approach. The only important difference that appears in this case is that the gas of scatterers is now moving as a whole with the drift velocity $-\boldsymbol{u}$ relative to the electron orbit center (see Fig. 2.3b), and the Green's function of scatterers is affected by the Doppler shift according to (1.46): for vapor atoms, we have $D_{\mathrm{a}}^{(0)}(q, \omega + \boldsymbol{q} \cdot \boldsymbol{u}_{\mathrm{f}}) \propto -\mathrm{i}N^{(\mathrm{a})}\delta(\omega + \boldsymbol{q} \cdot \boldsymbol{u}_{\mathrm{f}})$. As a result, in the SCBA theory, the relation between the electron self-energy $\Sigma_N(\varepsilon)$ and $G_N(\varepsilon)$ [(1.136)] becomes an integral equation

$$\Sigma_N(\varepsilon) = \frac{1}{4}\Gamma_{\mathrm{se}}^2 \sum_{N'} \int_0^{2\pi} \frac{\mathrm{d}\varphi}{2\pi} \int_0^{\infty} \mathrm{d}x_q \left|J_{N,N'}(x_q)\right|^2 G_{N'}(\varepsilon + \hbar \boldsymbol{q} \cdot \boldsymbol{u}) , \qquad (2.44)$$

where Γ_{se} is the single-electron Landau level broadening at $\boldsymbol{u} = 0$, which is the same for all Landau levels [see (1.143)], if electrons interact with short-range scatterers. The quantity $\hbar\boldsymbol{q} \cdot \boldsymbol{u}$ represents the correction to the energy exchange in a collision between the electron and a scatterer, arising in the moving reference frame. It should be remembered that the energy exchange for an elastic collision of the electron at an impurity is zero only in the frame where the impurity is at rest. In a moving frame, the impurity hits the electron, increasing or reducing its energy.

The difference between (2.44) and the conventional SCBA result concerns the argument of the electron Green's function entering the integrand. In the SCBA, the Doppler shift $\hbar\boldsymbol{q} \cdot \boldsymbol{u}$ is zero. If the energy exchange $\hbar\boldsymbol{q} \cdot \boldsymbol{u} \sim eEl_B$ is much less than Γ_N, then we can disregard it, as well as the mixing of different Landau levels. In this limit, (2.44) together with Dyson's equation yields the result of the SCBA. In the general case, (2.44) is an integral equation and the level shape differs from the simple semi-elliptic or Gaussian functions. Because the strict solution of the self-consistent equation is a very difficult problem, one can simplify it by fixing the Landau level shape to a Gaussian function. Then the broadening of the lowest Landau level $\Gamma_0 \propto \mathrm{Im}\Sigma_0$ can be found in an analytical form [89, 90]:

$$\Gamma_0^2 = \sqrt{\left[1 + C_0(x^*)\right]^2 \Gamma_{\mathrm{se}}^4 + 4(eEl_B)^4} - 2(eEl_B)^2 , \qquad (2.45)$$

where

$$C_0(x) = \frac{1}{\sqrt{\pi}} \sum_{n=1}^{\infty} \frac{1}{n!} \mathrm{e}^{-x^2 n^2} \int_0^{\infty} (t + x^2 n^2)^n \mathrm{e}^{-t} \frac{\mathrm{d}t}{\sqrt{t}} \qquad (2.46)$$

represents the effect of mixing between different Landau levels appearing when the parameter $x^* = \hbar\omega_c/(\sqrt{2}eEl_B)$ becomes of the order of unity or even less. A simple analytical interpolation of $C_0(x)$ can be written as $C_0(x) \simeq \mathrm{e}^{-x^2}(x^2 - 0.6 + 3/\sqrt{\pi}x)$.

Equation (2.45) meets the condition that, without the interaction with scatterers ($\Gamma_{\mathrm{se}} \to 0$), there is no level broadening at all ($\Gamma_0 \to 0$). In weak and intermediate electric fields E (when the parameter $x^* \gg 1$), one can disregard the mixing of different Landau levels in (2.45) ($C_0 \to 0$) and find a strong decrease in the collision broadening with E when $\sqrt{2}eEl_B > \Gamma_{se}$. The main effect of the external electric field on the electron density-of-state function broadened due to the scatterers is therefore a strong narrowing.

At stronger electric fields, when $C_0(x^*)$ becomes important, the effect of the electric field becomes the opposite and collision broadening increases. The typical dependencies of Γ_0/Γ_{se} and Γ_1/Γ_{se} on the dimensionless parameter $\lambda(E) = \sqrt{2}eEl_B/\Gamma_{se}$ are shown in Fig. 2.4 for three values of the magnetic field B. The Landau level broadening decreases fast when λ becomes comparable with unity. Then, at larger electric fields, it increases or approaches a constant, depending on the magnetic field value. The important point is that the decrease in both Γ_0 (continuous curves) and Γ_1 (dotted curves) is completely normalized by Γ_{se} – the curves corresponding to different magnetic fields merge in the region of small and medium values of λ. In contrast, the increase in collision broadening due to the electric field effect which appears at larger values of λ is not normalized by Γ_{se} because it depends on the other parameter $eEl_B/\hbar\omega_c$.

The physical reason for the narrowing of the Landau level width, induced by the electric field, can be differently presented in the laboratory and moving reference frames. In the first case, a drift of the electron orbit reduces the duration of the multiple elastic scattering of an electron with a chosen vapor atom ('impurity'). However, this picture is not transparent if there are many vapor atoms within the orbit area, as for the SEs on liquid helium. In the second instance, the electron orbit is motionless, while the vapor atoms are moving fast. They hit the electron orbit with velocity $-\boldsymbol{u}$ and cause an

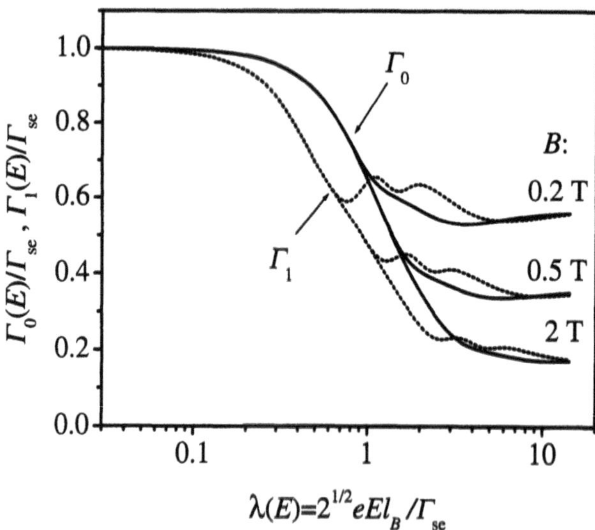

Fig. 2.4. Electric field effect on the collision broadening of Landau levels: $N = 0$ (*continuous lines*) and $N = 1$ (*dotted lines*)

additional energy exchange $\hbar \boldsymbol{q} \cdot \boldsymbol{u}$ at a collision. If this energy exchange is larger than the Landau level width, the electron has to scatter into the energy range with fewer or next to no states, which reduces the scattering probability. The scattering probability then increases if $\hbar \boldsymbol{q} \cdot \boldsymbol{u}$ becomes comparable with the Landau level separation $\hbar\omega_{\mathrm{c}}$. The treatment described above is also applicable to electron–ripplon scattering.

In the cumulant approach discussed in Sect. 1.9, the Doppler shift $\hbar \boldsymbol{q} \cdot \boldsymbol{u}$ enters the Green's function of scatterers in the second order cumulant [see (1.157)], and this affects the frequency argument of the function $\phi(t, \omega)$. Qualitatively, the electric field reduction in the collision broadening of the Landau levels in the cumulant approach is the same as that in the SCBA. It is the additional energy exchange $\hbar \boldsymbol{q} \cdot \boldsymbol{u}$ which cuts off the contributions of the scattering events with large q, when $\hbar \boldsymbol{q} \cdot \boldsymbol{u}$ becomes higher than Γ_{se}. Still, accurate evaluations are more complicated in this case. It is reasonable to use the field dependence $\Gamma_0(E)$ of the SCBA theory in the Gaussian forms of the density-of-state function as well, because both these approaches have the same broadening parameter for $E = 0$.

One should keep in mind that the substitution of the Gaussian form of $\mathrm{Im}G_N$ into the self-consistent equation is an approximation because, in the limit $\hbar \boldsymbol{q} \cdot \boldsymbol{u} \to 0$, the accurate solution of (2.44) has the semi-elliptic form. Nevertheless, this approximation does not cause any substantial error in the quantum conductivity equations because, as we shall see later, the DSF of the Coulomb liquid is more strongly affected by the same effect in a direct way which can be treated with much higher accuracy. Therefore, the approximate analytical solution for $\Gamma_0(E)$ given in (2.45) is quite sufficient for an accurate magnetotransport study. In Chap. 4 we shall discuss the similarities and differences in the electric field dependencies of the Landau level broadening $\Gamma_0(E)$ and the effective collision frequency $\nu(E)$.

2.3.2 Ensemble of Electrons with Ultra-Fast Orbit Centers

As noted above, the orbit center of an electron in the Coulomb liquid drifts fast with the velocity $u_{\mathrm{f}}^{(i)} = cE_{\mathrm{f}}^{(i)}/B$ with respect to the liquid center-of-mass frame owing to the quasi-uniform fluctuational electric field $\boldsymbol{E}_{\mathrm{f}}^{(i)}$. At high electron densities or in weak magnetic fields the drift velocity of the orbit center can be considered as ultra-fast because the Doppler shift $\hbar \boldsymbol{q} \cdot \boldsymbol{u}_{\mathrm{f}}$ strongly affects the probability of electron scattering. The effect of the fluctuational electric field on the single-electron density of states is equivalent to the effect of a uniform electric field discussed in the previous section.

To obtain the density dependence of the collision broadening of the Landau levels (in the reference frame moving along with the orbit center), we can just replace E by $E_{\mathrm{f}} \simeq E_{\mathrm{f}}^{(0)} = \mathcal{F}\sqrt{T} n_{\mathrm{s}}^{3/4}$ in (2.44) and (2.45). Typical density dependences of the broadening of Landau levels are shown in Fig. 2.5. Surprisingly, the strong Coulomb forces acting between electrons cause a narrowing of Landau levels. Even at high electron densities, where electron scattering between different Landau levels leads to an increase in Γ_N, the level broadening is less than in the case of noninteracting electrons.

Thus, in spite of the continuous correction to the Landau spectrum appearing in the laboratory frame, in local reference frames moving ultra-fast along with the electron orbit centers, the Landau levels of the Coulomb liquid become even narrower than in an ideal gas. As emphasized above, the only important difficulty is that such an electron excitation spectrum cannot be defined for all electrons in a single reference frame, in contrast to the case for conventional Fermi liquids. In the Coulomb liquid, for each electron, we have to introduce a particular reference frame where its spectrum coincides with the Landau spectrum. Nevertheless, a description of the transport properties in this system of strongly interacting particles appears to be possible by establishing a proper form for the more global property, the electron DSF.

The fluctuational forces acting on each electron can be treated as statistically independent. The many-particle problem is therefore reduced to the problem of a single electron in a fluctuational electric field. The whole Coulomb liquid under a magnetic field can be described as an ensemble of independent electrons subject to the quasi-uniform fluctuational electric field $\boldsymbol{E}_{\mathrm{f}}^{(i)}$. In this model, the system of strongly interacting electrons is replaced by a system of independent electrons with ultra-fast drift velocities $\boldsymbol{u}_{\mathrm{f}}^{(i)}$ distributed according to the fluctuational field, and the time average is replaced by an average over the ensemble.

The Coulombic reduction in Γ_N discussed above originates from the inelastic effect induced by the ultra-fast motion of the electron orbit centers.

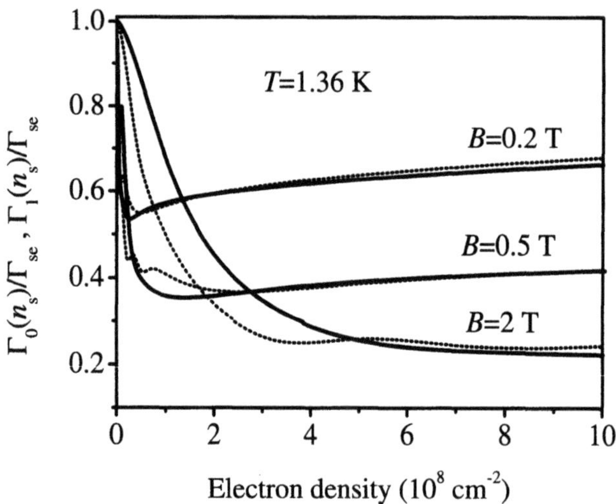

Fig. 2.5. Coulombic effect on the collision broadening of Landau levels vs. electron density: $N = 0$ (*continuous lines*) and $N = 1$ (*dotted lines*) [76]

It is very instructive to find its interplay with the intrinsic inelastic effect because of the finite energy exchange in the electron–phonon (ripplon) scattering events, which also decreases the level broadening. For example, from a naive point of view, the Coulomb reduction in level broadening should significantly strengthen the intrinsic inelastic effect, since the parameter $\hbar\omega/\Gamma_N$ becomes larger. Surprisingly, nothing like this happens, and, as we shall see in the following, the Coulombic effect actually reduces the inelastic effect.

Let us consider the low-temperature regime where electron scattering with 2D phonons (ripplons) dominates. Because the electron–ripplon coupling V_q is generally a complicated function of q, we cannot find an analytical form for $\Gamma_0(E_{\mathrm{f}})$ in this case. Keeping the semi-elliptic level shape, the self-consistent equation for Γ_0 can be found as

$$\Gamma_0^2 = \frac{T}{\pi\alpha}\int_{x_0}^{\infty}\frac{\mathrm{d}x_q}{x_q}\mathrm{e}^{-x_q}V_q^2 W\left(\frac{\hbar\omega_q}{\Gamma_0}, \sqrt{x_q y}\frac{\Gamma_{\mathrm{C}}}{\Gamma_0}\right), \tag{2.47}$$

where

$$W(\omega,\delta) = \int_0^{2\pi}\frac{\mathrm{d}\varphi}{2\pi}\sqrt{1-(\omega-\delta\cos\varphi)^2}, \tag{2.48}$$

$y = (E_{\mathrm{f}}/E_{\mathrm{f}}^{(0)})^2$ is the dimensionless parameter characterizing the fluctuational field, φ is the angle between the drift velocity and the wave vector $\boldsymbol{q}$, and $\Gamma_{\mathrm{C}} = \sqrt{2}eE_{\mathrm{f}}^{(0)}l_B$ is the quantity which we call the Coulomb broadening, even though its direct effect leads to the narrowing of Landau levels in the local moving reference frame. The reason for this name will become clear later, when we discuss the frequency dependence of the electron DSF. From (2.47) one can see that the level broadening is affected by the inelastic effect even at $E_{\mathrm{f}} \to 0$ because of the ripplon energy $\hbar\omega_q$ lost or gained in one-phonon scattering events. The form of the function $W(\omega,\delta)$ defined by (2.48) shows that there is indeed a strong interplay between the intrinsic inelastic effect of the electron–ripplon scattering and the inelastic effect which originates from the fluctuational electric field.

Now let us focus our attention on the physics of the discussed phenomenon and simplify evaluations as much as possible. (The accurate description of the interplay between the intrinsic inelastic effect and the many-electron effect will be given later for the electron DSF.) Previously, we analyzed a similar self-consistent equation assuming the Gaussian shape of Landau levels, which is equivalent to replacing $\sqrt{1-x^2}$ by $\exp(-2x^2)$ in (2.48). Such simplification even allows one to find an analytical solution for the many-electron reduction in the collision broadening [see (2.45)]. Still, numerically, the semi-elliptic form of the Landau levels is more consistent with the SCBA approach. Now we are looking for a better numerical simplification of the function $W(\omega,\delta)$. The comparison between the exact form $W(\omega,\delta)$ as a function of ω and the two approximate forms employing Gaussian approximations instead of the semi-elliptic shape is shown in Fig. 2.6. This figure indicates that, for

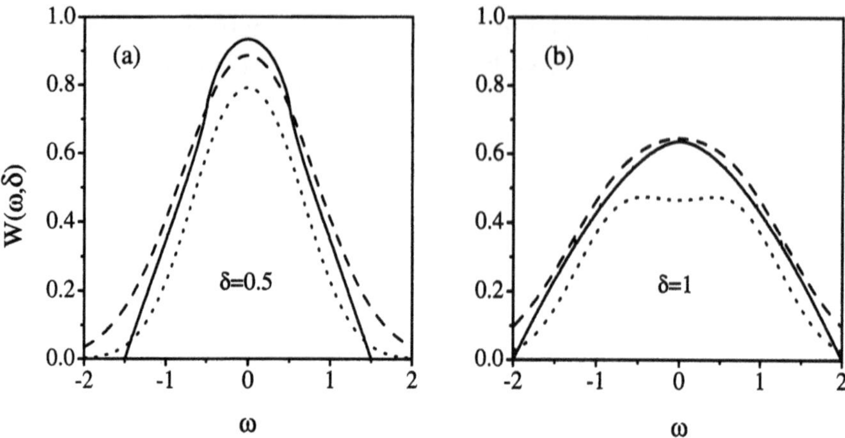

Fig. 2.6. Numerical comparison between the exact form of $W(\omega, \delta)$ given by (2.48) (*continuous line*) and two different forms using Gaussian approximations: $\exp(-2x^2)$ (*dotted line*) and $\exp(-x^2)$ (*dashed line*)

weak and intermediate fluctuational fields ($\delta \leq 1$), it is better to approximate the semi-elliptic function of (2.48) by $\exp(-x^2)$ (dashed curve) than by $\exp(-2x^2)$ (dotted curve). In the following, we shall take advantage of this numerical circumstance.

Searching for a simple solution to (2.47) which would describe the interplay of the many-electron effect with the intrinsic inelastic effect, we use the replacement $\sqrt{1-x^2} \rightarrow \exp(-x^2)$, proven to be numerically correct for the function $W(\omega, \delta)$ defined by (2.48). We then average (2.47) over the absolute values of the fluctuational electric field, assuming that $\Gamma_0^2(E_\mathrm{f})$ in the integrand W of the double integral on the right-hand side of this equation can be replaced by its average value $\langle \Gamma_0^2(E_\mathrm{f}) \rangle_\mathrm{f} \equiv \Gamma_\mathrm{av}^2$. For the Gaussian distribution of the fluctuational electric field,

$$\langle \cdots \rangle_\mathrm{f} = \int_0^\infty \mathrm{d}y \, \mathrm{e}^{-y} \int_0^{2\pi} \frac{\mathrm{d}\varphi}{2\pi} \cdots ,$$

where ϕ and y were defined in (2.48). The simplification made above yields a very instructive equation for Γ_av:

$$\Gamma_\mathrm{av}^2 = \frac{T}{\pi\alpha} \int_{x_0}^\infty \frac{\mathrm{d}x_q}{x_q \sqrt{1 + x_q \Gamma_\mathrm{C}^2 / \Gamma_\mathrm{av}^2}} V_q^2 \exp\left(-x_q - \frac{\hbar^2 \omega_q^2}{\Gamma_\mathrm{av}^2 + x_q \Gamma_\mathrm{C}^2} \right) . \tag{2.49}$$

This equation shows that the inelastic reduction in the Landau level broadening depends on the relation between the energy exchange $\hbar\omega_q$ and the effective broadening $\Gamma^{(*)} = \sqrt{\Gamma_\mathrm{av}^2 + x_q \Gamma_\mathrm{C}^2}$. Therefore, the many-electron decrease in Γ_av is compensated by the increase in $\Gamma_\mathrm{C} = \sqrt{2} e E_\mathrm{f}^{(0)} l_B$, and furthermore, the effective broadening increases with Γ_C.

It is interesting to note that, reducing the inelastic effect, the collision broadening of Landau levels and the Coulomb broadening parameter introduced above are combined in a way ($\sqrt{\Gamma_{\text{av}}^2 + x_q \Gamma_{\text{C}}^2}$) which is similar to combining the contributions from two different scattering mechanisms in a total broadening. In the following section, when analyzing the frequency dependence of the Coulomb liquid DSF, we shall see that this is a quite general rule, in spite of the fact that an increase in Γ_{C} generally leads to a decrease in the collision broadening in the local moving reference frames.

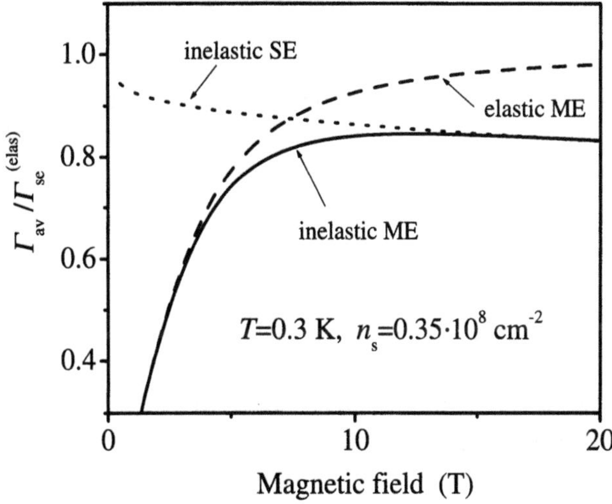

Fig. 2.7. Numerical solutions to (2.49) showing the interplay between the Coulombic and inelastic reductions in the collision broadening of the lowest Landau level

The normalized numerical solution of (2.49) is shown in Fig. 2.7 by the continuous curve. Two extreme limiting cases $\hbar\omega_q \ll \Gamma^{(*)}$ and $\Gamma_{\text{C}} \ll \Gamma_0$ are shown there by dashed and dotted curves, respectively. One can see that, in the low magnetic field range, the collision broadening of the lowest Landau level is practically the same as in the pure elastic case $\hbar\omega_q \to 0$, while at high magnetic fields it is close to the single-electron inelastic broadening. The Coulomb broadening correction in the exponent of the integrand of (2.49) makes the transition between these two regimes smooth in spite of the strong decrease in the average of Γ_0, which is seen in Fig. 2.7. In the high magnetic field range, the inelastic parameter $\hbar\omega_q/\Gamma^{(*)}$ increases with B because the Coulomb broadening parameter $\Gamma_{\text{C}}(B) \propto 1/\sqrt{B}$ can be disregarded in comparision with $\Gamma_0 \propto \sqrt{B}$ and the typical wave number $q \sim 1/l_B \propto \sqrt{B}$.

2.4 Coulomb Broadening of the Dynamical Structure Factor

The effect of Coulomb narrowing of the single-electron density of states described in local moving frames cannot completely reveal the equilibrium properties of the Coulomb liquid. For example, in the equation for the DSF of the nondegenerate electron gas [(2.38)], a careless replacement of Γ_N by the Landau level broadening, affected by the fluctuational electric field $\Gamma_N(n_s)$, would give a result quite opposite to the correct one.

The accurate form of the electron DSF can be found by describing the Coulomb liquid as an ensemble of independent electrons subject to the fluctuational electric field. For a particular electron, there is only one reference frame where it has the discrete Landau spectrum, namely, the frame which moves along with the orbit center with velocity $\boldsymbol{u}_{\mathrm{f}}$ relative to the center-of-mass reference frame. Only in this particular reference frame does the contribution of the electron to the DSF have the single-electron form $S_{\mathrm{se}}(q,\omega)$ given in (2.38). When changing the reference frame, the DSF obeys the transcription rule $S'(\boldsymbol{q},\omega) = S(q,\omega - \boldsymbol{q}\cdot\boldsymbol{u})$ because of Galilean invariance, as discussed above. Therefore, in the electron-liquid center-of-mass reference frame, the contribution of the electron to the DSF of the ensemble considered here can be written as $S_{\mathrm{se}}(q,\omega - \boldsymbol{q}\cdot\boldsymbol{u}_{\mathrm{f}}^{(i)})$. The proper many-electron approximation for the DSF of the ensemble of independent electrons subject to the fluctuational electric field can thus be found in the most simplified way as [76, 90]

$$S_{\mathrm{me}}(q,\omega) = \langle S_{\mathrm{se}}(q,\omega - \boldsymbol{q}\cdot\boldsymbol{u}_{\mathrm{f}})\rangle_{\mathrm{f}} \ , \tag{2.50}$$

where $\langle\cdots\rangle_{\mathrm{f}}$ denotes the average over the fluctuational electric field. The physics of this 'one-line' many-electron theory is very simple: the contribution from any electron to the DSF of the ensemble has a single-electron form, but in the center-of-mass reference frame, it acquires a Doppler shift because of the ultra-fast velocity of the electron orbit center.

The relation between the many-electron DSF and $S_{\mathrm{se}}(q,\omega)$ of (2.50) can also be found by evaluating the momentum loss per unit time of the collection of independent electrons subject to the quasi-uniform fluctuational electric field. One should take into account the fact that the energy exchange in a collision depends on the reference frame whilst the momentum exchange does not. Therefore one can calculate the momentum loss of each electron in the unique moving reference frame where it has the discrete energy spectrum and then just add contributions from different electrons. This procedure results in a momentum relaxation rate which is in agreement with the DSF form of (2.50).

The simple relation (2.50) also agrees with the conductivity evaluations performed previously [88] for the continuous electron spectrum in the fluctuational field, valid in the limiting case $\Gamma_0 \ll eE_{\mathrm{f}}l_B$. The important advantage of the form in (2.50) is that it allows one to find conductivity equations valid for any relation between the collision broadening of Landau levels and the many-electron effect, employing certain approximations for the single-electron DSF $S_{se}(q,\omega)$. An additional advantage is that one does not need to introduce the continuous correction to the Landau spectrum because $S_{se}(q,\omega)$ is evaluated for electrons without the fluctuational field. All conductivity evaluations can be performed for the discrete Landau spectrum, which greatly simplifies the many-electron theory. It should be emphasized that (2.50) does not stick to any particular approximation for the $S_{se}(q,\omega)$. For example, one can use the SCBA, or any other, even more accurate approximation.

For the Gaussian distribution of the fluctuational field E_f, the average of (2.50) can be written in the explicit form

$$S_{me}(q,\omega) = \int_0^\infty dy\, e^{-y} \int_0^{2\pi} \frac{d\varphi}{2\pi} S_{se}\left[q, \omega - \sqrt{x_q y}\cos(\varphi)\Gamma_C/\hbar\right] , \qquad (2.51)$$

where the dimensionless parameters of the fluctuational field y and φ are the same as those introduced in (2.48), and $\Gamma_C = \sqrt{2}eE_f^{(0)}l_B$. This equation can be evaluated numerically for particular transport considerations. Still, it is very instructive to begin by analyzing the case where the density dependence of $\Gamma_N(E_f)$, entering S_{se} of (2.38), is weak enough to be disregarded in the integral of (2.51). Strictly speaking, this assumption is valid only for rather high electron densities, according to Fig. 2.5, or in the limiting case $\Gamma_C \gg \Gamma_0$. Nevertheless, it is possible to prove that the integral of (2.51) is organized in such a way that the final result depends only weakly on the real behavior of $\Gamma_N(E_f)$ in the whole density range, so that even the replacement $\Gamma_0 \to \Gamma_{se}$ gives a quite accurate result. Substituting $\Gamma_N(E_f^{(0)})$ for $\Gamma_N(E_f)$ in the equation for S_{se} [(2.38)], the integrals of (2.51) can be taken and represented in the analytical form

$$S_{me}(q,\omega) = \sum_{N=0}^{\infty} \frac{2\sqrt{\pi}\hbar x_q^N}{N!\sqrt{\Gamma_{0,N}^2 + x_q\Gamma_C^2}} \exp\left[-x_q - \frac{\hbar^2(\omega - N\omega_c)^2}{\Gamma_{0,N}^2 + x_q\Gamma_C^2}\right] , \qquad (2.52)$$

which possesses very interesting features.

Like the DSF of noninteracting electrons [(2.38)], the many-electron DSF of (2.52) is a sum of Gaussian terms exhibiting resonant behavior with regard to $\omega - N\omega_c$ (with $N = 0, 1, 2, \ldots$). The important difference is that $\Gamma_{0,N}$ is replaced by the effective broadening parameter $\Gamma_{0,N}^{(*)}(q) = \sqrt{\Gamma_{0,N}^2 + x_q\Gamma_C^2}$. A similar quantity has already discussed in the preceding section, when analyzing the interplay between the inelastic and Coulombic effects on the collision broadening of Landau levels. One can see that $\Gamma_{0,N}^{(*)}(q)$ increases with n_s, despite the decrease in $\Gamma_{0,N}(n_s)$. For the particular case $N = 0$, the broadening of the many-electron DSF $\Gamma_{0,0}^{(*)} \simeq \sqrt{\Gamma_0^2 + \Gamma_C^2}$ is shown in Fig. 2.8 together with the collision broadening of the lowest Landau level evaluated according to (2.49). Thus the same many-electron effect that narrows Landau levels in the reference frame moving along with the electron orbit center broadens the electron DSF in the electron-liquid center-of-mass frame. Moreover, the rule for combining the collision broadening and the many-electron effect resembles the rule for combining contributions of two different scattering mechanisms: $\Gamma_{\mathrm{AB}}^2 = \Gamma_{\mathrm{A}}^2 + \Gamma_{\mathrm{B}}^2$. Therefore, we call Γ_{C} the Coulomb broadening of the electron dynamical structure factor.

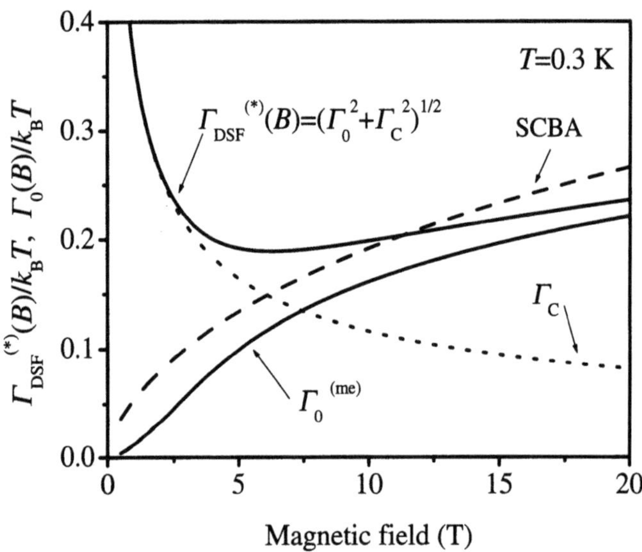

Fig. 2.8. Broadening of the Coulomb liquid DSF and collision broadening of the lowest Landau level vs. B for $n_{\mathrm{s}} = 0.35 \times 10^8\ \mathrm{cm}^{-2}$

It should be emphasized that the many-electron form $S_{\mathrm{me}}(q,\omega)$ was obtained for electrons with extremely narrow Landau levels. As discussed just after (2.38), for 'dirty' interfaces, the frequency shift $-\Gamma_0^2/4T\hbar$ in the Gaussian terms of $S_{\mathrm{se}}(q,\omega)$ should be taken into account. The above approach based on a formal averaging over the Doppler shifts is valid under the condition $\Gamma_{\mathrm{C}} \ll T$. This accuracy does not allow one to obtain the frequency shifts in the Gaussian terms of the Coulomb liquid DSF. One can find them using the general properties of the DSF as a function of frequency discussed above or by comparing (2.52) with the DSF obtained for the WS state.

It is interesting to note that in the limit $\Gamma_{\mathrm{C}} \gg \Gamma_{0,N}$, the DSF of the Coulomb liquid [(2.52)] appears to be the same as that of the Wigner solid, which we shall discuss in Chap. 7 dealing with properties of the electron solid. The only difference is that the corresponding Gaussians have small frequency shifts $-x_q\Gamma_{\mathrm{C}}^2/4T\hbar$. Comparing them with the frequency shifts $-\Gamma_N^2/4T\hbar$ in the expression for the single-electron DSF [(2.37)], it is possible to conclude that in the general case the Gaussians of the Coulomb liquid DSF should have similar frequency shifts $-(\Gamma_N^2 + x_q \Gamma_C^2)/4T\hbar$ [for (2.52) $N = 0$]. These also ensure the general properties of the DSF as a function of ω discussed above.

The DSF of a strongly correlated Coulomb liquid under a magnetic field can thus be found formally from the DSF of noninteracting electrons [(2.37)] simply by replacing the collision broadening of the single-electron states Γ_N by the effective broadening $\Gamma_N^{(*)}(q) = \sqrt{\Gamma_N^2 + x_q \Gamma_C^2}$. (For simplicity we assume that Γ_N does not depend much on N.) This is valid for both the broadening of the Gaussian terms and frequency shifts. For surface electrons on helium, the frequency shifts $-\Gamma_N^2(q)/4T\hbar$ are small because $\Gamma_{se} \ll T$. (The Coulomb broadening is also assumed to be smaller than the temperature.) They do not therefore affect the conductivity results very much. Still, taking these frequency shifts into account, it is remarkable that the simple model for the electron DSF presented in (2.52) will be valid across the whole range of the Coulomb coupling parameter, including the extreme cases of the electron gas state and the Wigner solid state. From a physical point of view it is remarkable that, according to (2.50), the DSFs of the Coulomb liquid and Wigner solid can be constructed from the DSF of the 2D gas of noninteracting electrons.

2.5 Plasmons and Magnetoplasmons in Reduced Dimensions

Interface electrons represent an ideal example for studying collective excitations in 2D electron systems. It is no wonder that the experimental observation of the unusual dispersion of 2D plasmons, described theoretically long ago [91–93], was first performed for surface electrons on liquid helium [47] and then reproduced for semiconductor 2D electron systems. Another example of collective excitations observed for SEs on liquid helium, the edge magnetoplasmon wave, was a completely unexpected phenomenon [94, 95]. Being basically a pure classical effect, it plays an important role in the study of the quantum Hall effect in semiconductor 2D electron structures [96, 97] and quantum magnetotransport phenomena in systems of SEs on liquid helium [98, 99].

When studying the long-wavelength excitations in a 2D system of charged particles, one cannot disregard the Coulomb potential energy, even if the plasma coupling parameter $\Gamma^{(\mathrm{pl})}$ is small. The reason is the same as in the 3D case: the Fourier transform of the Coulomb potential has a singularity as $k \to 0$. Although in the 2D case the singularity

$$V_k^{(2\mathrm{D})} = \frac{2\pi e^2}{k} \tag{2.53}$$

is weaker than in three dimensions ($V_k^{(3\mathrm{D})} \propto 1/k^2$), the long-wavelength excitations still possess a huge energy which determines the properties of such a one-component plasma. The difference in the long-wavelength asymptotic behavior of $V_k^{(2D)}$ and $V_k^{(3D)}$ causes crucial differences in the plasmon excitation spectrum. In three dimensions, the charge density oscillation spectrum has a huge gap (the plasma frequency), while in two dimensions the plasmon spectrum is gapless.

2.5.1 Interior Excitations

It is instructive to consider first the weak Coulomb coupling regime. It is well known that the approximation of noninteracting electrons [(2.21)] for the density–density response function is poor if one uses (2.13) for the dielectric function. Considering the electron response to the total field, we have to replace $V_k^{(2D)}$ by $V_k^{(2D)}/\varepsilon(k,\omega)$, which yields the following formula for the dielectric function:

$$\varepsilon(k,\omega) = 1 + V_k^{(2D)}\chi(k,\omega) \,. \tag{2.54}$$

In this equation, the use of the approximation $\chi(k,\omega) \simeq \chi_0(k,\omega)$ from (2.21) gives a reasonable result. This is the self-consistent field approximation. The spectrum of plasmons is determined by the equation $\varepsilon(k,\omega) = 0$, which means physically that the electric field perturbations $\varphi_{\text{total}}(k,\omega) = \varphi_{\text{ext}}(k,\omega)/\varepsilon(k,\omega)$ can exist even in the absence of the external field, i.e., when $\varphi_{\text{ext}}(k,\omega) \to 0$.

For the nondegenerate electron gas, the response functions were evaluated by Platzman and Tzor [100]:

$$\chi_0 = -\frac{n_s}{T}\left[R(x) + \mathrm{i}I(x)\right] \,, \tag{2.55}$$

where

$$x = \frac{\omega}{k}\sqrt{\frac{m}{2T}} \,, \qquad I(x) = \sqrt{\pi}x\,\mathrm{e}^{-x^2} \,,$$

$$R(x) = 1 - 2x\mathrm{e}^{-x^2}\int_0^x \mathrm{e}^{y^2}\mathrm{d}y \approx -\frac{1}{2x^2}\left(1 + \frac{3}{2x^2}\right) \,, \qquad x \gg 1 \,. \tag{2.56}$$

The real part $R(x)$ has a minimum at $x = x_{\min} \simeq 1.47$ with $R(x_{\min}) \simeq -0.284$.

In the long-wavelength limit, the dimensionless parameter x is large and the plasmon damping is small: $\omega = \Omega_k - \mathrm{i}\gamma_k$ with $\gamma_k \ll \Omega_k$. In this case, the plasmon dispersion and damping can be found as [30, 100]

$$\Omega_k^2 = \frac{n_s k^2 V_k^{(2D)}}{m}\left[1 + 3\frac{T}{n_s V_k^{(2D)}}\right] \,, \tag{2.57}$$

$$\gamma_k = \sqrt{\frac{2T}{m}}k\left[\sqrt{\pi}x\mathrm{e}^{-x^2}\left(\frac{\partial f}{\partial x}\right)^{-1}\right]_{x=x_k} \,, \tag{2.58}$$

where $x_k = \sqrt{n_s V_k^{(2D)}/T}$. In the limiting case $x_k \gg 1$, the plasmon damping is exponentially small. Equation (2.57) is valid if the second term in brackets is much smaller than unity.

The form of (2.57) establishes a simple relationship between the singularity of the interaction potential and the long-wavelength behavior of the excitation spectrum: $\Omega_k \propto k\sqrt{V_k}$. For example, if the interaction potential is of a short-range nature ($V_k^{(2D)} \to$ const.), as it is in typical neutral systems, the relationship yields the sound-like behavior $\Omega_k \propto k$. If the Fourier transform of the interaction potential is the same as in the 3D plasma $V_k \propto 1/k^2$, the excitation spectrum has the plasmon gap $\Omega_k \to$ const. For the 2D Coulomb liquid, we have an intermediate situation between these two extremes: $V_k^{(2D)} \propto 1/k$ and $\Omega_k \propto \sqrt{k}$. In the long-wavelength limit, the frequency of 2D plasmons approaches zero but at a much slower rate than in the case of particles with the short-range interaction.

For the Coulomb liquid with $\Gamma^{(\mathrm{pl})} \gg 1$, the use of the self-consistent field approximation is doubtful. Nevertheless, the main term of (2.57) describing the plasmon dispersion has a quite universal form which remains valid even for the longitudinal phonons of the electron solid ($\Gamma^{(\mathrm{pl})} > 140$). Therefore, in the intermediate range of the plasma coupling parameter, there is no reason for the interior plasmon dispersion to differ from the main term of (2.57) if $k \ll \sqrt{n_s}$.

As mentioned above, the first observation of the 2D plasmon dispersion was performed for the SEs on superfluid helium by Grimes and Adams [47]. The plasmon dispersion and damping was studied by exciting standing-wave resonances in the 2D plasma of SEs contained within a rectangular cell. The areal density of the electrons was modulated by an audio-frequency potential ($\sim 5\,\mathrm{kHz}$) in order to measure the derivative of the absorption with respect to the electron density $\mathrm{d}A/\mathrm{d}n_s$. The experimental plot of [47] is shown in Fig. 2.9. The most prominent feature in this plot are the first three standing-wave resonances corresponding to 1, 3, and 5 nodes contained within the width of the cell.

To calculate the frequencies of the standing-wave resonances in this experiment, it was necessary to take into account the screening of the Coulomb interaction due to the metal electrodes [47]:

$$\Omega^2 = \frac{2\pi n_s e^2}{m_e}\sqrt{k_x^2 F(k_x) + k_y^2 F(k_y)}\,, \tag{2.59}$$

where

$$F(k_i) = \frac{2\sinh(k_i d)\sinh[k_i(h-d)]}{\sinh(k_i h)}\,,$$

d is the helium depth, h is the cell height, $k_x = m\pi/W$, $k_y = n\pi/L$, m and n are integers, and W and L are the width and length of the rectangular cell, respectively. The resonance modes are identified by the two integers (m, n).

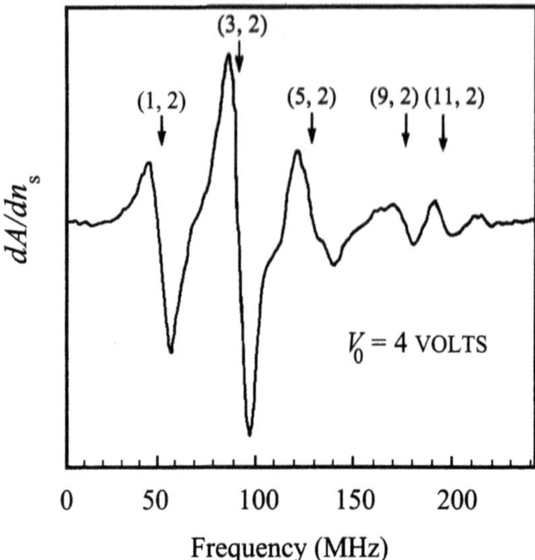

Fig. 2.9. Experimental plot of the derivative of the rf absorption with respect to electron density plotted vs. frequency [47]. The calculated positions of the standing-wave resonances are indicated by *arrows*

By varying the charging potential V_0 proportional to the average electron density, it was possible to check that the positions of the plasmon resonances change in accordance with the calculated dispersion of (2.59). For $m \geq 5$, the geometrical correction factor $F(k_i)$ is nearly unity, and the plasmon frequency varies as $\sqrt{k}$.

The measured damping of the standing-wave resonances behaves in accordance with the electron–ripplon momentum relaxation rate, indicating a strong increase with the holding electric field $E_\perp$. However, at weak holding fields and low temperatures, the linewidth of the these resonances was shown to become temperature independent, which disagrees with the low-frequency mobility data. Later, in similar measurements conducted in parallel with the low-frequency measurements of the electron mobility [101], it was proven that this temperature-independent plateau in the linewidth is caused by small deviations of the average electron density from the uniform distribution within the cell. Tilting the cell slightly shifts the linewidth plateau downwards.

The excitation spectrum of 2D plasmons is strongly affected by a magnetic field oriented normally to the system. Let us consider the hydrodynamic approach. The equation of motion for the Fourier transforms of the velocity field $\boldsymbol{v_k}$ can be written as

$$-\mathrm{i}\omega m_e \boldsymbol{v_k} = -\mathrm{i}\boldsymbol{k} V_k^{(2\mathrm{D})} n_{\boldsymbol{k}} - \frac{e}{c}\boldsymbol{v_k} \times \boldsymbol{B} , \qquad (2.60)$$

where the first term on the right-hand side is the self-consistent field and $n_{\boldsymbol{k}}$ are the electron density fluctuations. Conservation of matter yields the continuity equation, which we write in the linear approximation as

$$-\mathrm{i}\omega n_{\boldsymbol{k}} + n_{\mathrm{s}}\mathrm{i}\boldsymbol{k}\cdot\boldsymbol{v}_{\boldsymbol{k}} = 0\,. \tag{2.61}$$

The straightforward solution of these equations gives

$$\Omega_{\mathrm{mp}} = \sqrt{\Omega_k^2 + \omega_{\mathrm{c}}^2}\,, \qquad \Omega_k = \sqrt{\frac{n_{\mathrm{s}} V_k^{(2\mathrm{D})} k^2}{m_{\mathrm{e}}}}\,. \tag{2.62}$$

Thus the spectrum of magnetoplasmons has cyclotron frequency gap ω_{c} proportional to the magnetic field B. Therefore, the 'rigidity' of the electron liquid increases with the magnetic field.

2.5.2 Edge Waves

Edge magnetoplasmons (EMP) of a two-dimensional (2D) electron liquid were observed as a new and unanticipated perimeter mode in the system of electrons bound to a surface of superfluid helium and subject to a strong normal magnetic field. The observation of this mode was independently reported by two experimental groups [94, 95] at practically the same time. The EMP wave can propagate along the boundary of both rectangular and circular geometries in the direction determined by the sign of the z-component of the magnetic field. Of course, for the circular geometry, the excitation signal must not have axial symmetry.

In the EMP wave, the charge density oscillations are confined to the edge of the electron liquid within a quite narrow boundary strip $0 < x < b$, where b is much smaller than the wavelength, and the wave itself propagates along the boundary in one direction only, determined by the sign of the Hall conductivity σ_{xy} or by the direction of the magnetic field $\boldsymbol{B}$ applied normally to the electron layer. The width of the boundary strip b decreases with the magnetic field and, in the limit of strong fields, it becomes approximately equal to the length scale h of the equilibrium electron density profile at the edge $n(x) = n_0\tau(x/h)$, where $\tau(0) = 0$ and $\tau(\infty) = 1$.

In the last section we found that the normal magnetic field increases the frequency of the interior excitations of the 2D plasma by introducing the cyclotron frequency gap. The most important features of the edge magnetoplasmon wave are strikingly opposite: its frequency decreases with the magnetic field as $\Omega \propto 1/B$, and its spectrum is gapless $\omega \propto q_y$, where q_y is the wave vector component along the boundary. [For the pure Coulomb interaction, the excitation frequency also contains a logarithmic factor $\ln(1/|q_y|b)$.] The properties of the EMP wave mentioned above are very important for the semiconductor 2D electron systems in the regime of the quantum Hall effect (QHE) [96, 97, 102], because at low temperatures the electrical transport of these systems is confined to the so-called edge channels [103]. In the AC capacitive coupling method of the magnetotransport measurement used for SEs on liquid helium, this wave is actually a hindrance causing a strong spurious signal from the EMP excitation. However, at low temperatures the EMP damping itself contains important information about the quantum magnetoconductivity of the 2D electron system [98] and helps to measure it.

The theory of the EMP was primarily developed under the assumption that the boundary of the electron sheet is rigid or fixed. In this case, the edge excitations are just the density oscillations of the compressible electron liquid which are localized near the boundary and whose properties are strongly affected by the density profile at the edge $n(x)$. An accurate description of these waves requires a rather sophisticated mathematical treatment (for a review see [104]). It is remarkable that the main features of the EMP can be found by considering edge excitations of an incompressible electron liquid but with a movable boundary [102, 105, 106], which can be called magnetoripplons of the Coulomb liquid or boundary displacement waves (BDW). Although the 2D electron system on helium is not incompressible (this assumption is applicable to the semiconductor electron system under the fractional QHE regime), it is very instructive to begin with this simple model.

Let us firstly discuss the boundary properties of the electron pool in order to find the restoring force for a boundary displacement. 2D electrons are usually confined by an external electric field $\boldsymbol{E}_{\mathrm{G}}$ induced by the guard and cell capacitor electrodes. The equilibrium condition for the electron pool requires the total electric field $\boldsymbol{E} = \boldsymbol{E}_{\mathrm{G}} + \boldsymbol{E}_{\mathrm{el}}$ projected along the surface to be zero within the electron sheet and at its edge. Here $\boldsymbol{E}_{\mathrm{el}}$ is the long-range field of the interior electrons. Because the total force acting on a single electron at the edge satisfies $eE_x = 0$, the liquid boundary cannot be considered as truly rigid. Moreover, the confining field $\boldsymbol{E}_{\mathrm{G}}$ cannot even be considered as a linear restoring force for a small boundary displacement $\xi(y)$. Indeed, the single-electron potential energy $U(x)$ in the vicinity of the boundary (outside the electron pool, $x < 0$) has no terms proportional to the x-coordinate. In this range, one can expect $U(x) \propto x^2$ for a single electron. Therefore, the overall effect of the confining potential on the electron liquid potential energy should be proportional to ξ^3, which is a nonlinear effect for boundary oscillations.

The real restoring force for a liquid boundary displacement ξ originates from the electric field of the edge charges induced by this displacement. For a sharp edge, the induced charge per unit length is $en_0\xi(y)$. Any displacement of the boundary from the equilibrium position creates an electric field that acts on interior electrons in such a way as to restore the equilibrium. It is important to point out the difference between the pure Coulomb liquid and the electron layer which is close to a metal electrode of the parallel-plate condenser. In the first case, the electric potential of the displacement wave $\xi(y,t) = \xi_0 \exp(\mathrm{i}q_y y - \mathrm{i}\omega t)$ penetrates the interior of the electron liquid with a length-scale equal to the wavelength:

$$\varPhi_{\xi}(x,y) \simeq 2en_0\xi(y)K_0(|q_y x|) , \tag{2.63}$$

where $|x| \gg \xi, h$, and $K_0(x)$ is the modified Bessel function. Therefore, the electric potential Φ_ξ should enter the equation of motion of the electron liquid rather than the boundary conditions. In the case of strong screening by the electrodes, the electric field perturbations caused by $\xi(y)$ vanish in the interior of the electron liquid within the screening length which is of the same order of magnitude as the liquid helium depth d and the density profile width h. This means that the electric field induced by $\xi(y)$ should not enter the equations of motion of the electron liquid because it represents a pure edge quantity. These field perturbations should be treated in the same way as the van der Waals forces are treated in the theory of surface phenomena in ordinary liquids, by introducing a sort of edge tension.

Let us consider the unscreened system of charges first. As discussed above, assume that the velocity field $\boldsymbol{v}_\mathrm{e}$ of the electron liquid obeys the continuity equation of an incompressible liquid $\mathrm{div}(\boldsymbol{v}_\mathrm{e}) = 0$. In the linear theory, the Euler equation of motion can be written as

$$\mathrm{i}\omega v_\mathrm{e}^{(y)} = \frac{1}{mn_\mathrm{s}} \frac{\partial}{\partial y}(p_\mathrm{e} - en_\mathrm{s}\Phi_\xi) - \mathrm{sgn}\{B_z\}\omega_\mathrm{c} v_\mathrm{e}^{(x)} , \tag{2.64}$$

where the effect of the electric field of the edge charges $en_0\xi(y)$ is written as the correction $-en_\mathrm{s}\Phi_\xi$ to the 2D liquid pressure p_e. The magnetic field term changes its sign when $\boldsymbol{B} \to -\boldsymbol{B}$, which is taken into account by means of the factor $\mathrm{sgn}\{B_z\}$. The boundary conditions are

$$v_\mathrm{e}^{(x)}(x)\Big|_{x=0} = -\mathrm{i}\omega\xi , \qquad p_\mathrm{e}(x)|_{x=0} = 0 . \tag{2.65}$$

They represent the continuity conditions for the velocity field and for the pressure at the edge.

In a real system, Φ_ξ is a continuous function. Therefore, the boundary condition for the liquid pressure can be rewritten as a boundary condition for the function $\mathcal{P}_\mathrm{e}(x) = p_\mathrm{e} - en_\mathrm{s}\Phi_\xi$:

$$\mathcal{P}_\mathrm{e}(0) = -en_\mathrm{e}\Phi_\xi(0) . \tag{2.66}$$

For small displacements, we can write

$$\Phi_\xi(0) = \alpha_\mathrm{e} en_\mathrm{s}\xi ,$$

where α_e is a geometrical factor of the order of unity. In the limiting case $d \to \infty$, direct evaluation yields $\alpha_\mathrm{e} \simeq 2\ln(1/|q_y|h)$. In the 'screening' case, we find that

$$\alpha_\mathrm{e} = \int \tau'(t)\psi(t)\mathrm{d}t , \qquad \psi(t) = \ln\frac{4+t^2}{t^2} , \tag{2.67}$$

where $\tau'(t)$ is the derivative of the density profile function $\tau(x/h)$.

In (2.64), $\mathcal{P}_\mathrm{e} = p_\mathrm{e} - en_\mathrm{s}\Phi_\xi$ acts as an effective pressure in the 2D liquid and the boundary condition (2.66) appears to be similar to the pressure difference introduced for a typical liquid with surface tension. The right-hand side of (2.66) is analogous to the Laplace formula. The important difference in the new formula, induced by the long-range nature of the Coulomb forces, is that the Laplace edge pressure here is proportional to the amplitude of the boundary displacement. It is instructive to note that one could find this formula differently by removing the field of the edge charges from the equation of motion [(2.64)] and considering the energy of the electric field of the edge charges $\mathcal{E}(\xi) = (1/8\pi) \int (\nabla \Phi_\xi)^2 \mathrm{d}x\mathrm{d}z$ per unit length induced by a boundary displacement ξ. For small displacements

$$\mathcal{E}(\xi) \simeq \frac{1}{2}\gamma_{\mathrm{bdw}} e^2 n_{\mathrm{s}}^2 \xi^2 , \tag{2.68}$$

where γ_{bdw} is a dimensionless geometrical factor. Then one can introduce the edge-tension boundary condition for the electron liquid pressure:

$$p_{\mathrm{e}}(0) = -\frac{\partial \mathcal{E}(\xi)}{\partial \xi} = -\gamma_{\mathrm{bdw}} e^2 n_{\mathrm{s}}^2 \xi , \tag{2.69}$$

which is practically the same as the condition of (2.66). In this way the EMP excitations of the incompressible electron liquid were actually treated in [102]. For the 'screening' case, where this approach is fully applicable, straightforward evaluation gives [106]

$$\gamma_{\mathrm{bdw}} \simeq \int \mathrm{d}t_1 \int \mathrm{d}t_2 \tau'(t_1)\tau'(t_2)\psi(|t_1 - t_2|) . \tag{2.70}$$

If the boundary is rather sharp, one of the functions $\tau'(t)$ of the integrand can be replaced approximately by a δ-function. Then comparison with (2.67) leads to the relation $\gamma_{\mathrm{bdw}} \approx \alpha_{\mathrm{e}}$.

The easiest way to solve the above equations for $\boldsymbol{v}_{\mathrm{e}}$ is to introduce the velocity field potential φ_{e}, which satisfies the equation $\Delta\varphi_{\mathrm{e}} = 0$. We then find that $\varphi_{\mathrm{e}} \propto \exp(-|q_y|x)$. The interesting point is that the velocity field perturbation of the BDW penetrates deep into the interior of the 2D electron liquid causing no electric potential perturbations there. This is especially important for the 'screening' case, where the electric field perturbations cannot penetrate more deeply than the screening length $d \ll 1/|q_y|$. It should be noted that the Coulomb interaction between electrons is screened only at large distances $r \gg d$, while at small distances $r < d$, we still have the very strong Coulomb repulsion between electrons. Like the hard-core repulsion of typical neutral liquids, this repulsion is responsible for the transmission of the electron motion from the displaced boundary into the liquid interior. The latter is taken into account by the corresponding pressure (p_{e}) term of the Euler equation, as in typical neutral liquids.

In the absence of a magnetic field, the excitation spectrum of the edge wave can be found as

$$\Omega_0^2(q_y) = \frac{\alpha_{\mathrm{e}} e^2 n_{\mathrm{s}}}{m} |q_y| . \tag{2.71}$$

For the electron liquid with screened interaction due to the metal electrodes, the geometric factor α_e should be replaced by γ_{bdw} according to the condition (2.69). It is interesting to note that, at $B = 0$, the dispersion of the boundary displacement mode of an incompressible liquid is similar to the dispersion of the 2D plasmons: $\Omega_0 \propto \sqrt{|q_y|}$. If electrons are subject to a normal magnetic field, the solution of the dispersion equation depends on the sign of the product $q_y B_z$ [106]:

$$\Omega_{bdw}(q_y) = \frac{1}{2}\left[\sqrt{\omega_c^2 + 4\Omega_0^2(q_y)} - \text{sgn}\{q_y B_z\}\omega_c\right] . \tag{2.72}$$

For strong magnetic fields and at $q_y B > 0$, the frequency of the boundary displacement wave decreases with B as

$$\Omega_{bdw}(q_y) \simeq \frac{\Omega_0^2(q_y)}{\omega_c} \simeq \alpha_e \sigma_{yx} q_y . \tag{2.73}$$

If the wave propagates in the opposite direction ($\text{sgn}\{q_y B_z\} < 0$), the excitation spectrum has a high frequency gap ω_c.

Remarkably, at $\alpha_e \simeq 2\ln(1/|q_y|h)$, which corresponds to the case of the pure Coulomb liquid, the spectrum of the BDW [(2.73)] becomes equivalent to the spectrum of the EMP of a compressible electron liquid with a rigid boundary. This is not just a coincidence because the physics of these oscillations of the edge charge is the same: a charge density perturbation localized at the edge induces an electric field inside the liquid which acts as a restoring force.

The exact procedure for finding the EMP dispersion is much more complicated than the one described above for the BDW. Therefore, we consider here a simplified approach proposed by Shikin [107], which is based on the integral form of the charge conservation equation. Firstly, one should note that, under strong magnetic fields in the long-wavelength limit $q_y b \ll 1$, the electric field potential Φ of the edge wave satisfies the condition

$$\frac{\partial^2 \Phi}{\partial x^2} \ll \frac{\partial^2 \Phi}{\partial y^2} ,$$

which means that gradients are stronger for the perpendicular direction. In this case,

$$\text{div}\boldsymbol{j} \simeq -\sigma_{xx}\frac{\partial^2 \Phi}{\partial x^2} ,$$

and the integral form of the continuity equation can be written as [107]

$$-\sigma_{xx}\frac{\partial \Phi}{\partial x}\,|_{x=0} + \mathrm{i}\omega Q_{emp} = 0 , \tag{2.74}$$

where $Q_{emp} = -e\int \delta n dx$ is the linear charge density in the edge wave. According to the rigid boundary condition $j_x(0) = 0$, the first term of this equation can be replaced by

$$-\sigma_{yx} \mathrm{i} q_y \Phi(0) \equiv -\sigma_{yx} \mathrm{i} q_y \gamma_{\mathrm{emp}} Q_{\mathrm{emp}} ,$$

where γ_{emp} is a geometrical proportionality factor of the order of unity: $\Phi(0) = \gamma_{\mathrm{emp}} Q_{\mathrm{emp}}$. This replacement leads straightforwardly to the EMP dispersion

$$\Omega_{\mathrm{emp}}(q_y) = \gamma_{\mathrm{emp}} \sigma_{yx} q_y , \tag{2.75}$$

which is the same as the dispersion of the boundary displacement wave of the incompressible electron liquid [(2.73)]. As the edge charge is confined to the electron density transition strip near the edge ($0 < x < h \ll |q_y|^{-1}$), in the pure Coulomb liquid, the geometrical proportionality factor γ_{emp} is proportional to $\ln(1/|q_y|h)$ and is close to the corresponding factor α_{e} of the displacement wave.

The EMP damping is estimated as the Joule component of the energy dissipation divided by the total energy of the EMP [104]

$$\frac{1}{\tau_{\mathrm{emp}}} \sim \sigma_{xx} \frac{\int_b^{\infty} E^2 \mathrm{d}y}{\Phi(b) Q_{\mathrm{emp}}} , \tag{2.76}$$

where E and Φ are the electric field and potential of the EMP, respectively. Both E and Φ are proportional to the linear charge density Q_{emp} accumulated near the edge. In the case of strong magnetic fields (practically, at $B > 0.1\,\mathrm{T}$), the parameter b is equal to the electron density transition width h, which is independent of B and T. Therefore, at a fixed electron density, the EMP damping is proportional to σ_{xx}. The proportionality coefficient depends on the particular geometry of the experimental cell and the real shape of the electron density at the edge. The relation between the EMP damping and σ_{xx} was used in [98] to study the quantum magnetotransport of SEs at low temperatures.

As the frequencies of the EMP and the boundary displacement wave are quite close, in experiments with a free boundary, these modes interact to produce coupled edge modes. For the electron layer which is close to the bottom electrode ($qd \ll 1$), in the limit of strong magnetic fields, the coupling of these two edge waves can be approximately described by evaluating the energy of the electric field per unit length $\mathcal{E}(\xi, Q_{\mathrm{emp}})$ induced by the edge charges of the two kinds which interfere. Straightforward evaluations then yield the following coupled equations [106]:

$$(\omega - \Omega_{\mathrm{emp}}) Q_{\mathrm{emp}} = \alpha_{\mathrm{e}} \sigma_{yx} q_y e n_{\mathrm{s}} \xi , \tag{2.77}$$

$$(\omega - \Omega_{\mathrm{bdw}}) \xi = \beta_{\mathrm{e}} \sigma_{yx} q_y Q_{\mathrm{emp}} , \tag{2.78}$$

where β_{e} is the dimensionless coupling parameter

$$\beta_{\mathrm{e}} = \int \mathrm{d}t_1 \int \mathrm{d}t_2 \rho(t_1) \tau'(t_2) \psi \left(\left| \frac{b}{d} t_1 - t_2 \right| \right) , \tag{2.79}$$

the function $\rho(x/b)$ describes the distribution of the EMP charge across the edge, and b is the width of the EMP density-perturbation strip (in the limit of strong magnetic fields $b \to h \sim d$). Under the conditions considered here, the geometrical parameters γ_{bdw} and γ_{emp} do not contain the q-dependent logarithmic factor and are just numbers.

The set of coupled equations (2.77) and (2.78) has two similar solutions

$$\Omega_{\pm} = \gamma_{\pm}\sigma_{yx}q_y \ , \tag{2.80}$$

which differ only in dimensionless proportionality factors of the order of unity:

$$\gamma_{\pm} = \frac{\gamma_{\mathrm{emp}} + \gamma_{\mathrm{bdw}}}{2} \pm \sqrt{\left(\frac{\gamma_{\mathrm{bdw}} - \gamma_{\mathrm{emp}}}{2}\right)^2 + \alpha_{\mathrm{e}}\beta_{\mathrm{e}}} \ . \tag{2.81}$$

In the limit of weak coupling ($\alpha_{\mathrm{e}}\beta_{\mathrm{e}} \to 0$ when $b \gg d$), which is possible at rather low magnetic fields, (2.81) describes two independent modes. In the opposite limit $b \to h \approx d$, all parameters of (2.81) are just numbers of the same order of magnitude and the system is in the strong coupling regime.

In the high frequency mode Ω_+, the edge charges oscillate in phase, and the energy of the edge field perturbations is high. This can be seen directly from (2.77) and (2.78): at $\omega > \Omega_{\mathrm{bdw}}$ (or $\omega > \Omega_{\mathrm{emp}}$), the displacements ξ and the EMP charge Q have the same sign. The electron charge is defined as $-e$ and at $\xi > 0$, we have a deficiency of electrons or a positive edge charge of the BDW.

In the low frequency mode Ω_-, the edge charges oscillate out of phase and the edge field fluctuations are substantially screened. It follows from (2.77) that, at $\omega < \Omega_{\mathrm{bdw}}$, the displacement ξ and the EMP charge Q have opposite signs. This situation is sketched in Fig. 2.10. For the strong coupling regime, the Ω_- mode somewhat resembles the dipole edge acoustic plasmons of [108, 109]. In the edge acoustic plasmon wave, the electron density also oscillates across the edge within the transition strip $0 < x < h$. The substantial difference appears in the limit of weak fields, where the BDW uncouples from the EMP and has the characteristic behavior $\Omega \propto 1/B$ down to extremely weak magnetic fields, as long as $\omega_{\mathrm{c}} \gg \Omega_0$. In this limit, the edge acoustic plasmons have the opposite B-dependence: $\Omega \propto B$.

The boundary displacement wave was observed experimentally in [105, 110] for SEs on helium as the mode which has lower frequency than the conventional EMP. The new low-frequency mode has stronger damping than the EMP, and it appears only for sufficiently low temperatures $T < 0.9\,\mathrm{K}$, when electrons become mobile enough. This feature agrees qualitatively with the fact that the 'incompressible' electron motion of the BDW penetrates much more deeply into the interior of the electron liquid (on the scale $x \sim q_y^{-1}$) than the electron motion of the EMP, causing stronger damping. The magnetic field dependence $\Omega \propto 1/B$ observed under rather low magnetic fields ($B < 0.2\,\mathrm{T}$), and even the experimental value of the numerical proportion-

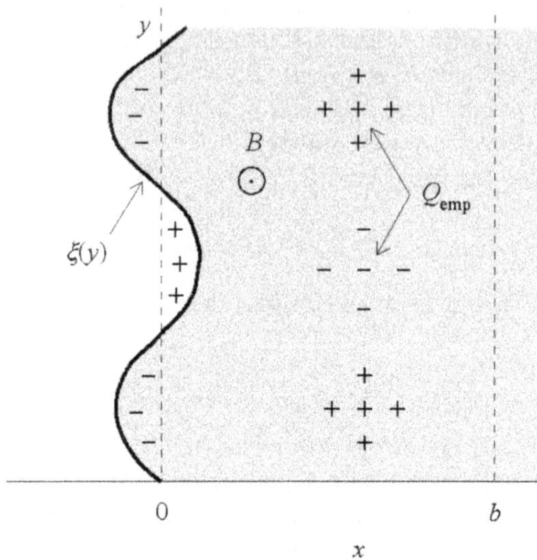

Fig. 2.10. Sketch of the interplay between the boundary displacement wave $\xi(y)$ and the EMP charge $Q_{\mathrm{emp}}(y)$. The charge distribution is shown by symbols + and −

ality factor $\gamma_{\mathrm{bdw}} \approx 2.7$, are in accordance with the concept of the boundary displacement wave of strongly correlated electrons, as shown in Fig. 2.11.

Under the experimental conditions in [105, 110], the conventional EMP mode is described by $\gamma_{\mathrm{emp}} \simeq 5.5$. Because the data of different magnetic field regimes were obtained for different electron density profiles and even for different experimental setups, it is impossible to make strong statements about the origin of the new mode shown in Fig. 2.11 by open circles. The relative shift of the lines that fit the data sets of the new mode for the strong and weak magnetic field regimes can be explained by the BDW–EMP coupling with $\alpha_{\mathrm{e}}\beta_{\mathrm{e}} = 3.57$. The theoretical curves that represent the high magnetic field asymptotes of $\Omega_{\pm}$ modes are shown in Fig. 2.11 by the dotted curves. It should be noted that the correction to the EMP spectrum caused by the coupling is substantially smaller than the correction to the BDW spectrum and therefore the Ω_{+} mode cannot be distinguished from the conventional EMP mode for the available data.

Among other interesting results concerning the edge excitations of the interface electrons, we would like to mention the following. The edge acoustic plasmons of [108, 109] were observed in the 2D pool of heavy ions confined under the free surface of liquid helium [112]. The edge excitations of the 2D Wigner solid under a magnetic field were described in [113]. It was shown that the electron solid with a sharp edge has only the BDW mode or a magneto-Rayleigh wave, which also has the field dependence $\Omega \propto 1/B$, but in a significantly lower frequency range. At finite temperature, the electron

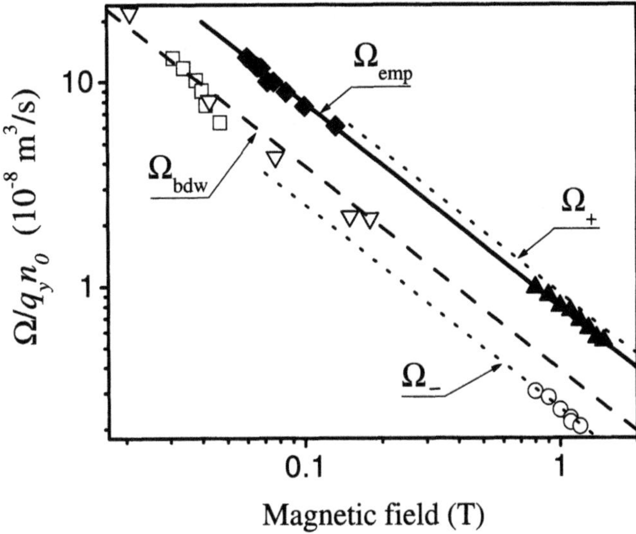

Fig. 2.11. Resonance frequencies of the conventional EMP (*solid symbols*) and novel edge excitations (*open symbols*) vs. the magnetic field; the uncoupled boundary displacement wave is shown by the *dashed line*, while the strongly coupled BDW–EMP modes are plotted by the *dotted lines* [111]. Conventional EMP data [105, 110] are shown by *solid symbols* and the novel edge excitation data of [105] are represented by *open symbols*

solid with a smooth density profile has a 'wet edge' of electrons in the liquid state. The EMP modes of this liquid strip are affected by the proximity of the Wigner solid boundary.

2.6 Electron Correlations and Binding Energy

The strong Coulomb repulsion acting between electrons is the main instability factor for the system of SEs on liquid helium. Without the electric field applied by the guard electrodes, they escape from the free surface. In addition, at high densities, the charged surface of the liquid dielectric becomes unstable for excitations with wavelength equal to the capillary length. In this context, it is remarkable that the strong Coulomb correlations inside the layer make SEs more stable with regard to electron evaporation from the ground surface level, which is important for the problem of the population of SE states at finite temperatures discussed in Sect. 1.2, and for escape experiments conducted with a fly-out external electric field applied to the layer [114–116].

Consider a 2D electron layer confined to the plane $z = 0$. If there are no correlations and the layer charge is evenly distributed along the plane (a sort of uniform negative background), then the electric field of the electron layer $E_\perp^{(\mathrm{el})}$ is discontinuous at $z = 0$ and behaves like a step function: $E_\perp^{(\mathrm{el})} = 2\pi e n_\mathrm{s}$ at $z < 0$ and $E_\perp^{(\mathrm{el})} = -2\pi e n_\mathrm{s}$ at $z > 0$. If we take into account the equilibrium holding electric field, which appears because of the potential difference applied to the cell capacitor electrodes V and because of the charges induced in the electrodes $E_\perp = E_\perp^{(0)} \equiv 2\pi e n_\mathrm{s}$, then the total electric field $E_\perp^{(0)} + E_\perp^{(\mathrm{el})}(z)$ is zero above the layer and equals $4\pi e n_\mathrm{s}$ below the layer, as sketched in Fig. 2.12a by the dotted line. The potential energy of an electron in the total field, viz.,

$$U_0(z) = e \int_0^z \left[E_\perp^{(0)} + E_\perp^{(\mathrm{el})}(z') \right] \mathrm{d}z' \,, \tag{2.82}$$

(excluding the image potential) is zero at $z > 0$ and equals $2eE_\perp^{(0)} z$ at $z < 0$, as shown in Fig. 2.12b by a line of the same kind.

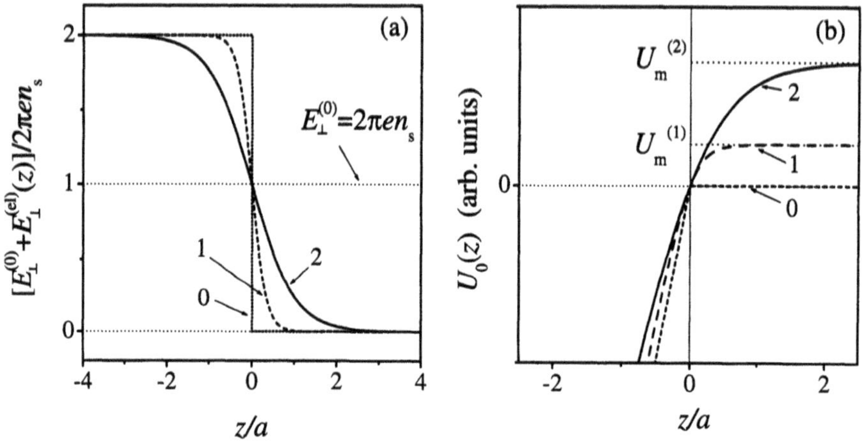

Fig. 2.12. Schematic view of the electric field (**a**) and potential energy (**b**) microstructures near the SE layer induced by Coulomb correlations in the presence of the equilibrium holding electric field $E_\perp^{(0)}$. Curves represent different stages of the interelectron interaction: (0) no correlations (*dotted line*), (1) weak correlations (*dashed line*), and (2) strong correlations (*continuous line*)

The situation changes significantly if Coulomb forces introduce a correlation hole of radius a_c in the vicinity of an electron. In this case, when leaving the layer, the electron feels the microstructure of the electric field of other electrons $E_\perp^{(\mathrm{el})}(z)$. In other words, the layer charge microstructure eliminates the discontinuity in the electric field $E_\perp^{(\mathrm{el})}(z)$ in the way sketched by the dashed and continuous lines in Fig. 2.12a (the dashed curve corresponds to weaker correlations). As a result, just above the layer, at distances $z \sim a_\mathrm{c}$, the total electric field is not zero because the holding field $E_{\perp}^{(0)}$ is not entirely compensated there. In the region below the layer $z < 0$, the correlation effect does not bring a significant change into the electron potential energy, since it goes to zero with $z \to 0$ anyway. A significant change occurs for the region above the layer $z > 0$, as indicated in Fig. 2.12b by the dashed and continuous lines. The potential energy of an electron above the layer increases linearly with z in the region $z \ll a_{\mathrm{c}}$ and then, at $z \gg a_{\mathrm{c}}$, approaches a saturation value U_{m} which corresponds to the area under the respective curves of Fig. 2.12a at $0 < z < \infty$. This limiting value of the electron potential energy should be added to its binding energy, which increases the ground level fraction of SEs, especially at high temperatures.

As an approximate estimate of the SE binding energy induced by correlations with $a_{\mathrm{c}} \sim n_{\mathrm{s}}^{-1/2}$, the area under the continuous curve can be replaced by the area of a triangle:

$$U_{\mathrm{m}} \approx \frac{eE_{\perp}^{(0)}}{2\sqrt{n_{\mathrm{s}}}} = \pi e^2 \sqrt{n_{\mathrm{s}}} = \sqrt{\pi}\Gamma^{(\mathrm{pl})}T \,. \tag{2.83}$$

Remarkably, this estimate is very close to the correct numerical result which we shall discuss later. Thus, the ratio U_{m}/T entering the exponential factors of the electron distribution function is approximately equal to $1.77\Gamma^{(\mathrm{pl})}$, and it becomes really large for the Coulomb liquid with $\Gamma^{(\mathrm{pl})} \gg 1$.

Let us consider a very simple model assuming that the pair distribution function of SEs is just a step function [$g(r) = 0$ at $r < a_{\mathrm{c}} \simeq (\pi n_{\mathrm{s}})^{-1/2}$], and that the correlation hole is frozen when an electron escapes the layer. For an electron above the hole with a fixed radius a_{c}, the field of other electrons $E_{\perp}^{(\mathrm{el})}(z)$ (an integral over the area $r \geq a_{\mathrm{c}}$) is easily evaluated:

$$E_{\perp}^{(\mathrm{el})}(z) = -2\pi n_{\mathrm{s}} \frac{z}{\sqrt{a_{\mathrm{c}}^2 + z^2}} \,. \tag{2.84}$$

At the same time, the total field $E_{\perp}^{(0)} + E_{\perp}^{(\mathrm{el})}(z)$ is actually the field of the electron sheet with a hole minus the field of the complete electron sheet, or plus the field of the positive background of the same density n_{s}. Therefore $E_{\perp}^{(0)} + E_{\perp}^{(\mathrm{el})}(z)$ is the field of the positively charged background circle of radius a_{c}. The potential energy $U^*(z)$ of an electron with this positively charged circle can be set to zero at $z \to \infty$. The energy $U(z)$, which is set to zero at the liquid helium surface, and $U^*(z)$ are related by $U(z) = U^*(z) + U_{\mathrm{m}}$. It is clear that the difference $U^*(0) - U^*(\infty) = U^*(0) = -U_{\mathrm{m}}$ is the interaction energy of an electron with the other electrons and the uniform positive background $\mathcal{E}_{\mathrm{I}} = -2\pi e^2 n_{\mathrm{s}} a_{\mathrm{c}} \simeq -2\Gamma^{(\mathrm{pl})}T$. [Here the second relation implies that $a_{\mathrm{c}} = (\pi n_{\mathrm{s}})^{-1/2}$.] This conclusion is confirmed by a direct evaluation of the electron potential energy in the total field which gives $U(\infty) = 2\pi e^2 n_{\mathrm{s}} a_{\mathrm{c}} = -\mathcal{E}_{\mathrm{I}}$.

The above result does not take into account the gradual rearrangement of the remaining electrons, or the shrinking of the correlation hole when an electron leaves the surface. Obviously, this effect should reduce $U(\infty)$. In this case, $U^*(0) = -U(\infty) \equiv -U_{\mathrm{m}}$ can be estimated as the energy it takes to add or remove an electron from the layer, which is known as the chemical potential μ. The simplest estimate of this quantity can be found using Seit's theorem (see [38]):

$$\mu(n_{\mathrm{s}}) = \mathcal{E}_{\mathrm{G}} + n_{\mathrm{s}} \frac{d\mathcal{E}_{\mathrm{G}}}{dn_{\mathrm{s}}} , \tag{2.85}$$

where $\mathcal{E}_{\mathrm{G}}$ is the ground state energy per particle. In our case $\mathcal{E}_{\mathrm{G}} = \mathcal{E}_{\mathrm{I}}/2$ (the 1/2 factor accounts for the double-counting of the electron interaction). According to the estimate given above, we have $\mathcal{E}_{\mathrm{G}} \simeq -\Gamma^{(\mathrm{pl})}T \propto -n_{\mathrm{s}}^{1/2}$ and $\mu \simeq -1.5\Gamma^{(\mathrm{pl})}T$. A more detailed analysis due to Nagano et al. [117] yields

$$\frac{\mu}{T} = -1.676\Gamma^{(\mathrm{pl})} + 3.19\left(\Gamma^{(\mathrm{pl})}\right)^{1/4} - 0.38\ln\left(\Gamma^{(\mathrm{pl})}\right) - 2.51 . \tag{2.86}$$

It is easy to see that, for large $\Gamma^{(\mathrm{pl})}$, this equation agrees with the simple estimates given above. Later we will see that this estimate is also confirmed by an accurate evaluation of the potential $U_0(z)$ and $U_{(0)}(\infty) = U_{\mathrm{m}}$ for the harmonic lattice model of the 2D electron liquid.

The simple analysis given above indicates that the correlation correction to the chemical potential evaluated and employed in [117] corresponds to the saturated case when the total field above the layer $E_\perp + E_\perp^{(\mathrm{el})}(z) \to 0$ at $z \to \infty$, or $E_\perp = E_\perp^{(0)} \equiv 2\pi e n_{\mathrm{s}}$. In an escape-rate measurement, an inverted potential is usually applied to the upper electrode to cause a current across the cell. In this case, the holding field $E_\perp$ is smaller than $E_\perp^{(0)} \equiv 2\pi e n_{\mathrm{s}}$. At large heights $z \gg a_{\mathrm{c}}$, the total electric field $E_\perp + E_\perp^{(\mathrm{el})}(z) \to E_\perp - E_\perp^{(0)} < 0$, and the potential energy of an electron decreases with z, as shown in Fig. 2.13. The maximum of the potential energy curve shown as U_{m} represents the additional activation energy for electron transport across the cell, which depends on the imbalance of the electric field

$$\delta = \frac{E_\perp^{(0)} - E_\perp}{E_\perp^{(0)}} = \frac{2d}{D} + \frac{V}{2\pi e n_{\mathrm{s}} D} , \tag{2.87}$$

where d is the liquid helium depth, D is the distance between the electrodes of the cell capacitor, and V is the potential difference between upper and lower electrodes. The correlation correction of (2.86) corresponds to the case $\delta \to 0$.

It should be noted that there is no need for a separate evaluation of the potential energy of an electron above the layer for a finite imbalance $\delta > 0$, because a very simple rearrangement of the holding field potential allows us to express $V(z)$ in terms of the potential $U_0(z)$ defined for the saturated case in (2.82):

$$V(z) = -\frac{\Lambda}{z} + U(z) , \qquad U(z) = -eE_\perp^{(0)} z\delta + U_0(z) . \tag{2.88}$$

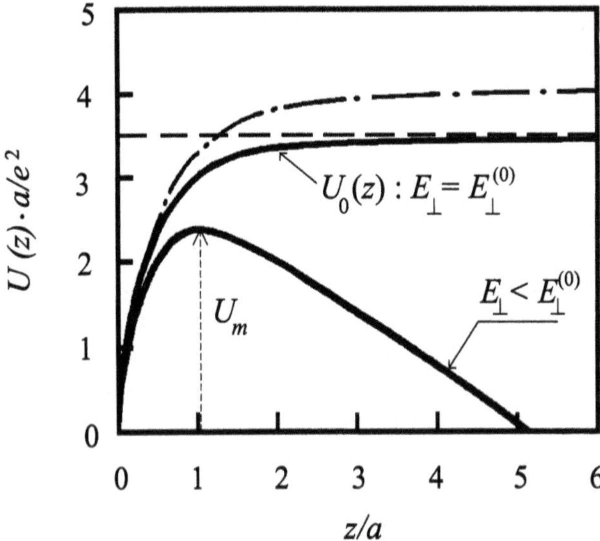

Fig. 2.13. Potential energy of an electron escaping from the electron lattice: frozen approximation (*dash-dotted curve*) and adiabatic approximation (*continuous curves*) [122]

Knowing the potential energy for the equilibrium case $U_0(z)$, one can easily plot the potential energy at any holding field $E_\perp$ and find the dependence $U_\mathrm{m}(\delta)$.

Accurate evaluation of $U_0(z)$ is quite a complicated task which can be performed only for certain simple models. A model of a correlation hole shrinking as an electron moves away from the surface was considered in [118]. The correlation radius a_c was defined by the condition $g(a_\mathrm{c}) = 1/2$, where $g(r)$ is the pair distribution function of the 2D electron liquid. According to the numerical experiment of [119], this function can be approximated in the short range by

$$g(r) = \exp\left[-\left(\frac{r_0}{r} - 1.18 + \frac{0.33r}{r_0}\right)\Gamma^{(\mathrm{pl})}\right] , \tag{2.89}$$

where $r_0 = (\pi n_\mathrm{s})^{-1/2}$. At large $\Gamma^{(\mathrm{pl})}$, (2.89) approaches unity around $r \sim r_0$. The potential energy $U(z)$ of an electron escaping from the fixed triangular electron lattice in the presence of an inverted electric field was evaluated in [114]. It was shown that the position of the potential maximum increases with the areal density n_s.

The harmonic lattice model of the Coulomb liquid proposed by Itoh, Ichimaru and Nagano [120, 121] was used in [122] to evaluate the functions $U_0(z)$ and $U_\mathrm{m}(\delta)$. As emphasized in this research, there are two extreme approximations with regard to the remaining electron lattice response to an electron escape: static (frozen) and adiabatic. In the first case, it is assumed that the escape is very fast and the remaining electrons do not have time to rearrange their positions near the escaping electron and to affect the potential $U_0(z)$. In the opposite limit, the escape is assumed to be slow and the positions of the remaining SEs are determined by the equilibrium condition for the electron system with an electron located at a height z above the surface. It is clear that this shrinking effect leads to a partial screening of the hole field and to a reduction of U_{m}.

Instead of a direct evaluation of the integral of (2.82), the escaping electron can be considered as a substitutional impurity in the 2D Wigner solid on the positive background. For the electron potential above the surface, it does not matter where the positive background, creating the holding field $E_{\perp}^{(0)}$, is actually placed. In practice this field is induced by charges at the cell capacitor electrodes. For simplicity, we can place the positive uniform background exactly at $z = 0$. In this case, the field below the layer is changed, but the electric field above remains the same. The energy $U_0(z)$ of such an impurity is equal to the difference $\Phi(z) - \Phi(0)$, where $\Phi(z)$ is the energy of the whole system (electrons plus background) with a single electron being removed from its lattice site and placed at a distance z just above the empty site.

For the frozen potential of the electron crystal with lattice spacing a, use of the Ewald summation method yields the result [122]

$$U_0(z) = \frac{e^2}{a}\left\{4.213 + \left(2b\tilde{z} - \frac{1}{\tilde{z}}\right)\mathrm{erf}\left(\sqrt{b}\tilde{z}\right)\right. \tag{2.90}$$

$$\left. -\sqrt{\frac{b}{\pi}}\left[2\mathrm{e}^{-b\tilde{z}^2} - 6\psi_{-1/2}(b\tilde{z}^2, b) + 6\psi_{-1/2}(0, b + b\tilde{z}^2)\right]\right\},$$

where $b = 2\pi/\sqrt{3}$ is a number, $\tilde{z} = z/a$ is the normalized height, and ψ_ν is the function defined by

$$\psi_\nu(x, y) = \int_1^\infty t^\nu \exp\left(-\frac{x}{t} - yt\right)\mathrm{d}t\,. \tag{2.91}$$

This potential is shown in Fig. 2.13 by the dash-dotted curve. Equation (2.90) allows us to find the asymptotic forms of $U_0(z)$ at large and small values of $\tilde{z}$:

$$U_0(z) \simeq 4.213\frac{e^2}{a} - \frac{e^2}{z}\,, \quad \text{if } z \gg a\,, \tag{2.92}$$

$$U_0(z) \simeq eE_{\perp}^{(0)} z - c_0 z^2 \frac{e^2}{a^3}\,, \quad \text{if } z \ll a\,, \tag{2.93}$$

where $c_0 \simeq 5.525$ is a numerical constant. At low heights, the potential of the electric field of the remaining electrons is proportional to z^2 and is much smaller than the potential of the holding field. On the other hand, under the saturation condition, we have $U_0(\infty) = U_m \simeq 4.21e^2/a \simeq 2.2\Gamma^{(pl)}T$, which agrees with the estimate $2\Gamma^{(pl)}T$ found for the frozen correlation hole of radius $a_c = (\pi n_s)^{-1/2}$. This value also agrees with the interaction energy of an electron in the Wigner solid (including the positive background) obtained by Bonsall and Maradudin [123]: $\mathcal{E}_I = -4.2134e^2/a$. It is larger than the correlation correction to the chemical potential of [(2.86)], because the adiabatic adjustment of the positions of the surrounding electrons was not taken into account.

The deformation correction to the potential $U_0(z)$ analyzed in [122] appears to be quite substantial. The numerical evaluation for the equilibrium case ($E_\perp = E_\perp^{(0)}$) is shown in Fig. 2.13 by a continuous line marked correspondingly. The resulting potential has the following asymptotic form at large distances:

$$U_0(z) \simeq 3.48\frac{e^2}{a} - \frac{e^2}{4z} . \tag{2.94}$$

The quantity $U_m = U_0(\infty)$ represents the total vacancy energy of the 2D Wigner solid. This vacancy energy was obtained previously in Monte Carlo simulations by Fisher et al. [124]: $U_v \simeq 3.43e^2/a$. The two results are quite close, giving the activation energy correction $\simeq 1.8\Gamma^{(pl)}T$, which agrees with the result of (2.86). It is remarkable that the second term of (2.94) behaves like an additional image potential due to the remaining electrons.

It is instructive to note that nearly the same value of $U_m = U_0(\infty)$ can be found for the adiabatic case even without the sophisticated evaluation of the lattice adjustment to the escaping electron. Knowing the interaction energy $\mathcal{E}_I$ of an electron in the Wigner solid given above [123], in order to obtain the chemical potential, we can simply use Seit's theorem [(2.85)] with $\mathcal{E}_G = \mathcal{E}_I/2$. Hence,

$$\mu = -1.5 \times \frac{1}{2} \times 4.21e^2/a \simeq -3.16e^2/a = -1.659\Gamma^{(pl)}T .$$

The absolute value of μ is a little bit less than the above energy $3.48e^2/a$ because a lattice with a vacancy is not a completely relaxed system. Once again, we note the agreement with (2.86) in the limit of large $\Gamma^{(pl)}$.

The dependence of the maximum value U_m of the potential of (2.88) on the imbalance parameter δ is shown in Fig. 2.14 by the continuous curve. At a small imbalance, we have a very strong contribution from the Coulomb correlations:

$$U_m(\delta) \simeq \left(3.48 - \sqrt{2b\delta}\right)\frac{e^2}{a} . \tag{2.95}$$

In the opposite limit ($\delta \gg 1$),

$$U_m(\delta) \simeq -\left[2\sqrt{2b(\delta-1)\lambda} + \frac{c_0\lambda}{2b(\delta-1)}\right]\frac{e^2}{a} , \tag{2.96}$$

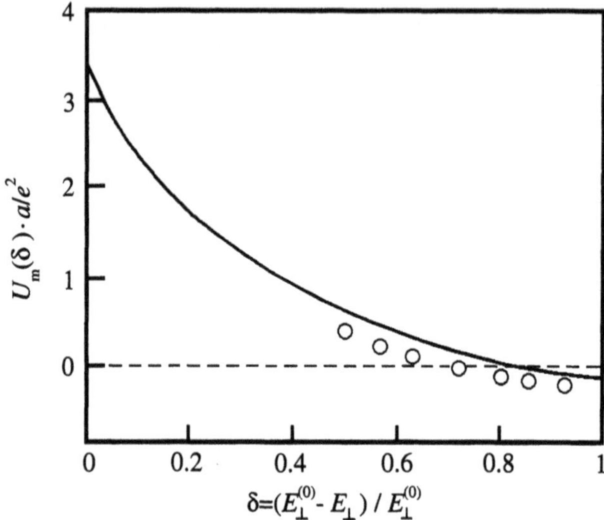

Fig. 2.14. Position of the maximum of the potential energy of an escaping electron as a function of the imbalance parameter $\delta = 1 - E_{\perp}/E_{\perp}^{(0)}$. The image potential $-\Lambda/z$ is included [122]

where $\lambda = \Lambda/e^2$. Here the first, dominated term is due to the image potential $-\Lambda/z$ and does not depend on electron correlations. Numerically, the second term already becomes negligibly small at $\delta > 1.5$. Therefore, in order to observe correlation effects on the escape current, one should make the imbalance parameter δ smaller. This is consistent with the observation that the correlation effect on the binding energy is stronger at smaller liquid helium depths d [116]. These data are shown in Fig. 2.15. It is clear that, at large values of the plasma parameter $\Gamma^{(\mathrm{pl})}$, the activation energy for the electron escape rate is proportional to $\Gamma^{(\mathrm{pl})}$ and increases with decreasing d.

The data of [118] were analyzed and compared with the harmonic lattice model in [122]. The corresponding experimental values of $U_{\mathrm{m}}(\delta)$ are shown in Fig. 2.14 as open circles. The activation energies are a bit smaller than in the case of the electron solid model because, under the experimental conditions, the plasma parameter $1 < \Gamma^{(\mathrm{pl})} < 15$ is much smaller than its value at the melting point. The results of the harmonic lattice model were also compared with experimental data of [125, 126] and a good agreement was reported at $n_{\mathrm{s}} > 0.4 \times 10^8\,\mathrm{cm}^{-2}$ [127]. At small values of the holding field imbalance $\delta \sim 0.1$, the lifetime of SEs with $\Gamma^{(\mathrm{pl})} > 47$ was shown to increase rapidly, exceeding several minutes [128].

The electric field microstructure at small heights shown in Figs. 2.12 and 2.13 also affects the electron transition frequencies between the ground surface level $l = 1$ and the excited states $l > 1$. Such local disorder in the 2D electron system of SEs was measured by Lambert and Richards [129] using

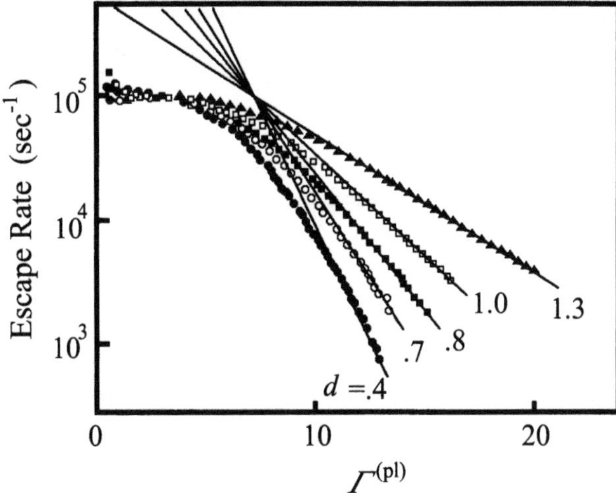

Fig. 2.15. Escape rate vs. plasma parameter $\Gamma^{(\mathrm{pl})}$ for different values of the helium depth d shown in units of mm [116]

the far-infrared technique. The experimental data were shown to agree with theoretical evaluations of the radial distribution function $g(r)$ given in [119].

An interesting phenomenon concerned with electron correlations in the 2D electron sheet on liquid helium has been reported by Zipfel, Brown and Grimes [130]. They observed that the linewidth for electronic transitions in the image-induced potential well is strongly affected by electron correlations if a magnetic field is applied parallel to the surface. Employing the gauge $\boldsymbol{A} = (zB, 0, 0)$, the kinetic energy terms of the electron Hamiltonian can be written as

$$K_{\mathrm{e}} = \frac{1}{2m_{\mathrm{e}}} \left(p_x - \frac{e}{c} Bz \right)^2 + \frac{1}{2m_{\mathrm{e}}} \left(p_y^2 + p_z^2 \right) , \tag{2.97}$$

where p_j are the components of the electron momentum. The electron motion in the x and y directions can be described by plane waves. The parallel applied magnetic field affects electron motion in the direction perpendicular to the layer. Expanding the first term of (2.97), we find two perturbation terms introduced by the magnetic field:

$$H^{(1)} = \omega_{\mathrm{c}} p_x z , \qquad H^{(2)} = \frac{1}{2} m_{\mathrm{e}} \omega_{\mathrm{c}}^2 z^2 , \tag{2.98}$$

where ω_{c} is the cyclotron frequency ($\omega_{\mathrm{c}} \propto B$). The important point is that the first term is proportional to the constant of motion p_x or to the electron velocity v_x. This term is equivalent to a shift in $E_{\perp}$. Electrons with different velocities have different shifts in the effective holding electric field. This must lead to a broadening of the transition linewidth if v_x remains constant during the transition.

In the free electron gas (FEG) model considered in [130], the probability of an electron having a velocity value v_x is proportional to $\exp\left(-m_e v_x^2/2T\right)$. This leads to a Gaussian broadening of the absorption line, if the interaction with scatterers is disregarded:

$$A(E_\perp) = A_0 \exp\left[-\frac{(E_\perp - E_\perp^{(0)})^2}{2w^2}\right] , \tag{2.99}$$

with

$$w = \omega_c \sqrt{m_e T}/e \simeq 1.5 \times B \text{ [GHz]} , \tag{2.100}$$

where B is in kilogauss. In the presence of scatterers, collision broadening makes the absorption line a convolution of the Lorentzian and the Gaussian. The absorption line in the FEG model thus becomes motionally broadened. On the other hand, collisions with other electrons and the scatterers cause changes in the sign of the electron velocity v_x. If such changes are quite fast, then the effective holding field correction due to $H^{(1)}$ should be reduced and one can expect a correlation-induced narrowing of the linewidth.

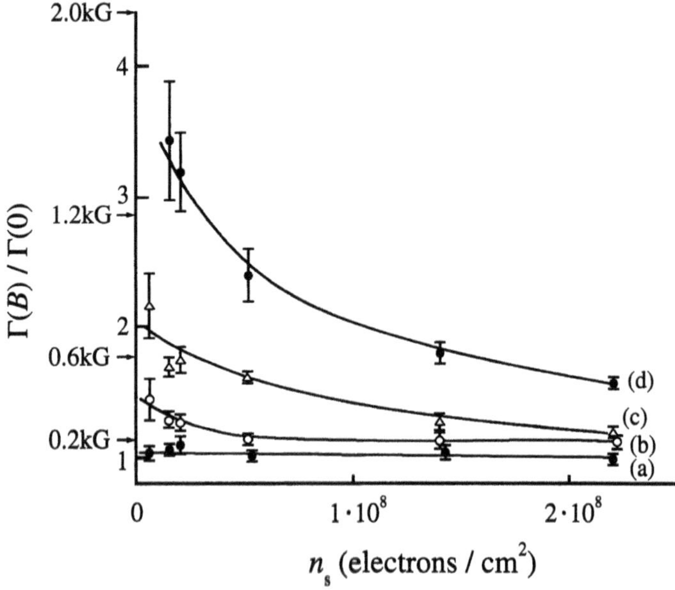

Fig. 2.16. Normalized linewidth vs. charge density n_s for several values of the magnetic field B: (**a**) 0.2 kG, (**b**) 0.6 kG, (**c**) 1.2 kG, (**d**) 2.0 kG [130]. Predictions from the FEG model are shown by *arrows* for each of these values of B

Experimental data from [130] are shown in Fig. 2.16 as the normalized linewidth $\Gamma(B)/\Gamma(0)$ vs. electron density. In the low electron density range, the parallel magnetic field causes a strong broadening of the absorption line in

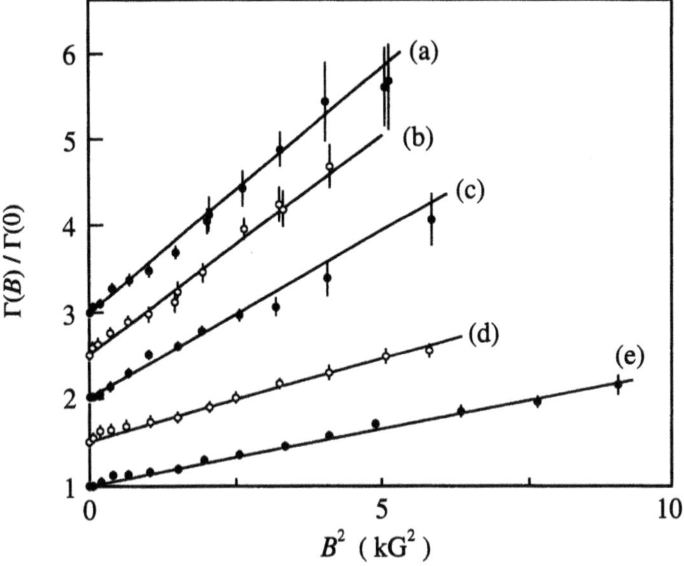

Fig. 2.17. Normalized linewidth vs. B^2 for several values of the charge density: (**a**) $0.15 \times 10^8\,\mathrm{cm}^{-2}$, (**b**) $0.2 \times 10^8\,\mathrm{cm}^{-2}$, (**c**) $0.51 \times 10^8\,\mathrm{cm}^{-2}$, (**d**) $1.4 \times 10^8\,\mathrm{cm}^{-2}$, (**e**) $2.2 \times 10^8\,\mathrm{cm}^{-2}$. For clarity, each successive curve has been displaced upward by one half unit [130]

qualitative accordance with the FEG model, although the predictions of this model are about twice as large as measured linewidths. With an increase in the electron density, the linewidths become strongly narrowed due to electron correlations.

In Fig. 2.17, the experimental data are shown as a function of B^2 for a few electron densities. The data were fitted by the expression

$$\frac{\Gamma(B)}{\Gamma(0)} = 1 + \frac{2(\delta\omega_0)^2\tau_{\mathrm{v}}}{\sqrt{3}\Gamma(0)} ,$$

found in analogy with the theory describing the motional narrowing of the nuclear magnetic resonance [131]. Here $\delta\omega_0$ is the change in the resonance frequency due to the magnetic field, viz.,

$$\delta\omega_0 = 2\pi \frac{v_x B}{c} \frac{\mathrm{d}\nu}{\mathrm{d}E} ,$$

$(\mathrm{d}\nu/\mathrm{d}E)$ is the Stark tuning rate, and τ_{v} is the electron velocity autocorrelation time. From the slope of each line, the authors of [130] deduced τ_{v}, which turned out to be surprisingly close to the reciprocal of the harmonic oscillator frequency in a 2D triangular lattice $\omega_0 = 2.1(e^2/m_{\mathrm{e}})n_{\mathrm{s}}^{3/4}$ in the range $9 \leq \Gamma^{(\mathrm{pl})} \leq 36$. For example, at $\Gamma^{(\mathrm{pl})} = 17$, it was found that $\tau_{\mathrm{v}} = 4.8 \times 10^{-11}\,\mathrm{s}$, while $\omega_0^{-1} = 5.0 \times 10^{-11}\,\mathrm{s}$. It is remarkable that in this case electron motion leads to both broadening and narrowing of the resonance linewidth: steady motion of the electrons with different but constant velocities broadens the linewidth because of the magnetic field, while dephasing introduced by collisions narrows it.

Concluding, SE escape experiments and autocorrelation narrowing of the electronic transition linewidth indicate that electron correlations in the Coulomb liquid can be well understood by means of the harmonic electron lattice model, even though the plasma parameter $\Gamma^{(\mathrm{pl})}$ is significantly smaller than its critical value $\Gamma_{\mathrm{c}}^{(\mathrm{pl})} \simeq 137$, representing the Wigner transition point.

3 Quantum Transport Framework for Highly Correlated Electrons

3.1 An Approach to Universality

Before discussing the quantum transport framework, which is the main topic of this chapter, we would like to underline the important points of the conventional treatment. Firstly, it should be noted that in the quantum transport theory the description of the electrical conductivity, even for the simplest cases, presents a substantially harder problem then in the kinetic equation method. The exact methods are usually based on evaluation of the current–current correlation function entering Kubo's formula. With this formula chosen as a starting point for the Feynman diagram treatment, in order to obtain even the well-known DC conductivity results of the semi-classical kinetic equation method, one needs to include the self-energy effects and vertex corrections, collecting contributions from the infinite series of diagrams (for examples, see [38]). The conductivity is found in an integral form containing the ratio $\Lambda(\varepsilon)/\Gamma(\varepsilon)$, where $\Lambda(\varepsilon)$ is the vertex function and $\Gamma(\varepsilon) = -2\,\mathrm{Im}\Sigma(\varepsilon) = \hbar\tau^{-1}$ is the scattering rate (the inverse average time between scattering events). Physically, the vertex function transforms the collision rate $\Gamma(\varepsilon)$ into the transport relaxation rate $\Gamma^{(\mathrm{tr})}(\varepsilon)$, by introducing a factor of $1-\cos\theta$ which favors scattering events at high-momentum transfer. Nevertheless, the accurate form must be found by solving an integral equation. Disregarding the vertex corrections and replacing $\Lambda(\varepsilon) \rightarrow 1$ is often a serious mistake. Therefore it seems that one cannot feel comfortable in this field without advanced knowledge and skills in the Feynman diagram technique.

Difficulties with the conventional quantum transport treatment multiply if one needs to describe AC conductivity under conditions $\hbar\omega \gg T$, DC magnetoconductivity or cyclotron resonance absorption. These cases conventionally require special considerations. For the Coulomb liquid, which is the subject matter of this book, the electron excitation spectrum changes drastically from the free electron spectrum of the ideal electron gas to the phonon spectrum of the Wigner solid. In the usual approaches, these extreme cases also require separate treatments. In the intermediate and strong Coulomb coupling regime, as discussed in the last chapter, one cannot even introduce a universal excitation spectrum for the Coulomb liquid in a single reference frame. For a long time, the situation was rather desperate (if not hopeless) for a universal quantum transport treatment applicable to the Coulomb liquid under different conditions and preferably comprehensible to a broad range of research.

A remarkably universal and reasonably simple approach to dynamic conductivity was proposed by Götze and Wölfle [132] and Götze and Hajdu [133, 134] using an interesting analogy with Dyson's equation for the single-electron Green's function. As discussed in Sect. 1.9, Dyson's equation, obtained by partial summation of the infinite series of diagrams, establishes the general relation between the one-particle Green's function $G_\alpha(\varepsilon)$ and the self-energy $\Sigma_\alpha(\varepsilon)$. The latter is in turn the sum of the infinite number of self-energy diagrams. The advantage of Dyson's form for the Green's function is that the simplest approximations for the self-energy usually provide us with quite accurate results, which must otherwise be found by collecting contributions from the infinite series of diagrams.

Hence, representing the perturbation theory for the Green's function as the perturbation theory for the self-energy gives a significant simplification: in most cases, accepting Dyson's form, we do not need to know much about Feynman diagrams. The conductivity form adopted in [132–134] looks like an extension of the classical Drude form: the conductivity has the resonance structure whose broadening and frequency shift are described by the imaginary and real parts of the memory function $M(\omega) \equiv w(\omega) + \mathrm{i}\nu(\omega)$, respectively. [Here $w(\omega)$ and $\nu(\omega)$ are real functions.] In the quantum transport theory, the memory function plays the same role as the self-energy plays for the single-particle propagator. The simplest approximations for the memory function are shown to be sufficient to reproduce the main results of the transport theory of metals at all relevant frequencies, without invoking Boltzman's equation or collecting the infinite series of Feynman diagrams.

Even more importantly, the memory function can be expressed in terms of the electron density–density correlation function $G^{(\mathrm{R})}_{n_q,n_{-q}}(\omega)$ or the density response function $\chi(q,\omega)$ [135, 136]. This means that, according to the fluctuation–dissipation theorem, the effective collision frequency $\nu(\omega)$ can be found in terms of the equilibrium dynamical structure factor (DSF) of the Coulomb liquid $S(q,\omega)$. This is an important step towards the necessary universality of the quantum transport theory. The situation becomes similar to that of the theory of thermal neutron (or X-ray) scattering by solids, where a flux of particles with well-known properties is used to study the properties of the solid or liquid target, and the scattering cross-section is expressed as an integral form of the DSF of the target. The important point is that, in the transport theory considered here, the scatterers play the role of the particle fluxes while the Coulomb liquid represents a moving target whose properties are to be studied.

An important imperfection in the memory function approach in its original form is that the approximate expressions for the memory function $M(\omega)$ and the effective collision frequency $\nu(\omega)$ are found from relations which are strictly valid only for high frequencies $\omega \gg \nu$ or $\omega_c \gg \nu$. There is no guarantee that they remain valid in the DC case if $\omega_c \ll \nu$. Even though the main results of the DC transport theory of metals are reproduced by means of these expressions, there are certain instances, which will be discussed in this chapter, where such approximation yields an incorrect numerical proportionality factor of the order of 2.

A similar problem is inherent in the force-balance method in the low electron density limit. This method, proposed and developed by Lei and Ting [137] and Cai, Lei and Ting [138], is very popular in studies of nonlinear transport phenomena in semiconductor systems. Still, it is not often emphasized that assumptions made when deriving the main result of the force-balance method are generally valid only for strongly correlated electron systems. A version of the force-balance method with an emphasis on this aspect was proposed by Vil'k and Monarkha [84]. In accordance with the memory function formalism, the DC conductivity and effective collision frequency of the force-balance method are found in terms of the electron DSF $S(q,\omega)$. A very important point is that these relations are valid only for highly correlated electron systems in which the electron–electron collision rate $\nu_{e\text{–}e}$ is much higher than the effective collision frequency with scatterers ν. Luckily, this is the case of particular interest for the Coulomb liquid. Combining the results of the memory function approach with those of the force-balance method, one can conclude that, for strongly correlated electrons, the approximate memory function proposed in the original version of the memory function formalism yields the correct conductivity equation in the whole frequency range, including the DC case. The analogous affinity between the high frequency approximation and the approximation of highly correlated electrons is well known in the semi-classical kinetic equation method (see discussion in Sect. 3.5).

Concerning the advantages of the memory function formalism for highly correlated electrons, we note first that in this framework the problem of quantum transport is reduced to evaluating the equilibrium electron DSF. In order to describe the electron transport of a particular system, it is sufficient to find a suitable approximation for the equilibrium DSF $S(q,\omega)$. This method can therefore be applied to any state of the electron system, liquid or solid. Secondly, in contrast with conventional methods, the simplest approximation for $S(q,\omega)$, disregarding self-energy effects and vertex corrections, reproduces the results of the kinetic equation method. The transport factor of $1-\cos\theta$ is pre-installed in the theory owing to the basic properties of the density–density correlation function. Only in the presence of the strong magnetic field, when the single-electron spectrum becomes purely discrete, does one need to include self-energy effects, because of the singular nature of the 2D electron system. Still, even in this instance, use of the conventional approximations for the one-electron propagator $G_N(\varepsilon)$ reproduces results which, in the usual treatment, are found by means of the Feynman diagram method, evaluating an infinite series of vertex terms. Thus, adopting the main results of the memory function formalism, one can perform practical evaluations for a particular case, whilst completely avoiding Feynman diagrams. In order to stress this important simplification, in this book we shall deliberately avoid drawing the Feynman diagrams, even though they do nicely decorate any theoretical study.

The 2D Coulomb liquid formed on the surface of liquid helium serves as a perfect model system for testing quantum transport theories and, in particular, the memory function framework adopted for highly correlated electrons. Because the system is pure and the properties of scatterers are known with high accuracy, the theory has no parameters to adjust and any conflict between experiment and theory usually means that important physics is missing. Besides the huge body of experimental data currently available, obtained under different conditions, we also have a large collection of theoretical methods introduced in parallel with experimental studies. Therefore we can compare the results given by the universal quantum transport framework with those found by other theories. For example, we shall see that this framework reproduces and extends the results of the method of moments applied to the 2D system of strongly interacting electrons [85]. Good agreement is also found with the path integral method describing the quantum magnetotransport of ripplonic polarons and the Wigner solid [139, 140]. As for the comparison with experiment, in Chaps. 4 and 5 which deal with DC magnetoconductivity and quantum cyclotron resonance in 2D electron systems, we shall see that the whole body of relevant experimental data is explained and described accurately (even numerically) by the universal quantum transport framework using the simple and rigorous model for the Coulomb liquid DSF discussed in the last chapter.

The relation between the memory function and the electron DSF is convenient also for studying the real part of the relaxation kernel of the dynamic conductivity $w(\omega) = \mathrm{Re}M(\omega)$. In Chaps. 7 and 8, we shall see that this relation and the analytical properties of the memory function help in describing the excitation spectrum of the Wigner solid under conditions of strong coupling with medium vibrations.

In this chapter, we shall not discuss quantum interference effects which result in localization of 2D carriers in the static disorder potential, because we are considering highly correlated Coulomb liquids: electron–electron collisions, dephasing single-electron states, have extremely high rate. Moreover, the natural scatterers for SEs formed on the surface of liquid helium (vapor atoms and capillary waves) induce intrinsically inelastic electron scattering with a substantial dephasing rate and, in order to observe the weak localization effects, special conditions are required. They are usually observed for electrons interacting with the quite dense gas of vapor atoms or with the pronounced roughness of the solid cryogenic substrate, as discussed in Sect. 1.7. We also note that there are a number of good textbooks which deal with this particular quantum transport phenomenon. Quantum transport effects, which are the subject matter of this chapter, originate mostly from quantization of the motion of 2D charge carriers subject to a normal magnetic field and its interplay with Coulombic effects.

This chapter is organized as follows. In Sect. 3.2, we discuss the most general form of the conductivity tensor which can be found by means of a phenomenological approach. Employing the formal structure of the linear response theory for the average value of any operator A [see (2.3)], it is possible to find quantum extensions of the classical Drude relations. In contrast with the classical case, the effective collision frequency ν of the quantum theory, appearing as a proportionality factor between the average friction and current, depends strongly on the frequency ω, magnetic field, and electron density. Then, in Sect. 3.3, we consider a version of the momentum-balance method. It establishes a relation between the effective collision frequency and the electron DSF for the DC conductivity, which cannot be covered by the memory function formalism. Section 3.4 provides a detailed discussion of the memory function formalism. Although this section is self-contained, those readers who are not interested in theoretical details and proofs are advised to read its conclusion with the formulation of the quantum transport framework for practical evaluations. Comparison with the kinetic equation method for particular scattering mechanisms is given in Sect. 3.5. This allows one to check the validity range of the memory function approach. The problem of energy relaxation between the electron layer and liquid helium is discussed in Sect. 3.6.

3.2 Phenomenological Analysis

The final result of the memory function formalism looks like a quantum extension of the classical Drude conductivity. The memory function $M(\omega)$, or the relaxation kernel, entering the denominator of the conductivity equation is usually a complex function. Its imaginary part $\mathrm{Im}M(\omega) \equiv \nu(\omega)$ serves as an effective collision frequency for the electron system, which depends on the frequency ω, magnetic field and electron density. The real part $\mathrm{Re}M(\omega) \equiv w(\omega)$, vanishing in the DC case, has the physical meaning of a shift in the position of the CR. This structure of the conductivity tensor, introduced in a remarkable analogy with Dyson's equation for the electron Green's function, is actually in accordance with the simple phenomenological treatment of quantum magnetotransport which we discuss in this section. It is instructive to note that 'improvements' on the classical Drude conductivity equation which follow from the quantum theory can be found and understood by means of very simple physical ideas which we discuss in this section before proceeding to a more sophisticated analysis.

Consider first a DC case assuming that a uniform infinite 2D electron system moves along the interface in crossed electric $\boldsymbol{E}_{\parallel}$ and magnetic $\boldsymbol{B}$ fields.

A magnetotransport experiment usually deals with average quantities such as the electric current or current density $\boldsymbol{j}$. Therefore, in order to understand basic properties of the electron transport, it is quite sufficient to analyze the average balance of forces which holds regardless of mechanics (quantum or classical) to be used to find the averages. It is easy to average the external force acting on the whole electron system:

$$\langle \boldsymbol{F}_{\text{ext}} \rangle = -N_{\text{e}} e \boldsymbol{E}_{\|} - N_{\text{e}} m_{\text{e}} \omega_{\text{c}} (\boldsymbol{v}_{\text{av}} \times \hat{\boldsymbol{z}}) , \qquad (3.1)$$

where $\boldsymbol{v}_{\text{av}}$ is the average velocity $\boldsymbol{v}_{\text{av}} = \langle \boldsymbol{v}_{\text{e}} \rangle$, and $\hat{\boldsymbol{z}}$ is the unit vector in the z-direction. In a steady regime, this external force should be compensated by the average frictional force $\boldsymbol{F}_{\text{scat}}$ acting on the electrons due to the interaction with scatterers. In order to find $\boldsymbol{F}_{\text{scat}}$, one should calculate the momentum relaxation rate or employ the quantum linear response theory. These approaches will be discussed in the following sections. Here we would like to show that the general structure of the conductivity tensor applicable to the quantum transport regime can be found by making use of a very simple physical assumption, namely, that the average kinetic friction $\boldsymbol{F}_{\text{scat}}$ acting on the electron system is antiparallel and proportional to the current $\boldsymbol{j}$, which can be written as

$$\boldsymbol{F}_{\text{scat}} = -N_{\text{e}} m_{\text{e}} \nu_{\text{eff}} \boldsymbol{v}_{\text{av}} , \qquad (3.2)$$

where $\nu_{\text{eff}}(B, n_{\text{s}})$ is an arbitrary proportionality factor. Then the balance of forces equation

$$\boldsymbol{F}_{\text{scat}} = - \langle \boldsymbol{F}_{\text{ext}} \rangle$$

gives a simple solution for the current ($\boldsymbol{j} = -e n_{\text{s}} \boldsymbol{v}_{\text{av}}$) described by the general form of the conductivity tensor

$$\sigma_{xx} = \frac{e^2 n_{\text{s}} \nu_{\text{eff}}}{m_{\text{e}} (\omega_{\text{c}}^2 + \nu_{\text{eff}}^2)} , \qquad \sigma_{yx} = \frac{\omega_{\text{c}}}{\nu_{\text{eff}}} \sigma_{xx} . \qquad (3.3)$$

It should be emphasized that this form is valid in both classical and quantum regimes and can be considered as an extension of the Drude equations.

In the conventional perturbation theory of quantum magnetotransport for σ_{xx}, we have the expansion

$$\frac{\nu_{\text{eff}}}{\omega_{\text{c}}^2 + \nu_{\text{eff}}^2} = \frac{\nu_{\text{eff}}}{\omega_{\text{c}}^2} \left(1 - \frac{\nu_{\text{eff}}^2}{\omega_{\text{c}}^2} + \cdots \right) , \qquad (3.4)$$

in which the first term is usually left as a high magnetic field approximation [81]. One should keep in mind that, in the quantum limit ($\hbar\omega_{\text{c}} > T$), the effective collision frequency ν_{eff} differs from the classical result, acquiring a strong magnetic field dependence. An increase in ν_{eff} with B can be even stronger than the linear increase in the cyclotron frequency, and the approximation $\sigma_{xx} \simeq e^2 n_{\text{s}} \nu_{\text{eff}} / m_{\text{e}} \omega_{\text{c}}^2$, used frequently as a high magnetic field asymptote, may fail for a particular electron system [141].

In the perturbation treatment, the conductivity tensor of (3.3) could be found by means of a partial summation of the infinite series of Feynmam diagrams. Knowing the exact form of the conductivity equation, we can evaluate the effective collision frequency in the simplest approximation or rearrange the perturbation theory so that it would be the perturbation theory for the effective collision frequency. In addition, we note that $\nu_{\rm eff}$ depends on the electron density when electron correlations become important. A detailed discussion of these properties of quantum magnetotransport in 2D electron systems will be given in Chap. 4.

It is instructive to note that, regardless of the actual field or density dependence of the effective collision frequency $\nu_{\rm eff}(B, n_{\rm s})$, inverting the conductivity tensor of (3.3) gives the universal Hall resistance

$$\rho_{yx} = \frac{B}{n_{\rm s}ec}\,, \qquad \rho_{xx} = \frac{m_{\rm e}\nu_{\rm eff}}{n_{\rm s}e^2}\,.$$

From this simple analysis one can conclude that experimental deviations of ρ_{yx} from the universal linear B-dependence can arise for the following reasons:

- the friction force is not parallel to the current, which is unlikely for isotropic systems,
- the density of carriers $n_{\rm s}$ depends on the magnetic field,
- edge effects are important.

The last two reasons are said to be responsible for the appearance of the ρ_{yx} plateaus in quantum Hall effect experiments [142, 143].

Consider now an alternating driving electric field $\boldsymbol{E}_{\parallel}$ with finite frequency ω. In this case, we have two characteristic vectors for the electron system: the current $\boldsymbol{j}$ and the average displacement vector of the electron liquid $\boldsymbol{u}_{\rm av}$, where $\boldsymbol{u}_{\rm av}$ is defined by $\boldsymbol{v}_{\rm av} = {\rm d}\boldsymbol{u}_{\rm av}/{\rm d}t$. The average force $\boldsymbol{F}_{\rm scat}(t)$ acting on the electron system is generally a retarded function depending on $\boldsymbol{j}(t')$ and $\boldsymbol{u}_{\rm av}(t')$ at all previous times ($t' \leq t$). Once again, we can assume that, in the linear transport regime, the retarded reaction of scatterers $\boldsymbol{F}_{\rm scat}(t)$ is antiparallel to the current and furthermore is proportional to the average displacement vector $\boldsymbol{u}_{\rm av}$ of the electron liquid (both vectors being taken at the appropriate time t'):

$$\boldsymbol{F}_{\rm scat}(t) = -N_{\rm e}m_{\rm e}\int_{-\infty}^{t}\left[\chi_v(t-t')\boldsymbol{v}_{\rm av}(t') + \chi_u(t-t')\boldsymbol{u}_{\rm av}(t')\right]{\rm d}t'\,, \qquad (3.5)$$

where $\chi_v(t)$ and $\chi_u(t)$ are arbitrary functions. This form of the force acting on the electron system due to the interaction with scatterers is consistent with the general equation of quantum linear response theory for an average value of an operator A discussed in the last chapter [see (2.3)].

The first term of (3.5) is the conventional friction. The effective collision frequency is defined as

$$\nu(\omega) = \int_0^\infty \chi_v(\tau) e^{i\omega\tau} d\tau \ ,$$

where we omit the subscript 'eff' from now on. The second term of (3.5) usually describes the response of the medium polarization cloud induced by electrons. This term should vanish in the limit $\omega \to 0$, which means that the medium polarization cloud follows the slow electron. Therefore, the analogous time integral containing $\chi_u(\tau)$ can be defined as

$$\int_0^\infty \chi_u(\tau) e^{i\omega\tau} d\tau = \lambda(0) - \lambda(\omega) \equiv -\omega w(\omega) \ , \tag{3.6}$$

in order to satisfy the zero-frequency limiting condition. In the opposite limit of high frequencies, $\lambda(\omega)$ usually decreases and we have an oscillatory response from the medium, viz., $\boldsymbol{F}_{\text{scat}} \simeq -N_e m_e \lambda(0) \boldsymbol{u}_{\text{av}}$, with characteristic frequency $\omega_0 = \sqrt{\lambda(0)}$. The functions $\nu(\omega)$ and $w(\omega)$ are related to each other because they originate from the imaginary and real parts of the same force:

$$\boldsymbol{F} = -\sum_e \frac{\partial V_{\text{int}}}{\partial \boldsymbol{r}_e} \ , \tag{3.7}$$

averaged according to the linear response theory.

The AC magnetoconductivity tensor can be found from the balance-of-force equation which includes the reaction of scatterers [see (3.5)] and the inertia term:

$$\sigma_{xx} = \frac{e^2 n_s}{m_e} \frac{\nu(\omega) - i[\omega + w(\omega)]}{\omega_c^2 + \{\nu(\omega) - i[\omega + w(\omega)]\}^2} \ , \qquad \sigma_{yx} = \frac{\omega_c \sigma_{xx}}{\nu(\omega) - i[\omega + w(\omega)]} \ . \tag{3.8}$$

With the notation $M(\omega) = w(\omega) + i\nu(\omega)$, this equation can be transformed into the general conductivity formula adopted in the memory function formalism [132, 133] (see also the discussion in Sect. 3.4). The structure of the function $w(\omega)$, viz.,

$$w(\omega) = [\lambda(\omega) - \lambda(0)] / \omega \ , \tag{3.9}$$

is also consistent with the corresponding property of the memory function. The effective collision frequency $\nu(\omega)$, equal to the imaginary part of the memory function, is responsible for a finite resistivity ρ_{xx} or the broadening of the CR. The real part of the memory function $w(\omega)$ leads to a shift in the position of the CR. For the Wigner solid state of SEs, $w(\omega)$ also describes the coupling between WS phonons and medium excitations [see Sect. 7.4]. It should be emphasized that, in contrast to the classical Drude equation, the effective collision frequency of the quantum transport theory generally has a strong frequency dependence if $\hbar\omega > T$. For example, the pronounced frequency dependence of the dynamical conductivity of simple metals was analyzed in [132]. The frequency dependence of the conductivity of the 2D electron gas and Wigner solid will be discussed in Sect. 8.2.

Thus, even without a substantial use of quantum theory, a simple phenomenological analysis based on reasonable physical assumptions allows us to retrieve the main general features of quantum magnetotransport.

3.3 Force-Balance Method (DC Case)

The functions $\nu(\omega)$ and $w(\omega)$ which determine the electron conductivity tensor in the phenomenological theory [see (3.8)] were introduced using the most general form of the average force exerted by scatterers on the whole moving electron system. In order to calculate these functions, one should be more specific when considering the interaction force $\boldsymbol{F} = -\sum_e (\partial V_{\mathrm{int}}/\partial \boldsymbol{r}_e)$. Let us consider first the DC case, which allows us to find an accurate expression for the effective collision frequency of highly correlated electrons. This result will be used in the following section to formulate the framework for practical evaluations of the electron conductivity, valid over the whole frequency range for strongly interacting systems. An important advantage of the force-balance method discussed here as compared to the memory function formalism is the possibility of describing nonlinear transport phenomena, because the kinetic friction $\boldsymbol{F}(\boldsymbol{j})$ can generally be found for arbitrary currents.

Thanks to the general formula for the electron conductivity tensor given in (3.8), it is reasonable to formulate the transport theory as a certain approximation for the effective collision frequency ν, so that any conductivity or resistivity component can be found in the usual way. As mentioned in the last section, in the extended theory, the effective collision frequency is actually the proportionality factor between the kinetic friction and the average electron velocity. Therefore, in order to find ν, one should calculate the momentum loss of the electron system per unit time for a particular interaction Hamiltonian. Because we wish to present the final result in a general form without making use of any specific assumption about the electron excitation spectrum, it is convenient to evaluate the momentum gained by scatterers instead. In the Born approximation, we obtain

$$\boldsymbol{F}_{\mathrm{scatt}} = -\frac{2\pi}{\hbar}\left\langle \sum_{\boldsymbol{n}',\boldsymbol{j}'} \left(\boldsymbol{p}_{\boldsymbol{n}'}^{(\mathrm{b})} - \boldsymbol{p}_{\boldsymbol{n}}^{(\mathrm{b})}\right) |\langle \boldsymbol{n}', \boldsymbol{j}' | V_{\mathrm{int}} | \boldsymbol{n}, \boldsymbol{j}\rangle|^2 \, \delta(\mathcal{E}_{\boldsymbol{n}',\boldsymbol{j}'} - \mathcal{E}_{\boldsymbol{n},\boldsymbol{j}}) \right\rangle , \tag{3.10}$$

where $\mathcal{E}_{\boldsymbol{n},\boldsymbol{j}}$ is the energy of the entire system with electrons in the many-body state $|\boldsymbol{j}\rangle$ and scatterers in the state $|\boldsymbol{n}\rangle$. Scatterers are assumed to be bosons (helium vapor atoms or ripplons), where $\boldsymbol{p}_{\boldsymbol{n}}^{(\mathrm{b})}$ stands for the in-plane (x, y) projection of the momentum of the whole Bose system and $\langle\ \rangle$ denotes the Gibbs average over the states $|\boldsymbol{n}, \boldsymbol{j}\rangle$.

The following evaluations of (3.10) are similar to those in the theory of thermal neutron scattering by solids. The interaction Hamiltonian V_{int} is generally a linear function of the Fourier transforms of the electron density operator $n_{\boldsymbol{q}}$ [see (1.74)]. Therefore, the force acting on electrons can finally be expressed in terms of the electron DSF $S(\boldsymbol{q},\omega)$. For example, if the scatterers are 2D phonons (ripplons), using basic properties of the phonon creation and destruction operators,

$$\langle \boldsymbol{n}' | b_{\boldsymbol{q}} | \boldsymbol{n} \rangle = N_{\boldsymbol{q}} \delta_{\boldsymbol{n}',\boldsymbol{n}-\boldsymbol{e}_{\boldsymbol{q}}} , \qquad \langle \boldsymbol{n}' | b_{\boldsymbol{q}}^{\dagger} | \boldsymbol{n} \rangle = (N_{\boldsymbol{q}}+1) \delta_{\boldsymbol{n}',\boldsymbol{n}+\boldsymbol{e}_{\boldsymbol{q}}} ,$$

we find that

$$|\langle \boldsymbol{n}', \boldsymbol{j}' | V_{\mathrm{int}} | \boldsymbol{n}, \boldsymbol{j} \rangle|^2 = \frac{1}{S_{\mathrm{A}}} \sum_{\boldsymbol{q}} V_q^2 Q_q^2 |\langle \boldsymbol{j}' | n_{-\boldsymbol{q}} | \boldsymbol{j} \rangle|^2 \times \left[N_{\boldsymbol{q}} \delta_{\boldsymbol{n}',\boldsymbol{n}-\boldsymbol{e}_{\boldsymbol{q}}} + (N_{-\boldsymbol{q}}+1) \delta_{\boldsymbol{n}',\boldsymbol{n}+\boldsymbol{e}_{-\boldsymbol{q}}} \right] , \tag{3.11}$$

where $N_{\boldsymbol{q}}$ is the ripplon occupation number and $\boldsymbol{e}_{\boldsymbol{q}}$ is the unit vector in the occupation number space ($\boldsymbol{n} = \sum_{\boldsymbol{q}} N_{\boldsymbol{q}} \boldsymbol{e}_{\boldsymbol{q}}$). The required relation between $\boldsymbol{F}_{\mathrm{scat}}$ and $S(\boldsymbol{q},\omega)$ is easily obtained by combining $|\langle \boldsymbol{j}' | n_{-\boldsymbol{q}} | \boldsymbol{j} \rangle|^2$ with the energy conservation δ-function in order to assemble $S(\boldsymbol{q},\omega)$ according to the form presented in (2.17):

$$\boldsymbol{F}_{\mathrm{scat}} = \frac{N_{\mathrm{e}}}{\hbar S_{\mathrm{A}}} \sum_{\boldsymbol{q}} \boldsymbol{q} V_q^2 Q_q^2 \left[N_{\boldsymbol{q}}^{(\mathrm{r})} S(\boldsymbol{q},\omega_q) + \left(N_{-\boldsymbol{q}}^{(\mathrm{r})} + 1 \right) S(\boldsymbol{q},-\omega_q) \right] , \tag{3.12}$$

where $N_{\boldsymbol{q}}^{(\mathrm{r})}$ is the ripplon distribution Bose function, depending only on the absolute value of the wave vector.

If the electrons are in equilibrium in the laboratory reference frame, then the electron DSF is isotropic $S(\boldsymbol{q},\omega) = S_0(q,\omega)$ and the frictional force of (3.12) disappears as a result of the summation over all wave vectors $\boldsymbol{q}$. This is similar to the vanishing of the collision integral in the usual kinetic equation theory, when the equilibrium Fermi distribution function is used. To proceed with (3.12), one should evaluate the non-equilibrium DSF which is not isotropic because of the driving electric field. Generally, this is quite a complicated task. Luckily, for highly correlated electrons, there is an easy way. One can assume that the highly correlated electron liquid is in equilibrium in the center-of-mass reference frame, where it can be described by the isotropic equilibrium DSF $S_0(q,\omega)$. The DSF $S(\boldsymbol{q},\omega_q) \equiv S_{\mathrm{lab}}(\boldsymbol{q},\omega_q)$ entering (3.12) describes the electron system in the laboratory frame. In order to find it from $S_0(q,\omega)$ evaluated in the moving frame, one can use Galilean invariance, which leads to the transcription rule of (2.41). When changing the reference frame from the center-of-mass frame moving with velocity $\boldsymbol{v}_{\mathrm{av}}$ to the motionless laboratory frame, the frequency argument of the DSF acquires the Doppler shift $S_{\mathrm{lab}}(\boldsymbol{q},\omega) = S(\boldsymbol{q},\omega - \boldsymbol{q}\cdot\boldsymbol{v}_{\mathrm{av}})$. If the electrons are in equilibrium in the center-of-mass frame as assumed above, the proper expression for the electron DSF to be used in (3.12) has the very simple form

$$S(\boldsymbol{q},\omega) = S_0(q,\omega - \boldsymbol{q}\cdot\boldsymbol{v}_{\mathrm{av}}) . \tag{3.13}$$

In the semi-classical regime, which we discuss in Sect. 3.5, this treatment is equivalent to the description of electron transport by means of the shifted Fermi distribution function $f(\varepsilon_k - \hbar \boldsymbol{q} \cdot \boldsymbol{v}_{\rm av})$, which is valid when the electron–electron collision rate is much higher than the electron collision frequency $\nu_{\rm eff}$ due to the scatterers.

Instead of expanding the DSF in powers of $\boldsymbol{q} \cdot \boldsymbol{v}_{\rm av}$ and finding its anisotropic correction, it is convenient to make some rearrangements in the basic equation for the frictional force. We note first that the second term of (3.12) can be transformed to

$$\left(N_q^{(\rm r)} + 1\right) S_0(q, -\omega_q - \boldsymbol{q} \cdot \boldsymbol{v}_{\rm av}) = N_q^{(\rm r)} {\rm e}^{-\hbar \boldsymbol{q} \cdot \boldsymbol{v}_{\rm av}/T} S_0(q, \omega_q + \boldsymbol{q} \cdot \boldsymbol{v}_{\rm av}) ,$$

because of the property of the equilibrium DSF

$$S_0(q, -\omega) = \exp(-\hbar\omega/T) S_0(q, \omega) .$$

Then replacing the summation vector $\boldsymbol{q} \to -\boldsymbol{q}$ for this term only, the friction force can be rewritten in the form [36, 84]

$$\boldsymbol{F}_{\rm scat} = \frac{N_{\rm e}}{\hbar S_{\rm A}} \sum_{\boldsymbol{q}} V_q^2 Q_q^2 N_q^{(\rm r)} \boldsymbol{q} \left(1 - \exp \frac{\hbar \boldsymbol{q} \cdot \boldsymbol{v}_{\rm av}}{T}\right) S_0(q, \omega_q - \boldsymbol{q} \cdot \boldsymbol{v}_{\rm av}) . \quad (3.14)$$

This equation is a first-principles presentation of the average frictional force acting on the electron system in a DC experiment, which is valid for arbitrary values of the average electron velocity $\boldsymbol{v}_{\rm av}$.

In the linear transport regime, the form of (3.14) allows us to disregard the Doppler shift in the argument of the electron DSF because of the factor in the large round brackets. The frictional force obtained above satisfies the assumptions used in the phenomenological treatment of quantum magneto-transport given in the last section: $\boldsymbol{F}_{\rm scat}$ is a function of the electron current density or the average electron velocity $\boldsymbol{v}_{\rm av}$, and in the linear regime it is antiparallel and proportional to the current.

The parameter $\hbar \boldsymbol{q} \cdot \boldsymbol{v}_{\rm av}/T$ is usually the smallest parameter of the non-linear theory. Therefore, we can expand the exponent of (3.14) in powers of $\hbar \boldsymbol{q} \cdot \boldsymbol{v}_{\rm av}/T$, which makes the expression in large round brackets proportional to $-\boldsymbol{q} \cdot \boldsymbol{v}_{\rm av}$ and builds in the transport factor of $1 - \cos\theta$: $q^2 = (2k)^2 \sin^2(\theta/2) \propto 1 - \cos\theta$. Here we would like to note that, in the conventional quantum transport theory, this factor appears as a result of the summation of the infinite series of vertex diagrams. Here it originates from the basic property of the equilibrium DSF $S_0(q, \omega)$. Disregarding the Doppler shift in $S_0(q, \omega_q - \boldsymbol{q} \cdot \boldsymbol{v}_{\rm av})$, one can find the first-principles presentation for the effective collision frequency ν defined by the relation $\boldsymbol{F}_{\rm scat} = -N_{\rm e} m_{\rm e} \nu \boldsymbol{v}_{\rm av}$ in the phenomenological theory of the last section:

$$\nu = \frac{1}{2 m_{\rm e} T S_{\rm A}} \sum_{\boldsymbol{q}} q^2 V_q^2 Q_q^2 N_q^{(\rm r)} S_0(q, \omega_q) . \quad (3.15)$$

Thus, in order to find the magnetoconductivity tensor for highly correlated electrons, one should find a suitable approximation for the equilibrium DSF of the Coulomb liquid, instead of solving the kinetic equation for the system with a complicated energy spectrum.

Consider now the electron interaction with impurity-like scatterers. According to the analysis given in Sect. 1.4, this sort of electron scattering can be described as the scattering induced by heavy atoms forming a dilute Bose gas. For electrons on liquid helium, this is the real Bose gas of vapor atoms. The static impurity scattering of other systems can be considered in the same way as scattering induced by fictitious heavy ($M_{\mathrm{a}} \to \infty$) atoms interacting with the electron via a given potential $V(\boldsymbol{R}_{\mathrm{e}} - \boldsymbol{R}_{\mathrm{a}})$. For this kind of electron scattering, the interaction Hamiltonian is proportional to the electron density operator $n_{-\boldsymbol{q}}$ and the operator $A_{\mathrm{a},\boldsymbol{q}}$, which represents a sort of projection of the 3D impurity system onto the plane of the 2D electron system according to (1.39). The latter operator annihilates a Bose atom in the state described by a 3D wave vector $\boldsymbol{K}'$ and creates an atom in another state defined as $\boldsymbol{K}' - \boldsymbol{K}$, which describes a scattering event. The momentum exchange between the electron and atom caused by this scattering is denoted by $\boldsymbol{K} = \{\boldsymbol{q}, k\}$, where $\boldsymbol{q}$ is a 2D vector in the plane of the system and $k \equiv K_z$. The matrix elements entering the frictional force of (3.10) can be written in the form

$$\langle \boldsymbol{n}' | A_{\mathrm{a},\boldsymbol{q}} | \boldsymbol{n} \rangle = \sum_{k} \eta_k \sum_{\boldsymbol{K}'} \sqrt{N_{\boldsymbol{K}'} (N_{\boldsymbol{K}'-\boldsymbol{K}} + 1)} \delta_{\boldsymbol{n}', \boldsymbol{n} - \boldsymbol{e}_{\boldsymbol{K}'} + \boldsymbol{e}_{\boldsymbol{K}'-\boldsymbol{K}}} , \quad (3.16)$$

where $N_{\boldsymbol{K}}$ is the occupation number for the Bose atoms. Because we consider the dilute Bose gas, $N_{\boldsymbol{K}'-\boldsymbol{K}}$ can be disregarded in (3.16) in comparison with unity.

Using (3.16), the average frictional force can be evaluated in a straightforward manner:

$$\boldsymbol{F}_{\mathrm{scat}} = \frac{N_{\mathrm{e}}}{\hbar \Omega_{\mathrm{v}}^2} \sum_{\boldsymbol{K}} \boldsymbol{q} \left| \eta_k V_{\boldsymbol{K}}^{(\mathrm{a})} \right|^2 \sum_{\boldsymbol{K}'} N_{\boldsymbol{K}'}^{(\mathrm{a})} S(\boldsymbol{q}, \Delta_{\boldsymbol{K},\boldsymbol{K}'}) , \quad (3.17)$$

where $\hbar \Delta_{\boldsymbol{K},\boldsymbol{K}'} = \varepsilon_{\boldsymbol{K}'}^{(\mathrm{a})} - \varepsilon_{\boldsymbol{K}'-\boldsymbol{K}}^{(\mathrm{a})}$ is the energy exchange in a single collision, $N_{\boldsymbol{K}}^{(\mathrm{a})}$ is the Bose distribution function of the vapor atoms, and $\Omega_{\mathrm{v}} = S_{\mathrm{A}} L_z$. Once again, replacing $S(\boldsymbol{q}, \omega) \to S_0(q, \omega - \boldsymbol{q} \cdot \boldsymbol{v}_{\mathrm{av}})$, we find that the form of (3.17) is not convenient for the expansion in $\boldsymbol{q} \cdot \boldsymbol{v}_{\mathrm{av}}$. In order to arrive at a convenient form for $\boldsymbol{F}_{\mathrm{scat}}$, consider first the pure elastic scattering when $\Delta_{\boldsymbol{K},\boldsymbol{K}'} = 0$. In this case, we can use the transformation

$$\sum_{\boldsymbol{q}} \left| V_{q,k}^{(\mathrm{a})} \right|^2 \boldsymbol{q} S_0(q, -\boldsymbol{q} \cdot \boldsymbol{v}_{\mathrm{av}}) = \frac{1}{2} \sum_{\boldsymbol{q}} \left| V_{q,k}^{(\mathrm{a})} \right|^2 \boldsymbol{q} \left[S_0(q, -\boldsymbol{q} \cdot \boldsymbol{v}_{\mathrm{av}}) - S_0(q, \boldsymbol{q} \cdot \boldsymbol{v}_{\mathrm{av}}) \right]$$
$$= \frac{1}{2} \sum_{\boldsymbol{q}} \left| V_{q,k}^{(\mathrm{a})} \right|^2 \boldsymbol{q} \left(1 - \exp \frac{\hbar \boldsymbol{q} \cdot \boldsymbol{v}_{\mathrm{av}}}{T} \right) S_0(q, -\boldsymbol{q} \cdot \boldsymbol{v}_{\mathrm{av}}) , \quad (3.18)$$

which employs the basic properties of the equilibrium DSF discussed above. In the general case $\Delta_{\boldsymbol{K},\boldsymbol{K}'} \neq 0$, to obtain the second term with $S_0(q,-\omega)$, one should use a more complicated change in the summation vectors $\boldsymbol{K}' - \boldsymbol{K} \to \boldsymbol{K}'$ and $\boldsymbol{K} \to -\boldsymbol{K}$, which changes the sign of $\Delta_{\boldsymbol{K},\boldsymbol{K}'}$. As a result, (3.17) transforms to

$$\boldsymbol{F}_{\mathrm{scat}} = \frac{N_{\mathrm{e}}}{2\hbar\Omega_{\mathrm{v}}^2} \sum_{\boldsymbol{q}} \boldsymbol{q} \left(1 - \exp\frac{\hbar \boldsymbol{q}\cdot\boldsymbol{v}_{\mathrm{av}}}{T}\right) \sum_{k,\boldsymbol{K}'} \left|\eta_k V_K^{(\mathrm{a})}\right|^2 N_{\boldsymbol{K}'}^{(\mathrm{a})} S_0(q, \Delta_{\boldsymbol{K},\boldsymbol{K}'} - \boldsymbol{q}\cdot\boldsymbol{v}_{\mathrm{av}}), \tag{3.19}$$

which allows us to disregard the Doppler shift $\boldsymbol{q}\cdot\boldsymbol{v}_{\mathrm{av}}$ in the frequency argument of the DSF.

In the case of elastic scattering ($\Delta_{\boldsymbol{K},\boldsymbol{K}'} \to 0$) at the short-range impurity potential with $V_K^{(\mathrm{a})} = V^{(\mathrm{a})}$ from (1.37), the effective collision frequency of the linear transport regime has the simplest form:

$$\nu = \frac{\hbar^3 \nu_0^{(\mathrm{a})}}{4m_{\mathrm{e}}^2 T S_{\mathrm{A}}} \sum_{\boldsymbol{q}} q^2 S_0(q,0), \tag{3.20}$$

where we have introduced

$$\nu_0^{(\mathrm{a})} = \frac{3(V^{(\mathrm{a})})^2 n_{3\mathrm{D}}^{(\mathrm{a})} m_{\mathrm{e}} \gamma}{8\hbar^3}, \tag{3.21}$$

using the relations

$$\frac{1}{\Omega_{\mathrm{v}}} \sum_{\boldsymbol{K}'} N_{\boldsymbol{K}'}^{(\mathrm{a})} = n_{3\mathrm{D}}^{(\mathrm{a})}, \qquad \sum_k |\eta_k|^2 = \frac{3\gamma}{8} L_z. \tag{3.22}$$

The latter relation is found for the electron wave function of the ground surface level $f_1(z) = 2\gamma^{3/2} z \exp(-\gamma z)$ with the localization parameter γ depending on the holding electric field, as discussed in Chap. 1. Under zero magnetic field, the DSF $S_0(q,0)$ given in (2.22) for the nondegenerate 2D electron gas yields $\nu = \nu_0^{(\mathrm{a})}$, a result which could be found by means of the kinetic equation method. The important point is that the general forms of (3.15) and (3.20) are valid for strongly correlated electrons and under a quantizing magnetic field when the excitation spectrum of the electron system is far from the simple parabolic spectrum $\varepsilon_{\boldsymbol{k}} = \hbar^2 k^2 / 2m_{\mathrm{e}}$ of free electrons.

The magnetoconductivity treatment discussed above represents a version of the quantum force-balance method developed by Cai, Lei and Ting [138]. This method is frequently used in studies of nonlinear magnetoconductivity under a strong magnetic field in semiconductor 2D electron systems [144, 145], without emphasis on the importance of the approximation in (3.13), which is valid only for highly correlated electrons. In [137, 138], the basic transport equations are found by introducing the center-of-mass and relative electron coordinates. The frictional force experienced by the center of mass due to electron–phonon and electron–impurity interactions is found by averaging $-(\mathrm{i}/\hbar)[\boldsymbol{P}, V_{\mathrm{int}}]$ in the framework of the quantum linear response theory with regard to interactions [see (2.3)]. (Here $\boldsymbol{P}$ is the momentum operator of the center of mass.) Because the perturbation V_{int} and the operator $[\boldsymbol{P}, V_{\mathrm{int}}]$ to be averaged are both proportional to the electron density fluctuation operator $n_{-\boldsymbol{q}}$, the final result can be found in terms of the density–density correlation function or the retarded Green's function for the density fluctuation operator:

$$G^{(\mathrm{R})}_{n_{\boldsymbol{q}}, n^{\dagger}_{\boldsymbol{q}}}(\omega) \equiv \langle\langle n_{\boldsymbol{q}} | n_{-\boldsymbol{q}} \rangle\rangle_{\omega} , \tag{3.23}$$

discussed in Sect. 2.1. For example, the frictional force induced by impurities was found in the form [138]

$$\boldsymbol{F}_{\mathrm{scat}} = \frac{n^{(\mathrm{i})}}{\hbar} \sum_{\boldsymbol{q}} |u(q)|^2 \, \boldsymbol{q} \mathrm{Im} G_{n_{\boldsymbol{q}}, n^{\dagger}_{\boldsymbol{q}}}(\boldsymbol{q} \cdot \boldsymbol{v}_{\mathrm{av}}) , \tag{3.24}$$

where $n^{(\mathrm{i})}$ is the impurity density and $u(q)$ is the electron–impurity interaction in momentum space. (A more detailed analysis of the impurity distribution and the interaction matrix elements is given in [145].) Recalling the fluctuation–dissipation theorem,

$$\mathrm{Im} G^{(\mathrm{R})}_{n, n^{\dagger}}(\boldsymbol{q} \cdot \boldsymbol{v}_{\mathrm{av}}) = -\frac{1 - \mathrm{e}^{-\hbar \boldsymbol{q} \cdot \boldsymbol{v}_{\mathrm{av}}/T}}{2} N_{\mathrm{e}} S(q, \boldsymbol{q} \cdot \boldsymbol{v}_{\mathrm{av}}) , \tag{3.25}$$

one can see the equivalence of the final results of (3.24) and (3.19), if the energy exchange $\Delta_{\boldsymbol{K},\boldsymbol{K}'}$ at a collision is disregarded and the summation vector is taken with the opposite sign: $\boldsymbol{q} \to -\boldsymbol{q}$.

3.4 Memory Function Formulation (AC Case)

The phenomenological analysis of the electron conductivity tensor results in (3.8). With the notation

$$\sigma_{\pm} \equiv \sigma_{xx} \pm \mathrm{i}\sigma_{xy} , \qquad M(\omega) = w(\omega) + \mathrm{i}\nu(\omega) ,$$

this form of the conductivity tensor can be rewritten as

$$\sigma_{\pm} = \frac{\mathrm{i}e^2 n_{\mathrm{s}}}{m\left[\omega \mp \omega_{\mathrm{c}} + M(\omega)\right]} . \tag{3.26}$$

This conductivity equation looks very like Dyson's equation for the single-electron propagator. Götze and Wölfle [132] and Götze and Hajdu [134] adopted the view that one can avoid solving kinetic equations and summing Feynman diagrams by starting from this presentation of the conductivity tensor with the built-in resonance structure. In order to arrive at a conductivity equation applicable at all frequencies ω, it seems to be enough to use the simplest approximation for the memory function $M(\omega)$ which is finite in the hydrodynamic limit.

The use of the form (3.26) in the quantum transport theory is assumed to be equivalent to a partial summation of an infinite series of diagrams including the self-energy and vertex corrections. This situation is reminiscent of the DC magnetoconductivity case discussed in the last section: the simplest approximation for the effective collision frequency $\nu(B)$ gives a conductivity equation valid for any relation between ν and ω_c, while the conventional procedure in the same approximation gives the conductivity equation $\sigma_{xx} \propto \nu(B)/\omega_c^2$, valid only for $\nu \ll \omega_c$.

Although Shiwa and Isihara [136] found that the relaxation kernel of the electron conductivity $M(\omega)$ has a more complicated structure than the one presented in [132], depending on the sign $\pm$ chosen in (3.26), it agrees with the previous result for small impurity effects. Moreover, we would like to note that, in the general case, one should be very careful when applying the approximation for the memory function of [132] to the low frequency case with $B = 0$, as we shall discuss later.

We shall now find the general form of the conductivity tensor, similar to (3.26), starting from Kubo's conductivity formula

$$\sigma_{\alpha\beta}(\omega) = \frac{\mathrm{i}}{\hbar} \int_0^{\infty} \mathrm{d}t\, \mathrm{e}^{\mathrm{i}\omega t - \delta t} \left\langle [J_\alpha(t), D_\beta(0)] \right\rangle , \tag{3.27}$$

where $\boldsymbol{D} = e \sum_e \boldsymbol{r}_e$ and $\boldsymbol{J} = \mathrm{d}\boldsymbol{D}/\mathrm{d}t$. This equation is a straightforward result of the linear response theory for the average current $\boldsymbol{J}$ [see the general form of (2.3)] when the perturbation Hamiltonian is due to an external driving electric field $V = -\boldsymbol{D} \cdot \boldsymbol{E}(t)$.

The particular form of (3.27) is not very convenient because it contains the correlation function of different operators. A suitable transformation of (3.27) into a form containing the current–current correlation function can be found by moving the time dependence from the operator J_α to the operator D_β according to the rule $\langle A(t)B\rangle = \langle AB(-t)\rangle$ and then integrating by parts with respect to the time:

$$\sigma_{\alpha\beta}(\omega) = \frac{\mathrm{i}}{\hbar\omega S_{\mathrm{A}}} G^{(\mathrm{R})}_{J_\alpha, J_\beta}(\omega) + \frac{\mathrm{i}e^2 n_{\mathrm{s}}}{m_{\mathrm{e}}\omega} \delta_{\alpha,\beta} , \tag{3.28}$$

where $G^{(\mathrm{R})}_{J_\alpha, J_\beta}(\omega) \equiv \langle\langle J_\alpha | J_\beta \rangle\rangle_\omega$ is the retarded current–current Green's function defined in Sect. 2.1. The surface area S_{A} entering the denominator of (3.28) and the areal electron density n_{s} appear because we are considering 2D electron systems. (In the 3D case, we have to replace S_{A} by Ω_{v}). The form of (3.28) is often used as the starting point for quantum transport theories. The relation between $\sigma_{\alpha\beta}$ and $G^{(\mathrm{R})}_{J_\alpha, J_\beta}(\omega)$ shows that the conductivity of the linear regime is an intrinsic property of the equilibrium state of the system.

In the presence of a magnetic field, it is convenient to introduce a scalar current operator

$$J = \frac{J_x + \mathrm{i}J_y}{\sqrt{2}} . \tag{3.29}$$

Then (3.28) can be rewritten as

$$\sigma_+(\omega) = \frac{\mathrm{i}}{\hbar\omega S_\mathrm{A}} \left[G^{(\mathrm{R})}_{J^\dagger,J}(\omega) - G^{(\mathrm{R})}_{J^\dagger,J}(0) \right] , \tag{3.30}$$

$$\sigma_-(\omega) = \frac{\mathrm{i}}{\hbar\omega S_\mathrm{A}} \left[G^{(\mathrm{R})}_{J,J^\dagger}(\omega) - G^{(\mathrm{R})}_{J,J^\dagger}(0) \right] , \tag{3.31}$$

where we have used the relation

$$G^{(\mathrm{R})}_{J^\dagger,J}(0) = G^{(\mathrm{R})}_{J,J^\dagger}(0) = -\frac{e^2\hbar N_\mathrm{e}}{m} , \tag{3.32}$$

which guarantees that the conductivity tends to a finite value as $\omega \to 0$.

The scalar current operator introduced above satisfies the commutation relations

$$\left[J^\dagger, J \right] = \frac{e^2 N_\mathrm{e} \hbar\omega_\mathrm{c}}{m} , \qquad [H_\mathrm{K}, J] = \hbar\omega_\mathrm{c} J , \tag{3.33}$$

where H_K is the kinetic energy term of the electron Hamiltonian. Employing the second relation, the time evolution equation for the electron current can be written as

$$\frac{\mathrm{d}J}{\mathrm{d}t} - \mathrm{i}\omega_\mathrm{c} J = \frac{\mathrm{i}}{\hbar} [H_\mathrm{int}, J] , \tag{3.34}$$

where the interaction Hamiltonian $H_\mathrm{int} = V_\mathrm{C} + V_\mathrm{int}$ is a sum of the mutual Coulomb interaction V_C and the interaction with scatterers V_int. The right-hand side of (3.34) is proportional to the force acting on the whole electron system,

$$F \equiv \frac{F_x + \mathrm{i}F_y}{\sqrt{2}} = \frac{m}{\hbar e} \mathrm{i} [J, H_\mathrm{int}] . \tag{3.35}$$

It is important to note that the Coulomb potential term V_C cannot contribute to this force because

$$\frac{m}{\hbar e} \mathrm{i} [\boldsymbol{J}, V_\mathrm{C}] = \frac{1}{2} \sum_{s,s'} \left[-\frac{\partial}{\partial \boldsymbol{r}_s} V\left(|\boldsymbol{r}_s - \boldsymbol{r}_{s'}| \right) \right] \to 0 . \tag{3.36}$$

Here we have used the well known commutation relation between the momentum p and coordinate q:

$$[p, f(q)] = -\mathrm{i}\hbar \frac{\partial f}{\partial q} .$$

Thus, if the interaction with scatterers V_int is disregarded, we have $J(t) = \exp(\mathrm{i}\omega_\mathrm{c} t) J(0)$ and consequently,

$$G^{(R)}_{J^\dagger,J}(\omega) = \frac{e^2 N_e \hbar\omega_c}{m(\omega-\omega_c)} , \qquad G^{(R)}_{J,J^\dagger}(\omega) = -\frac{e^2 N_e \hbar\omega_c}{m(\omega+\omega_c)} , \tag{3.37}$$

$$\sigma_\pm(\omega) = \frac{\mathrm{i}e^2 n_s}{m(\omega \mp \omega_c)} , \tag{3.38}$$

which means that Coulomb forces acting between electrons cannot affect the conductivity and CR absorption directly. This is the statement of the famous Kohn theorem [146]. As we shall see later, the Coulomb forces can affect the CR absorption from electrons and the DC magnetoconductivity in an indirect way through the term $[J, V_{\text{int}}]$ on the right-hand side of (3.34), by changing the probability of electron scattering due to impurities and phonons.

In order to include the interaction with scatterers V_{int} in the current–current correlation function, we have to consider the equations of motion for the Green's function $G^{(R)}_{A,B}(\omega) \equiv \langle\langle A|B\rangle\rangle_\omega$ introduced in (2.11) of the last chapter. The Hamiltonian of electrons H entering these equations contains the term V_{int} describing interaction with scatterers. For the particular current–current Green's function, we have $A = J^\dagger$ and $B = J$. Using the commutation relation

$$[H, J^\dagger] = -\hbar\omega_c J^\dagger + \mathrm{i}\frac{\hbar e}{m} F^\dagger , \tag{3.39}$$

the first line of (2.11) can be rewritten

$$(\omega - \omega_c) G_{J^\dagger,J}(\omega) = \frac{e^2 N_e}{m}\omega_c - \mathrm{i}\frac{e}{m} G_{F^\dagger,J}(\omega) , \tag{3.40}$$

where the operator F describes the frictional force acting on the electron system. [It has the form (3.35), with H_{int} replaced by V_{int}.]

We now represent $G_{F^\dagger,J}(\omega)$ of the right-hand side of (3.40) in terms of the force–force Green's function $G_{F^\dagger,F}(\omega)$. This can be done using the second line of (2.11) with $A = F^\dagger$, $B = J$, and the commutation relation for $[J, H]$ which is similar to (3.39). Finally, we have

$$G^{(R)}_{F^\dagger,J}(\omega) = \mathrm{i}\frac{e}{m(\omega-\omega_c)} \left[G_{F^\dagger,F}(\omega) - \mathrm{i}\frac{m}{\hbar e} \langle [F^\dagger, J] \rangle \right] . \tag{3.41}$$

The zero-frequency limiting condition of (3.32), (3.40) and (3.41) results in the relation

$$\mathrm{i}\frac{m}{e\hbar} \langle [F^\dagger, J] \rangle = G^{(R)}_{F^\dagger,F}(0) , \tag{3.42}$$

which can be used in (3.41) to simplify the expression in square brackets.

For an isotropic system, $\langle F^\dagger(t)F(0)\rangle = \langle F_x(t)F_x(0)\rangle$, and we therefore introduce the function

$$M_F(\omega) = \frac{1}{m\omega N_e} \left[G^{(R)}_{F_x,F_x}(0) - G^{(R)}_{F_x,F_x}(\omega) \right] . \tag{3.43}$$

This function is also sometimes called the memory function. The current–current Green's functions entering equations for $\sigma_{\pm}$ can be expressed in terms of the force–force Green's function or the memory function:

$$G^{(\mathrm{R})}_{J^{\dagger},J}(\omega) = \frac{e^2 N_{\mathrm{e}} \hbar}{m} \left[\frac{\omega_{\mathrm{c}}}{\omega - \omega_{\mathrm{c}}} - \frac{\omega M_F(\omega)}{(\omega - \omega_{\mathrm{c}})^2} \right] , \tag{3.44}$$

$$G^{(\mathrm{R})}_{J,J^{\dagger}}(\omega) = \frac{e^2 N_{\mathrm{e}} \hbar}{m} \left[-\frac{\omega_{\mathrm{c}}}{\omega + \omega_{\mathrm{c}}} - \frac{\omega M_F(\omega)}{(\omega + \omega_{\mathrm{c}})^2} \right] . \tag{3.45}$$

We have thus succeeded in establishing the relations between the current–current and force–force Green's functions. This simplifies evaluations, although we still do not have the proper resonance structure of the conductivity tensor.

In order to obtain the proper form of the conductivity equation in the presence of the magnetic field, we introduce two auxiliary functions $M_+(\omega)$ and $M_-(\omega)$:

$$M_{\pm}(\omega) = \frac{M_F(\omega)}{1 - M_F(\omega)/(\omega \mp \omega_{\mathrm{c}})} . \tag{3.46}$$

Inverting this equation, we find two forms of $M_F(\omega)$:

$$M_F(\omega) = \frac{M_{\pm}(\omega)}{1 + M_{\pm}(\omega)/(\omega \mp \omega_{\mathrm{c}})} , \tag{3.47}$$

which we shall use in the equations for the corresponding current–current Green's functions entering $\sigma_{\pm}(\omega)$ [see (3.41), (3.44) and (3.45)]. After simple rearrangements, we arrive at the final conductivity equation

$$\sigma_{\pm}(\omega) = \frac{\mathrm{i} e^2 n_{\mathrm{s}}}{m \left[\omega \mp \omega_{\mathrm{c}} + M_{\pm}(\omega) \right]} , \tag{3.48}$$

which was our goal. This kind of conductivity equation was first found by Shiwa and Isihara [136] in a different way, using Mori's projection operator method [147].

Firstly, we note that the exact conductivity form of (3.48) has the right resonance structure, similar to (3.26) found in the phenomenological treatment. Still, the relaxation kernel $M_{\pm}$ of the first-principles approach depends on the sign label of the conductivity component $\sigma_{\pm}$, and more importantly it vanishes when $\omega \to \omega_{\mathrm{c}}$ (if M_F is finite), which looks rather frustrating. Nevertheless, when $\omega_{\mathrm{c}} \gg \nu$ and the frequency is not in the vicinity of the CR, the resonant term in the denominator of (3.46) can be disregarded for small impurity effects, and we have $M_{\pm}(\omega) \simeq M_F$. In this case, the relaxation kernel is independent of the sign label and can be taken as $M(\omega)$ of the phenomenological formula (3.26). Thus, at $\omega_{\mathrm{c}} \gg \nu$, the memory function approach for DC magnetotransport yields an accurate conductivity form which also agrees with the results found in other methods, including the force–balance method discussed above.

In the absence of the magnetic field $B = 0$, the approximation $M_{\pm}(\omega) \simeq M_F(\omega)$ is strictly valid only for high enough frequencies $\omega \gg \nu$. The limiting case $\omega \to 0$ requires a special treatment. The point adopted by Götze and Wölfle [132] is that, knowing the right structure of the conductivity form, one can get $M(\omega)$ by expanding (3.26) and (3.48) in $M(\omega)/\omega$ and $M_F(\omega)/\omega$, respectively, and equating the linear terms, which gives $M(\omega) \simeq M_F(\omega)$. There is a certain good sense in such a treatment and in many cases it gives an accurate result, even for the DC case. Still, one should remember that the approximation $M(\omega) \simeq M_F(\omega)$ may generally give an incorrect numerical proportionality factor of the order of 2, if it is applied to the zero frequency case. A highly correlated electron system is an important exception. In this case, the approximation $M(\omega) \simeq M_F$ gives a numerically correct result even at $\omega = 0$, which is confirmed by the force-balance method. As we shall see in the following, this approximation appears to be accurate for both high and low frequency regimes, if the electron collision rate $\nu_{\text{e–e}}$ due to mutual interactions is much higher than the collision rate ν due to scatterers.

Consider the particular scattering mechanisms. For surface electrons on liquid helium, the interaction Hamiltonian has the universal form (1.74), which can be used in a similar way for both electron–ripplon (2D phonon) and electron–vapor atom ('impurity') scattering. The force $\boldsymbol{F}$ experienced by electrons due to these interactions can be written as a linear functional of the electron density operator:

$$\boldsymbol{F} = -\mathrm{i} \sum_{\mathrm{s=r,a}} \sum_{\boldsymbol{q}} \boldsymbol{q} U_{\mathrm{s}} A_{\mathrm{s},\boldsymbol{q}} n_{-\boldsymbol{q}} \ , \tag{3.49}$$

where the scatterer operator $A_{\mathrm{s},\boldsymbol{q}}$ was defined in (1.39) and (1.56). Then the relation between $\nu_{\text{eff}}(\omega) \simeq \mathrm{Im}[M_F(\omega)]$ and the electron DSF $S(q,\omega)$ becomes evident. According to (3.42), the evaluation of the memory function in the approximation of [132, 133] requires knowledge of the force–force Green's function $G^{(\mathrm{R})}_{F_x,F_x}(\omega)$, which has the following spectral representation:

$$G^{(\mathrm{R})}_{F_x,F_x}(\omega) = \int_{-\infty}^{+\infty} \langle F_x(t) F_x(0) \rangle_{\omega'} \frac{1 - \mathrm{e}^{-\hbar\omega'/T}}{\omega - \omega' + \mathrm{i}0} \frac{\mathrm{d}\omega'}{2\pi\hbar} \ . \tag{3.50}$$

This is a particular case of the general form (2.9) discussed in the last chapter.

According to (1.39), for electron scattering by vapor atoms, the particular form of the operators $A_{\mathrm{a},\boldsymbol{q}}$ contains the vapor atom density fluctuation operator

$$\rho_{\mathrm{a},\boldsymbol{K}} = \sum_{\boldsymbol{K}'} a^{\dagger}_{\boldsymbol{K}'-\boldsymbol{K}} a_{\boldsymbol{K}'} \ ,$$

reducing the momentum of the system by $\hbar\boldsymbol{K}$. Therefore, in the absence of interactions, its time evolution is described by

$$A_{a,\boldsymbol{q}}(t) = \sum_k \eta_k \sum_{\boldsymbol{K}'} a^{\dagger}_{\boldsymbol{K}'-\boldsymbol{K}} a_{\boldsymbol{K}'} \exp(-\mathrm{i}\Delta_{\boldsymbol{K},\boldsymbol{K}'} t) \,, \tag{3.51}$$

where $\Delta_{\boldsymbol{K},\boldsymbol{K}'}$ is the energy exchange defined in (1.42). Evaluating the force–force correlation function in the simplest approximation, one can disregard the effect of interaction on the scatterers and describe them by the equilibrium density matrix. Then straightforward evaluation yields

$$\langle F_\alpha(t)F_\beta(0)\rangle_\omega = \frac{N_e}{S_A} \sum_{\boldsymbol{q}} q_\alpha q_\beta \frac{1}{L_z} \sum_k \left|\eta_k V_K^{(a)}\right|^2 \times \frac{1}{\Omega_v} \sum_{\boldsymbol{K}'} N_{\boldsymbol{K}'}^{(a)} \left\langle S\left(\boldsymbol{q}, \omega + \Delta_{\boldsymbol{K},\boldsymbol{K}'}\right)\right\rangle_{sc} \,, \tag{3.52}$$

where $\langle\,\rangle_{sc}$ denotes the average over the scatterer variables. This anticipates the fact that, for a 2D electron system under a magnetic field, the collision broadening of Landau levels should be taken into account. For the scattering induced by the short-range potential of vapor atoms, $V_K^{(a)} = V^{(a)} = \text{const.}$, according to (1.37). The approximation used above can be written as the following assumption:

$$\langle A_{a,\boldsymbol{q}}(t) n_{-\boldsymbol{q}}(t) A_{a,\boldsymbol{w}} n_{-\boldsymbol{w}}\rangle \simeq \langle A_{a,\boldsymbol{q}}(t) A_{a,\boldsymbol{w}}\rangle \left\langle \langle n_{-\boldsymbol{q}}(t) n_{-\boldsymbol{w}}\rangle_{el}\right\rangle_{sc} \,. \tag{3.53}$$

Below we shall omit the brackets $\langle\,\rangle_{sc}$ implying that this sort of averaging is included in the definition of the average $\langle\,\rangle$ of the electron DSF $S(\boldsymbol{q},\omega)$.

For elastic scattering ($\Delta_{\boldsymbol{K},\boldsymbol{K}'} = 0$) induced by the short-range electron–impurity interaction, (3.52) and the spectral representation of (3.50) yield

$$G^{(R)}_{F_x,F_x}(\omega) = \frac{\hbar^3 \nu_0^{(a)}}{m_e S_A} \sum_{\boldsymbol{q}} q_x^2 G^{(R)}_{n_{\boldsymbol{q}},n^{\dagger}_{\boldsymbol{q}}}(\omega) \,. \tag{3.54}$$

Then according to the fluctuation–dissipation theorem, the effective collision frequency $\nu(\omega) = \mathrm{Im}\left[M_F(\omega)\right]$ of the memory function approach can be written as

$$\nu(\omega) = \left(1 - \mathrm{e}^{-\hbar\omega/T}\right) \frac{\hbar^2 \nu_0^{(a)}}{4 m_e^2 \omega S_A} \sum_{\boldsymbol{q}} q^2 S_0(q,\omega) \,. \tag{3.55}$$

Remarkably, in the zero frequency limit $\omega \to 0$, this equation agrees with the result of the force-balance method (3.20) found for highly correlated electrons. It therefore extends the result of (3.20), making it applicable to the AC case as well. In the high-frequency regime $\hbar\omega \gg T$, the effective collision frequency deviates from the usual semi-classical result, acquiring a strong frequency dependence, even if $S_0(q,\omega)$ can be approximated by $S_0(q,0)$. This distinguishes the electron conductivity of the quantum theory from the semi-classical result.

At $\omega \approx \omega_c$, the quantity $\gamma_{CR}(\omega) = 2\nu(\omega)$ describes the CR linewidth. When $\omega \to \omega_c$ the denominator $\omega - \omega_c$ entering the auxiliary function $M_-(\omega)$

of (3.46) leads to the same problem as that discussed in the case $\omega = 0$ and $B = 0$. Therefore, when considering the CR absorption from highly correlated electrons, we shall use the approximation $M(\omega) = M_F(\omega)$, disregarding the nonphysical high order term in M_-. The same approximation was used in [135] describing the quantum CR absorption in a semiconductor inversion layer. Additionally, we should remember that, at $\omega \approx \omega_c$, the DSF of 2D electrons under a magnetic field, and consequently $\nu(\omega)$, have resonance structures themselves. The validity of the above approximation for the conductivity relaxation kernel will be checked in Chap. 5, where we compare the different approaches introduced to describe the CR absorption from 2D electron systems.

For the electron–ripplon scattering regime, the force–force correlation function can be found in a similar way [148, 149]:

$$\langle F_\alpha(t) F_\beta(0) \rangle_\omega = n_s \sum_{\boldsymbol{q}} q_\alpha q_\beta V_q^2 Q_q^2 \times \left[(N_q^{(r)} + 1) S(-\boldsymbol{q}, \omega - \omega_q) + N_q^{(r)} S(\boldsymbol{q}, \omega + \omega_q) \right] . \tag{3.56}$$

Using this form and $S(\boldsymbol{q}, \omega) = S_0(q, \omega)$, the effective collision frequency $\nu(\omega) = \mathrm{Im} M_F(\omega)$ can be found as

$$\nu(\omega) = \frac{1 - \mathrm{e}^{-\hbar\omega/T}}{4\hbar\omega m_e S_A} \sum_{\boldsymbol{q}} q^2 V_q^2 Q_q^2 \left[(N_q^{(r)} + 1) S_0(q, \omega - \omega_q) + N_q^{(r)} S_0(q, \omega + \omega_q) \right] . \tag{3.57}$$

The two terms in the brackets of this equation represent contributions from one-ripplon creation and destruction processes. We note that, in the zero frequency limit, (3.57) reproduces the effective collision frequency of the DC case found in the force–balance method [see (3.15)], because

$$(N_q^{(r)} + 1) S_0(q, -\omega_q) = N_q^{(r)} S_0(q, \omega_q) .$$

In the following, we shall omit the subscript 0 for the electron DSF, implying that $S(q, \omega)$ is an equilibrium DSF if it does not depend on the direction of the vector $\boldsymbol{q}$. At typical frequencies of the CR experiment ($\omega \gg \omega_q$), one can ignore the ripplon frequency ω_q in the argument of the DSF and find a direct and simple relation between $\nu(\omega)$ and $S(q, \omega)$, viz.,

$$\nu(\omega) = \frac{1 - \mathrm{e}^{-\hbar\omega/T}}{4\hbar\omega m_e S_A} \sum_{\boldsymbol{q}} q^2 V_q^2 Q_q^2 \left[2N_q^{(r)} + 1 \right] S(q, \omega) , \tag{3.58}$$

which can be used to study the CR absorption from both the 2D electron liquid and the Wigner solid.

We are now ready to formulate the quantum transport framework for practical evaluations of the electron conductivity in a strongly correlated system. Firstly, we adopt the general structure of the conductivity tensor of (3.26) consistent with the phenomenological analysis. The relaxation kernel of this equation $M(\omega) = w(\omega) + i\nu(\omega)$ determines the effective collision frequency $\nu = \mathrm{Im}M$ and the frequency shift $w = \mathrm{Re}M$. For highly correlated electrons, the relaxation kernel can be expressed in terms of the force–force correlation function $G_{F_x,F_x}(\omega) \equiv \langle\langle F_x|F_x\rangle\rangle_\omega$, using the approximation $M(\omega) = M_F(\omega)$, where $M_F(\omega)$ is given in (3.43).

For the two kinds of scatterers usually considered in solid state physics (impurities and phonons), the frictional force $\boldsymbol{F}$ can be quite generally expressed as the linear form of the electron density fluctuation operator $n_{\boldsymbol{q}}$ [see (3.49)]. Therefore the effective collision frequency can be found in terms of the equilibrium dynamical structure factor $S(q,\omega)$ of the electron system, as illustrated in (3.55) and (3.57). The frequency shift w can be obtained from the fluctuation–dissipation theorem (3.50). Thus, in the framework considered here, the description of quantum transport is reduced to finding the right approximation for the equilibrium electron DSF $S(q,\omega)$, which is very convenient for systems with a complicated electron excitation spectrum. Once the DSF is found, the quantum transport properties of the system can be obtained by taking certain integrals over the momentum exchange $\boldsymbol{q}$. This transport framework will be used frequently throughout this book, especially in Chaps. 4 and 5 which deal with the description of quantum magnetotransport and CR absorption in 2D Coulomb liquids.

3.5 Comparison with the Kinetic Equation Method

In the last section, we found a remarkable affinity between the high-frequency approximation ($\omega \gg \nu$) for the relaxation kernel in the memory function formalism and the approximation of strongly correlated electrons ($\nu_{\mathrm{e-e}} \gg \nu$) for the DC collision frequency in the force-balance method. It is instructive to track down this affinity in the framework of the conventional kinetic equation method for particular scattering mechanisms.

Let us consider quasi-classically a 2D system of electrons with the simple energy spectrum $\varepsilon_k = \hbar^2 k^2/2m_{\mathrm{e}}$. In order to obtain the current density $\boldsymbol{j} = en_{\mathrm{s}}\sum_{\boldsymbol{k}} \boldsymbol{v} f(\boldsymbol{k})$, we have to evaluate the deviation of the electron distribution function $\delta f(\boldsymbol{k}) = f(\boldsymbol{k}) - f_0(\varepsilon_k)$ from its equilibrium form $f_0(\varepsilon_k)$ caused by the driving electric field $\boldsymbol{E}_\parallel$. The kinetic equation for the electron distribution function $f(\boldsymbol{k})$ is usually written as a balance equation for changes caused by the field and changes induced by collisions:

$$\frac{\partial f}{\partial t} - e\boldsymbol{E}_\parallel \frac{1}{\hbar}\frac{\partial f}{\partial \boldsymbol{k}} = \hat{I}_{\mathrm{es}}\{f\} + \hat{I}_{\mathrm{ee}}\{f\}\,, \tag{3.59}$$

where $\hat{I}_{\mathrm{ee}}\{f\}$ is the collision integral for electron–electron scattering and $\hat{I}_{\mathrm{es}}\{f\}$ is the collision integral for the interaction with scatterers. If the electron–electron collision rate is low ($\hat{I}_{\mathrm{ee}}\{f\} \to 0$) and the static ($\omega = 0$) electric field is weak, then $\delta f(\boldsymbol{k})$ can be found from (3.59), by using the relation

$\hat{I}_{\text{es}}\{f\} = -\nu(\varepsilon_k)\delta f$, where $\nu(\varepsilon_k)$ is the collision frequency with scatterers:

$$\delta f(\boldsymbol{k}) \simeq (\hbar \boldsymbol{k} \cdot \boldsymbol{E}_{\parallel}) \frac{e}{m_{\text{e}}\nu(\varepsilon_k)} \frac{\partial f_0}{\partial \varepsilon_k} . \tag{3.60}$$

This linear correction to the equilibrium distribution function is responsible for the appearance of the electric current.

For example, according to (1.101), the electron–ripplon collision frequency as a function of energy $\nu(\varepsilon_k)$ can be written in the form

$$\nu(\varepsilon_k) \simeq \frac{T}{4\alpha\hbar\varepsilon_k} \left\langle V_q^2 \right\rangle_{\varphi} ,$$

where φ is the scattering angle. Then, averaging the electron velocity over the new distribution [see (3.60)], the conductivity of SEs can be found in the form [27, 43]

$$\sigma = \frac{e^2 n_{\text{s}}}{m\nu_{\text{eff}}^{(\text{se})}} , \tag{3.61}$$

where

$$\frac{1}{\nu_{\text{eff}}^{(\text{se})}} = \int_0^{\infty} \frac{x}{\nu(Tx)} \mathrm{e}^{-x} \mathrm{d}x . \tag{3.62}$$

In the limit of strong holding fields, $V_q \simeq eE_{\perp} = \text{const.}$ and therefore the collision frequency has the strong energy dependence

$$\nu(\varepsilon_k) \simeq \frac{(eE_{\perp})^2 T}{4\alpha\hbar\varepsilon_k} \propto \varepsilon_k^{-1} , \tag{3.63}$$

and the effective collision frequency defined above, viz.,

$$\nu_{\text{eff}}^{(\text{se})} \simeq \frac{(eE_{\perp})^2}{8\alpha\hbar} , \tag{3.64}$$

is reduced by the additional factor 2 because of the averaging given in (3.62).

If the electron–electron collision rate is high $\nu_{\text{e–e}} \gg \nu_{\text{eff}}$, one can conclude that the shape of the electron distribution function is mainly formed by electron–electron collisions. In the first approximation, it can be found as a solution of the equation [150]

$$\hat{I}_{\text{ee}}\{f\} = 0 . \tag{3.65}$$

The general solution of this equation is a shifted Fermi function:

$$f = \frac{1}{\exp\left[(\varepsilon_k - \hbar \boldsymbol{k} \cdot \boldsymbol{v}_{\text{av}} - \mu_F)/T_{\text{e}}\right] + 1} , \tag{3.66}$$

where the parameters T_{e} and $\boldsymbol{v}_{\text{av}}$ are to be found from the energy and momentum balance equations. This regime of electron transport is usually called the complete control regime, which means that the redistribution of the momentum and energy within the electron layer is governed by electron–electron collisions. If the energy relaxation rate $\tilde{\nu}$ is much lower than the momentum relaxation rate ν, there is also a partial control regime, in which electron–electron collisions govern only the redistribution of energy: $\tilde{\nu} \ll \nu_{\text{e–e}} \ll \nu$.

In the linear approximation $T_{\text{e}} = T$, we have

$$f = f_0(\varepsilon_k - \hbar \boldsymbol{k} \cdot \boldsymbol{v}_{\text{av}}) \simeq f_0(\varepsilon_k) - (\hbar \boldsymbol{k} \cdot \boldsymbol{v}_{\text{av}}) \frac{\partial f_0}{\partial \varepsilon_k} . \tag{3.67}$$

The vector parameter $\boldsymbol{v}_{\text{av}}$ is the drift velocity of the whole electron system caused by the driving electric field, because $\langle \boldsymbol{v} \rangle = \boldsymbol{v}_{\text{av}}$. Using the relation $-en_{\text{s}}\boldsymbol{v}_{\text{av}} = (e^2/m\nu_{\text{eff}}^{(\text{me})})\boldsymbol{E}_{\parallel}$, the deviation of the distribution function can be written as

$$\delta f(\boldsymbol{k}) \simeq (\hbar \boldsymbol{k} \cdot \boldsymbol{E}_{\parallel}) \frac{e}{m_{\text{e}}\nu_{\text{eff}}^{(\text{me})}} \frac{\partial f_0}{\partial \varepsilon_k} , \tag{3.68}$$

which is similar to (3.60). The important difference is that the average quantity $\nu_{\text{eff}}^{(\text{me})}$ (the effective collision frequency) enters the distribution function instead of the collision frequency $\nu(\varepsilon_k)$. This quantity does not depend on the electron energy, by definition. The electron–electron collisions thus affect the shape of the distribution function, in particular its asymmetric correction induced by the driving electric field.

In order to obtain $\boldsymbol{v}_{\text{av}}$ or equally $\nu_{\text{eff}}^{(\text{me})}$ for highly correlated electron systems, one can multiply (3.59) by $\hbar\boldsymbol{k}$ and perform the summation over all $\boldsymbol{k}$. Because the electron–electron collisions preserve the total momentum, the term containing $\hat{I}_{\text{ee}}\{f\}$ does not contribute to the momentum balance equation

$$e\boldsymbol{E}_{\parallel} = \sum_{\boldsymbol{k}} \hbar \boldsymbol{k} \hat{I}_{\text{er}}\{f\} . \tag{3.69}$$

Using the relation $\hat{I}_{\text{er}}\{f\} = -\nu(\varepsilon_k)\delta f$ and (3.68), we find the new equation for the effective collision frequency of the nondegenerate electron gas:

$$\nu_{\text{eff}}^{(\text{me})} = \sum_{\boldsymbol{k}} \frac{\varepsilon_k}{T} \nu(\varepsilon_k) f_0(\varepsilon_k) = \int_0^{\infty} x\nu(Tx)\text{e}^{-x}\text{d}x . \tag{3.70}$$

It it important to note that this equation differs from the analogous equation found by means of the single-electron approximation [see (3.62)]. The difference consists in the way the thermal averaging is done: in the single-electron theory, we have to average the relaxation time $1/\nu(\varepsilon_k)$, while for highly correlated electrons it is necessary to average the collision frequency.

It is clear that the two extreme regimes discussed above yield the same effective collision frequency $\nu_{\text{eff}}^{(\text{se})} = \nu_{\text{eff}}^{(\text{me})}$ only if the collision frequency ν determined by the collision integral $\hat{I}_{\text{er}}\{f\}$ does not depend on the electron energy. For electron–ripplon scattering, the holding field term $eE_{\perp}$ brings a strong dependence of the collision frequency on the electron energy, as discussed above. In the limit of strong fields ($V_q \simeq eE_\perp$), the collision frequency $\nu(\varepsilon_k) \propto \varepsilon_k^{-1}$ and, for highly correlated electrons, (3.70) yields

$$\nu_{\mathrm{eff}}^{(\mathrm{me})} = \frac{(eE_\perp)^2}{4\alpha\hbar}, \tag{3.71}$$

which is exactly two times larger than the asymptote of (3.64) found in the single-electron approximation. Remarkably, this factor of 2 for the mobility data of SEs on superfluid helium was reported in the plasmon resonance experiment of Grimes and Adams [47]. A detailed experimental study of the low-frequency conductivity of SEs under the complete control regime was conducted by Buntar' et al. [151, 152].

In contrast, the polarization term of the electron–ripplon coupling eE_q [see (1.70)] yields a collision frequency which has a very weak dependence on the electron energy. As a result, under the saturation condition $E_\perp = 2\pi e n_s$ and in the low electron density limit ($E_\perp \to 0$), both definitions of the effective collision frequency [see (3.62) and (3.70)] give numerically close results. With an increase in the electron density and $E_\perp$, the effective collision frequency term, proportional to $(eE_\perp)^2$, increases twice, causing a strong density dependence in the electron mobility.

The density dependent mobility of SEs on superfluid helium was observed by Mehrotra et al. [50]. At low densities, experimental data are in good numerical agreement with the theory of electron–ripplon scattering, as shown in Fig. 1.12 of Chap. 1. At high densities, the data shown in Fig. 3.1 deviate from the single-electron mobility (dashed curve). Still, only part of the observed increase in the effective collision rate can be explained by the result found for highly correlated electrons in the framework of the kinetic equation method (solid curve). As shown in [84], the Coulomb liquid corrections to the effective collision frequency are proportional to a rather small parameter $g_1/2k_T$, where $g_1 = 2\pi(2/\sqrt{3}^{1/2} n_s^{1/2})$ is the first reciprocal lattice vector of the 2D Wigner solid and $k_T = \sqrt{2m_e T}/\hbar$. At large values of the plasma parameter, this correction has the right sign, but its absolute value found in the local field approximation cannot explain the observed effect.

An additional increase in the effective collision frequency may also be caused by the zero-point corrections to the mean kinetic energy of electrons due to 2D plasmons or solid-like behavior. This affects the DSF factor of the electron liquid of (2.22) in such a way that the parameter T in it should be replaced by the mean kinetic energy of electrons. This coincides with the temperature only at $T \gg \hbar\Omega_{\mathrm{pl}}$, where Ω_{pl} is the typical plasmon frequency. As a result, the collision rate increases.

The transition from the single-electron behavior of (3.62) to the result of the complete control regime (3.70) was analyzed in [153]. According to this analysis, the effective collision frequency has the form

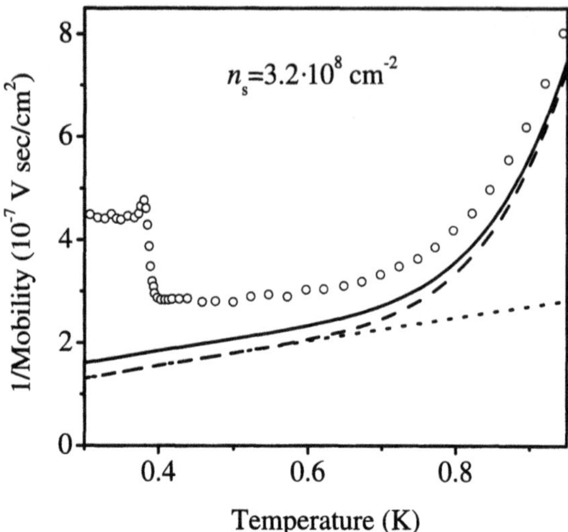

Fig. 3.1. Mobility of SEs along the surface of liquid ^{4}He for two extreme regimes with regard to electron–electron collisions: firstly, the single-electron approximation (*dashed curve*) and secondly, when the shape of the electron distribution function is controlled by electron–electron scattering (*solid curve*). The result of electron scattering by ripplons alone is shown by the *dotted line*. Data (*symbols*) are from Mehrotra et al. [50]

$$\nu_{\text{eff}} = \nu_{\text{eff}}^{(\text{me})} \frac{1 + \nu_{\text{eff}}^{(\text{se})}/\nu_{\text{e–e}}}{1 + \nu_{\text{eff}}^{(\text{me})}/\nu_{\text{e–e}}} , \tag{3.72}$$

which transforms into the asymptotic results of the extreme cases [$\nu_{\text{eff}}^{(\text{se})}$ and $\nu_{\text{eff}}^{(\text{me})}$] when the electron–electron collision rate $\nu_{\text{e–e}}$ changes from the regime $\nu_{\text{e–e}} \ll \nu_{\text{eff}}$ to the regime $\nu_{\text{e–e}} \gg \nu_{\text{eff}}$.

As noted above, the expressions for $\nu_{\text{eff}}^{(\text{se})}$ and $\nu_{\text{eff}}^{(\text{me})}$ give the same result only if the electron collision frequency does not depend on the electron energy. Electron scattering on short-range impurities or vapor atoms represents another important example of such behavior. Consider electron scattering on a single impurity (vapor atom). In two dimensions, the outgoing wave function of the electron scattered by an atom with a fixed position ($\boldsymbol{r} = 0$) and the Green's function of the Schrödinger equation are proportional to the Hankel function $H_0(kr)$, which has the asymptotic behavior

$$H_0(kr) \simeq \sqrt{\frac{2}{\pi kr}} \exp\left[\mathrm{i}\left(kr - \pi/4\right)\right] \tag{3.73}$$

at large distances. Therefore the 2D 'cross-section' for electron scattering by a single atom, viz.,

$$\sigma_{\text{cs}} = \frac{4\pi^2 a_0^2}{k} f_1^4(z_{\text{a}}) , \tag{3.74}$$

is inversely proportional to the electron velocity $\hbar k/m$. Here $f_1(z)$ is the wave function of the ground SE state, which determines the dependence of the 'cross-section' on the vertical position of the scatterer. The collision frequency

$$\nu = \frac{\hbar k}{m} \int \mathrm{d}z_{\mathrm{a}} n_{3\mathrm{D}}^{(\mathrm{a})} \sigma_{\mathrm{cs}}(z_{\mathrm{a}}) = \frac{3m_{\mathrm{e}}(V^{(\mathrm{a})})^2 n_{3\mathrm{D}}^{(\mathrm{a})}\gamma}{8\hbar^3} \equiv \nu_0^{(\mathrm{a})} \tag{3.75}$$

does not depend on k.

The same result can be found in a more rigorous way by evaluating

$$\nu(\varepsilon_{\boldsymbol{k}}) = \frac{2\pi}{\hbar} \sum_{\boldsymbol{k}} (1 - \cos\theta) \left|\langle \boldsymbol{k}| V_{\mathrm{int}} |\boldsymbol{k}'\rangle\right|^2 \delta\left(\varepsilon_{\boldsymbol{k}'} - \varepsilon_{\boldsymbol{k}}\right) , \tag{3.76}$$

where θ is the angle between $\boldsymbol{k}$ and $\boldsymbol{k}'$, which is valid for elastic scattering. Using $V_{\mathrm{int}} = V^{(\mathrm{a})} \sum_a \delta(\boldsymbol{R} - \boldsymbol{R}_a)$ and disregarding the interference effects, one can easily arrive at $\nu(\varepsilon_{\boldsymbol{k}}) = \nu_0^{(\mathrm{a})}$.

We are now ready to compare the conductivity results found for SEs by means of the conventional kinetic equation method and the universal quantum transport framework discussed in the preceding sections. In the latter approach, the effective collision frequency of SEs is completely determined by the equilibrium electron DSF $S_0(q,\omega)$ entering (3.55) and (3.57). The single-electron form of $S_0(q,\omega)$ [see (2.22)] represents the simplest approximation. For the electron–ripplon scattering regime, the typical energy of ripplons $\hbar\omega_q$ involved in scattering events ($q \leq 2k$) is much lower than the thermal energy, and one can disregard ω_q in the frequency argument of $S(q,\omega)$. In the DC case ($\omega = 0$), substituting the single-electron approximation for the DSF into the equation for the effective collision frequency [see (3.57)] yields

$$\nu_{\mathrm{eff}} = \frac{1}{2\sqrt{2\pi m_{\mathrm{e}} T} T} \int_0^\infty q^2 \left|V_q^{(\mathrm{r})}\right|^2 Q_q^2 N_q^{(\mathrm{r})} \exp\left(-\frac{\hbar^2 q^2}{8 m_{\mathrm{e}} T}\right) \mathrm{d}q . \tag{3.77}$$

As noted above, the same equation can be found using the result (3.15) of the force-balance method. At first glance, (3.77) differs from both the kinetic equation results (3.62) found for an ideal electron gas and for highly correlated electrons (3.70). Still, it is possible to prove that, changing the order of integrals in the definition (3.70) [assuming that $\nu(\varepsilon_{\boldsymbol{k}})$ has the integral form (1.101)] and evaluating one of the integrals analytically, the result found for highly correlated electrons in the kinetic equation method can be transformed into the form (3.77).

We thus find that, in the simplest approximation for the electron DSF, the memory function formalism with relaxation kernel $M(\omega) = M_F(\omega)$ yields the DC conductivity result that corresponds to the result of the kinetic equation method which is valid only for highly correlated electrons ($\nu_{\mathrm{e\text{-}e}} \gg \nu_{\mathrm{er}} \equiv \nu_{\mathrm{eff}}$). For the force-balance method, this could be anticipated from the very beginning, because of the condition (3.13), which assumes that the Coulomb interaction of the electrons is strong enough. The comparison made above shows that the memory function formalism with the relaxation kernel $M(\omega)$ found from the high frequency condition $[M(\omega) = M_F(\omega)]$ can give numerically incorrect DC conductivity results if the electron–electron collision rate $\nu_{\text{e–e}}$ is low and the electron collision frequency $\nu(\varepsilon_k)$ depends strongly on the electron energy. In this case, the memory function approach can only provide us with qualitative results, valid up to a numerical proportionality factor of the order of 2. At the same time, for highly correlated electrons ($\nu_{\text{e–e}} \gg \nu_{\text{eff}}$), the memory function formalism provides us with numerically accurate DC conductivity results.

The origin of the remarkable affinity between the high-frequency approximation for the memory function proposed in [133] and the approximation of highly correlated electrons used in the force-balance method for DC transport can be explained by means of the kinetic equation method applied to the AC case: $\boldsymbol{E}_{\|}(t) = \boldsymbol{E}_{\|}^{(0)} \cos(\omega t)$. For an ideal electron gas, the oscillating correction to the electron distribution function can be written as

$$\delta f(\boldsymbol{k}, t) \simeq (\hbar \boldsymbol{k} \cdot \boldsymbol{E}_{\|}^{(0)}) \frac{e}{m_{\text{e}}[\nu^2(\varepsilon_k) + \omega^2]} \frac{\partial f_0}{\partial \varepsilon_k} \left[\nu(\varepsilon_k) \cos(\omega t) + \omega \sin(\omega t) \right] . \tag{3.78}$$

From this equation, one can see that, in the high frequency limit $\omega \gg \nu$, the term of the distribution function oscillating in phase with the driving field is proportional to $\nu(\varepsilon_k)/\omega^2$. After averaging, this ratio transforms into $\nu_{\text{eff}}/\omega^2$, where ν_{eff} coincides with the definition of $\nu_{\text{eff}}^{(\text{me})}$ given in (3.70) for highly correlated electrons.

The high-frequency condition changes the way of averaging the collision frequency $\nu(\varepsilon_k)$. [Formally, it moves $\nu(\varepsilon_k)$ from the denominator to the numerator.] This in turn affects the numerical proportionality factor of the effective collision frequency, as noted in [154, 155]. Extrapolating the high-frequency approximation for ν_{eff} into the low frequency range in the framework of both the kinetic equation method and the memory function formalism generally gives an incorrect numerical proportionality factor of the order of 2. For highly correlated electrons, the corresponding correction to the distribution function $\delta f(\boldsymbol{k}, t)$ has the form (3.78) with $\nu(\varepsilon_k)$ replaced by $\nu_{\text{eff}}^{(\text{me})}$. Therefore, the high frequency form of the effective collision frequency ν_{eff} gives a conductivity result applicable for highly correlated electrons in the whole frequency range. This explains the remarkable accuracy of the high-frequency approximation for the memory function in the whole frequency range in the case of highly correlated electrons.

A high magnetic field applied normally to the system also changes the way of averaging the collision frequency $\nu(\varepsilon_k)$ in the semi-classical treatment if $\omega_{\text{c}} \gg \nu$. In this case, the effective collision frequency ν_{eff} of the single-electron approximation coincides with that found for highly correlated electrons $\nu_{\text{eff}}^{(\text{me})}$. Incidentally, this is the regime where the memory function formalism gives an accurate result even for the DC case. The important point is that, for highly correlated electrons, the effective collision frequency remains the same for any relationship between ω_c and ν, while for an ideal gas it is valid only in the high-cyclotron-frequency range.

3.6 Energy Relaxation Rate

As mentioned above, the wave numbers of surface excitations of liquid helium taking part in scattering events of SEs are limited by the electron wave numbers $q \leq 2k$. The thermal wave numbers of electrons and ripplons have different scales: at $T \sim 1\,\mathrm{K}$, we estimate $k_T = \sqrt{2Tm_e}/\hbar \sim 5 \times 10^5\,\mathrm{cm}^{-1}$ and $q_T = (\rho/\alpha)^{1/3}(T/\hbar)^{2/3} \sim 2 \times 10^7\,\mathrm{cm}^{-1}$. Therefore thermal ripplons cannot be emitted or absorbed by an electron with $k \sim k_T$. In other words, momentum and energy cannot both be conserved at the same time for such processes. As a result, the energies of ripplons giving the main contribution to the momentum relaxation rate ν are very low ($\hbar\omega_q \sim 10^{-2}\,\mathrm{K}$ at $T \sim 1\,\mathrm{K}$), and energy exchange between the electron layer and the liquid helium is suppressed. This means that the SEs can easily be overheated by the driving electric field.

For SEs on liquid helium, electron–electron correlations are usually very strong and the electron system is at least under the conditions of the partial control regime: $\nu_{e\text{-}e} \gg \tilde{\nu}$. In this case, the symmetrical part of the distribution function $f_0(\varepsilon_k)$ has the usual equilibrium shape, although the effective electron temperature T_e entering this function can be substantially higher than T. The energy loss of the electron system per unit time due to electron–ripplon scattering can be written as

$$\dot{\mathcal{E}}(T_e) = -N_e \tilde{\nu}_{\mathrm{eff}}(T_e)\,(T_e - T)\ , \tag{3.79}$$

where

$$\tilde{\nu}_{\mathrm{eff}}(T_e) = \int_0^\infty x\tilde{\nu}(xT_e)\mathrm{e}^{-x}\mathrm{d}x\ , \tag{3.80}$$

and $\tilde{\nu}(\varepsilon_k)$ is the energy relaxation rate as a function of electron energy. For one-ripplon scattering events, $\tilde{\nu}(\varepsilon_k)$ is determined as [27]

$$\tilde{\nu}(\varepsilon_k) = \frac{m_e}{2\pi\hbar^3}\int_0^{2\pi}\mathrm{d}\varphi\frac{(\hbar\omega_q)^2}{2T\varepsilon_k}\frac{\hbar V_q^2}{2\mu_q\omega_q}(2N_q+1)\ , \tag{3.81}$$

where $q = 2k\sin(\varphi/2)$. As compared to the momentum collision frequency ν defined in (1.101), the energy relaxation rate contains an additional small parameter $(\hbar\omega_q)^2/2T\varepsilon_k \ll 1$. Therefore, for one-ripplon processes, $\tilde{\nu}_{\mathrm{eff}}$ is approximately four orders of magnitude smaller than ν_{eff}. At $T \sim 1\,\mathrm{K}$ and $E_\perp \to 0$, we estimate $\tilde{\nu}_{\mathrm{eff}} \sim 10^4\,\mathrm{s}^{-1}$.

The electron energy relaxation rate can be significantly faster for scattering processes involving two ripplons [44]. For example, consider the energy conservation for scattering events in which an electron emits a couple of short-wavelength ripplons with total momentum $\boldsymbol{s} = \boldsymbol{q} + \boldsymbol{q}'$. By momentum conservation $\boldsymbol{k}' = \boldsymbol{k} - \boldsymbol{s}$, it can be rewritten in the form

$$2\sqrt{\varepsilon_k \varepsilon_s} \cos\theta - \varepsilon_s = \hbar\omega_{\boldsymbol{q}} + \hbar\omega_{\boldsymbol{q}-\boldsymbol{s}} \,, \tag{3.82}$$

where θ is the angle between $\boldsymbol{k}$ and $\boldsymbol{s}$. It is clear that this condition can be fulfilled for $\hbar\omega_q \sim \varepsilon_k$ if $s \sim k \ll q$. This means that an electron can emit pairs of short-wavelength surface excitations propagating in opposite directions: $\boldsymbol{q}' \simeq -\boldsymbol{q}$. The important point is that the energies of these ripplons and the electron energy are of the same order of magnitude. Therefore the energy relaxation rate caused by two-ripplon processes can be faster in spite of the smaller probability.

Thermal ripplons usually have large wave numbers: $q \gg 1/\langle z \rangle$. Therefore one has to find out which part of the interaction Hamiltonian is responsible for these two-phonon processes. Naturally, it can be the same term of V_{int} (the linear term in the surface displacement operator ξ_q) used in the next order of the perturbation treatment. Still, the bound electrons represent a special case. Firstly, we note that, for nonlinear interaction terms proportional to ξ^2, two-phonon processes can appear in the usual order of the perturbation treatment. According to the results discussed in Sect. 1.5.2 for SEs on liquid helium, the most important nonlinear term in the interaction Hamiltonian $V_{\text{int}}^{(2)}$ is proportional to the second derivative of the potential energy above the flat surface $V_{\text{e}}(z)$ [see (1.76)]. Because the 1D potential well is very sharp in the vicinity of the interface, one can expect this term to give the major contribution to the probability of emission of the short-wavelength ripplons by the SEs. This assumption is confirmed by more detailed analysis.

As discussed in Sect. 1.5.2, when evaluating the second derivative of the electron potential, one should take into account the fact that the 1D potential well is highly asymmetric and that the contribution from the repulsive part dominates:

$$V_{\text{int}}^{(2)} \simeq V_0 \chi_1(0) \chi_1'(0) \xi^2 = \kappa_0 V_0 \chi_1^2(0) \xi^2 \equiv W_0 \xi^2 \,, \tag{3.83}$$

where $\chi_1(0)$ is the SE wave function at the surface $z = 0$ of the liquid helium and κ_0^{-1} is the penetration length for the electron wave function inside the liquid helium, according to (1.8). We have used the relation $\chi_1(0) = \kappa_0^{-1} \chi_1'(0)$. The value $\kappa_0 \simeq 5.1 \times 10^7\ \text{cm}^{-1}$ is nearly two orders of magnitude larger than γ. Therefore, the interaction potential of (3.83), which is approximately proportional to $|\varepsilon_1^{(\perp)}| 4(\xi/\langle z \rangle)^2 \kappa_0/\gamma$, is larger than the estimate accounting for the polarization attraction term by the factor $\kappa_0/\gamma \sim 10^2$.

In the adiabatic theory of electron–ripplon scattering discussed in Sect. 1.5.3, the electron wave function was chosen to be the shifted function $\chi_l(z - \xi) \exp(i\boldsymbol{k} \cdot \boldsymbol{r})$. In this case, the perturbations induced by the repulsive step function of the electron potential above the uneven surface are included in the adiabatic wave function and they do not cause any scattering. The scattering Hamiltonian for this approach has a nonlinear term proportional to ξ_q^2, which appears because of the kinetic energy perturbation

$$\delta K_{\mathrm{e}}^{(2)} = \frac{p_z^2}{2m_{\mathrm{e}}} (\nabla \xi)^2 . \tag{3.84}$$

This term has a strong q-dependence. A simple estimate indicates that, at $q > 3 \times 10^7\,\mathrm{cm}^{-1}$, the perturbation term of (3.84) is approximately one order of magnitude larger than the corresponding interaction term found with the modified Bloch method [see (3.83)]. Physically, the adiabatic approximation should not cause larger scattering because the adiabatic wave function adjusts to the interaction potential. Therefore, the estimate found above means that the shifted wave function is a bad approximation for the adiabatic electron wave function if $q > 3 \times 10^7\,\mathrm{cm}^{-1}$, which could be anticipated from the beginning. In contrast, for long-wavelength ripplons, the approximation of (3.83) fails and one should use the adiabatic treatment to describe two-ripplon processes.

The energy transfer from the electron system to the liquid helium per unit time can be found using a relation similar to (3.10):

$$\dot{\mathcal{E}}_{\mathrm{e}} = -\frac{2\pi}{\hbar} \left\langle \sum_{\boldsymbol{n}',\boldsymbol{j}'} \left(E_{\boldsymbol{n}'}^{(\mathrm{r})} - E_{\boldsymbol{n}}^{(\mathrm{r})} \right) \left| \langle \boldsymbol{n}', \boldsymbol{j}' | V_{\mathrm{int}} | \boldsymbol{n}, \boldsymbol{j} \rangle \right|^2 \delta \left(\mathcal{E}_{\boldsymbol{n}',\boldsymbol{j}'} - \mathcal{E}_{\boldsymbol{n},\boldsymbol{j}} \right) \right\rangle , \tag{3.85}$$

where $E_{\boldsymbol{n}}^{(\mathrm{r})}$ is the energy of the phonon system in the many-body state $\boldsymbol{n}$ and $\mathcal{E}_{\boldsymbol{n},\boldsymbol{j}}$ is the energy of the whole system. Substituting here the interaction Hamiltonian of (3.83) and using the condition $s \ll q$, the energy transfer rate can be found in the form [84]

$$\dot{\mathcal{E}}_{\mathrm{e}} = -\frac{N_{\mathrm{e}} \hbar W_0^2}{4\pi^2 \rho^2} \int_0^\infty \mathrm{d}q \frac{q^3}{\omega_q} (N_q + 1)^2 \left(\mathrm{e}^{-2\hbar\omega_q/T_{\mathrm{e}}} - \mathrm{e}^{2\hbar\omega_q/T} \right) \int_0^\infty s S(s, 2\omega_q) \mathrm{d}s . \tag{3.86}$$

In spite of the fact that the ripplons involved belong to the short-wavelength range, the electron DSF corresponds to the long-wavelength range ($s \ll q$). This is because the nonlinear perturbation $\xi^2(\boldsymbol{r})$ becomes effectively of long-wavelength nature at $\boldsymbol{q}' \to -\boldsymbol{q}$.

If electrons can be described by the nondegenerate gas (NG) form of the DSF [see (2.22)], the energy transfer rate can be simplified to

$$\dot{\mathcal{E}}_{\mathrm{e}} = -\frac{N_{\mathrm{e}} W_0^2 m_{\mathrm{e}}}{2\pi \rho^2} \int_0^\infty \frac{q^3}{\omega_q} (N_q + 1)^2 \left(\mathrm{e}^{-2\hbar\omega_q/T_{\mathrm{e}}} - \mathrm{e}^{2\hbar\omega_q/T} \right) \mathrm{d}q . \tag{3.87}$$

Here we have taken into account the fact that the integral

$$\int_0^\infty \exp\left[-\left(x - a/x \right)^2 \right] \mathrm{d}x = \sqrt{\pi}/2 \tag{3.88}$$

does not depend on the parameter a.

In the short-wavelength range, the ripplon spectrum may deviate substantially from the hydrodynamic asymptotic behavior $\omega_q = \sqrt{\alpha/\rho} q^{3/2}$. If it remains the same for large q, then straightforward evaluation of the integral over q yields

$$\dot{\mathcal{E}}_{\mathrm{e}} = -N_{\mathrm{e}} \tilde{\nu}_{\mathrm{hot}}(T_{\mathrm{e}}) T_{\mathrm{e}} \Phi(T_{\mathrm{e}}/T) , \qquad \tilde{\nu}_{\mathrm{hot}}(T_{\mathrm{e}}) = \frac{\Gamma(5/3) W_0^2 m_{\mathrm{e}} T_{\mathrm{e}}^{2/3}}{2^{2/3} 6\pi \rho^{2/3} \alpha^{4/3} \hbar^{5/3}} , \tag{3.89}$$

where we have introduced the energy transfer function

$$\Phi(x) = \frac{1}{\Gamma(5/3)} \int_0^\infty y^{2/3} \frac{(\mathrm{e}^{-y} - \mathrm{e}^{-xy})}{(1 - \mathrm{e}^{-xy/2})^2} \mathrm{d}y . \tag{3.90}$$

The numerical graph of this function is shown in Fig. 3.2. At large x, it has the simple asymptotic behavior $\Phi(x) \to 1$, which states that at $T_{\mathrm{e}} \gg T$ the energy transfer rate is proportional to $T_{\mathrm{e}} \tilde{\nu}_{\mathrm{hot}}(T_{\mathrm{e}}) \propto T_{\mathrm{e}}^{5/3}$. In the opposite limit $T_{\mathrm{e}} - T \ll T$, we have $\Phi(x) \to 0$, which means that there is no energy transfer at $T_{\mathrm{e}} = T$. The accurate analysis of this asymptotic behavior gives

$$\frac{\dot{\mathcal{E}}_{\mathrm{e}}}{N_{\mathrm{e}}} = -A \tilde{\nu}_{\mathrm{hot}}(T_{\mathrm{e}})(T_{\mathrm{e}} - T) , \qquad A = 2^{8/3} \frac{5}{3} \left[\zeta(5/3) - \zeta(8/3)\right] \simeq 8.88 . \tag{3.91}$$

Therefore the warm electrons ($T_{\mathrm{e}} - T \ll T$) have an additional numerical factor A, increasing their energy relaxation rate as compared to hot electrons.

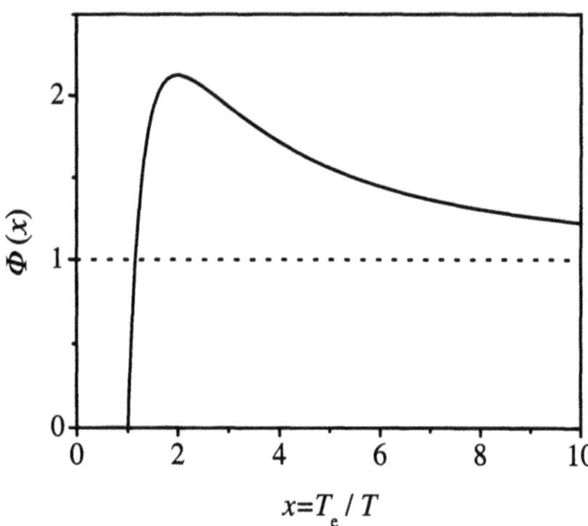

Fig. 3.2. Energy transfer function

For the interaction parameter W_0 defined in (3.83), simple estimates indicate that the energy relaxation rate caused by two-ripplon scattering processes is higher than that found for one-ripplon processes by approximately two orders of magnitude:

$$\tilde{\nu}_{\text{hot}}(0.5\,\text{K}) \sim 6 \times 10^5\,\text{s}^{-1}\,, \qquad \tilde{\nu}_{\text{warm}}(0.5\,\text{K}) \sim 5.3 \times 10^6\,\text{s}^{-1}\,. \tag{3.92}$$

This estimate agrees with numerous experiments in which SEs were heated by the driving electric field or MW radiation. We would like to emphasize that, unlike the linear interaction Hamiltonian, the interaction potential for short-wavelength ripplons of (3.83) only weakly depends on the holding electric field. The holding field dependence of $\chi_1(0)$ was analyzed in Sect. 1.2 and shown in Fig. 1.6.

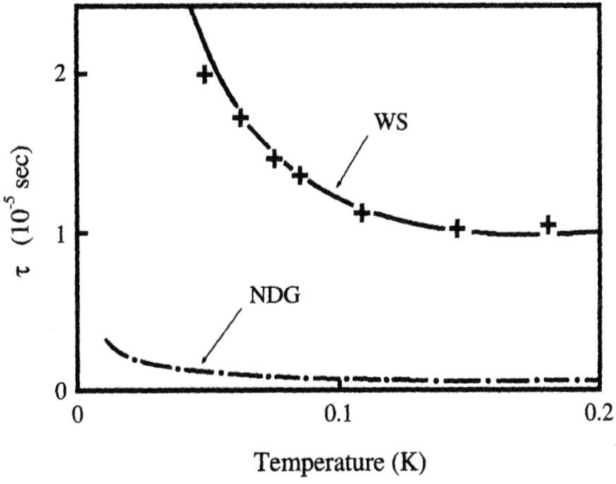

Fig. 3.3. Energy relaxation time of the 2D Wigner solid (WS) for $n_s = 1.02 \times 10^8\,\text{cm}^{-2}$: theory (*solid curve*) [157] and data (*crosses*) [158, 159]. The *dash-dotted curve* represents the calculation performed for the nondegenerate gas (NDG) of SEs

The energy relaxation rate decreases significantly if the electron system undergoes the Wigner solid transition. For this case, the analysis of (3.86) was given in [156, 157] and the results were compared with the WS energy relaxation time measured in [158, 159]. This comparison is shown in Fig. 3.3 displaying the energy relaxation time vs. temperature. The interaction parameters were chosen to be $V_0 = 1\,\text{eV}$ and $\chi_1(0) = 58.8\,\text{cm}^{-1/2}$. The solid curve calculated for the WS state of SEs is very close to the experimental data. At the same time, the dash-dotted curve evaluated for the nondegenerate gas state is far below the experimental data.

A high magnetic field applied normally to the system should strongly affect the energy relaxation rate of the 2D electron gas. In this case, the electron energy is confined to a very narrow range near the Landau level position $-\Gamma_0 \leq (\varepsilon - \varepsilon_0) \leq \Gamma_0$. Therefore, emitting a couple of short-wavelength ripplons can be inconsistent with the Landau level width, when $2\hbar\omega_q > \Gamma_0$. Energy transfer from the electron layer into the liquid helium caused by the two-ripplon processes is then suppressed by the quantizing magnetic field, because the Landau levels of SEs are usually extremely narrow $\Gamma_0 \ll T$.

4 Unconventional Hall Effect

4.1 Transport of Electrons with Discrete Energy Spectrum

A 2D electron system subject to a strong magnetic field applied normally to it represents a special case for the quantum transport theory. Contrary to the semi-classical picture which yields a single-electron density of states $\Delta_{2D}(\varepsilon) = m_e/\pi\hbar^2$ independent of energy, in the quantum mechanical treatment, the electron density of states is squeezed into the set of discrete Landau levels, as discussed in Sect. 1.9 [see (1.129)]. We thus have a unique situation: charge carriers are characterized by a purely discrete energy spectrum.

The singular nature of a 2D electron gas under a magnetic field causes many interesting phenomena including the quantum Hall effect (QHE). Instead of the universal linear increase with B, under the conditions of the QHE, the transverse resistance ρ_{yx} of the electron layer develops plateaux at values $\rho_{yx} = \rho_0/i$, where $\rho_0 = h/e^2$ is a simple combination of fundamental constants, and i is an integer [5] or simple fraction [6]. Important issues of the QHE are the localization caused by impurities [160] and edge effects [142, 143]. In the framework of the phenomenological analysis presented in Sect. 3.2, the appearance of such plateaux might be explained by the appropriate field dependence of the density $n_s(B)$ of mobile electrons compensating the linear increase B/ec, if the QHE is caused by interior electrons of the 2D electron liquid. The integer QHE model with a fixed chemical potential and field-dependent electron density was discussed by Yennie [161]. Later developments of the QHE theory revealed the importance of the current-carrying edge states [162, 163].

Regarding the QHE itself, there are a number of good books and reviews on this subject. Therefore we shall not go into the details of these aspects of quantum magnetotransport. Here we shall focus our attention on the quantum transport of interior electrons under conditions where localization effects are small and the influence of edges can be disregarded. We would like to emphasize that, besides the quantum plateaux of ρ_{xy} and minima of ρ_{xx} which are the essential features of the QHE observed in the degenerate electron layers of semiconductors, the quantization of the electron spectrum under a magnetic field introduces other important changes in the conventional picture of the Hall effect observed in nondegenerate interface electron systems. The extremely narrow Landau levels of SEs on superfluid helium cause many interesting quantum transport phenomena and this is the subject matter of this chapter.

Even in three dimensions, the quantization of the electron spectrum causes a divergence in the longitudinal conductivity component induced by elastic scattering [81, 164]. In this case, the conductivity formula contains an additional integral over p_z, and therefore the divergence is only logarithmic. Physically, the divergence appears because an electron encounters the same scatterer many times for small velocity component v_z in the direction of the magnetic field, and the duration of the collision becomes effectively infinite if $v_z \to 0$. In this picture, a 2D electron system characterized by $\langle 1|v_z|1\rangle = 0$ represents an extreme case. The origin of the divergence is the overlap of two level densities which appears because of energy conservation in elastic scattering. Therefore, the most important factors which can eliminate the divergence are improved level density and inelastic scattering.

Qualitatively, the magnetoconductivity of the 2D electron system can be described by the Einstein relation [78]

$$\sigma_{xx} = \frac{e^2 n_s D}{T} , \tag{4.1}$$

where D is the diffusion constant. The diffusion constant is usually estimated by $D \sim (\Delta X)^2/\tau_B$, where we have introduced the mean jumping distance ΔX and the lifetime τ_B under a magnetic field. The form of (4.1) does not show the conductivity divergence mentioned above: the conductivity vanishes as $\tau_B \to \infty$, which means that in the absence of scatterers the electrons just drift in the perpendicular direction. The important point is that the quantity τ_B differs from the lifetime under zero magnetic field τ_0. Because the electron states are squeezed into the extremely narrow Landau levels, the scattering rate is enhanced by a factor $\hbar\omega_c/\Gamma_0$:

$$\frac{1}{\tau_B} \simeq \frac{1}{\tau_0}\frac{\hbar\omega_c}{\Gamma_0} , \tag{4.2}$$

where Γ_0 is the Landau level broadening. Equation (4.2) shows the divergence of the collision rate and σ_{xx} for extremely narrow levels: $\sigma_{xx} \propto 1/\Gamma_0 \to \infty$ if $\Gamma_0 \to 0$.

The broadening parameter Γ_0 and the lifetime under a magnetic field are related by $\Gamma_0 \sim \hbar/\tau_B$. Together with (4.2), this relation represents a set of self-consistent equations whose solution can be written in the form

$$\frac{\hbar}{\tau_B} \sim \Gamma_0 \sim \Gamma_{se} \equiv \sqrt{\frac{2}{\pi}(\hbar/\tau_0)\hbar\omega_c} , \tag{4.3}$$

which is in accordance with the result of the SCBA theory discussed in Sect. 1.9. The parameter Γ_{se} was introduced previously in (1.143). In scattering events, $\Delta X = X' - X = q_y l_B^2$, where q_y is the component of the momentum exchange. The latter quantity is limited by the condition $q \sim 1/l_B$. We therefore estimate $\Delta X \sim l_B$. Then, in the high-cyclotron-frequency approximation $\sigma_{xx} \simeq e^2 n_s \nu_{\text{eff}}/m\omega_c^2$, (4.1) and (4.3) result in the approximate relation

$$\nu_{\text{eff}} \sim \frac{\omega_c \Gamma_{\text{se}}^2}{T\Gamma_0} \sim \frac{\omega_c \Gamma_{\text{se}}}{T} . \tag{4.4}$$

The important consequence of this form is that $\nu_{\text{eff}}(B) \propto B\Gamma_0(B) \propto B^{3/2}$ increases with B faster than the cyclotron frequency $\omega_c \propto B$. This means that, in some cases, the high-cyclotron-frequency approximation $\omega_c \gg \nu_{\text{eff}}$ can be inconsistent with the limit of extremely strong magnetic fields, and it is reasonable to formulate the quantum transport theory for the effective collision frequency rather than for the conductivity component. Another interesting consequence of (4.4) is that, for the ripplon scattering regime $[\Gamma_0(T) \propto \sqrt{T}]$, the effective collision frequency and $\sigma_{xx}(T) \propto 1/\sqrt{T}$ increase as the temperature falls, even though the number of scatterers $N_q^{(\text{r})} \propto T$ becomes smaller.

We now consider the rigorous treatment of the galvanomagnetic effect. In the DC case, according to the quantum transport framework discussed in the last chapter, the effective collision frequency of electrons interacting with impurity-like scatterers can be written as an integral form of the electron DSF:

$$\nu_{\text{eff}} = \frac{\Gamma_{\text{se}}^2 \omega_c}{8T\hbar} \int_0^\infty x_q R(x_q) S(q,0) \mathrm{d}x_q , \tag{4.5}$$

where $x_q = q^2 l_B^2/2$ is a dimensionless parameter and $R(x_q)$ is the range factor which equals $\exp(-x_q d^2/l_B^2)$ for the Gaussian interaction potential of (1.145). [For short-range scatterers, $R(x_q) = 1$.] This form of ν_{eff} follows directly from (3.19) and (3.52). It should be noted that, in this equation, Γ_{se}^2 does not originate from the electron density-of-states function and just represents a combination of the interaction potential $V^{(\text{a})}$ and the parameters of the interface electron states [see (1.143)]. The electron density-of-states function enters the equation for the DSF $S(q,0)$. Hence, all the delicate features of the quantum magnetotransport of the 2D Coulomb liquid are hidden in its equilibrium DSF.

Using the DSF of noninteracting electrons from (2.36) with the self-energy effects taken into account, (4.5) can be rewritten as

$$\nu_{\text{eff}} = \frac{m_e \omega_c^2}{4\pi^2 \hbar n_s} \sum_{N=0}^{\infty} \left(\Gamma_N^{\text{tr}}\right)^2 \int \left(-\frac{\partial f}{\partial \varepsilon}\right) \left[\text{Im} G_N(\varepsilon)\right]^2 \mathrm{d}\varepsilon , \tag{4.6}$$

where

$$\left(\Gamma_N^{\text{tr}}\right)^2 = \frac{1}{2}\Gamma_{\text{se}}^2 \int_0^\infty x_q R(x_q) \left|J_{N.N}(x_q)\right|^2 \mathrm{d}x_q . \tag{4.7}$$

This equation is valid regardless of the particular form of $\text{Im}G_N(\varepsilon)$ to be employed. In the SCBA theory [74], the density-of-states function has the semi-elliptic shape, which yields

$$\nu_{\text{eff}} = \frac{m_e \omega_c^2}{\pi^2 \hbar n_s} \sum_N \left(\frac{\Gamma_N^{\text{tr}}}{\Gamma_N} \right)^2 \int d\varepsilon \left(-\frac{\partial f}{\partial \varepsilon} \right) \left[1 - \left(\frac{\varepsilon - \varepsilon_N}{\Gamma_N} \right)^2 \right] , \tag{4.8}$$

where Γ_N is the true broadening of Landau levels. Assuming the relation $\sigma_{xx} \simeq (e^2 n_s / m_e) \nu_{\text{eff}} / \omega_c^2$ (the high-cyclotron-frequency approximation, $\omega_c \gg \nu$), we arrive at the well-known conductivity equation of Ando and Uemura. Thus, the general conductivity form (3.3) and (4.8) extend the SCBA so that it is applicable for any relation between ω_c and ν_{eff}. This is very important for nondegenerate 2D electrons because of the fast increase in the effective collision frequency with B (see also discussions in Sect. 4.3).

For the short-range impurity potential $[R(x_q) = 1]$, (4.7) can be evaluated analytically:

$$\left(\Gamma_N^{\text{tr}} \right)^2 = \left(N + \frac{1}{2} \right) \Gamma_{\text{se}}^2 . \tag{4.9}$$

Here we have used the formula

$$\int_0^\infty x \left| J_{N.N'}(x) \right|^2 dx = N + N' + 1 .$$

Under the conditions employed $\Gamma_N = \Gamma_{\text{se}}$.

It is instructive to study the range dependence of Γ_N^{tr}. For the Gaussian interaction potential discussed above $[R(x_q) = \exp(-x_q d^2 / l_B^2)]$ and $N = 0$, we have

$$\left(\Gamma_0^{\text{tr}} \right)^2 = \frac{\Gamma_{\text{se}}^2}{2 \left(1 + d^2 / l_B^2 \right)^2} \equiv \frac{\Gamma_0^2}{2 (1 + d^2 / l_B^2)} , \tag{4.10}$$

which is the result of the Ando and Uemura theory. The last relation uses the range dependence of the level broadening Γ_0 from (1.146). The important point is that the conventional SCBA theory for the electron conductivity includes vertex corrections which are not small for the long-range impurity potential. Thus, in order to arrive at the same result within the quantum transport framework discussed in the last chapter, it is quite sufficient to include only the self-energy effects in the DSF. The equivalence of the range dependencies of Γ_0^{tr} in the above-mentioned approaches justifies the approximation (2.36) for the electron DSF discussed in Sect. 2.1.3.

For the degenerate electron gas, the effective collision frequency has a series of maxima as a function of ε_f according to (4.8) and (4.9). In the high-cyclotron-frequency limit, the peak value of the conductivity, viz.,

$$\sigma_{xx}^{(\text{peak})} = \frac{e^2}{\pi^2 \hbar} (N + 1/2) , \tag{4.11}$$

depends only on the Landau level number and natural constants. In order to avoid the sharp edges of the conductivity, Ando [77] employed the many-site approximation. In this case, the level density of states is a sort of mixture of the Gaussian and semi-elliptic shapes. For the same purpose, Gerhardts [75] proposed to use the electron density of states following from the cumulant approach, which we discussed in Sect. 1.9.2. In the latter case, from (4.6) and (1.166), one can conclude that the effective collision frequency ν_{eff} is a sum of Gaussians.

As mentioned above, in the degenerate 2D electron system of semiconductors, electron localization and the current-carrying edge states strongly affect the electron transport and yield the resistivity features typical of the QHE, which differ from those given by the SCBA. It is very interesting to apply the extended SCBA theory to the nondegenerate electron system formed on the free surface of superfluid helium, which is usually free from strong localization effects. Under typical helium temperatures $T < 2\,\text{K}$, the electron interaction with vapor atoms and ripplons is weak and edge effects are small. (The possibility of polaronic self-trapping in this system is discussed in Chap. 6.) Therefore the SCBA theory seems to be applicable to this 2D electron system, at least in the range of low electron densities.

In Sect. 2.1.3 we found that, even under zero magnetic field, the main properties of the electron DSF introduced by quantum mechanics remain for the nondegenerate statistics as well (see Fig. 2.1). In the presence of a strong magnetic field ($\hbar\omega_c > T$), practically all SEs populate the lowest Landau level, and they are therefore confined to the very narrow energy range near the position of the Landau level because of the extremely small broadening $\Gamma_0 \ll T$. For the electron distribution function $f(\varepsilon) \simeq \exp\left[(\mu - \varepsilon)/T\right]$ and the semi-elliptic level shapes, (4.6) can be evaluated including all Landau levels [165]:

$$\nu_{\text{eff}} = \frac{2\omega_c \Gamma_{\text{se}}^2}{\pi \Gamma^2 I_1(\Gamma/T)} \coth\left(\frac{\hbar\omega_c}{2T}\right) \left[\cosh\left(\frac{\Gamma}{T}\right) - \frac{T}{\Gamma}\sinh\left(\frac{\Gamma}{T}\right)\right] , \tag{4.12}$$

where $I_1(x)$ is the modified Bessel function of first order. The mixing of different Landau levels has been disregarded here, which means that, in the limit of weak magnetic fields where $\Gamma > \hbar\omega_c$, the above formula does not transform into the semi-classical result.

For the Gaussian shape of $\text{Im}G_N(\varepsilon)$, we have a slightly different result [165]:

$$\nu_{\text{eff}}(B) = \frac{\sqrt{\pi}\omega_c \Gamma_{\text{se}}^2}{4\Gamma T} \exp\left[-\left(\frac{\Gamma}{4T}\right)^2\right] \coth\left(\frac{\hbar\omega_c}{2T}\right) . \tag{4.13}$$

The numerical comparison of these two equations indicates that they are quite close, because, for extremely sharp Landau levels ($\Gamma \ll T$), electrons are evenly distributed within the level density of states and the sharp edges of the original SCBA theory are only of minor importance. In the ultra-quantum limit $\hbar\omega_c > T$, the effective collision frequency has a very strong magnetic field dependence which we shall discuss in Sect. 4.3. In the high-cyclotron-frequency regime $\sigma_{xx} \simeq (e^2 n_s/m)\nu_{\text{eff}}/\omega_c^2$, (4.12) and (4.13) lead to conductivity equations which coincide with the early results of [166–168] if $\Gamma = \Gamma_{\text{se}}$.

Here we found it instructive to distinguish Γ_{se}, which just represents the matrix elements of the interaction Hamiltonian for short-range scatterers, from the true level broadening $\Gamma = \Gamma_N$ which originates from the electron density of states. In this case, the forms of (4.12) and (4.13) with $\nu_{eff} \propto \Gamma_{se}^2/\Gamma$ reveal the singular nature of the 2D electron gas under a magnetic field, which agrees with the qualitative form of (4.3): the limiting case $\Gamma \to 0$ gives us an infinite collision rate ($\nu_{eff} \to \infty$) because of multiple electron scattering. The cancellation of Γ in the denominator of (4.13) with Γ_{se} of the numerator would give a misleading asymptote: $\nu_{eff} \to 0$ if $\Gamma \to 0$. Of course, if there is only one scattering mechanism, $\Gamma = \Gamma_{se}$ and the limiting behavior $\nu_{eff} \propto \Gamma_{se} \to 0$ just reflects the trivial fact that, in the absence of scatterers, there is no scattering.

4.2 Experimental Techniques

In this chapter we mainly discuss the DC magnetoconductivity of the system of SEs formed on the free surface of liquid helium, although nearly all conductivity measurements of SEs are done in a contactless way by means of the AC capacitive coupling technique or a resonance method. The real DC measurement in this system presents a hard problem because it is very difficult to attach leads to the electron layer on the free surface of liquid helium. Time-of-flight experiments are not very convenient because the electron pool produces a macroscopic dimple on the helium surface and the electron mobility is restricted by the very slow mobility of the macroscopic dimple. Therefore, when talking about the DC measurement for SEs on liquid helium, we imply that the frequency of the driving signal is much lower than the effective collision frequency ν_{eff} and the frequency of plasmons Ω_p which can be excited in a finite cell.

The capacitive detection method [46, 169] is frequently used for conductivity measurements in the system of SEs on liquid helium. This method employs an electrode array in the form of a Hall bar or a Corbino disk placed below the liquid helium surface, as shown in Fig 4.1. One electrode of the array is usually driven with an AC voltage V_{in} which induces a current in the electron layer above the surface. The response of the 2D electron system to the oscillating electric field is monitored by another electrode. The induced currents in and out of phase with respect to the driving voltage V_{in} are measured by a current amplifier in conjunction with a dual-phase lock-in amplifier. In order to account for stray resistance and capacitance, currents measured with the uncharged surface are usually subtracted from the data.

There are two different treatments for analyzing the currents measured and extracting the electron conductivity. The simplified method uses the equivalent circuit model $C_1 - G_e - C_2$, where C_1 and C_2 denote the capacitance between bottom electrodes and the electron sheet (the upper plates of

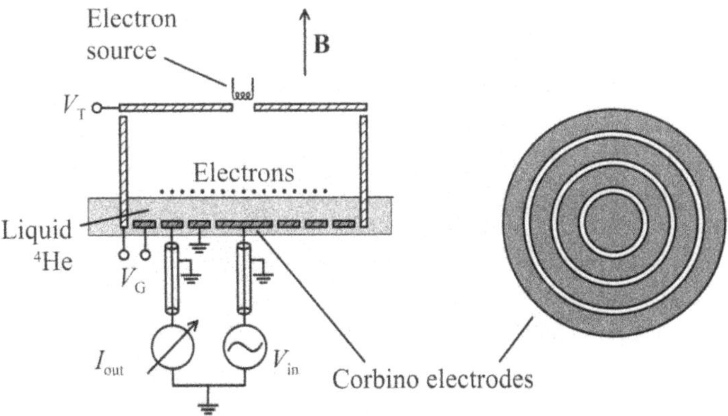

Fig. 4.1. Schematic diagram of the experimental cell for the magnetoconductivity measurements in the system of SEs on liquid helium [170]

these capacitors are formed by the electrons). The conductance G_e of the electron layer is evaluated by circuit analysis, which includes the direct coupling between electrodes [165]. At low frequencies, the experimental system behaves as an RC circuit, so that the convention is to write the measured current as $I_{\text{out}} = ZV_{\text{in}}$, where the complex admittance is defined as $Z = Z_G + \mathrm{i}Z_C$. The phase shift φ is defined as $\varphi = \arctan(Z_G/Z_C)$. The zero phase shift corresponds to a purely capacitive coupling.

Another treatment is based on the transmission line approach. This method is applicable for thin layers of liquid helium above the metal electrode: $d \ll R_{\text{el}}$, where R_{el} is the radius of the electron pool. In this model, the electrons and electrodes form a 2D transmission line with distributed parameters such as resistance and capacitance. The response of such a system can be calculated analytically and expressed as a function of the electron conductivity σ_{xx}. For the electrode array of Fig. 4.1 with excitation electrode (radius r_0), ground electrode (radius r_1), and receiving electrode (radius r_2), the current can be found as

$$\frac{I}{V_{\text{in}}^{(0)}} = \pi^2 r_0 r_1 \mathrm{i}\omega C \frac{J_1(z_0)}{J_1(z_2)} \left[J_1(z_1)N_1(z_2) - J_1(z_2)N_1(z_1)\right] , \tag{4.14}$$

where $J_1(z)$ and $N_1(z)$ are the Bessel and Neumann functions, $V_{\text{in}}^{(0)}$ is the amplitude of the applied voltage,

$$z_n = r_n \left(-\mathrm{i}\frac{\omega C}{\sigma_{xx}}\right)^{1/2} , \tag{4.15}$$

C is the effective capacitance per unit area between the electrons and the electrodes [169],

$$C = \varepsilon_0 \left(\frac{1}{h} + \frac{\varepsilon}{d}\right) , \tag{4.16}$$

d is the depth of the liquid helium and h is the distance from the electrons to the top plate. Equation (4.14) establishes the relation between I and the magnetoconductivity of electrons. Thus, knowing the experimental signal $I_{\rm out}$ induced in the outer electrode, it is possible to find the conductivity of the electron layer. More detailed information concerning the application of the transmission line model for the conductivity measurement in the system of SEs on liquid helium is given in [73, 171, 172].

The magnetoconductivity measurements are usually done using Corbino electrodes. In this geometry (the excitation field is directed along the radius), no Hall voltage is induced, and the radial current density is proportional to the radial electric field $j_r = \sigma_{xx} E_r$. The main advantage of the experiment with axial symmetry is that the edge magnetoplasmon (EMP) wave is not excited. The important property of this wave discussed in Sect. 2.5.2 is that its spectrum is gapless, and its frequency decreases strongly with the magnetic field $\Omega_{\rm emp} \propto 1/B$. The latter means that an AC measurement with a low excitation frequency $\omega \ll \nu_{\rm eff}$ and with Hall bar electrodes cannot give accurate DC magnetoconductivity data in the limit of strong magnetic fields because $\Omega_{\rm emp}$ eventually reaches the resonance condition $\omega = \Omega_{\rm emp}$ and this spoils the circuit analysis.

Even a Corbino experiment can be affected by the excitation of the EMP wave at low temperatures and high B because of small deviations from the radial symmetry or a slight tilt. In [173], Corbino electrodes of very small radius were introduced to avoid excitation of the EMP wave. This method was successfully used to study many-electron effects on the DC magnetoconductivity in the relatively high temperature regime ($T > 0.7\,{\rm K}$) [174, 175], where electron scattering was dominated by vapor atoms.

In recent measurements of [76, 89, 90], the experimental cell was equipped to study the DC magnetoconductivity and the CR absorption in the same experiment. The bottom plate with the Corbino electrodes was mounted on a movable plunger so that the height of the resonator could be changed in order to tune its resonance frequency in situ. In a certain range of the experimental parameters, simultaneous measurements of the CR absorption and the low-frequency admittance allow one to control the magnetoconductivity and the Landau level broadening independently in the same experiment in order to check the theoretical concepts underlying the quantum magnetotransport of strongly interacting electrons.

Another interesting method for measuring σ_{xx} at low temperatures ($T < 1\,{\rm K}$) was realized in [98, 99]. This method uses the fact that, under medium and high magnetic fields, the damping coefficient of the EMP is proportional to σ_{xx} with the prefactor depending on the geometry of the layer [104] [see (2.76)]. In the case of strong magnetic fields (practically, $B \gg 0.1\,{\rm T}$), the thickness b of the edge strip where the EMP charge is confined is equal to the electron density transition width, which is independent of B and T. Thus, the magnetic field and temperature dependencies of the electron magnetoconductivity can be obtained from the respective dependencies of the EMP damping coefficient. Remarkably, the effect that spoils the conventional method for measuring σ_{xx} now serves as a good tool for studying electron magnetoconductivity. A number of important aspects of quantum magnetotransport under strong magnetic fields and low temperatures were uncovered using this method [36, 99]. It should also be noted that a real DC experiment with SEs on liquid helium has been carried out recently in [176].

4.3 Quantization-Induced Decrease in the Hall Angle

In the classical picture of the Hall effect, the normal magnetic field deflects the current vector $\boldsymbol{j}$ from the direction of the applied electric field $\boldsymbol{E}$ forming the Hall angle φ_{H}. Because the collision frequency ν of electrons with the semi-classical spectrum does not depend on the magnetic field, the Hall angle increases with B as $\tan\varphi_{\mathrm{H}} = \omega_{\mathrm{c}}/\nu \propto B$, reaching 90° as $B \to \infty$. The quantization of the electron spectrum introduces a strong magnetic field dependence of the effective collision frequency $\nu_{\mathrm{eff}}(B)$ which competes with the classical Hall effect. The important point is that, in a particular case, the dependence $\nu_{\mathrm{eff}}(B)$ can be stronger than the linear dependence of $\omega_{\mathrm{c}} \propto B$, which leads to a decrease in the Hall angle, and more importantly to a breakdown of the high-cyclotron-frequency approximation ($\omega_{\mathrm{c}} \gg \nu_{\mathrm{eff}}$) with increasing magnetic field. Thus, as noted in Sect. 4.1, the high-cyclotron-frequency approximation can be inconsistent with the high magnetic field limit.

Consider a nondegenerate gas of SEs interacting with helium vapor atoms above a free surface of liquid helium. According to (4.12) for the semi-elliptic level shape, in the ultra-quantum limit ($\hbar\omega_{\mathrm{c}} > T$) and at $\Gamma_{\mathrm{se}} \ll T$, the effective collision frequency of the DC case is

$$\nu_{\mathrm{eff}}(B) \simeq \frac{4\omega_{\mathrm{c}}\Gamma_{\mathrm{se}}}{3\pi T} \propto B^{3/2} . \tag{4.17}$$

For the Gaussian shape of Landau levels, (4.13) gives nearly the same result:

$$\nu_{\mathrm{eff}}(B) \simeq \frac{\sqrt{\pi}\omega_{\mathrm{c}}\Gamma_{\mathrm{se}}}{4T} . \tag{4.18}$$

The numerical difference between the proportionality factors of (4.17) and (4.18) is about 4%. As anticipated, in both cases the effective collision frequency increases with B faster than the cyclotron frequency because $\Gamma_{\mathrm{se}} \propto \sqrt{B}$. This means that the Hall angle decreases with B which is opposite (!) to the classical Hall effect. For intermediate magnetic fields, $\tan\varphi_{\mathrm{H}}$ vs. B is plotted in Fig. 4.2 for three typical temperatures. Instead of the straight lines representing classical behavior (dashed lines), the extended SCBA theory gives nonlinear curves (continuous curves) with maxima. Using the Gaussian level shape (4.13) yields quite close curves. This result indicates that, at

$T > 1\,\mathrm{K}$, the high-cyclotron-frequency approximation ($\omega_c \gg \nu_{\rm eff}$) frequently used in the literature is inconsistent with the strong magnetic field limit, and one should use the extended SCBA equations [(3.3) and (4.12)]. Here we should remember that (4.12) does not transform into the classical result as $B \to 0$ because it was obtained by disregarding the mixing of different Landau levels which becomes important at low fields.

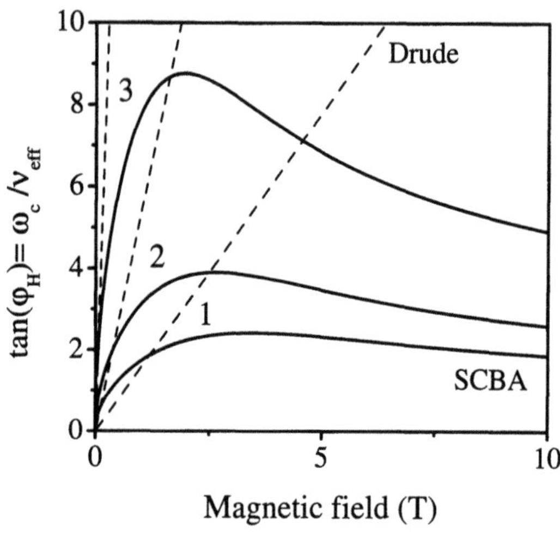

Fig. 4.2. Unconventional Hall effect caused by the quantization of electron orbital motion: SCBA theory (*continuous lines*), semi-classical regime (*dashed lines*) for $T = 2\,\mathrm{K}$ (1), $T = 1.6\,\mathrm{K}$ (2), $T = 1.2\,\mathrm{K}$ (3) [170]

In the magnetoconductivity experiment of [141], the phase shift φ (away from $\pi/2$) between the AC excitation voltage and the current was measured under magnetic fields up to 20 T. The data were compared with both the conventional SCBA established using the high-cyclotron-frequency approximation and the extended SCBA resulting in (4.12). With increasing magnetic field, the measured phase shifts designated in Fig. 4.3 by e–h deviate strongly from the high-cyclotron-frequency SCBA (dash-dotted lines) marked by e1–h1 because of the unconventional decrease in the Hall angle. At the same time, they agree well with the extended SCBA based on the universal quantum framework discussed in the last chapter (dashed lines). It should be remembered that, under the conditions $\nu_{\rm eff} \sim \omega_c$, the conventional approximation for the memory function $M(\omega) \simeq M_F(\omega)$ is not strictly proven within the framework of the memory function approach. This approximation was shown to be accurate only for highly correlated electrons when $\nu_{\rm e\text{–}e} \gg \nu_{\rm eff}$. The good agreement between experiment and theory favors the framework formulated for highly correlated electrons and the above-mentioned approximation for the memory function, even though under the experimental conditions [141] ($T \sim 2\,\mathrm{K}$) we have $\nu_{\rm eff} > \nu_{\rm e\text{–}e}$. This is because, for short-range impurity scat-

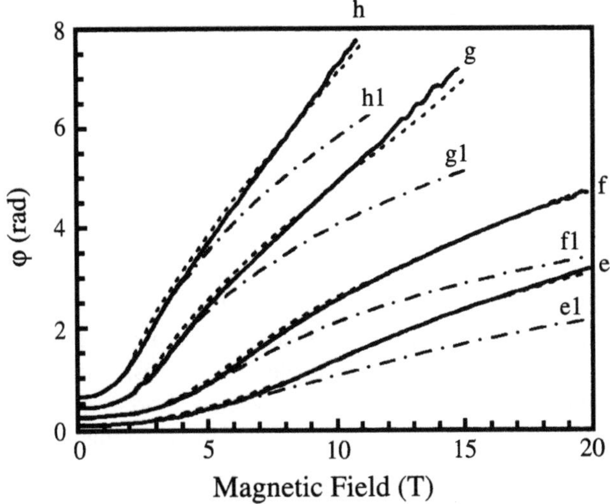

Fig. 4.3. Measured phase shifts $\varphi = \arctan(Z_G/Z_C)$ for a fixed electron density $n_s = 0.58 \times 10^8\,\mathrm{cm}^{-2}$ and different frequencies of the signal: 3.0 kHz (e), 5.0 kHz (f), 10.0 kHz (g), and 15.0 kHz (h). *Dashed curves* are the calculated phase shifts for the extended SCBA and *dash-dotted curves* designated by e1–h1 are calculated according to the original SCBA [141]

tering, the effective collision frequency does not depend on $\nu_{e\text{–}e}$, as discussed in Sect. 3.5. In the next section, we shall see that the anomalous Hall effect discussed here is strongly enhanced by internal forces.

The same strong field dependence of the effective collision frequency of SEs can be found in the ripplon scattering regime. The ratio $\omega_c/\nu_{\mathrm{eff}}$ decreases with B, causing an anomalous decrease in the Hall angle. Regarding the consistency of the high-cyclotron-frequency approximation and the strong field limit, at low temperatures, ν_{eff} is very small and, under typical conditions (for reasonable magnetic fields), the high-cyclotron-frequency approximation is valid with good accuracy.

4.4 Many-Electron Effects

The history of physical ideas concerning the influence of electron–electron interactions on electron transport is very interesting and instructive. In the semi-classical picture of electron transport, internal forces can affect the conductivity only by changing the shape of the electron distribution function, as discussed in Sect. 3.5. This can change the conductivity by introducing a new numerical proportionality factor of the order of 2, if the magnetic field is weak $\omega_c \ll \nu_{\mathrm{eff}}$. In the high-cyclotron-frequency limit typical of the DC magnetotransport experiment, electron correlations do not affect σ_{xx} in the semi-classical theory. Furthermore, if electrons are scattered by the short-range impurity potential, the effective collision frequency does not depend on ν_{e-e}.

In the quantum treatment of many-electron transport, the situation changes a great deal. Historically, Ryzhii [177] was apparently the first to note that the electron–electron interaction should broaden the Landau levels and affect the conductivity. The broadening of the Landau levels was introduced via the inverse phenomenological lifetime of single-electron states. Coulomb broadening was discussed and studied for the degenerate 2D electron gas by Fukuyama, Kuramoto and Platzman [178]. Introducing the concept of the quasi-uniform fluctuational internal field, Dykman and Khazan [85] noted that, rather than the lifetime of single-electron states, it is the characteristic time of interaction of the drifting electron with a short-range defect (the time of flight) τ_f that cuts off the divergence of σ_{xx} in two dimensions. An electron subject to a fluctuational electric field drifts away from the impurity, which reduces the number of multiple scattering events with the same scatterer to $N_{sc} = \omega_c \tau_f$. The quantity τ_f has no direct relationship with the lifetime of the single-electron state (for the many-electron theory τ_B can even be infinitely long), and it can be estimated as $\tau_f^{-1} = u_f / l_B = eE_f^{(0)} l_B / \hbar$.

The relation between the lifetime limited by scatterers and the time of flight was found by Monarkha, Teske and Wyder [90], who considered an electron in the reference frame moving along with the orbit center (see the discussion in Sect. 2.3). In this picture, the electron orbit center is still and impurities drift with the velocity $-\boldsymbol{u}_f$, causing inelastic scattering with the energy exchange $\hbar q u_f \sim \hbar u_f / l_B = \hbar / \tau_f$ which relates directly to the time of flight near an impurity in the laboratory reference frame. Internal forces thus eliminate the divergence of σ_{xx} for impurity scattering by introducing the second factor discussed in Sect. 4.1, namely, inelastic scattering. The intriguing point is that the lifetime τ_B of single-electron states becomes longer as a result of the Coulomb interaction, and instead of Coulomb broadening of Landau levels, we have Coulomb narrowing, as discussed in Sect. 2.3.

Knowing the behavior of the collision broadening Γ_N as a function of electron density n_s, previously discussed in Sect. 2.3, it might be concluded that the replacement $\Gamma \to \Gamma_0(n_s)$ in (4.13) would give us a suitable many-electron approximation for the effective collision frequency in the quantum limit $\nu_{eff} \simeq \sqrt{\pi}\omega_c \Gamma_{se}^2 / 4T\Gamma_0(n_s)$. However, this replacement would give an incorrect density dependence of the electron conductivity. Indeed, $\Gamma_0(n_s)$ in the denominator is a mainly decreasing function of density, while Γ_{se} in the numerator increases slowly with n_s due to the fact that $\nu_0 \propto \gamma(E_\perp)$ and $E_\perp = 2\pi e n_s$. Therefore, this naive many-electron treatment would lead to an increase in the collision rate which contrasts with the idea of a cutoff in the multiple electron scattering induced by the fluctuational field and contradicts numerous experimental observations. This could be anticipated from the beginning, because the density dependence of the Landau level broadening is not the main many-electron effect for the fluctuational field concept underlying the DC magnetoconductivity: the broadening of single-electron states and inelastic scattering are two different factors limiting the divergence of σ_{xx}, and the many-electron conductivity σ_{xx} is finite even for the singular level density due to the inelastic cutoff. The solution of the above-mentioned paradox is that the single-electron approximation for the DSF $S(q,0)$ is not the right one for describing Coulombic effects [even substituting $\Gamma_N \to \Gamma_N(n_s)$].

In the simplified treatment, the many-electron magnetoconductivity can be found from the Einstein relation (4.1) and the estimate $D \sim l_B^2/\tau_B$ for the diffusion constant. In the single-electron approach, we used the relation $1/\tau_B = 1/\tau_0(\hbar\omega_c/\Gamma_0)$ of (4.2) with Γ_0 found by means of the self-consistent procedure. The enhancement factor $\hbar\omega_c/\Gamma_0$ can also be considered as the number of multiple scattering events within the lifetime $\omega_c\tau_B$. In the opposite limit of an extremely strong many-electron effect, one can assume that the quantum scattering rate $1/\tau_B$ is enhanced by the number of multiple electron scattering events within the time of flight $N_{sc} = \omega_c\tau_f$ in the presence of the fluctuational electric field. We thus have

$$\frac{1}{\tau_B} \simeq \frac{1}{\tau_0} N_{sc} \simeq \frac{1}{\tau_0}\omega_c\tau_f \sim \frac{1}{\tau_0}\frac{\hbar\omega_c}{\Gamma_C} . \tag{4.19}$$

The parameter $\Gamma_C = \sqrt{2}eE_f^{(0)}l_B$ was introduced in Sect. 2.4 as the characteristic broadening of the electron DSF. Unlike the single-electron broadening Γ_0, the many-electron broadening Γ_C of the DSF is not determined by the self-consistent procedure and relates simply to the time of flight in the fluctuational field and to the average energy exchange in the moving frames: $\Gamma_C \sim \hbar/\tau_f \sim \hbar q u_f$.

Comparing (4.2) and (4.19), one can see that in the case of an extremely strong many-electron effect ($\Gamma_C \gg \Gamma_0$), the collision broadening of (4.2) is just replaced by the Coulomb broadening Γ_C of the DSF. Therefore, for the two extreme regimes of quantum magnetotransport, the effective collision frequency can be written as

$$\nu_{eff} = \frac{\sqrt{\pi}\omega_c\Gamma_{se}^2}{4T\Gamma^{(*)}} , \tag{4.20}$$

where $\Gamma^{(*)}$ equals Γ_{se} in the single-electron approximation, and it approximately equals Γ_C in the extreme many-electron regime. The proportionality factor of the order of unity is fixed by (4.18).

In the intermediate case $\Gamma_C \sim \Gamma_{se}$, one should establish a way of combining the lifetime τ_B and the time of flight τ_f to form the interaction time τ_i which determines the number of multiple scattering events and enters the relations

$$\frac{1}{\tau_B} \simeq \frac{1}{\tau_0}\omega_c\tau_i , \tag{4.21}$$

$$\nu_{eff} \simeq \frac{\sqrt{\pi}\omega_c\tau_i\Gamma_{se}^2}{4T\hbar} . \tag{4.22}$$

In a simplified manner, one can just add the inverse quantities according to the semi-classical rule for two different scattering mechanisms: $\tau_{\mathrm{i}}^{-1} = \tau_B^{-1} + \tau_{\mathrm{f}}^{-1}$. Nevertheless, under a quantizing magnetic field, the level broadening is proportional to $\sqrt{\nu_0} \propto \sqrt{V_{\mathrm{int}}^2}$ and we have to add squares when combining two different scattering mechanisms: $\Gamma_{\mathrm{AB}} = \sqrt{\Gamma_{\mathrm{A}}^2 + \Gamma_{\mathrm{B}}^2}$. It is therefore reasonable to use the rule

$$\tau_{\mathrm{i}}^{-2} = \tau_B^{-2} + \tau_{\mathrm{f}}^{-2} . \tag{4.23}$$

In this case, substituting (4.23) for τ_{i} in (4.21), we have a self-consistent equation whose solution can be written as

$$\tau_B^{-2} = \sqrt{(\Gamma_{\mathrm{se}}/\hbar)^4 + \tau_{\mathrm{f}}^{-4}/4} - \tau_{\mathrm{f}}^{-2}/2 . \tag{4.24}$$

This relation is in accordance with the equation for collision broadening affected by a uniform electric field, discussed in Sect. 2.3.1 [see (2.45)].

The lifetime which follows from (4.24) increases (the level broadening decreases) with an increase in the fluctuational field. In contrast, the interaction time

$$\tau_{\mathrm{i}} = \frac{1}{\sqrt{\tau_B^{-2} + \tau_{\mathrm{f}}^{-2}}} = \frac{1}{\sqrt{\sqrt{(\Gamma_{\mathrm{se}}/\hbar)^4 + \tau_{\mathrm{f}}^{-4}/4} + \tau_{\mathrm{f}}^{-2}/2}} \tag{4.25}$$

and the effective collision frequency of (4.22) decrease with an increase in the average fluctuational field (in spite of the increase in τ_B), which explains the above-mentioned paradox. One can use the even simpler interpolation formula

$$\Gamma^{(*)} = \sqrt{\Gamma_{\mathrm{se}}^2 + b^2 \Gamma_{\mathrm{C}}^2} , \tag{4.26}$$

with a numerical parameter $b = 2/\sqrt{\pi} \simeq 1.128$ of the order of unity chosen to give an accurate conductivity result in the limiting case $\Gamma_{\mathrm{C}} \gg \Gamma_0$. We have already met this kind of relation in Sect. 2.3.2 when analyzing the interplay between the intrinsic inelastic effect and the effect of internal forces, and in Sect. 2.4 which dealt with Coulomb broadening of the electron DSF.

The qualitative analysis of the DC magnetoconductivity given above interprets the many-electron effect in the simple diffusion picture. The more accurate analysis is based on the exact relation (4.5) given by the universal quantum transport framework discussed in Chap. 3 and the many-electron approximation for the DSF of the Coulomb liquid. In Chap. 2, we found that (2.50) establishes a simple relation between the many-electron DSF $S_{\mathrm{me}}(q,\omega)$ and its single-electron form $S_{\mathrm{se}}(q,\omega)$. Therefore, when accepting the general form of (4.5) for the effective collision frequency, we actually reduce the many-electron transport theory to a 'one-line' theory based on the simple relation of (2.50).

In Sect. 2.4, we noted that the Coulomb liquid can be described by a model in which independent electrons are exposed to a quasi-uniform fluctuational electric field. The universal electron excitation spectrum cannot be introduced for all electrons of this system in a single reference frame, because the fluctuational electric fields $E_{\rm f}^{(i)}$ acting on electrons differ. Instead, we can introduce the Landau spectrum for each electron in its particular reference frame moving along with the orbit center, which is different for different electrons. In the laboratory reference frame, one can only introduce a global property of the Coulomb liquid, namely, the dynamical structure factor. In Sect. 2.4, we found that the many-electron fluctuational electric field broadens the electron DSF $S(q,\omega)$ [see (2.52)] in spite of the Coulomb narrowing of Landau levels in the local moving reference frames. This is the main many-electron effect which significantly affects all transport properties of the Coulomb liquid. This effect exists even for $\Gamma_N = 0$. In the general case, we have to compare the broadening $\Gamma_{N,N'} \sim \Gamma_N$ of the DSF caused by the finite lifetime of the Landau states with the Coulomb broadening $\Gamma_{\rm C}$ of the DSF induced by the fluctuational field.

In order to reveal the influence of the Coulomb broadening of the electron DSF on quantum magnetotransport at low temperatures $T < \hbar\omega_{\rm c}$, we separate the DSF of the 2D electron liquid $S_{\rm me}(q,0) = \langle S_{\rm se}(q, -\boldsymbol{q}\cdot\boldsymbol{u}_{\rm f})\rangle_{\rm f}$ into two parts:

$$S(q,0) = S_{\rm ins}(q,0) + S_{\rm out}(q,0)\ , \tag{4.27}$$

representing contributions from electron scattering 'inside' the lowest Landau level and electron scattering on higher Landau levels ('out' of the lowest Landau level). The first term $S_{\rm ins}(q,0)$ on the right-hand side of (4.27) originates from the first term ($N = 0$) of the single-electron DSF $S_{\rm se}(q,\omega)$ [see (2.38)] to be employed in the many-electron form $S_{\rm me}(q,0)$ given above. The other term $S_{\rm out}(q,0)$ originates formally from the other terms ($N > 0$) of $S_{\rm se}(q,\omega)$. Equations (4.5) and (4.27) lead to a corresponding separation of the effective collision frequency which we denote as

$$\nu_{\rm me} = \nu_{\rm me}^{(\rm ins)} + \nu_{\rm me}^{(\rm out)}. \tag{4.28}$$

Here and below we omit the subscript 'eff' for the effective collision frequency.

The separation of the DSF and effective collision frequency into two parts has the following physical meaning. The terms $S_{\rm ins}(q,0)$ and $\nu_{\rm me}^{(\rm ins)}$ represent and describe the effect discussed above in the diffusion picture: the number of multiple electron scatterings is reduced because of the shortening of the time of flight $\tau_{\rm f}$ and the interaction time $\tau_{\rm i}$ with an increase in the fluctuational field. The appearance of the other terms $S_{\rm out}(q,0)$ and $\nu_{\rm me}^{(\rm out)}$ can be explained differently, depending on the reference frame. In the moving local reference frames where the pure discrete Landau spectrum can be introduced, the electron scattering out of the lowest Landau level is induced by the ultrafast drift velocities $-\boldsymbol{u}_{\rm f}$ of 'impurities', causing the energy exchange $\hbar q u_{\rm f} \sim \hbar\omega_{\rm c}$. In the laboratory frame, the scattering is elastic, but an electron has a spectrum with tilted levels $\varepsilon_N(X) = \varepsilon_N^{(0)} - eE_{\rm f}X$. Therefore an elastic scattering event with $\Delta X \sim l_B$ can bring the electron into the next Landau level, if $eE_f l_B \sim \hbar\omega_c$. It is clear that the above condition is the same as the one found in the inelastic picture for moving frames.

The contribution $\nu_{me}^{(ins)}$ can be calculated using two different approximations. Firstly, in the character of S_{ins}, we can use the main term ($N = 0$) of the approximation given in (2.52), where the Landau level broadening $\Gamma_0(E_f)$ as a function of the fluctuational electric field is replaced by $\Gamma_0(E_f^{(0)})$. This substitution for $S_{ins}(q,0)$ in the effective collision frequency equation [(3.20) and (4.5)] yields

$$\nu_{me}^{(ins)} \approx \frac{\sqrt{\pi}\Gamma_{se}^2\omega_c}{4T}\int_0^\infty \frac{x\exp(-x)}{\sqrt{\Gamma_0^2 + x\Gamma_C^2}}dx \,. \tag{4.29}$$

Remarkably, this equation is close to the result found above in the qualitative diffusion picture. The important difference is the additional averaging over the momentum exchange ($x = q^2 l_B^2/2$). The result of the diffusion picture can be obtained just by replacing x in the denominator of the integrand by its average value of the order of unity. Equation (4.29) also justifies the way of combining the lifetime τ_B and the time of flight τ_f to form the interaction time τ_i chosen in the qualitative analysis [see (4.23)].

A more accurate result can be found without the replacement $\Gamma_0(E_f) \to \Gamma_0(E_f^{(0)})$ by integrating over x_q before averaging over the absolute value of the fluctuational field. This approach gives [90]

$$\begin{aligned}\nu_{me}^{(ins)} &= \frac{\sqrt{\pi}\Gamma_{se}^2\omega_c}{4T}\int_0^\infty \frac{\Gamma_0^2(y) + \Gamma_C^2 y/2}{[\Gamma_0^2(y) + \Gamma_C^2 y]^{3/2}}e^{-y}dy \\ &= \nu_{se}\int_0^\infty \frac{\sqrt{1+\lambda^4 y^2} - \lambda^2 y/2}{(1+\lambda^4 y^2)^{3/4}}e^{-y}dy \,, \end{aligned} \tag{4.30}$$

where we have introduced the dimensionless parameter $\lambda = \Gamma_C/\Gamma_{se}$ showing the measure of the Coulombic effect, and $y = (E_f/E_f^{(0)})^2$. In spite of the seemingly cumbersome form, (4.30) is suitable for numerical analysis.

The many-electron reduction in the effective collision frequency as a function of the dimensionless parameter λ which follows from (4.30) is shown in Fig. 4.4 by the continuous curve. Remarkably, this accurate result is very close to what is found in the phenomenological treatment of Coulomb broadening of the average density of states in the laboratory frame $\Gamma \to \sqrt{\Gamma_{se}^2 + b^2\Gamma_C^2}$ used in [36]:

$$\frac{\nu_{me}}{\nu_{se}} = \frac{1}{\sqrt{1+b^2\lambda^2}} \,, \tag{4.31}$$

with the numerical parameter b fixed at $2/\sqrt{\pi} \simeq 1.128$ (dotted curve). Another interesting numerical coincidence is that the exact form of (4.30) is very close to the Coulombic reduction in the collision broadening of Landau levels discussed in Sect. 2.3, taken at the average fluctuational field (dash-dotted curve):

$$\frac{\Gamma(\lambda)}{\Gamma_{se}} = \sqrt{\sqrt{1+\lambda^4} - \lambda^2}\,. \tag{4.32}$$

Therefore, the relation $\nu_{eff} \propto \Gamma_0 = \Gamma(\lambda)$ remains approximately valid in the many-electron theory as well. This numerical result agrees with the diffusion picture of the many-electron effect given above: $\sigma_{xx} \propto 1/\tau_B$, with τ_B defined by (4.24). The important point is that the many-electron reduction of ν_{me} in the strict treatment [see (4.30)] is not due to the dependence $\Gamma_0(\lambda)$ but happens rather despite it, because leaving the Landau level broadening unchanged $\Gamma_0(\lambda) \equiv \Gamma_{se} =$ const. would give the even stronger reduction shown in Fig. 4.4 by the dashed curve. Inclusion of the actual dependence $\Gamma_0(\lambda)$ in (4.30) acts in the opposite way and shifts the many-electron curve slightly upwards, as shown in Fig. 4.4 (continuous curve). It is really a piece of good luck that the theory is not very 'sensitive' to the actual dependence $\Gamma_0(\lambda)$ whose derivation was the most difficult and approximate part of the many-electron theory.

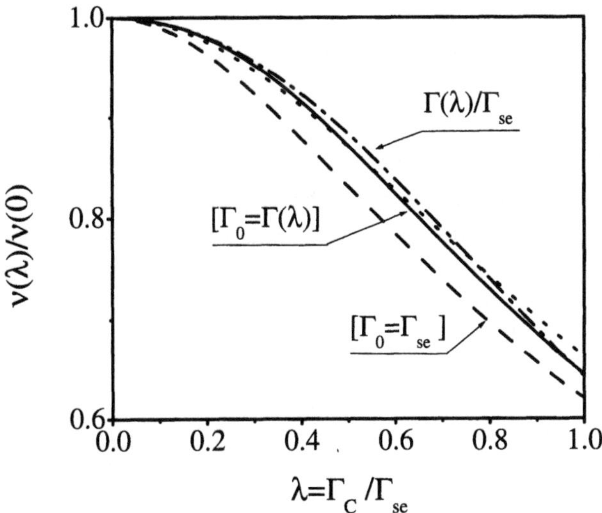

Fig. 4.4. Coulomb suppression of the effective collision frequency of SEs interacting with vapor atoms in the presence of a quantizing magnetic field: accurate evaluation based on (4.30) (*continuous curve*), approximation $\Gamma_0 = \Gamma_{se}$ (*dashed curve*), and analytical approximation of (4.31) (*dotted curve*). The Coulomb reduction in the collision broadening of Landau levels (4.32) is shown by the *dash-dotted curve*

It should be noted that it is only the exact equation (4.30) for ν_{me} which is, remarkably, not very sensitive to the actual dependence $\Gamma_0(\lambda)$. The approximate form of (4.29) found assuming $\Gamma_0(\lambda) \simeq$ const. would deviate from the continuous curve at $\lambda \sim 1$ if Γ_0 were replaced by (4.32). Therefore, when employing the approximate form of (4.29), it is better to keep $\Gamma_0 = \Gamma_{se}$

because in this case we have a consistent and quite accurate result which, according to Fig. 4.4, is close to the exact result based on numerical evaluation of (4.30).

Comparing (4.29) and (4.31) with the single-electron result of (4.13), one can see that the many-electron transport equations look approximately like the single-electron equation with the Landau level broadening Γ replaced by the effective broadening of the electron dynamical structure factor $\sqrt{\Gamma_{\mathrm{se}}^2 + b^2\Gamma_{\mathrm{C}}^2}$. Therefore an approximate but quite accurate way of treating the many-electron effects in the magnetotransport theory of the Coulomb liquid is just to replace Γ by $\sqrt{\Gamma_{\mathrm{se}}^2 + b^2\Gamma_{\mathrm{C}}^2}$ in the single-electron conductivity equations. This approximate treatment is also in accordance with the analysis based on the diffusion picture.

The broadening of the electron DSF thus leads to a strong decrease in the collision rate for electron scattering inside the lowest Landau level, just because of the decrease in the proportionality factor $1/\sqrt{\Gamma_0^2 + x_q\Gamma_{\mathrm{C}}^2}$ of the Gaussian function entering $S_{\mathrm{ins}}(q,0)$. Physically, this decrease in the electron collision rate can be explained as follows. When describing a scattering event in the reference frame moving along with the electron orbit center, any collision is accompanied by an energy exchange $\hbar \boldsymbol{q} \cdot \boldsymbol{u}_{\mathrm{f}}$ between the electron and a scatterer. If the energy exchange is larger than the broadening of the single electron density of states, the electron has to scatter in the energy range with fewer or next to no states. This causes a strong decrease in $\nu_{\mathrm{me}}^{(\mathrm{ins})}$. In the laboratory frame, it is the decrease in the interaction time $\tau_{\mathrm{i}} \sim 1/\sqrt{\tau_B^{-2} + \tau_{\mathrm{f}}^{-2}}$ due to the ultra-fast drift velocity that makes the collision frequency smaller.

For electron scattering out of the lowest Landau level, Coulomb broadening of the DSF produces the opposite effect, namely, the collision rate $\nu_{\mathrm{me}}^{(\mathrm{out})}$ increases with Γ_{C}. At high electron densities, the effective broadening $\Gamma_{0N}^{(*)} = \sqrt{\Gamma_0^2 + x_q\Gamma_{\mathrm{C}}^2}$ becomes large, and the terms of the DSF with $N > 0$ start to contribute substantially to the effective collision frequency. The contribution of electron scattering out of the lowest Landau levels can be found using (2.52) for the many-electron DSF. Because this kind of electron scattering becomes important when Γ_{C} is substantially larger than Γ_{se}, one can disregard Γ_0 as compared with $\sqrt{x_q}\Gamma_{\mathrm{C}}$. Then straightforward evaluation gives

$$\nu_{\mathrm{me}}^{(\mathrm{out})} = \frac{\sqrt{\pi}\Gamma_{\mathrm{se}}^2\omega_{\mathrm{c}}}{4T\Gamma_{\mathrm{C}}} A_{\mathrm{DC}}(\beta_{\mathrm{f}}) , \tag{4.33}$$

$$A_{\mathrm{DC}}(\beta_{\mathrm{f}}) = 2\sum_{N=1}^{\infty} \frac{1}{N!} \left(\frac{N}{\beta_{\mathrm{f}}}\right)^{N+3/2} K_{N+3/2}\left(2N/\beta_{\mathrm{f}}\right) , \tag{4.34}$$

where $\beta_{\mathrm{f}} = \Gamma_{\mathrm{C}}/\hbar\omega_{\mathrm{c}}$ and $K_\nu(x)$ is the modified Bessel function. For a simple estimate of the effect of stimulation of electron scattering between different Landau levels, one can use the interpolation formula

$$A_{\rm DC}(\beta_{\rm f}) \simeq 1.85\beta_{\rm f}^{2.36} \exp\left[-\left(\frac{0.3}{\beta_{\rm f}}\right)^3\right] ,$$

which is very close to the exact equation in the important range $0.3 \le \beta_{\rm f} \le 1.7$, where this function is restricted to $0.04 \le A_{\rm DC}(\beta_{\rm f}) \le 6.5$.

Numerically, the strong decrease in $\nu_{\rm me}^{(\rm ins)}$ occurs even before the Coulomb broadening $\Gamma_{\rm C}$ becomes equal to $\Gamma_{\rm se}$, as shown in Fig. 4.4. Because of this strong reduction in the main term, the contribution of $\nu_{\rm me}^{(\rm out)}$ cannot be disregarded even at intermediate electron densities when $\Gamma_{\rm C}$ is still substantially smaller than $\hbar\omega_{\rm c}$ and the fluctuational electric field can be considered approximately as quasi-uniform. The fast increase in $\nu_{\rm me}^{(\rm out)}$ changes the sign of the many-electron effect: the decrease in $\nu_{\rm me}$ with $n_{\rm s}$ turns into an increase. For a fixed magnetic field, the effective collision frequency vs. electron density is shown in Fig. 4.5. Under the conditions of this figure, electrons predominantly populate the lowest Landau level. The extended SCBA curve increases slowly with $n_{\rm s}$, owing to the holding field dependence of the electron wave function parameter $\gamma(E_\perp)$ under the saturation condition $E_\perp = 2\pi e n_{\rm s}$. In contrast, the many-electron effective collision frequency $\nu_{\rm me}(n_{\rm s})$ [see (4.28)] displays a strong dip due to the suppression of electron scattering, and then, at higher densities, the collision rate increases with $n_{\rm s}$ due to electron scattering to higher Landau levels caused by the fluctuational electric field. Thus, the same many-electron effect, the Coulomb broadening of the electron DSF, leads to a successive decrease and increase in the electron collision frequency with $n_{\rm s}$.

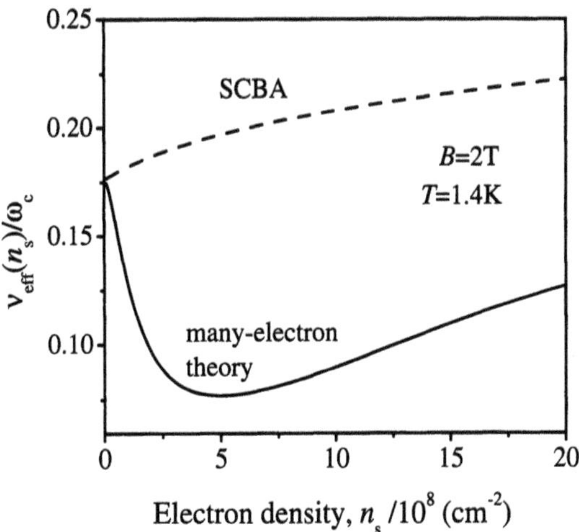

Fig. 4.5. The effective collision frequency vs. $n_{\rm s}$ shows the strong many-electron effect (*continuous curve*). The SCBA curve (*dashed curve*) increases because $E_\perp = 2\pi e n_{\rm s}$

Magnetoconductivity data are conventionally plotted against the magnetic field B at a fixed electron density and hence at a fixed average fluctuational electric field $E_{\rm f}^{(0)}$. In this case, the energy exchange $\hbar \boldsymbol{q} \cdot \boldsymbol{u}_{\rm f} \sim (\hbar/l_B)cE_{\rm f}/B \propto 1/\sqrt{B}$ increases when the magnetic field becomes lower, which causes the Coulombic effects discussed above. The typical magnetic field dependence of the effective collision frequency is shown in Fig. 4.6, together with field dependencies of the broadening of the two lowest Landau levels. One can see that the fluctuational electric field first reduces the normalized quantities $\nu_{\rm me}/\nu_{\rm se}$ and $\Gamma_0/\Gamma_{\rm se}$ in a similar way. Then, at lower magnetic fields, the many-electron increase in the collision rate becomes significantly stronger than the corresponding increase in the level broadening, because of scattering events out of the lowest Landau level. It should be noted that the presentation of the Coulombic effect as an effect caused by a decrease in the magnetic field is not very convenient, because decreasing the magnetic field also causes a breakdown in the quantum transport regime.

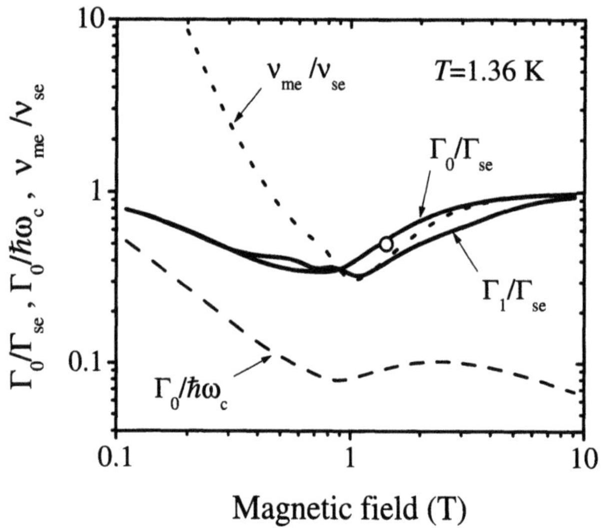

Fig. 4.6. Magnetic field dependence of the effective collision frequency (*dotted curve*) and normalized collision broadening of Landau levels (*continuous* and *dashed curves*) for $n = 10^8\,{\rm cm}^{-2}$[90]. The *circle* is the CR linewidth datum normalized to $\sqrt{2}\Gamma_{\rm se}$

The many-electron increase in the broadening of the DSF enhances the anomalous Hall effect because the magnetic field dependence

$$\nu_{\rm me}^{\rm (ins)}(B) \propto \omega_{\rm c}\Gamma_{\rm se}^2/\Gamma_{\rm C} \propto B^{5/2} \tag{4.35}$$

is even stronger than in the single-electron theory. The strict evaluation of $\tan\varphi_H = \omega_{\rm c}/\nu(B)$ according to the many-electron theory is shown in Fig. 4.7

by the continuous curve. The many-electron effect under the chosen conditions is very strong. Electron scattering to higher Landau levels caused by the Coulombic effect keeps the many-electron curve close to the classical Drude curve at weak and intermediate magnetic fields. Then, at approximately 1 T, the classical Hall behavior breaks down and, within a narrow magnetic field range, the field dependence of the many-electron curve changes sign, slowly approaching the SCBA curve (dash-dotted curve) in the limit $B \to \infty$. The sharp maximum of the many electron curve (continuous) placed far above the extended SCBA curve indicates the region where the many-electron effect on the electron collision frequency changes sign: the Coulomb stimulation of electron scattering to higher Landau levels changes to the suppression of electron scattering within the ground Landau level. The circles in the scatter graph in Fig. 4.7 show early magnetoconductivity data from [168], transformed into the ratio $\omega_c/\nu(B)$ according to the extended Drude form (3.3). The data show the same behavior as the many-electron curve. The numerical difference between experiment and theory may be caused by not including ripplon scattering in the theoretical curves and by an error in determination of the electron density in the experiment.

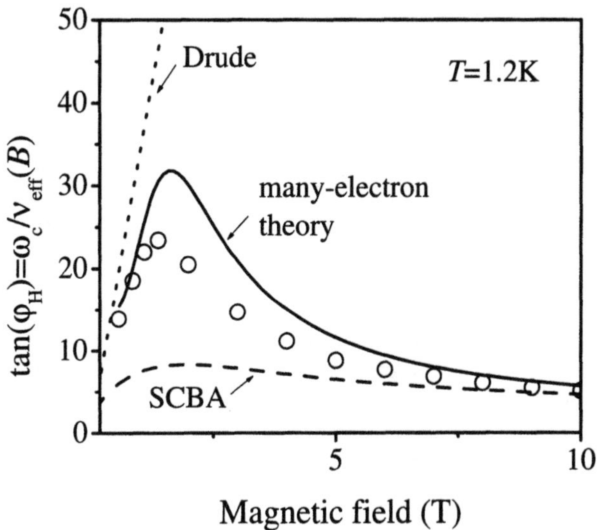

Fig. 4.7. Magnetic field dependence of the ratio $\omega_c/\nu_{\text{eff}}$ influenced by strong internal forces at $n_s = 3.2 \times 10^8\,\text{cm}^{-2}$. The many-electron theory (*continuous curve*) shows an even sharper decrease in the Hall angle than the SCBA (*dashed curve*), which deviates from the Drude result (*dotted curve*) at stronger fields. Data (*circles*) are taken from [168]

Conductance data from [170], analyzed by the equivalent circuit model, and the corresponding theoretical curves are shown in Fig. 4.8a as the ratio $\sigma_{xx}(0)/\sigma_{xx}(B)$ for two fixed electron densities. Note that, because of the holding field dependence of the SE localization parameter $\gamma(E_{\perp})$, the SCBA theory has different curves for each electron density shown in Fig. 4.8a by dashed lines. Therefore, for different n_s, the many-electron theory has different asymptotic behavior at $B \to \infty$. The Coulombic effect at low B and the dependence $\nu_{se}(E_{\perp})$ at high B affect the many-electron curve in opposite ways. As a result, the data for different densities cross at intermediate magnetic fields ($B \sim 6$–$7\,\mathrm{T}$).

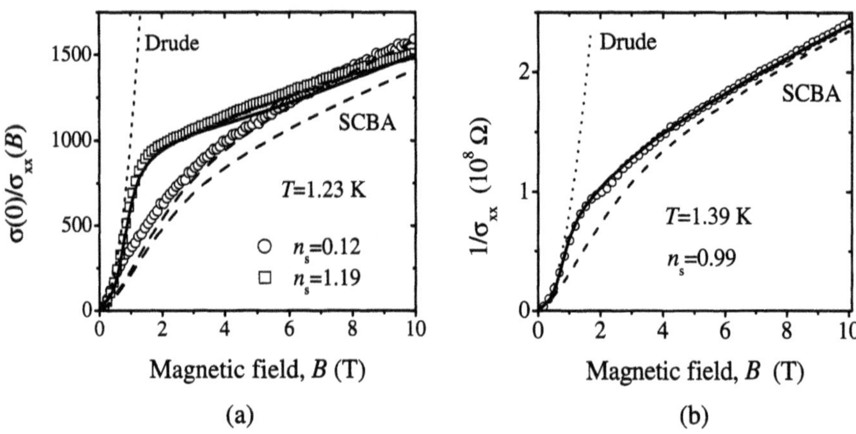

Fig. 4.8. Inverse conductivity vs. magnetic field found using the equivalent circuit (**a**) [170] and transmission line (**b**) [90] models. Electron densities are shown in units of $10^8\,\mathrm{cm}^{-2}$. The many-electron theory is plotted as *continuous curves*. The single-electron SCBA theory is shown by *dashed curves*, while the semi-classical Drude result is plotted as *dotted curves*

The equivalent circuit model used to analyze the conductance data provides the experimental conductivity data up to a numerical proportionality factor of the order of unity. The latter is usually found by fitting the data to the SCBA in the regime of extremely strong magnetic fields where the Coulombic effect is extremely small. Then at lower magnetic fields the data display the Coulombic effects in accurate agreement with the many-electron theory. The absolute value of the magnetoconductivity data can also be found using the transmission line model. The data of [90] found by means of the transmission line model and theory are shown in Fig. 4.8b for the intermediate electron density $n_s = 9.93 \times 10^7\,\mathrm{cm}^{-2}$. The data agree (even numerically) with the rigorous evaluation of the effective collision frequency (continuous curve), without any adjusting parameter.

It is instructive to note that the Coulombic effects discussed here occur without smearing of Landau levels. Despite the extremely strong internal forces, each electron has a well-defined Landau spectrum in its local frame (different for different electrons) because the forces are quasi-uniform at high magnetic fields. According to Fig. 4.6, illustrating the magnetic field dependencies of the most important magnetotransport parameters, the many-electron effect makes the Landau levels sharper. Even the strong increase in the collision rate (dotted curve) restoring the Drude conductivity behavior occurs at narrow levels $\Gamma_0/\hbar\omega_c \ll 1$. The CR linewidth datum (indicated by a circle) measured under the conditions of Fig. 4.8b does not show any smearing of Landau levels and is reasonably placed between two continuous curves describing the collision broadening of the ground ($N = 0$) and first excited ($N = 1$) levels. Thus, the 2D Coulomb liquid remarkably preserves the quantum properties of the 2D electron gas under a quantizing magnetic field.

It is interesting to compare the treatment of quantum magnetotransport discussed here with other theories. One can see that the quite clear and simple approximation (2.52) for the many-electron DSF used in the final equations of the universal quantum transport framework reproduces the results of the theory of Dykman and Khazan [85] and those of its sophisticated extension [88] as the extreme limiting case $\Gamma_C \gg \Gamma_{se}$. It should also be noted that the same many-electron decrease in the DC magnetoconductivity was found by Saitoh [140] in the framework of the path integral formalism for the Wigner solid interacting with ripplons. The important advantage of the discussed many-electron treatment consists in the possibility of combining self-energy effects (collision broadening) with Coulombic effects (inelastic scattering in the moving reference frames), which was previously impossible.

In the electron–ripplon scattering regime, the Coulombic effect on the magnetoconductivity of SEs is analogous to that discussed above for impurity-like scattering, if the scattering can be treated as quasi-elastic: $\hbar\omega_q \ll \Gamma_0^{(*)} = \sqrt{\Gamma_0^2 + b^2\Gamma_C^2}$. Because this kind of electron scattering eventually becomes inelastic in the limit of high magnetic fields, we discuss the low temperature magnetotransport properties of SEs in the following section, where we treat the problem of inelastic magnetotransport in 2D electron systems.

4.5 Inelastic Magnetotransport

In the low temperature regime $T < 0.7\,\mathrm{K}$, the vapor atom density is negligibly small and electron scattering is due to the interaction with capillary wave quanta (ripplons). The momentum exchange in one-ripplon processes is restricted by the electron magnetic length $q \sim 1/l_B$. Therefore, ripplons contributing to the effective collision frequency have quite low energies $\hbar\omega_q$ in comparison with T and their distribution function can be approximated as $N_q^{(r)} \simeq T/\hbar\omega_q$. As mentioned in Sect. 1.9.1, this results in a temperature dependence $\Gamma_0 \propto \sqrt{T}$ of the Landau level broadening. It is interesting that the Coulomb broadening Γ_C has the same temperature dependence and therefore that the total broadening of the DSF has the universal temperature dependence $\Gamma_0^{(*)} \simeq \sqrt{\Gamma_0^2 + b^2\Gamma_C^2} \propto \sqrt{T}$. In spite of this slow decrease in $\Gamma_0^{(*)}(T)$ with cooling, the Gaussians of the electron DSF are extremely narrow $\Gamma_{0,N}^{(*)} \ll T$ down to $T \sim 0.1\,\mathrm{K}$, which causes an interesting inelastic effect.

Let us first analyze the low temperature magnetotransport qualitatively, assuming that the electron–ripplon scattering is quasi-elastic. (The energy exchange at a collision $\hbar\omega_q$ is set to zero in the argument of the DSF.) Then, using the analogy with the vapor atom scattering case [see (4.29)], one can write $\nu_{\mathrm{eff}} \sim \omega_c(\Gamma_{\mathrm{se}}^{\mathrm{tr}})^2/(T\Gamma_0^{(*)})$. We estimate $\Gamma_{\mathrm{se}}^{\mathrm{tr}} \sim \Gamma_{\mathrm{se}}^{(\mathrm{elas})}$, where $\Gamma_{\mathrm{se}}^{(\mathrm{elas})}$ is the level broadening in the quasi-elastic approximation, which differs from the simple relation of (1.143) because of the long-range nature of the interaction. The quantity $(\Gamma_{\mathrm{se}}^{\mathrm{tr}})^2$ is proportional to the Bose distribution function of ripplons $N_q^{(\mathrm{r})}$. Since $N_q^{(\mathrm{r})} \simeq T/\hbar\omega_q$, the ratio $(\Gamma_{\mathrm{se}}^{\mathrm{tr}})^2/T$ does not depend on temperature and we conclude that the effective collision frequency increases with cooling as $\nu_{\mathrm{eff}} \propto 1/\sqrt{T}$. The unique temperature dependence $\sigma_{xx} \propto \nu_{\mathrm{eff}} \propto 1/\sqrt{T}$ (the collision rate increases when the number of scatterers $N_q^{(\mathrm{r})}$ decreases!), first predicted in the single-electron treatment [139] and in the extreme many-electron regime [85], represents the singular nature of elastic magnetotransport in two dimensions: $\nu_{\mathrm{eff}} \propto 1/\Gamma_0^{(*)} \to \infty$, if $\Gamma_0^{(*)} \propto \sqrt{T} \to 0$. This temperature dependence was first observed by Ito, Shirakhama and Kono [98] (see also [99]).

When considering the regime of extremely narrow maxima of the electron DSF, one should be very careful in treating the electron–ripplon scattering, because the limit $\Gamma_0^{(*)} \to 0$ leads to an uncertain conductivity for quasi-elastic scattering. Unlike the semi-classical treatment, where scattering is quasi-elastic if $\hbar\omega_q \ll T$, the quantum limit requires the more severe condition $\hbar\omega_q \ll \Gamma_0^{(*)}$ to be fulfilled for the scattering to be quasi-elastic. This requirement comes from (2.52) for the electron liquid DSF and it means that the energy exchange at a collision should be consistent with the broadening of the maxima of the DSF. Moreover, one can see that the unusual dispersion $\omega_q \propto q^{3/2}$ of capillary waves makes electron scattering eventually inelastic in the limit of strong magnetic fields because the energy exchange $\hbar\omega_q \propto 1/l_B^{3/2} \propto B^{3/4}$ increases with B faster than the broadening of the DSF: $\Gamma_0^{(*)} \to \Gamma_0 \propto \sqrt{B}$.

In the ripplon scattering regime, the DC effective collision frequency $\nu^{(\mathrm{r})}$ is related to the electron DSF by

$$\nu^{(\mathrm{r})} = \frac{\omega_c}{8\pi\alpha\hbar}\int_0^{\infty} \mathrm{d}x_q V_q^2 S(q,\omega_q)\,, \tag{4.36}$$

where $x_q = q^2 l_B^2/2$ is the dimensionless variable used previously. As the electron–ripplon coupling V_q is the complicated function (1.69), evaluation of (4.36) with the exact form of the many-electron DSF given in (2.51) is reasonably easy only if ω_q in the argument of the DSF can be disregarded (quasi-elastic approximation). In this case, we find that

$$\nu^{(r)} = \frac{\omega_c}{4\sqrt{\pi}\alpha\Gamma_{se}^{(elas)}} \int_0^\infty dx_q V_q^2 e^{-x_q} \int_0^\infty \frac{dy}{g(y)} e^{-y} \times \exp\left[-\frac{\lambda^2}{2g^2(y)} x_q y\right] I_0 \left[\frac{\lambda^2}{2g^2(y)} x_q y\right] , \tag{4.37}$$

where $I_0(x)$ is the modified Bessel function and we have introduced the normalized broadening of the lowest Landau level

$$g(y) \equiv g(y, \lambda) = \Gamma_0(y, \lambda)/\Gamma_{se}^{(elas)} ,$$

which depends on the dimensionless fluctuational field parameter $y = (E_f/E_f^{(0)})^2$ and also on $\lambda = \Gamma_C/\Gamma_{se}^{(elas)}$.

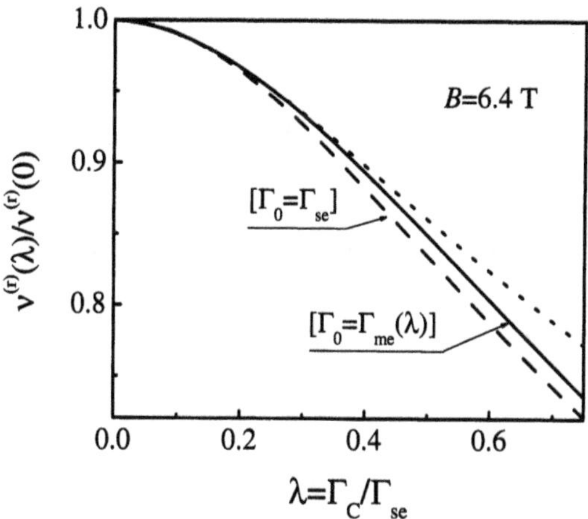

Fig. 4.9. Numerical graph showing the weak dependence of the many-electron decrease in the effective collision frequency on the Coulomb narrowing of Landau levels in the ripplon dominated scattering regime: strict evaluation with $\Gamma_0 = \Gamma_{me}(\lambda)$ (*continuous curve*) and with $\Gamma_0 = \Gamma_{se}$ (*dashed curve*). The approximate Gaussian form of the many-electron DSF yields a result which coincides with the *dashed curve* if $\Gamma_0 = \Gamma_{se}$ and deviates from both the *continuous* and *dashed curves* if $\Gamma_0 = \Gamma_{me}(\lambda)$ (*dotted curve*)

It is important to note that, in the many-electron theory, the exact form of $\nu^{(r)}$ shown in (4.37) is not very sensitive to the actual dependence $g(y, \lambda)$, as was also proven for electron–atom scattering. In Fig. 4.9, we show that the direct Coulombic reduction in the effective collision frequency is much stronger than the effect produced by the Coulomb narrowing of Landau levels. In addition, the latter acts in the opposite way. Therefore, any reasonable approximate equation for $\Gamma_0(E_f)$ would give a quite accurate result for the electron collision rate. Even the replacement $g(y,\lambda) \to 1$ [$\Gamma_0(\lambda) = \Gamma_{se} = \Gamma_{se}^{(elas)}$] works remarkably well. This approximation is shown in Fig. 4.9 by the dashed curve, which is quite close to the continuous curve taking into account the Coulomb narrowing of Landau levels. For $\Gamma_0 = \Gamma_{se}^{(elas)}$, the approximate form for the DSF of (2.52) yields the same conductivity result as that found using the exact formula (dashed curve), which could be anticipated from the condition $\Gamma_0(n_s) = \text{const.}$ used to obtain this approximate form. At the same time, the use of the more accurate expression for $\Gamma_0(E_f)$ in the approximate form of the DSF (dotted curve) deflects this approximation substantially from the result of the accurate evaluation (continuous curve). Considering the numerical results of Fig. 4.9, we can conclude that, in the approximate form of the DSF, it is sometimes better to keep the level broadening unchanged in order to stay close to the accurate result.

At high magnetic fields, the energy exchange $\hbar\omega_q$ can be larger than the Landau level width. In this case, use of the exact form of the DSF [see (2.51)] leads to numerical difficulties in evaluating the effective collision frequency. Nevertheless, the main inelastic effect on the collision rate can be understood and even described using the approximate Gaussian form for the DSF presented in (2.52). [We recall that this approximation concerns the dependence of $\Gamma_0(E_f)$ and it becomes an exact form if $\Gamma_0(E_f) \approx \text{const.}$] Using this form and disregarding electron transitions on higher Landau levels, the effective collision frequency of the low-temperature regime can be found as

$$\nu^{(r)} = \frac{\omega_c}{4\sqrt{\pi}\alpha} \int_0^\infty \frac{dx_q}{\sqrt{\Gamma_0^2 + x_q \Gamma_C^2}} V_q^2 \exp\left(-x_q - \frac{\hbar^2 \omega_q^2}{\Gamma_0^2 + x_q \Gamma_C^2} \right) . \tag{4.38}$$

As discussed above, it is reasonable in this equation to approximate $\Gamma_0 = \Gamma_{se}^{(r)}$ in order to keep the approximation close to the result which follows from the exact form of the DSF.

Equation (4.38) shows that the many-electron effect suppresses the reduction in the collision rate induced by the energy exchange $\hbar\omega_q$, and in order to estimate the inelastic effect, one should compare $\hbar\omega_q$ with the effective broadening $\Gamma_0^{(*)} \simeq \sqrt{\Gamma_0^2 + \Gamma_C^2}$ of the DSF. The same conclusion was found previously regarding the interplay of the intrinsic inelastic and the Coulombic effects for the average broadening of Landau levels [see (2.49)]. If the inelastic effect is small, the energy exchange $\hbar\omega_q$ does not depend on the temperature because it is fixed by the estimate $x_q \sim 1$. Under these conditions, the inelastic parameter $(\hbar\omega_q/\Gamma_0^{(*)})^2$ entering the exponent increases as the temperature decreases: $(\hbar\omega_q/\Gamma_0^{(*)})^2 \propto 1/T$. When it becomes of the order of unity, the collision rate falls significantly. As a result, the increase in $\nu_{eff}(T) \propto 1/\sqrt{T}$ with falling T eventually changes to a decrease, and $\nu_{eff}(T)$ attains a maximum when $\Gamma_0^{(*)}(T) \simeq \hbar\omega_q$. This shows how the self-energy effects eliminate the uncertainty in the quasi-elastic magnetotransport in the 2D electron system.

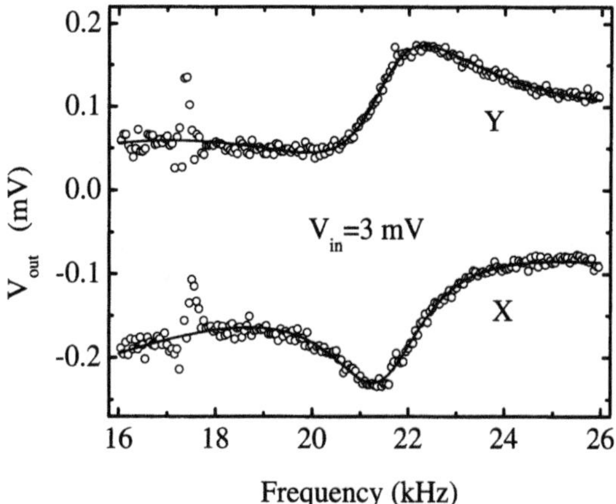

Fig. 4.10. Experimental signal from the biphase lock-in amplifier (X in phase, Y out of phase) as a function of frequency for $T = 0.3\,\mathrm{K}$, $n_s = 3.5 \times 10^7\,\mathrm{cm}^{-2}$, and $B = 1.8\,\mathrm{T}$ [99]. *Continuous curves* represent results obtained by fitting to Lorentzians

Employing the EMP damping method, Ito, Shirahama and Kono [98] measured the low temperature magnetoconductivity of SEs on superfluid helium. The output signal of the EMP resonance was fitted to the Lorentzian line shape with certain corrections added to eliminate the baseline offset, as shown in Fig. 4.10. The magnetoconductivity was found from the EMP damping using the relation (2.76). The proportionality factor in this relation was fixed in the high temperature range where electron–vapor atom scattering predominates, and it was not changed in the low temperature range. The conductivity vs. temperature data measured using the EMP damping method [99] are shown in Fig. 4.11 for two different values of the magnetic field. In the low temperature range ($T < 0.7\,\mathrm{K}$), the conductivity data of the lowest magnetic field $B_1 = 1.84\,\mathrm{T}$ (Fig. 4.11a) and the theoretical result (continuous curve) show a temperature dependence which is characteristic of quasi-elastic electron–ripplon scattering: $\sigma_{xx} \propto 1/\sqrt{T}$. As mentioned above, this unusual temperature dependence of the magnetoconductivity of SEs originates directly from the singular nature of the 2D electron system subject to a magnetic field: $\sigma_{xx}(T) \propto 1/\Gamma_0^{(*)}(T) \propto 1/\sqrt{T}$. Therefore, the data set of Fig. 4.11a is the most convincing manifestation of the quantum magnetotransport of 2D electrons with extremely narrow density of states.

The data of Fig. 4.11a also show the strong many-electron reduction in magnetoconductivity σ_{xx}, as compared to the single-electron SCBA theory (dotted curve marked 'elastic SE'). At the same time, the low temperature

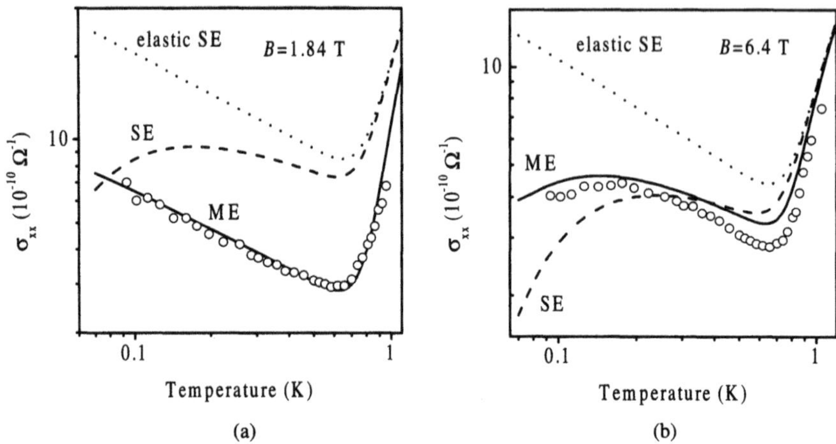

Fig. 4.11. Magnetoconductivity vs. temperature for two values of the magnetic field [99]. (**a**) At $B = 1.84\,\mathrm{T}$, the many-electron effect suppresses the intrinsic inelastic effect (*continuous curve*), which is expected to be strong in the single-electron theory (*dashed curve*), and makes the temperature dependence similar to the elastic SCBA result (*dotted curve*). (**b**) At $B = 6.4\,\mathrm{T}$, the many-electron broadening Γ_{C} is smaller and the inelastic effect becomes prominent at low temperatures. Experimental data are shown as *circles*

reduction of σ_{xx}, induced by the intrinsic inelastic effect in the single-electron theory and shown by the dashed curve marked 'SE', is clearly suppressed by internal Coulomb forces, in accurate agreement with the concept of the many-electron effect discussed above (continuous curve marked 'ME').

With a substantial increase in the magnetic field up to $B_2 = 6.4\,\mathrm{T}$, the temperature dependence of the conductivity data changes drastically, as shown in Fig. 4.11b. The data (circles) and theory (continuous curve) agree well but deviate strongly from the dependence $\sigma_{xx} \propto 1/\sqrt{T}$ and exhibit successively a maximum and a strong decrease with further cooling. The conductivity maximum relates to the condition $\hbar\omega_q \simeq \Gamma_0^{(*)}$, which can also be used as a probe for the broadening of the electron DSF. Both the Coulombic and inelastic effects are described by the same Coulomb broadening of the DSF calculated for the fluctuational electric field value, reduced slightly from its average interior value $E_{\mathrm{f}}^{(0)}$ by the proportionality factor 0.71 to cater for the reduced electron density at the edge strip of the electron pool.

4.6 Cold Nonlinear Effect

In the semi-classical treatment of electron transport, nonlinear conductivity usually appears through heating of the electron gas by the external electric field. This happens when the energy gained by an electron from the field within its mean-free path becomes large enough. In the 2D electron system with extremely narrow Landau levels ($\Gamma_0 \ll T$), a new kind of nonlinear effect becomes possible owing to the sharp structure of the electron density of states under a strong magnetic field [36, 179].

As discussed in Sect. 2.3, the driving electric field affects the probability of electron scattering and the collision broadening of Landau levels when the Doppler shift correction $\hbar \boldsymbol{q} \cdot \boldsymbol{u}$ becomes larger than Γ_0. Here $\boldsymbol{u}$ is the drift velocity of the whole system in the crossed magnetic and driving electric fields. In this case, the driving electric field $\boldsymbol{E}_{\parallel}$ itself reduces the interaction time τ_{i} and the number of multiple scattering events, which affects the collision frequency ν_{eff} making it dependent on $E_{\parallel}$. In order to describe such a nonlinear effect on the electron magnetoconductivity, one can use the momentum balance method discussed in Sect. 3.3.

The system considered here is characterized by an extremely narrow density-of-states function ($\Gamma_0 \ll T$), so that the exponents of (3.14) and (3.19) can be expanded in the parameter $\hbar \boldsymbol{q} \cdot \boldsymbol{u}/T$. At the same time, another nonlinear parameter $\hbar \boldsymbol{q} \cdot \boldsymbol{u}/\Gamma_0$ can be substantially larger than unity, and this strongly affects the collision frequency. Therefore, when describing the nonlinear magnetoconductivity as a function of $\hbar \boldsymbol{q} \cdot \boldsymbol{u}/\Gamma_0$, we can use the linear approximation with regard to the smaller parameter $\hbar \boldsymbol{q} \cdot \boldsymbol{u}/T$. This approximation yields

$$\nu(B, u) = \frac{1}{m_{\mathrm{e}} T_{\mathrm{e}} S_{\mathrm{A}}} \sum_{\boldsymbol{q}} q_u^2 \left[V_q^2 Q_q^2 S(q, \omega_q - \boldsymbol{q} \cdot \boldsymbol{u}) + \frac{\hbar^3 \nu_0^{(\mathrm{a})}}{2 m_{\mathrm{e}}} S(q, -\boldsymbol{q} \cdot \boldsymbol{u}) \right] , \tag{4.39}$$

where q_u is the projection of the vector $\boldsymbol{q}$ onto the direction of $\boldsymbol{u}$. The inelastic effect induced by the driving electric field thus affects the frequency argument of the DSF of the Coulomb liquid. In the single-electron treatment, the collision rate decreases when $\hbar \boldsymbol{q} \cdot \boldsymbol{u}$ becomes comparable with Γ_0.

Generally, in order to determine the conductivity tensor, we have to solve the nonlinear balance of forces equation. In the case of low temperatures, even under a high magnetic field, the Hall drift velocity is much larger than the average migration velocity along the electric field, i.e., $\sigma_{xy} \gg \sigma_{xx}$. Therefore, on the right-hand side of (4.39), we can replace $\boldsymbol{q} \cdot \boldsymbol{u}$ by $q(cE_{\parallel}/B) \cos \varphi$, where φ is the angle between $\boldsymbol{q}$ and the direction of the drift velocity in the absence of scattering. Then equation (4.39) directly gives the driving field dependence of the conductivity tensor caused by the cold nonlinear effect. Taking into account the fact that the electron DSF is not isotropic in the laboratory reference frame under a strong driving electric field, the effective collision frequency as a function of $E_{\parallel}$ can be found as

$$\nu^{(\mathrm{r})}(E_{\parallel}) = \frac{\omega_{\mathrm{c}}}{4\sqrt{\pi}\alpha} \int_0^{\infty} \frac{\mathrm{d}x_q}{\sqrt{\Gamma_0^2 + x_q \Gamma_{\mathrm{C}}^2}} V_q^2 \mathrm{e}^{-x_q} \tag{4.40}$$
$$\times \int \frac{\mathrm{d}\varphi}{2\pi} \cos^2\varphi \exp\left[-\frac{\left(\hbar\omega_q - \sqrt{2x_q} e E_{\parallel} l_B \cos\varphi\right)^2}{\Gamma_0^2 + x_q \Gamma_{\mathrm{C}}^2} \right] .$$

Here we have used the Gaussian approximation (2.52) for the DSF of the Coulomb liquid.

In the many-electron theory, the broadening of the electron DSF is determined by $\Gamma_0^{(*)} \simeq \sqrt{\Gamma_0^2 + \Gamma_{\mathrm{C}}^2}$. Equation (4.40) shows that the electron transport changes drastically if $\sqrt{2} e E_{\parallel} l_B$ becomes larger than $\Gamma_0^{(*)}$. The probability of electron scattering is significantly reduced because the energy exchange between an electron and a scatterer in the moving center-of-mass frame cannot be adopted by the narrow density-of-states function.

This cold nonlinear effect also reduces the Landau level broadening because of the electric field effect on the density of states discussed in Sect. 2.3.1 and described by (2.45), where E denotes the external electric field $E_{\parallel}$. Still, recalling the results of numerical evaluations in Sects. 4.4 and 4.5 in the approximate Gaussian form (2.52) of the DSF, it is numerically sufficient to approximate $\Gamma_0 = \Gamma_{\mathrm{se}}^{(\mathrm{r})} \simeq \Gamma_{\mathrm{se}}^{(\mathrm{elas})}$, because the direct field decrease in the effective collision frequency is much stronger than the change induced by the reduction in the collision broadening.

Generally speaking, the Coulombic effect on quantum magnetotransport discussed in Sect. 4.4 can also be considered as a sort of cold nonlinear effect with respect to the strong internal fluctuational electric field E_{f}. The important difference between the external driving field and the internal fluctuational field is that the latter cannot heat the system, by definition. In contrast, in the first case, heating interferes with the quantum cold nonlinear effect. Because the energy exchange in one-phonon processes is very small ($\hbar\omega_q \ll T$), the energy relaxation rate of SEs is also small. Under a strong magnetic field, the two-ripplon emission of thermal excitations is suppressed owing to the extremely narrow lowest Landau level. As a result, in the usual Corbino experiment, the cold nonlinear effect is rather difficult to observe against the usual background of heating. The situation changes a great deal when magnetoconductivity is measured by the edge magnetoplasmon method because, in this case, only a very narrow strip of the electron system near the edge is under the nonlinear regime and absorbs energy from the strong electric field. At the same time, high electron correlations redistribute this energy among all electrons, and all of them take part in the energy relaxation due to the electron–ripplon interaction. This increases the effective energy relaxation rate by approximately one order of magnitude and reduces heating.

In the nonlinear regime, the experimental signal deviates from the simple Lorentzian shape [36], as shown in Fig. 4.12a. The physics of this shape

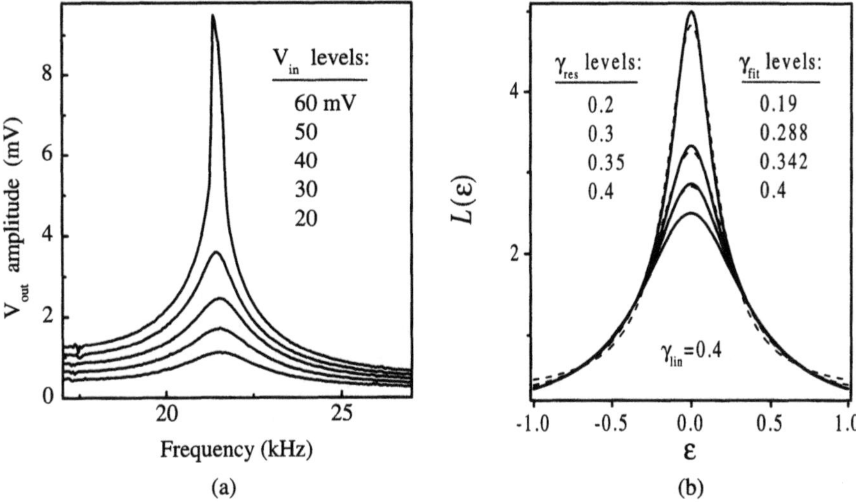

Fig. 4.12. (**a**) Amplitude of the experimental signal as a function of frequency for different input voltages. The *lowest curve* corresponds to the smallest value of V_{in}. (**b**) Nonlinear resonance curve from the frequency-dependent width model described in the text (*continuous curves*) for different strengths of the nonlinear effect $\gamma_{\text{lin}} - \gamma_{\text{res}}$. The *lowest curve* corresponds to the linear regime $\gamma_{\text{res}} = \gamma_{\text{lin}} = 0.4$. *Dashed curves* represent the result of fitting to Lorentzians with the frequency independent half-width γ_{fit} [36]

transformation can be understood in the following way. The collision rate and σ_{xx} are reduced by the cold nonlinear effect. For small nonlinear changes, to which we mainly confine our study, the correction to σ_{xx} is proportional to $-E_{\|}^2$. At the same time, the edge charge and the electric field in the EMP wave $E_{\|}$ exhibit resonant behavior as functions of frequency. Therefore, since the dimensionless EMP damping $\gamma = 1/\omega\tau$ is proportional to σ_{xx}, it has a strong frequency dependence due to the field $E_{\|}$, which can be described by

$$\gamma(\epsilon) = \gamma_{\text{lin}} - (\gamma_{\text{lin}} - \gamma_{\text{res}}) \frac{\gamma_{\text{lin}}^4}{(\epsilon^2 + \gamma_{\text{lin}}^2)^2} , \tag{4.41}$$

where $\epsilon = (\omega - \omega_{\text{emp}})/\omega_{\text{emp}}$, γ_{lin} is the normalized linear EMP damping or the half-width of the linear resonance curve, and γ_{res} represents the EMP damping at the resonance condition $\epsilon = 0$.

The frequency-dependent half-width of (4.41) is used in the usual Lorentzian function instead of γ_{lin}: $L(\epsilon) = \gamma(\epsilon)/[\epsilon^2 + \gamma^2(\epsilon)]$. Fig. 4.12b shows the nonlinear resonance curves of this frequency-dependent width model for different levels of the nonlinear effect described by $\gamma_{\text{res}} \leq \gamma_{\text{lin}}$. The difference $\gamma_{\text{lin}} - \gamma_{\text{res}}$ is a measure of the nonlinear effect. Comparing Fig. 4.12a and Fig. 4.12b, one can see that the frequency-dependent width model describes the nonlinear narrowing of the EMP resonances quite well. The observed nonlinear effect makes the resonance curves narrower and higher, mainly in the vicinity of the maximum, while the tails remain unchanged, in accordance with theory. Another important numerical consequence of Fig. 4.12b is that fitting the nonlinear curves to the usual Lorentzians gives the frequency independent half-width γ_{fit}, which is very close to γ_{res}.

EMP waves of the 2D electron liquid are excited by an AC voltage V_{in} applied to one of the outer electrodes. Thus the external AC electric field $\boldsymbol{E}_{\parallel}^{(0)} = -\nabla\Phi^{(0)}$ is mainly localized near the gap between two neighboring electrodes and can be estimated as V_{in}/d, where d is the helium depth. The real electric field $\boldsymbol{E}_{\parallel} = -\nabla\Phi$ responsible for EMP damping is produced by electron density perturbations δn. In general, it is very difficult to find a direct relationship between this field $E_{\parallel}$ and V_{in}. It is nevertheless possible to establish the frequency and magnetic field dependencies of the proportionality coefficient between $E_{\parallel}$ and V_{in}, which we use when comparing experimental data and theory. For this purpose we use the qualitative analysis of the EMP dispersion given in Sect. 2.5.2. Taking into account the presence of the external potential, the relation between the EMP charge Q and V_{in} can be found as

$$Q \propto \frac{\sigma_{yx}}{\omega - \omega_{\text{emp}}} V_{\text{in}} \,. \tag{4.42}$$

This equation just represents the resonance structure for the linear EMP charge.

At the resonance condition $\omega - \omega_{\text{emp}} \to 0$, the denominator of (4.42) should be replaced by $1/\tau_{\text{emp}} \propto \sigma_{xx}$, and the relation between the amplitude of the electric field in the EMP wave and the applied voltage acquires the form

$$E_{\parallel} \propto \frac{\sigma_{yx}}{\sigma_{xx}} V_{\text{in}} \,. \tag{4.43}$$

Because of this relation, the driving electric field in the EMP wave depends on σ_{xx}. Therefore, it is only for sufficiently small nonlinear changes that the nonlinear narrowing of the EMP damping is proportional to the nonlinear change in the SE conductivity. This condition is characterized by the linear relation between $E_{\parallel}$ and V_{in}. Strong nonlinear effects produce more rapid narrowing of the EMP damping than the change in σ_{xx}, which agrees with observations discussed below.

Equation (4.43) results in the following dependence of the nonlinear parameter $\lambda_{\parallel} = \sqrt{2}eE_{\parallel}l_B/\Gamma_0^{(*)}$ on the input voltage:

$$\lambda_{\parallel} \propto V_{\text{in}} \frac{l_B}{B\sigma_{xx}\Gamma_0^{(*)}} \,. \tag{4.44}$$

In the qualitative analysis of the linear transport regime presented at the beginning of Sect. 4.5, we found that $\nu_{\text{eff}} \sim \omega_c(\Gamma_{\text{se}}^{(\text{r})})^2/(T\Gamma_0^{(*)})$ and therefore $\sigma_{xx}\Gamma_0^{(*)} \propto (\Gamma_{\text{se}}^{(\text{r})})^2/\omega_c \approx \text{const.}$, where we have used the fact that $\Gamma_{\text{se}}^{(\text{r})} \propto \sqrt{B}$ for relatively low densities and $T \sim 0.5\,\text{K}$. This leads to $\lambda \propto V_{\text{in}}B^{-3/2}$, and therefore for the same nonlinear change, we have the following relation between the input voltage and magnetic field: $V_{\text{in}}B^{-3/2} \approx$ const. This relation is close to the experimentally observed result $V_{\text{in}}B^{-2} \approx$ const. We attribute the difference to the heating effect which becomes more important with decreasing magnetic field.

The magnetoconductivity data vs. the input voltage V_{in} found from the edge magnetoplasmon damping [36] are presented in Fig. 4.13 together with theoretical curves. The data represent the results of measurements at four different values of the magnetic field. The data sets and curves of the strongest magnetic field are marked by the number 1. The continuous curves take into account both the cold nonlinear and heating effects. Dashed curves are calculated disregarding the usual heating. For small nonlinear changes in σ_{xx}, the data follow the continuous curves. The sharp fall of the experimental data away from the continuous curves in the strong nonlinear regime is due to the fact that the electric field of the EMP and the drift velocity of electrons become nonlinear functions of V_{in}, while the theory assumes linear relations between them. In Fig. 4.13 one can see that, at strong magnetic fields, heating is very small and the major nonlinear change in the magnetoconductivity is due to the quantum cold nonlinear effect. This is because the highest magnetic field ($B \simeq 3.6\,\text{T}$) presented in Fig. 4.13 is close to the position of the minimum of the effective broadening of the electron DSF $\Gamma_0^{(*)} = \sqrt{\Gamma_0^2 + b^2\Gamma_C^2}$ as a function of B.

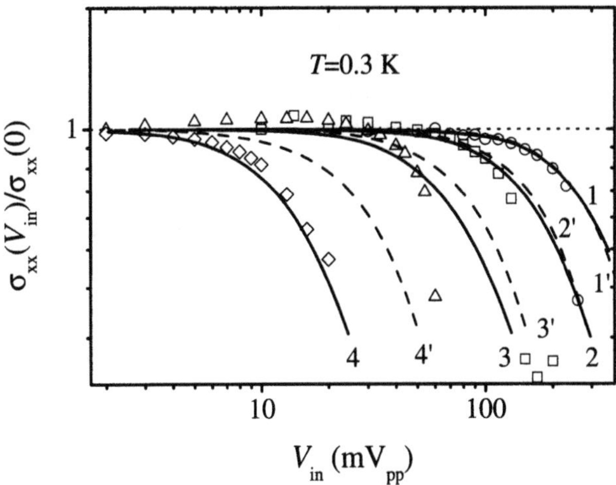

Fig. 4.13. SE magnetoconductivity vs. input voltage for four values of the magnetic field: 3.6 T (*circles* and *curves* 1 and 1′), 2.7 T (*squares* and *curves* 2 and 2′), 1.8 T (*triangles* and *curves* 3 and 3′), and 0.91 T (*diamonds* and *curves* 4 and 4′) [36]. *Continuous curves* represent the many-electron nonlinear theory. *Dashed curves* are plotted disregarding the heating effect

The effective broadening $\Gamma_0^{(*)}$ increases with falling magnetic field, which makes the parameter $\hbar \boldsymbol{q} \cdot \boldsymbol{u}/\Gamma_0^{(*)}$ smaller. At the same time, the heating effect becomes relatively stronger, causing the shift between the dashed curves and the corresponding data and continuous curves. The electron temperature is found from the energy balance equation

$$m_e u^2 \nu(T_e) \Delta N_e = (T_e - T) \tilde{\nu}(T_e) N_e \ , \tag{4.45}$$

where $\tilde{\nu}(T_e)$ is the energy relaxation rate and ΔN_e is the number of electrons in the edge strip which are heated by the EMP electric field. We use the approximate relation $\Delta N_e \approx 2hN_e/R_{el}$, with the electron sheet radius $R_{el} \simeq$ 15 mm and the width of the edge strip (the transition region, where the electron density changes) $h \simeq 0.3$ mm.

Hence, the universal quantum transport framework and the model for the DSF of the Coulomb liquid discussed in the preceding chapters provide a quite effective description of the nonlinear narrowing of the EMP damping coefficient and the dependence of the SE magnetoconductivity on the amplitude of the input voltage V_{in}. According to (4.43), the theoretical curves in Fig. 4.13 are strictly valid only for small nonlinearities because σ_{xx} in this equation is actually a function of λ. For small λ, it is possible to use the linear relation between λ and V_{in}. The decrease in σ_{xx} with λ sharply enhances the effective driving electric field, and this causes the more rapid fall in the experimental data with V_{in}, if the nonlinear change exceeds 30%.

In conclusion, for the surface electrons on liquid helium at low temperatures, there is a physically interesting regime where the nonlinear decrease in the edge magnetoplasmon damping and σ_{xx} with an increase in the input voltage is due to the extremely narrow maxima of the dynamical structure factor of the 2D Coulomb liquid under a strong magnetic field. Such an effect is a pure quantum phenomenon which does not exist in a classical electron liquid. It relates physically to the many-electron reduction in the electron magnetoconductivity: both effects are due to the electric-field-induced decrease in the interaction time τ_i and the number of multiple scattering events of an electron with the same scatterer. Therefore, the direct observation of the cold nonlinear effect nicely completes the physical picture of the unconventional Hall effect in the nondegenerate quantum electron liquid.

5 Quantum Cyclotron Resonance

5.1 Early Achievements

The basic idea and the phenomenon of cyclotron resonance (CR) in ionized gases have been known for quite some time. In the semi-classical picture, a charged particle subject to a magnetic field gains energy resonantly from the alternating electric field and its orbit radius increases if the frequency of the field satisfies $\omega = \omega_c$. In the quantum theory, under these conditions, a single particle absorbs energy from the electric field by moving to higher energy levels which are equally spaced in the absence of interactions. In solids, the CR absorption method has proved to be a powerful tool for studying electronic properties of matter.

Considerable research has been performed on cyclotron resonance in two-dimensional electron systems created in semiconductor structures and on the free surface of superfluid helium. Even raw data for CR absorption usually have a quite visual and convincing form. For example, the first observation of CR absorption from surface-bound electrons on liquid helium performed by Brown and Grimes [180] was a very important event for experimental progress in studying this two-dimensional electron system. Just after the release of the negative result due to Ostermeier and Schwarz [181], questioning the existence of the image-potential-induced surface states on helium, Brown and Grimes conclusively demonstrated the existence of electrons in 2D states outside of liquid helium by tilting the magnetic field under CR conditions. The evidence was based on the fact that for 3D electrons the CR condition $\omega = \omega_c$ does not depend on the direction of the magnetic field, while for 2D electrons it is only the component B_z which contributes to the cyclotron frequency for the in-plane motion. Therefore, at a fixed signal frequency ω, tilting the magnetic field to an angle θ to the vertical requires one to apply a stronger magnetic field $B = B_0/\cos\theta$ in order to reach the resonance condition.

Examples of CR absorption as a function of magnetic field from [180] are shown in Fig. 5.1 for two different orientations of the field. The shift in the resonance peak owing to the tilting $\boldsymbol{B}$ clearly seen in this figure is the most compelling evidence for the two-dimensional character of electron motion. Measuring at different angles shows that the peak position is really proportional to $\sec\theta$, as indicated in Fig. 5.2. An interesting extension of this field-tilting effect was observed in the nonlinear CR experiment of Edel'man

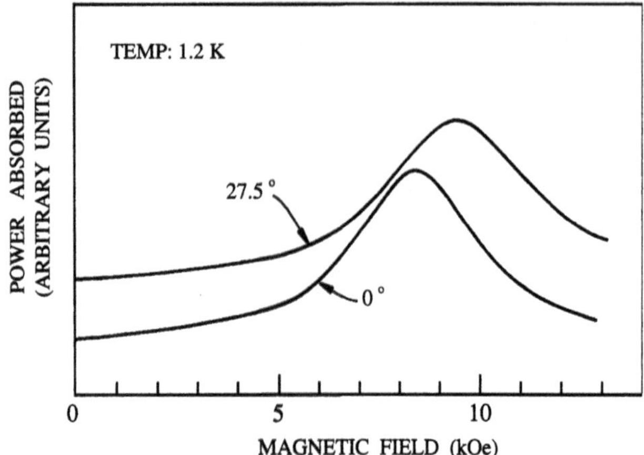

Fig. 5.1. Typical examples of the CR absorption data for SEs on liquid helium in tilted and untilted fields [180]

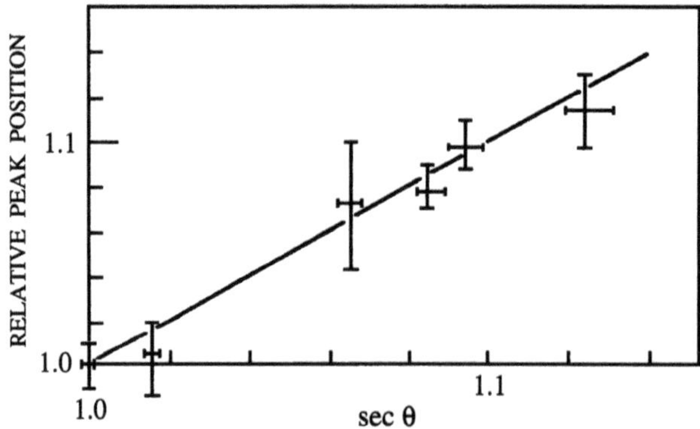

Fig. 5.2. Relative peak shift vs. sec θ. *Error bars* are $\pm 0.5°$ horizontally and represent maximum uncertainty vertically. The *straight line* is the expected behavior for an electron in a purely 2D state [180]

[24]. In the linear regime, the CR absorption line has the usual resonance shape (the lower line in Fig. 5.3 marked $P = P_0$). At higher powers, the CR line has an additional peak owing to electrons evaporating into 3D states because of power heating, as shown in Fig. 5.3 by the upper absorption lines. When the magnetic field is tilted, the broad shape of the CR absorption shifts leftward according to the condition $\omega = \omega_c \cos\theta$. At the same time, the sharp peak for the 3D electrons does not change its position.

The position of the CR peak provides information about the effective mass of charged particles and about their coupling with medium vibrations. Experiments [182, 183] performed at low temperatures ($\approx 0.4\,\mathrm{K}$) showed that, in the limit $E_\perp \to 0$, the mass of SEs coincides with the free electron mass m_e to an accuracy of 10^{-4}. If the holding electric field increases, the position of the CR shifts approximately as $\Delta B \propto -E_\perp^2$, which indicates that the

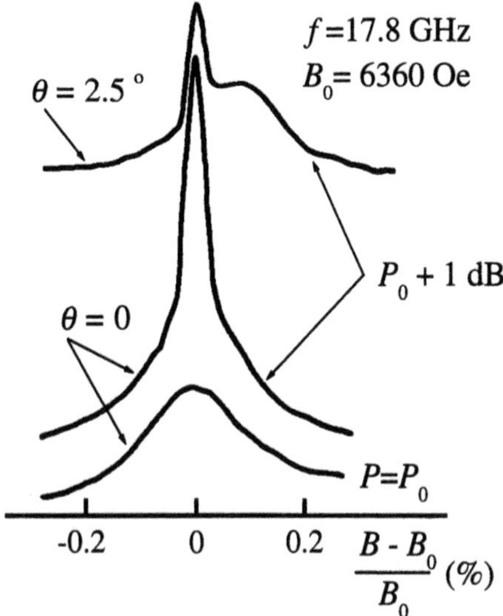

Fig. 5.3. Amplitude of the reflected MW vs. magnetic field for two different power levels and tilting angles θ [24]

response of the medium to the high-frequency motion of SEs becomes stronger with $E_\perp$, in accordance with the electron–ripplon coupling theory ($V_q \propto E_\perp$) discussed in Chap. 1. At this point, we would like to pay attention to the sign of the CR shift. From the negative sign of ΔB, it might be naively concluded that the interaction with the medium reduces the electron effective mass. In reality, the character of the medium response depends crucially on the frequency of the signal: at low frequencies, the medium excitation cloud follows the electron motion, increasing its effective mass, while at high frequencies, the cloud is motionless and the interaction induces an oscillatory potential for the electron. The origin of this sign is easily understood using the result of the phenomenological treatment of electron conductivity discussed in Sect. 3.2. According to (3.8) and (3.9), the resonance condition for the magnetic field can be written as

$$B_{\mathrm{res}} = \frac{m_e c}{e}\left[\omega + w(\omega)\right] \equiv \frac{m_e c}{e}\left[\omega + \frac{\lambda(\omega) - \lambda(0)}{\omega}\right] . \tag{5.1}$$

As mentioned in Sect. 3.2, $\lambda(\omega)$ represents the response of the medium. This function usually has a resonance structure due to medium vibrations, which we describe by the excitation spectrum ω_q:

$$\lambda(\omega) \propto \frac{\omega_q^2}{\omega_q^2 - \omega^2} . \tag{5.2}$$

Generally, we should have the sum over all wave vectors $\boldsymbol{q}$ here. At low frequencies $\omega \ll \omega_q$, we have $\lambda(\omega) - \lambda(0) \propto +\omega^2$ and $w(\omega) \propto +\omega$, which corresponds to an increase in the electron effective mass due to the polarization cloud and confirms the sign chosen in (5.2). At high frequencies $\omega \gg \omega_q$, corresponding to CR conditions, the response function $\lambda(\omega)$ strongly decreases and we find that

$$B_{\rm res} - B_0 = -\frac{m_e c \lambda(0)}{e\omega} < 0 \,, \tag{5.3}$$

which is in accordance with the observed effect. Thus the negative sign in (5.3) is caused by the oscillatory response of the medium polarization cloud to the high-frequency motion of electrons: $\boldsymbol{F}_{\rm scat} \simeq -N_{\rm e} m_{\rm e} \lambda(0) \boldsymbol{u}_{\rm av}$. In order to explain the experimentally observed shift of the CR peak, Cheng and Platzman introduced a magnetopolaron model [184] which we discuss later.

Most importantly, the linewidth of the CR relates directly to the effective collision frequency of electrons. Therefore, in the CR experiment, $\nu_{\rm eff}(\omega)$ can be measured directly without any uncertainty in proportionality factors which are typical for the low-frequency magnetoconductivity measurement in the system of SEs on liquid helium. This allows one to perform a direct experimental verification of the magnetotransport theory of strongly interacting electrons and to check the quantum transport framework discussed in Chap. 3. Figure 5.4 shows how an increase in the electron–ripplon coupling affects the linewidth and position of the CR observed in [183] for SEs on liquid ^{3}He and ^{4}He. At weak holding fields, the CR absorption line is narrow. An increase in the holding field makes it substantially broader and shifts the extremum of the recordings into the lower magnetic field range.

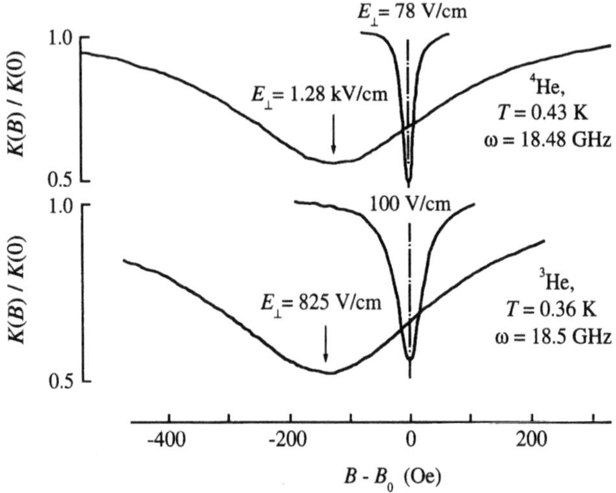

Fig. 5.4. Typical recordings showing the dependence of the CR line (a signal that has passed through the cavity vs. B) on the holding electric field [183]. An increase in $E_\perp$ broadens the CR line and shifts the peak position to the left

Most of the CR measurements in [9, 183] were performed under the conditions of quantum ($\hbar\omega_c \sim T$) or ultra-quantum ($\hbar\omega_c \gg T$) transport regimes. Unfortunately, the quantum transport theory of the 2D Coulomb liquid had not been developed by that time. Therefore the comparison of the linewidth data with the semi-classical theory of the CR given in [9] revealed many contradictions, especially in the low density limit. It was even concluded that the electron–ripplon interaction was not well understood [9, 183]. Typical examples of these data and curves for the semi-classical theory (dotted) are shown in Fig. 5.5. For nondegenerate electrons, one may expect a strong deviation of experimental data from the single-electron semi-classical theory in the opposite limit of high densities or strong holding electric fields. Nevertheless, in this regime the data were in qualitative agreement with theory and the strong deviation occurs for low electron densities.

The important insight into the problem of strongly interacting 2D electrons under a quantizing magnetic field was made by Dykman and Khazan [85, 86], who introduced the concept of a quasi-uniform fluctuational electric field. The low-density range of Edel'man's experiments was actually a rather high density range for the theory. (In Fig. 5.5, Dykman's theory is shown by dashed lines.) This theory explained the decrease in the CR linewidth with the holding electric field in the low density range. Still, at higher holding fields, the increase in the linewidth data is substantially stronger than that given by the quantum theory. In 1988, the direct verification of the many-electron theory of [86] performed by Wilen and Giannetta [185] down to $n_s \sim 0.2 \times 10^8\,\mathrm{cm}^{-2}$ showed a strong qualitative conflict between experiment and theory. For fixed holding fields, the CR linewidth was shown to have opposite density dependencies in theory and experiment. This conflict 'froze' experimental and theoretical research into the quantum CR in the 2D Coulomb liquid for a long while.

A significant advance in the study of the quantum CR of highly correlated 2D electron systems on liquid helium was achieved quite recently in [76, 89] by revising both experiment and theory. The first step was the new treatment of the fluctuational electric field based on the description of electron scattering events in local, moving reference frames, where $E_f' \to 0$, as discussed in Chap. 2. This procedure simplifies the many-electron theory to such an extent that one can study the interplay between Coulombic and self-energy effects and incorporate the effect of electron scattering between different Landau levels due to Coulomb broadening of the electron DSF. It should also be noted that the quantum transport framework discussed in the Chap. 3 appears to be very practical for the description and understanding of the quantum CR of that strongly interacting electron system. For the present, all aspects of the CR experiment such as line broadening, line shape transformation and even power dependence of the linewidth are well described by this theory.

We would like to stress important differences between CR studies of semiconductor 2D electron systems and electrons on a liquid helium surface. In

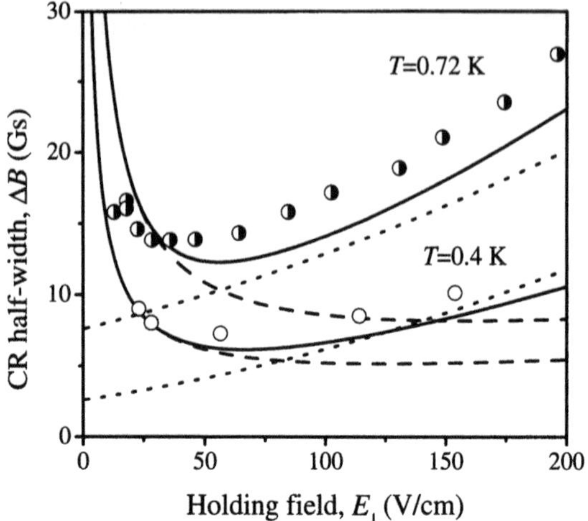

Fig. 5.5. CR half-width vs. holding electric field $E_{\perp} = 2\pi e n_s$. Data obtained by Edel'man (*circles*), semi-classical theory (*dotted curves*), Dykman's result obtained for the Wigner crystal state (*dashed curves*), quantum transport framework employing the Coulomb liquid DSF (*continuous curves*)

degenerate 2D electron systems, the single-electron approach is applicable in the high electron density regime, and lowering n_s introduces Coulomb correlations. Interesting many-electron effects on the CR from Si inversion layers were observed in [186–188] when the lowest Landau level is partially occupied. At the smallest electron densities $n_s \sim 5 \times 10^9\,\mathrm{cm}^{-2}$ achieved in these studies, the CR linewidth is much narrower than it can be according to any single-electron theory or the semi-classical Drude formula. As for SEs on liquid helium, this nondegenerate electron system is on the other side of the phase diagram, and lowering n_s makes the single electron approximation more applicable, while an increase in the electron density enhances many-electron effects. Thus, in contrast to the semiconductor systems, SEs on liquid helium can be used to study the appearance of Coulombic effects on CR absorption with increasing n_s, starting from the smallest densities $n_s \sim 10^7\,\mathrm{cm}^{-2}$, where the single electron approximation does not interfere with the effect of completion of the lowest Landau level. Comparing experimental data obtained from such a pure system with different theoretical models reveals the origin of the strong CR narrowing induced by electron–electron interactions.

5.2 Single-Electron Approaches

Before proceeding to discuss CR absorption in the strongly interacting 2D electron system, it is instructive to compare the results of the main single-electron approaches. Two major dynamical conductivity treatments can be used to describe the quantum CR. The conventional approach developed by Ando [189] is based on the perturbation theory for electron conductivity with self-energy effects and vertex corrections taken into account. The single-electron Green's function entering the conductivity terms is treated in the framework of the SCBA. Another approach is based on the memory function formulation of the dynamical conductivity discussed in Chap. 3 as an essential part of the universal quantum transport framework. In this method, line broadening relates directly to the electron DSF. At first glance, these two approaches should lead to substantially different absorption line shapes. Therefore it is very important to check the consistency of the results which follow from the universal quantum transport framework discussed in Chap. 3 and that found in the conventional perturbation treatment.

Consider electron scattering by the short-range vapor atom potential, which dominates at $T > 1\,\mathrm{K}$. In this case, direct application of the Ando theory for the CR line shape [189] developed for noninteracting 2D electrons yields the result

$$\mathrm{Re}\left[\sigma_{xx}(\omega)\right] = \frac{e^2\omega_c}{4\pi^2\hbar^2}\int \mathrm{d}\varepsilon\left[f(\varepsilon) - f(\varepsilon+\hbar\omega)\right]\mathrm{Im}G_0(\varepsilon)\mathrm{Im}G_1(\varepsilon+\hbar\omega)\ , \quad (5.4)$$

where $f(\varepsilon)$ is the Fermi distribution function and $\mathrm{Im}G_N(\varepsilon)$ represents the density of states of Landau levels. In the quantum limit, one can disregard $f(\varepsilon+\hbar\omega)$ in (5.4), and even replace $f(\varepsilon)$ by $f(\varepsilon_0)$ because of the extremely narrow Landau levels $\Gamma_0 \ll T$. Then the absorption line shape becomes proportional to the integral

$$I_G(\hbar\omega) = \Gamma_0 \int \mathrm{Im}G_0(\varepsilon)\mathrm{Im}G_1(\varepsilon+\hbar\omega)\mathrm{d}\varepsilon\ , \quad (5.5)$$

which is a measure of the overlap of the lowest Landau level and the first excited level shifted by $\hbar\omega$ due to the microwave radiation (MR).

As discussed in Chap. 1, the SCBA theory results in the semi-elliptic Landau level shape, while the cumulant expansion method yields the Gaussian level shape with the same broadening parameter Γ_N. According to Fig. 1.20 of Chap. 1, these shapes of the density-of-states function are quite different, and one may expect the CR line shapes obtained using these forms of $\mathrm{Im}G_N(\varepsilon)$ to be substantially different as well. It is therefore remarkable that the integral in (5.5) does not much depend on what kind of Landau level shape is used, be it semi-elliptic or Gaussian. Numerical evaluation of $I_G(\hbar\omega)$ for the semi-elliptic shape of $\mathrm{Im}G_N(\varepsilon)$ is shown in Fig. 5.6 by the continuous curve. Surprisingly, it is rather close to that found for the Gaussian shape (dashed curve):

$$I_G(\hbar\omega) = \pi^{3/2}\exp\left[-\hbar^2(\omega-\omega_c)^2/\Gamma_{0,1}^2\right]\ , \quad (5.6)$$

where $\Gamma_{0,1} = \sqrt{(\Gamma_0^2+\Gamma_1^2)/2}$ is the average level broadening, which equals Γ_0 for the interaction with short-range scatterers. Therefore, the CR line shape is very close to Gaussian regardless of the actual shape of the Landau levels, with the exception of the region $|\omega - \omega_c| / \Gamma_{se} > 2$, where the absorption tails originating from the semi-elliptic functions vanish while the Gaussian tails are still present but negligibly small.

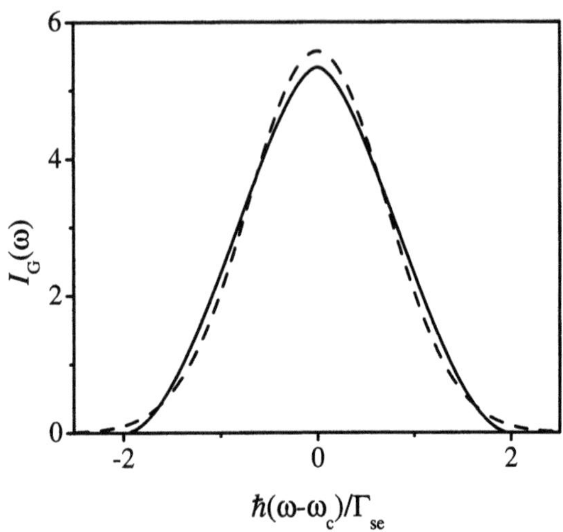

Fig. 5.6. CR line shape which follows from the semi-elliptic (*continuous curve*) and Gaussian (*dashed curve*) forms of $\mathrm{Im}G_N(\varepsilon)$, according to (5.5)

Taking into account the above conclusion and the fact that the Gaussian level shape usually provides very simple analytical results, here and below we shall mainly employ the Gaussian shape for the imaginary part of the electron Green's function. In this case, (5.4) can be rewritten in the simple Gaussian form

$$\mathrm{Re}\left[\sigma_{xx}(\omega)\right] = \frac{\sqrt{\pi} e^2 n_s \hbar}{4 m_e \Gamma_{0,1}} \exp\left[-\frac{\hbar^2 (\omega - \omega_c)^2}{\Gamma_{0,1}^2}\right] . \tag{5.7}$$

In the pure single-electron treatment, $\Gamma_{0,1} = \Gamma_0 = \Gamma_1 = \Gamma_{se}$. Still, we prefer to distinguish these quantities, intending an application to the many-electron case.

The framework for practical evaluations based on the memory function formulation seems to provide us with a different CR line shape which follows from (3.8) and (3.26). Near the resonance frequency, the real part of the electron conductivity can be written as

$$\mathrm{Re}\left[\sigma_{xx}(\omega)\right] \simeq \frac{e^2 n_s}{2 m_e} \frac{\nu_{\mathrm{eff}}(\omega)}{(\omega - \omega_c)^2 + \nu_{\mathrm{eff}}^2(\omega)} . \tag{5.8}$$

The usual replacement $\omega \to \omega_c$ in the effective collision frequency $\nu_{\mathrm{eff}}(\omega)$ would give us the Lorentzian shape of the CR absorption line with full width equal to $\gamma_{\rm CR} = 2\nu_{\rm eff}(\omega_{\rm c})$. It should be noted that this is not an entirely correct procedure for 2D electrons because, as a function of ω, $\nu_{\rm eff}(\omega)$ changes on the same scale as the main conductivity equation. For example, recalling that in the quantum limit

$$\nu_{\rm eff}(\omega) = \frac{\Gamma_{\rm se}^2 \omega_{\rm c}}{8\hbar^2 \omega} \int_0^\infty x_q S(q,\omega) {\rm d}x_q \,, \tag{5.9}$$

and inserting here the DSF for non-interacting electrons in (2.38), we find that

$$\gamma_{\rm CR}(\omega) \simeq \frac{\sqrt{\pi}\Gamma_{\rm se}^2}{\Gamma_{0,1}\hbar} \exp\left[-\frac{\hbar^2(\omega-\omega_{\rm c})^2}{\Gamma_{0.1}^2}\right] . \tag{5.10}$$

Here we have taken into account the fact that the high frequency electric field ($\omega \sim \omega_{\rm c}$) mixes two nearest Landau levels, and that at $\Gamma_0 \ll \hbar\omega_{\rm c}$ the main contribution to $S(q,\omega)$ comes from the $N=1$ term, which we shall call the resonance term. We note first that in the high frequency case, the effective collision frequency has a different temperature dependence to the DC case. The factor $1/T$ in the collision frequency for the DC case, originating from the general equation factor

$$\left[1 - \exp(-\hbar\omega/T)\right]/\omega \,,$$

is replaced by $1/\hbar\omega$ under the condition $\hbar\omega \gg T$.

The most important feature of the effective collision frequency of (5.10) is that it has a resonance structure itself. The width of the resonance form of the effective collision frequency is practically the same as that of the original conductivity form because $\gamma_{\rm CR}(\omega_{\rm c}) \sim \Gamma_{\rm se}/\hbar$. This does not allow one to conclude that the CR line shape is Lorentzian. The sharp resonance structure of the effective collision frequency affects the CR line shape, making it narrower in the tail region $[\hbar(\omega-\omega_{\rm c}) \sim \Gamma_{\rm se}]$. Moreover, the CR line shape is affected in such a way that the initial Lorentzian form of (5.8) is transformed into a form which is close to Gaussian, in accordance with the result of the conventional theory [see (5.7)].

The simple numerical comparison shown in Fig. 5.7 is very instructive. In terms of the dimensionless parameter x, the structure of (5.8) and (5.10) is equivalent to a generalized Lorentzian function

$$\mathcal{L}(x) = \frac{\gamma(x)}{4(x-x_0)^2 + \gamma(x)} \,, \tag{5.11}$$

with the frequency-dependent width parameter

$$\gamma(x) = \sqrt{\frac{\pi}{2}} w_0 \exp\left[-\frac{2(x-x_0)^2}{w_0^2}\right] , \tag{5.12}$$

where $w_0 = \sqrt{2}\Gamma_0/\hbar\omega_{\rm c}$. If one disregards the frequency dependence of γ, assuming $\gamma(x) \simeq \gamma(x_0) \equiv \sqrt{\pi/2}w_0$, then the Lorentzian function (continuous curve L) will be substantially removed from the corresponding Gaussian function

$$G(x) = \sqrt{\frac{2}{\pi w_0^2}} \exp\left[-\frac{2(x - x_0)^2}{w_0^2}\right] \tag{5.13}$$

(continuous curve G) in the tail region (we set $w_0 = 0.085$ and $x_0 = 1$), although the width at half-height for these functions is nearly the same. Here we compare functions with the same maximum position. Conversely, the generalized Lorentzian function of (5.11) with frequency-dependent width (dashed curve) is quite close to the proper Gaussian function. The difference between these two curves may be attributed to the accuracy of the approximation. It is interesting to note that the model with $\gamma(x) = \gamma(x_0) \exp\left[-2(x - x_0)^2/\gamma^2(x_0)\right]$ leads to a curve (dotted) which is very difficult to distinguish from the Gaussian curve. Thus, both the conventional conductivity treatment performed by Ando and the memory function formalism give nearly the same CR absorption curves, if Coulombic effects are disregarded.

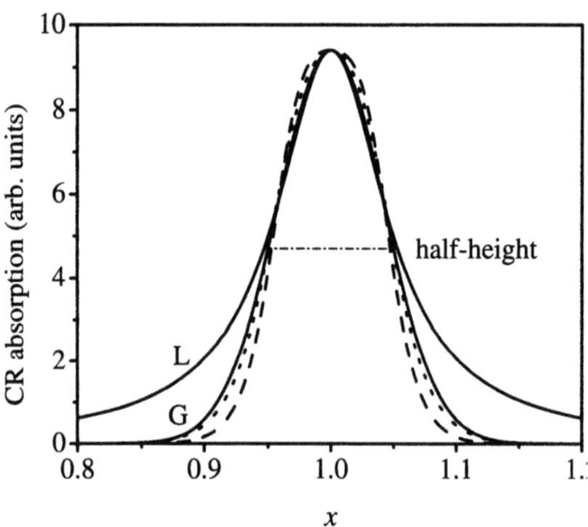

Fig. 5.7. Typical CR absorption curves: Gaussian (*continuous curve* G), Lorentzian (*continuous curve* L), frequency-dependent width models (*dashed* and *dotted curves*) presented according to the memory function formalism, as described in the text [76]

It should also be noted that the semi-classical approach usually results in a Lorentzian CR absorption shape. Therefore, a Gaussian line shape of non-degenerate free 2D electrons is one of the distinctive features of the quantum CR theory. The broadening of the CR absorption line is determined by the broadening Γ_{se} of the Landau levels. In the following section, we shall see that mutual Coulomb forces strongly affect the shape and broadening of the CR absorption line.

5.3 Cyclotron Resonance and Internal Forces

Before discussing an accurate description of the many-electron theory of the CR, let us have a further look at the result given by the single-electron quantum theory [see (5.4)]. This contains the density-of-states functions of the two nearest Landau levels $\mathrm{Im}G_0(\varepsilon)$ and $\mathrm{Im}G_1(\varepsilon + \hbar\omega)$ broadened due to the interaction with scatterers. If internal forces introduced a sort of broadening of the single-electron states described by $\mathrm{Im}G_N(\varepsilon)$, then (5.4) would give a broadening of the CR line existing even in the absence of scatterers, and this would be very difficult to explain because of Konh's theorem. In reality, the Coulombic effect discussed in this book is not a self-energy effect and it does not lead to any broadening of the CR line in the absence of scatterers. It can only reduce or increase the broadening of the CR line induced by scatterers.

Considering the Coulomb liquid as an ensemble of independent electrons exposed to a quasi-uniform fluctuational electric field, one can introduce two different approaches for incorporating many-electron effects in the quantum theory of the CR. We note first that the drift velocities $\boldsymbol{u}_\mathrm{f}$ of the electron orbits caused by the fluctuational electric field E_f are ultra-fast only with regard to electron scattering by vapor atoms or ripplons, because they affect the probability of scattering events if $\boldsymbol{u}_\mathrm{f} > \Gamma_N/\hbar q$. Concerning the electron transitions induced by MW radiation under CR conditions, the velocities $\boldsymbol{u}_\mathrm{f}$ of the electron orbit centers are too slow to be taken into account in a direct way because of the high value of the velocity of light. For microwaves, the corresponding Doppler shift $\boldsymbol{k} \cdot \boldsymbol{u}_\mathrm{f}$ is extremely small because $k \approx 0$.

This means that, within the Coulomb liquid model mentioned above, one can describe the major many-electron effect on CR absorption by considering independent electrons and using the Ando formula (5.7). The important point is that, in this pure single-electron formula, the Landau level broadening $\Gamma_N = \Gamma_\mathrm{se}$ of free electrons should be replaced by $\Gamma_N(E_\mathrm{f})$, defined in reference frames moving along with the electron orbit centers according to (2.45). Here E_f is the fluctuational electric field acting on the chosen electron in the Coulomb liquid center-of-mass frame. In the local moving frames, this field E'_f is zero and the level broadening actually depends on the drift velocity $\boldsymbol{u}_\mathrm{f}$, which relates directly to E_f. Then the average MW absorption is approximately proportional to

$$\langle \mathrm{Re}\left[\sigma_{xx}(\omega)\right]\rangle_\mathrm{f} = \frac{\sqrt{\pi}e^2 n_\mathrm{s}\hbar}{4m_\mathrm{e}} \left\langle \frac{1}{\Gamma_{0,1}} \exp\left[-\frac{\hbar^2(\omega-\omega_\mathrm{c})^2}{\Gamma_{0,1}^2}\right]\right\rangle_\mathrm{f} , \qquad (5.14)$$

where the broadening parameter $\Gamma_{0,1}(E_\mathrm{f}) = \sqrt{(\Gamma_0^2 + \Gamma_1^2)/2}$ depends on the fluctuational electric field according to (2.45) and Fig. 2.4. This treatment can be called the density-dependent Landau level width approach.

Averaging over the fluctuational electric field changes the line shape of the CR, making the absorption curve broader in the tail region. Although the shape of the CR line is no longer Gaussian, the width at half-height can be written approximately as

$$\gamma(n_s) \simeq \frac{\sqrt{\Gamma_0^2(n_s) + \Gamma_1^2(n_s)}}{0.849\hbar} . \quad (5.15)$$

Because of the Coulomb narrowing of Landau levels discussed in Sect. 2.3, the linewidth of the CR decreases with electron density. Later we shall see that this equation gives a good description of the Coulomb narrowing of the CR observed experimentally and agrees with the approach based on the memory function formalism. At the same time, (5.15) cannot describe the broadening effect on the CR due to Coulomb stimulation of electron scattering to higher Landau levels. This is because the probability of such scattering was not taken into account in a direct way (i.e., the system leaves the SCBA regime). This probability does not depend on the Landau level width, which is much smaller than the level spacing. The effect of Coulomb broadening of the CR linewidth is better described by means of the memory function formalism.

The convenience of the quantum transport framework based on the memory function approach is that the CR half-width ν_{eff} is directly related to the equilibrium DSF $S(q,\omega)$ of the Coulomb liquid (5.9), for which one can use certain approximations. At high MW radiation frequencies $\omega \sim \omega_c$, the general form of the electron DSF in (2.50) and (2.52), which includes summation over all Landau level numbers N, can be used to separate the effective collision frequency into a sum of resonant ($N = 1$) and nonresonant ($N \neq 1$) terms:

$$\nu_{\text{eff}}(\omega) = \nu_{\text{R}}(\omega) + \nu_{\text{NR}} \equiv [\gamma_{\text{R}}(\omega) + \gamma_{\text{NR}}]/2 . \quad (5.16)$$

The second term ν_{NR} (or γ_{NR}) is practically independent of $\omega - \omega_c$ and taken alone would give a pure Lorentzian CR absorption line. The nonresonant term can be found in a similar way to that discussed in Sect. 4.4 [76] in relation to DC magnetotransport:

$$\gamma_{\text{NR}} = \frac{\sqrt{\pi}\Gamma_{\text{se}}^2}{2\hbar^2\omega_c} A(x_f^{(0)}) , \quad (5.17)$$

$$A(x) = 2\sum_{N\neq 1} \frac{|N-1|^{N+3/2} x^{N+5/2}}{N!} K_{N+3/2}(2x|N-1|) , \quad (5.18)$$

where $x_f^{(0)} = \hbar\omega_c/(\sqrt{2}eE_f^{(0)}l_B)$ and $K_\nu(z)$ is the modified Bessel function. For an approximate analysis, one can use the interpolation formula $A(x) \simeq 3.2\exp[-(0.3x)^3]/x^{0.94}$, which is valid for $x > 0.9$.

The resonant term $\nu_{\text{R}}(\omega)$ can be evaluated numerically [76]. Still, in order to explain Coulombic effects on the CR linewidth, it is convenient to use an approximate but more transparent equation based on the simplification $\sqrt{\Gamma_{0,N}^2 + x_q\Gamma_{\text{C}}^2} \to \sqrt{\Gamma_{\text{se}}^2 + \Gamma_{\text{C}}^2}$. In this approximation, the resonant term can be written in the very instructive form

$$\gamma_{\mathrm{R}}(\omega) \simeq \frac{\sqrt{\pi}\Gamma_{\mathrm{se}}^2}{\hbar\sqrt{\Gamma_{\mathrm{se}}^2+\Gamma_{\mathrm{C}}^2}} \exp\left[-\frac{\hbar^2(\omega-\omega_{\mathrm{c}})^2}{\Gamma_{\mathrm{se}}^2+\Gamma_{\mathrm{C}}^2}\right] , \tag{5.19}$$

similar to the single-electron result of (5.10). This term has the Gaussian resonance form with respect to $\omega - \omega_{\mathrm{c}}$. Taken at $\omega = \omega_{\mathrm{c}}$, (5.19) explains the Coulomb narrowing of the CR linewidth:

$$\hbar\gamma_{\mathrm{R}}(\omega_{\mathrm{c}}) \simeq \frac{\sqrt{\pi}\Gamma_{\mathrm{se}}^2}{\sqrt{\Gamma_{\mathrm{se}}^2+\Gamma_{\mathrm{C}}^2}} ,$$

in a similar way to that discussed above (Sect. 4.4), when we considered the unconventional Hall effect: the Coulomb broadening of the electron DSF reduces the collision rate, because it substitutes for the broadening of the single-electron density of states.

The presence of Γ_{C} in the broadening parameter of the Gaussian function in (5.19) affects the shape of the CR line, so that, in the limiting case $\Gamma_{\mathrm{C}} \gg \Gamma_{\mathrm{se}}$, the linewidth parameter of the Lorentzian function $\gamma_{\mathrm{R}}(\omega) \simeq$ const. is practically independent of $\omega - \omega_{\mathrm{c}}$, and therefore the CR line shape becomes a pure Lorentzian function. The Coulombic effect on the CR linewidth should thus be accompanied by the line shape transformation from a pure Gaussian (at $\Gamma_{\mathrm{C}} \ll \Gamma_{\mathrm{se}}$) to a pure Lorentzian (at $\Gamma_{\mathrm{C}} \gg \Gamma_{\mathrm{se}}$). This line shape transformation was observed experimentally in [76, 89] for SEs on liquid helium. The raw data represented in Fig. 5.8 show the gradual change in the CR absorption for several electron densities. The absorption line for the lowest density is substantially broader than the others and its shape can be well fitted with a Gaussian function. An increase in the electron density causes the absorption line to become narrower in the vicinity of the maximum, while the tails are left almost unchanged. The upper curves are pure Lorentzian functions. The CR line shape of the intermediate densities is a sort of mixture of Gaussian and Lorentzian shapes. Figure 5.8 also shows the successive narrowing and broadening of the CR linewidth with increasing electron density.

The simplification $\sqrt{\Gamma_{0,N}^2 + x_q\Gamma_{\mathrm{C}}^2} \to \sqrt{\Gamma_{\mathrm{se}}^2+\Gamma_{\mathrm{C}}^2}$ made above is not really necessary for evaluation of the CR linewidth. Recalling that the linewidth at half-height is nearly the same for both extreme shapes (see Fig. 5.7), one can determine the linewidth as $\gamma_{\mathrm{CR}}(\omega_{\mathrm{c}})$. In this case, the contribution of the resonance term can be found directly using the strict relation of (2.51):

$$\gamma_{\mathrm{R}}(\omega_{\mathrm{c}}) = \frac{\sqrt{\pi}\Gamma_{\mathrm{se}}}{2\hbar}\int_0^\infty \mathrm{e}^{-y} W_{0,1}(\lambda^{(0)}\sqrt{y})\mathrm{d}y , \tag{5.20}$$

where

$$W_{0,1}(\lambda) = \frac{3(g_{0,1}^2+\lambda^2/2)^2}{(g_{0,1}^2+\lambda^2)^{5/2}} - \frac{g_{0,1}^2}{(g_{0,1}^2+\lambda^2)^{3/2}} ,$$

$\lambda^{(0)} = \sqrt{2}eE_{\mathrm{f}}^{(0)}l_B/\Gamma_{\mathrm{se}}$ and $g_{0,1} = \Gamma_{0,1}/\Gamma_{\mathrm{se}} \equiv \sqrt{(g_0^2+g_1^2)/2}$ is a function of $\lambda = \sqrt{2}eE_{\mathrm{f}}l_B/\Gamma_{\mathrm{se}} = \lambda^{(0)}\sqrt{y}$, according to (2.45). The nonresonant term γ_{NR}

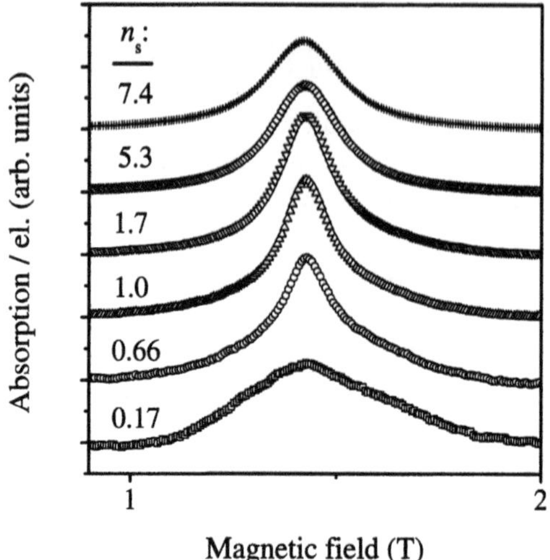

Fig. 5.8. CR absorption data at $T = 1.36\,\mathrm{K}$ for six electron densities shown in units of $10^8\,\mathrm{cm}^{-2}$ [76]

remains the same as in (5.17), because it does not depend on $\omega - \omega_c$ near the resonance condition.

It should be noted that the accurate analysis of the limiting case $\Gamma_C \gg \Gamma_{se}$ based on $\sqrt{\Gamma_{0,N}^2 + x_q \Gamma_C^2} \to \sqrt{x_q}\Gamma_C$ slightly changes the numerical proportionality factor in the simplified formula (5.19) and agrees with the result of the Dykman–Khazan theory:

$$\gamma_{CR} \simeq \frac{3\pi\Gamma_{se}^2}{8\hbar\Gamma_C} \propto \frac{1}{n_s^{3/4}} . \tag{5.21}$$

This can also be seen directly from (5.20), assuming $g_{0,1} \to 0$. The result of (5.21) is restricted at both ends: in the low density region, γ_{CR} is finite according to the self-energy effects; at high densities the nonresonant terms dominate and change the sign of the many-electron effect.

It is instructive to compare the two ways of describing Coulombic effects on the quantum CR: the density-dependent Landau level width model and the framework based on the memory function formalism. The numerical comparison is given in Fig. 5.9 for three typical values of the magnetic field. It is clear that the lowest curves of the memory function formalism (continuous) and the density-dependent width model (dashed) plotted for the limiting case $\hbar\omega_c \gg \Gamma_C$ are remarkably close over a wide range of values of the parameter $eE_f l_B/\Gamma_{se}$. The small relative increase in the continuous curve in the limit $\lambda_f^{(0)} \to 0$ reflects the shaping effect: in this range the CR shape of the memory function approach is not Lorentzian, but rather is close to a Gaussian with width at half-height equal to $\sqrt{2/\pi}\gamma_L/0.849 \simeq 0.94\gamma_L$. There is no doubt

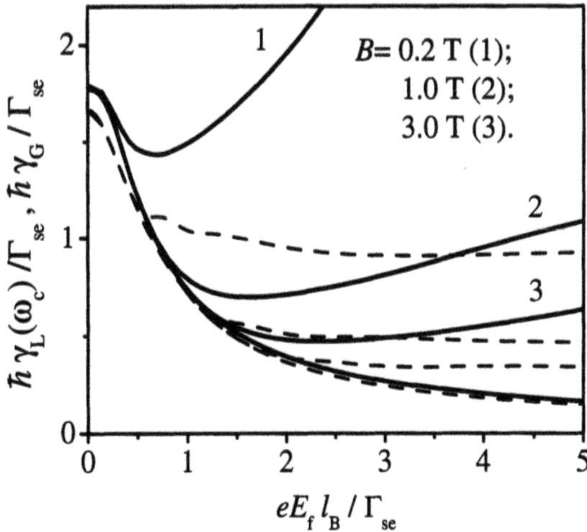

Fig. 5.9. CR linewidth vs. the many-electron parameter $eE_f l_B/\Gamma_{se}$ for three typical resonant magnetic fields and $T = 1.36\,\text{K}$. *Continuous curves*: framework based on the memory function formalism. *Dashed curves*: the density-dependent Landau level width approach. The lowest curves (*dashed* and *dotted*) are plotted for the limiting case $\hbar\omega_c \gg \Gamma_C$

that both theoretical approaches result in the same Coulomb narrowing of the CR linewidth.

Unlike the Coulomb narrowing, the broadening of the CR induced by internal forces is described differently by the two approaches introduced above. The increase in the CR linewidth with n_s found for the memory function formalism is substantially bigger than that found for the density-dependent width model, as shown in Fig. 5.9 for three values of the magnetic fields $B = 0.2, 1$ and $3\,\text{T}$ (continuous curves 1, 2 and 3). One can see that the small increase in the broadening of Landau levels caused by the mixing effect (dashed curves) cannot be a measure of the many-electron broadening of the CR. This is reasonable, because the probability of electron scattering between different Landau levels induced by the high energy exchange $\hbar\boldsymbol{q}\cdot\boldsymbol{u}_f \sim \hbar\omega_c$ does not actually depend on the Landau level broadening if $\Gamma_N \ll \hbar\omega_c$. Moreover, Fig. 5.9 shows that the magnetic field dependence of the Coulomb narrowing can be normalized by $\Gamma_{se} \propto \sqrt{B}$, while the Coulomb broadening cannot. The normalized linewidth is different for different B, approaching the lowest continuous curve only in the limit $B \to \infty$, because the Landau level separation $\hbar\omega_c \propto B$ increases faster with B than Γ_{se}.

Figure 5.9 also shows that the minimum of the CR linewidth appears at rather small values of the parameter $eE_f^{(0)} l_B/\hbar\omega_c$ because of the fast decrease in the resonant term $\gamma_L^{(R)}$ with $\lambda^{(0)}$. This means that the CR narrowing and most of the CR broadening of Fig. 5.9 occur under conditions in which the fluctuational electric field can be approximately considered as quasi-uniform. Beyond this regime, the approach discussed above takes into account only the uniform part of the fluctuational field, which induces drift velocities $\boldsymbol{u}_{\mathrm{f}}^{(i)}$. Because the main effect discussed here originates from the ultra-fast fluctuational motion of electron orbit centers, one may assume that this approach will be qualitatively valid even for nonuniform fluctuational fields.

The conductivity formula (5.8) itself has only one resonant frequency $\omega = \omega_{\mathrm{c}}$. At the same time, the effective collision frequency $\nu_{\mathrm{eff}}(\omega)$ has a resonance structure at all excitation frequencies $\omega = N\omega_{\mathrm{c}}$. This is clearly seen from the approximations for the electron DSF used here [(2.38) and (2.52)]. In Chap. 7 we shall see that $S(q,\omega)$ and $\nu_{\mathrm{eff}}(\omega)$ have the same resonance structure even in the Wigner solid state. Because the real part of σ_{xx} is proportional to $\nu_{\mathrm{eff}}(\omega)$, one can expect the appearance of subharmonics of the CR at lower magnetic fields such that $\omega_{\mathrm{c}} = \omega/N$. In contrast with the basic resonance at $\omega = \omega_{\mathrm{c}}$, the sub-harmonic resonances should not have the Coulomb narrowing, since the broadening of the maxima of the electron DSF $\Gamma_{0,N}^{(*)} = \sqrt{\Gamma_{0,N}^2 + x_q \Gamma_{\mathrm{C}}^2}$ increases steadily with the electron density. Of course, the intensity of subharmonic resonances is much smaller than that of the basic resonance and special precautions are required to observe them on the background of the experimental noise. For degenerate 2D electrons, the subharmonic structure of the quantum CR was predicted by Ando [189] and observed experimentally in [190].

In the low temperature regime, the main scatterers are ripplons. In this case, the linewidth equation has a different form which follows directly from (3.58). At low temperatures, the collision broadening of Landau levels is small and at medium and high electron densities it can be disregarded in comparison with the Coulomb broadening of the DSF. Moreover, in the low temperature regime, the system is close to the Wigner solid (WS) state. As we shall see in Chap. 7, the WS DSF coincides with the Coulomb liquid DSF of (2.52) if Γ_0 is set to zero. According to the analysis of [148] based on (3.58), this electron DSF leads to the following result for the effective collision frequency of the electron–ripplon scattering regime:

$$\nu_{\mathrm{eff}}(\omega_{\mathrm{c}}) = \frac{(eE_{\perp})^2 T}{4\sqrt{\pi}\hbar\alpha\Gamma_{\mathrm{C}}} \sum_{N=0}^{\infty} \frac{1}{N!} F_N\left(\hbar\omega_{\mathrm{c}} \left|N-1\right| / \Gamma_{\mathrm{C}}\right) , \tag{5.22}$$

where

$$F_N(y) = \int_0^{\infty} x^{N-1/2} V_{\mathrm{S}}^2(x) \mathrm{e}^{-x-y^2/x} \mathrm{d}x , \tag{5.23}$$

$$V_{\mathrm{S}}(x) = 1 + \frac{\Lambda x}{eE_{\perp} l_B^2} w_{\mathrm{C}}\left(x/2\gamma^2 l_B^2\right) , \tag{5.24}$$

and the coupling function $w_{\mathrm{C}}(x)$ is determined by the interaction potential introduced in Sect. 1.5.2. Equation (5.22) is reminiscent of the result found for the vapor atom scattering regime. In Sect. 7.5.4, we shall see that the same sum over N represents all terms of the expansion series of the WS DSF in powers of the cyclotron motion factor $\exp(-i\omega_c t)$, similar to the expansion series introduced previously for the DSF of simple oscillators [see (2.35)]. At low electron densities, the main contribution comes from the resonance term with $N = 1$. Because $F_1(0)$ is independent of Γ_C, the CR linewidth decreases with n_s according to $\gamma_{CR} = 2\nu_{eff} \propto 1/\Gamma_C$, as predicted in [85].

For larger values of Γ_C, the nonresonant terms with $N \neq 1$ become important, changing the sign of the Coulombic effect. Therefore, the same Coulombic effect (broadening of the electron DSF) leads to the successive narrowing and broadening of the CR linewidth. Even though at high densities the system leaves the regime where the fluctuational field can be considered as quasi-uniform, the sign change of the many-electron effect described by these nonresonant terms is very important for understanding the available experimental data.

5.3.1 Many-Electron Effects in the Linear Regime

According to the numerical results plotted in Fig. 5.9, in order to reach the Coulomb narrowing regime of the quantum CR for SEs on a free surface of liquid helium, the magnetic field should be strong enough and the sensitivity of the detection system should make it possible to conduct measurements down to very low electron densities. In the experiments reported in [76, 89], the authors were using an MW spectrometer operating at frequencies of 40–60 GHz. Figure 5.10 shows a schematic view of the low temperature part of the setup. It consists essentially of a metallic cavity in the form of an upright cylinder, acting as a resonator for microwaves in the TE_{011} mode. The bottom plate of the cavity is mounted on a movable plunger to enable the height of the resonator to be changed and hence to tune its resonance frequency in situ. The cavity has diameter 10.8 mm and height between 7 mm (at 40 GHz) and 3 mm (at 60 GHz). The resonator was operated in reflection mode through a single rectangular waveguide, ending above a coupling hole in the top plate. For signal detection, a phase-sensitive heterodyne system with high sensitivity was employed. To avoid possible line shape distortions owing to heating of the electron system, the MW input power was kept at an estimated level below 10^{-18} watt per electron. At this level, the line shape was in the equilibrium regime.

Comparing the data with theory, let us first check whether the discussed many-electron theory can describe the transformation of the CR line shape observed with increasing electron density. As already mentioned, the qualitative agreement with theory is obvious. For a quantitative comparison, the data and theory are shown in Fig. 5.11 by two typical data plots and corresponding theoretical curves calculated according to (5.8) and (5.9), using the DSF of the Coulomb liquid in (2.51). The results of the density-dependent Landau level width model are shown by dashed curves. For the lowest electron density, the data plot has a slightly wider linewidth than the theoretical

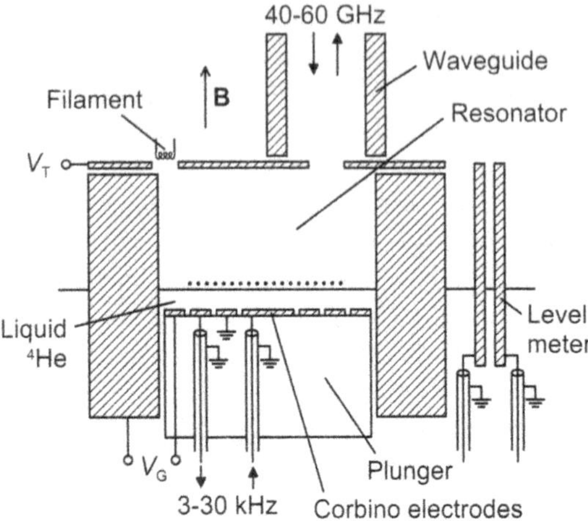

Fig. 5.10. Schematic diagram of the experimental cell for CR absorption measurements [76]

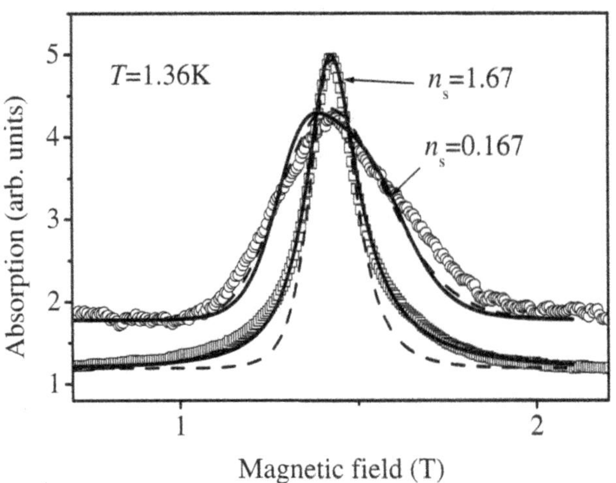

Fig. 5.11. CR absorption for two electron densities. Data are represented by *circles* and *squares*. The quantum transport framework based on the memory function formalism is shown by *continuous curves*, whilst the density-dependent Landau level width approach is illustrated by *dashed curves* [76]

curve, while for the higher density, the continuous theoretical curve fits the CR data perfectly with no adjusting parameter.

The linewidth data vs. electron density are shown in Fig. 5.12 for two values of the magnetic field. It should be noted that low electron densities and high magnetic fields represent the most difficult experimental ranges for measuring the CR from SEs on liquid helium. Therefore, experimental errors usually increase in these limits. On the contrary, the most accurate data are found for high electron densities where the experimental signal is strong. The absorption curves were fitted to both Gaussian and Lorentzian forms in order to extract the linewidth data. The CR linewidth data vs. electron density (Fig. 5.12) displays a sharp fall in the low electron density range which is in a good accordance with both theoretical models: the density-dependent Landau level width approach (dashed curves) and the quantum transport framework based on the memory function formalism (continuous curves). In this range, both models give very close results. Generally, the linewidth data and theoretical curves are reminiscent of the density dependence of the effective collision frequency of the DC case shown in Fig. 4.5. In the high density range, the CR linewidth data increase with n_s in a good (even numerical) accordance with the continuous curve, without any adjusting parameter. In this regime, the density-dependent Landau level width model cannot explain the data, because the electron scattering to higher Landau levels does not depend on the level broadening. The result of Dykman–Khazan theory [85], which is equivalent to (5.21) (dotted curves), is close to the data only at intermediate electron densities.

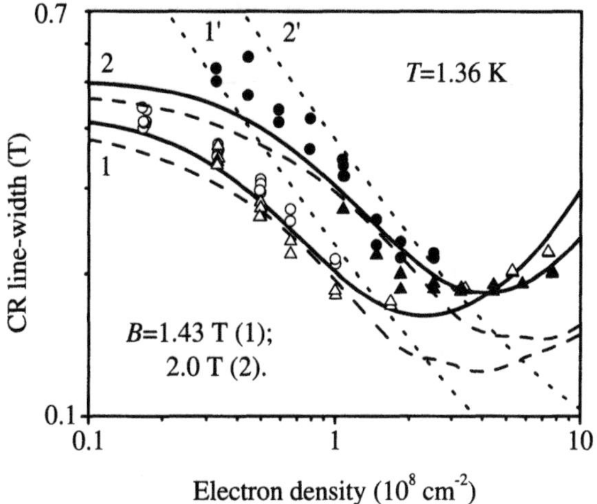

Fig. 5.12. CR linewidth vs. n_s for two resonant values of the magnetic field [76]: $B = 1.43\,\mathrm{T}$ (*open symbols*), $B = 2.0\,\mathrm{T}$ (*filled symbols*). *Circles*: Gaussian fitting data. *Triangles*: Lorentzian fitting data. The quantum transport framework based on the memory function formalism with the many-electron approximation for the DSF is shown by *continuous curves*, whilst the density-dependent Landau level width approach is illustrated by *dashed curves*, and the theory of Dykman and Khazan by *dotted curves*

At low and intermediate densities, with an increase in the magnetic field, all theoretical curves and data acquire upward shifts caused by the increased Landau level broadening $\Gamma_{\mathrm{se}}(B) \propto \sqrt{B}$. At high densities, the data and continuous curves behave in the opposite way, acquiring downward shifts because the probability of electron scattering to higher Landau levels caused by the fluctuational motion of orbit centers decreases with the level spacing $\hbar\omega_{\mathrm{c}} \propto B$. As a result, the two continuous curves and both data plots cross at high n_{s}, practically at the same point.

The relationship between the Coulomb effect on the CR absorption linewidth and the Landau level width is shown in Fig. 5.13 by means of the normalized plots. Here Γ_{se}, chosen as a normalization parameter for both theory and experiment, is calculated for each experimental density according to the SCBA of the electron–vapor atom scattering regime. To adjust the theory to the data, all theoretical curves are shifted slightly upward by 11%. Figure 5.13 clearly shows that the effect of the magnetic field on the Coulomb narrowing is perfectly normalized by $\Gamma_{\mathrm{se}}(B) \propto \sqrt{B}$, which proves that it depends on the ratio $\Gamma_{\mathrm{C}}/\Gamma_{\mathrm{se}}$. In contrast, the Coulomb broadening of the CR cannot be normalized by $\Gamma_{\mathrm{se}}(B)$ because it depends on another parameter $\Gamma_{\mathrm{C}}/\hbar\omega_{\mathrm{c}}$ decreasing with the magnetic field faster than $\Gamma_{\mathrm{C}}/\Gamma_{\mathrm{se}}$. The agreement between theory and experiment shown in Fig. 5.13 indicates that Coulombic effects on the CR linewidth are well understood.

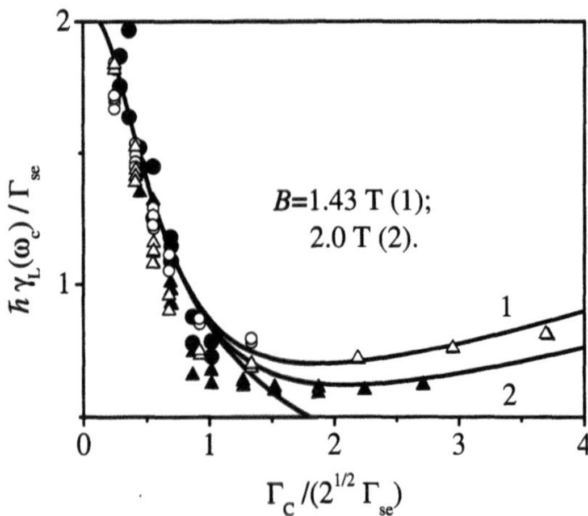

Fig. 5.13. Normalized CR linewidth data and theory vs. $\Gamma_{\mathrm{C}}/\sqrt{2}\Gamma_{\mathrm{se}} = eE_{\mathrm{f}}l_B/\Gamma_{\mathrm{se}}$ for two resonant frequencies [76]. The notation is the same as in Fig. 5.12

It is important to prove that the observed changes in the CR linewidth are not related to a change in the electron–vapor atom coupling and represent solely the many-electron effect. The above-mentioned data were found under saturation conditions with regard to the holding electric field $E_{\perp} = 2\pi e n_{\mathrm{s}}$

and an increase in n_s also means an increase in $E_\perp$ and the parameter γ of the electron wave function $f_1(z)$. To check that the observed effects do not come from this simple change in the electron–vapor atom coupling, some measurements were conducted at fixed electron densities, varying only the holding electric field [76]. Two distinctive electron densities were chosen in different regimes of the Coulombic effect, as shown in Fig. 5.14. In this case, the data show only a very slight increase in the linewidth with the holding electric field, which is in accordance with the increase in the electron localization length $\gamma^{-1}(E_\perp)$ entering $\nu_0^{(a)}$ and Γ_{se}. The corresponding theoretical results are shown in Fig. 5.14 by dashed curves. Thus, neither the narrowing, nor the broadening of the CR observed with the increase in electron density and described by the continuous curve in Fig. 5.14 can be caused by the change in the holding electric field, and so represent pure Coulombic effects.

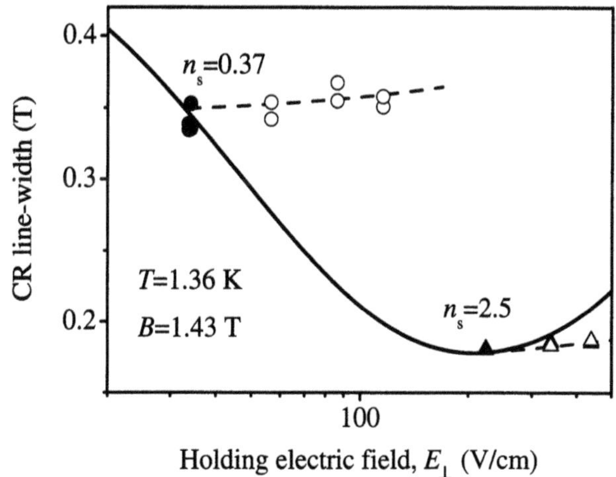

Fig. 5.14. CR linewidth data vs. $E_\perp$ for two fixed electron densities: $n_s \simeq 0.37 \times 10^8\,\mathrm{cm}^{-2}$ (*circles*), $n_s \simeq 2.5 \times 10^8\,\mathrm{cm}^{-2}$ (*triangles*). *Solid symbols* represent the saturation condition $E_\perp = 2\pi e n_s$ [76]. For the CR linewidth given by the memory function formalism: saturation condition (*continuous curves*), fixed electron density conditions (*dashed curves*)

It is interesting to compare the theory of the CR absorption in the low temperature regime ($T < 1\,\mathrm{K}$) with the old linewidth vs. $E_\perp$ data of Edel'man [9, 183] presented earlier in Fig. 5.5, where the electron density n_s was varied according to the saturation condition $E_\perp = 2\pi e n_s$. Two sets of the CR half-width curves evaluated by means of (5.22) for different temperatures are shown in the same figure (Fig. 5.5). Dashed curves represent the contribution of the resonant term ($N = 1$) to the effective collision frequency (the approximation of [85]). The continuous curves include all terms of (5.22). For comparably high temperatures ($T = 0.72\,\mathrm{K}$), in addition to the electron–ripplon scattering, the electron interaction with vapor atoms is taken into account, while for the low temperature regime ($T = 0.4\,\mathrm{K}$), the interaction with vapor atoms can be omitted.

The important thing is that the resonant term itself cannot explain a strong increase in the CR linewidth observed in the region $E_\perp > 50\,\mathrm{V/cm}$. The reference to the holding field dependence of the interaction Hamiltonian used previously is misleading. Indeed, the semi-classical curves for the effective collision frequency (dotted curves of Fig. 5.5) have a proper rate of increase with $E_\perp$. But the resonant term (in its denominator) also contains the Coulomb broadening Γ_C, which depends strongly on $n_\mathrm{s} = E_\perp/2\pi$. It is easy to check that combining the field dependence of the semi-classical result (dotted curves) with $1/\Gamma_\mathrm{C}(E_\perp)$ gives the holding field dependence of the resonant term (dashed curves), which has a much weaker increase with $E_\perp$, and a minimum shifted significantly in the range of strong holding fields. In contrast, the inclusion of the nonresonant terms resulting in continuous curves agrees well with the experimental data.

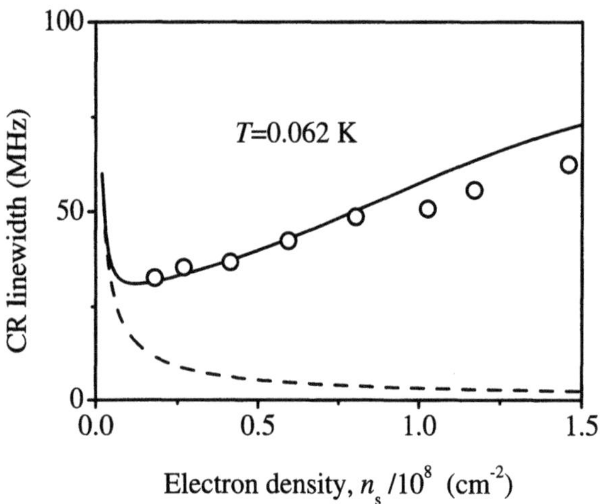

Fig. 5.15. CR linewidth vs. n_s. Data of Wilen and Giannetta [185] are represented by *circles*. *Continuous curve*: the quantum transport framework employing the DSF of the Wigner solid (5.22) [148]. *Dashed curve*: Dykman's approximation, equivalent to keeping only the resonant term

Let us now consider the conditions of the experiment of Wilen and Giannetta [185], where SEs were in the Wigner solid state. The linewidth data and the corresponding theoretical curves are shown in Fig. 5.15. While varying the electron density n_s, the holding field was fixed at $E_\perp = 275\,\mathrm{V/cm}$. It is clear that the contribution from the resonant term alone, which corresponds to the approximation of [86] (dashed curve), is in distinct (even qualitative) contradiction with the experimental data. In contrast with the graphs of [185], the curve representing the contribution from the resonance term does not cross the data, which is physically more reasonable. This is due to the use of the exact form of the electron–ripplon interaction Hamiltonian in (5.22), rather than an asymptote. Figure 5.15 shows that all experimental data are in the regime of Coulomb broadening of the CR and that the theory of CR absorption from the Wigner crystal (continuous curve), including nonresonant terms ($N \neq 1$), describes this many-electron effect well. It should be noted that under the conditions of this experiment, at higher densities, the fluctuational electric field cannot be considered as quasiuniform and this causes deviations of the experimental data from the discussed model. In any case, the nonresonant terms give a quite reasonable explanation of the puzzling many-electron effect observed by Wilen and Giannetta. Thus the Coulomb stimulation of electron scattering to other Landau levels eliminates the discrepancy between theory and two independent experiments [183, 185].

5.3.2 Power-Induced Coulomb Narrowing

As mentioned in the early CR studies by Brown and Grimes [180] and by Edel'man [9, 183], the SEs on liquid helium are easily overheated by MW radiation at very low powers. The nonlinear experimental data shown above in Fig. 5.3 were explained theoretically by Saitoh and Aoki [191]. The nonlinear change in the CR line shape which was calculated taking into account electron transitions on higher surface levels is shown in Fig. 5.16 for different levels of the driving electric field. The sharp peak in the vicinity of the resonance is due to electron evaporation from the ground surface levels under conditions $T_{\mathrm{e}} > \left|\varepsilon_1^{(\perp)}\right|$. The reason for the level narrowing is that, in the ripplon-dominated regime, electron scattering is much stronger on the ground surface level than on other levels and in 3D states. The theoretical curves are in accordance with the experimental graph of Edel'man shown previously.

If heating is low and electron transitions on higher surface levels can be disregarded, then the broadening of the CR usually increases with power. In such cases, there is a convention to obtain the equilibrium data as an extrapolation to zero power input. The power broadening of the CR absorption from SEs on liquid helium was already reported in experiments by Brown and Grimes [180]. In later studies [89], it was shown that the Coulombic effect makes the power dependence substantially non-monotonic with a deep minimum and, therefore, the extrapolation to zero power input can give an incorrect result. These experimental data correspond to the vapor-atom scattering regime. The non-monotonic behavior of the CR linewidth, in this instance, originates from the strong dependence of the fluctuational electric field E_{f} on electron temperature $T_{\mathrm{e}} > T$.

Because SEs on liquid helium form a highly correlated system, in a nonequilibrium case one can introduce the electron temperature parameter T_{e}, which can be substantially higher than T. The MW radiation increases T_{e} and consequently the fluctuational electric field $E_{\mathrm{f}} \propto \sqrt{T_{\mathrm{e}}}$. At the same time, Γ_{se} does not depend on electron temperature, unless it changes the temperature of vapor atoms (which is unlikely). Therefore, the many-electron

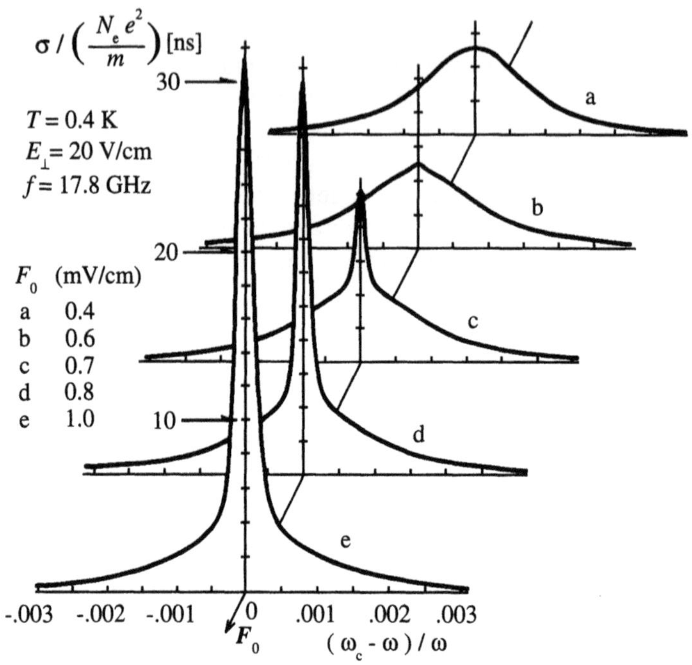

Fig. 5.16. Normalized absorption line-shape of the CR for different input MW field strengths F_0 [191]. Here σ is the effective conductivity

parameters $\Gamma_C(T_e)/\Gamma_{se}$ and $\Gamma_C(T_e)/\hbar\omega_c$ responsible for Coulomb narrowing and broadening of the CR linewidth increase rapidly with the power. As a result, the power dependence of the CR linewidth should reflect its density dependence and have a deep minimum at certain powers.

The CR linewidth vs. input power data of [89] are shown in Fig. 5.17 for three typical electron densities. The power narrowing is strong for the low density limit. Increasing the electron density makes the minimum shallow until it disappears at some density corresponding to the minimum of the density dependence of the equilibrium CR linewidth (Fig. 5.12). At high powers, the whole data sets show the same power broadening in accordance with previous results [180, 183]. It is clear that the power dependence of Fig. 5.17 represents the same Coulombic effect as that shown in Fig. 5.12 under equilibrium conditions. It is observed with E_f being varied by changing T_e instead of n_s.

5.4 Peak Shift

As discussed above, the increase in the electron–ripplon coupling with the holding electric field $E_\perp$ leads to a shift in the CR peak. The formal reason for this was given in Sect. 5.1 by means of the qualitative analysis of the

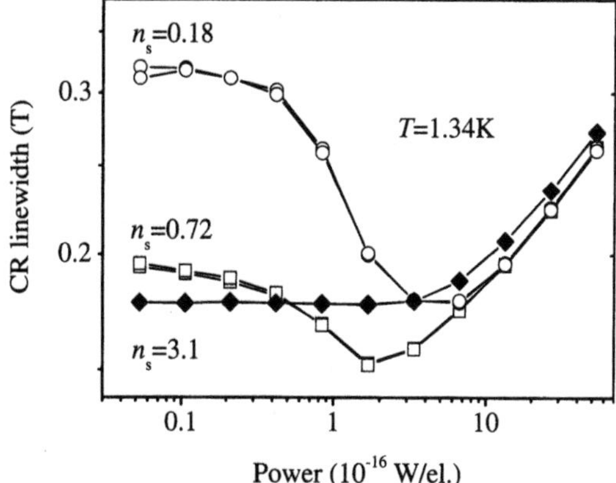

Fig. 5.17. CR linewidth data vs. absorbed power at resonance $\omega = \omega_c$ for three values of n_s shown in units of $10^8\ \text{cm}^{-2}$ [89]. *Lines* connect data points for magnetic field sweep upwards and downwards, respectively

real part of the memory function $w(\omega) = \text{Re}(\omega)$. To calculate this shift, one needs to perform the following procedure. Firstly, one must determine the most important terms in the DSF. These terms should then be substituted into the fluctuation–dissipation equation in order to determine the Green's functions $G_{n_q,n_q^\dagger}(\omega)$ and $G_{F_x,F_x}(\omega)$ according to (3.54). Finally, the peak shift can be found using the relation between the memory function and these Green's functions given by (3.43). The result of the formal evaluation of the peak shift will be discussed in Chap. 7, which deals with the properties of the electron solid. Historically, experimental shifts were explained using a simple magnetopolaron model [184], and the model of phonon–ripplon coupling of the Wigner solid under a magnetic field [192]. We would like to discuss these models briefly here because they introduce important physical ideas.

In the presence of a high magnetic field, instead of the usual Landau set of eigenstates $|N, X\rangle$, one can use a set of purely localized electron states determined by the angular momentum m and N for the symmetrical gauge. For $N = 0$, these states are described by the following wave functions

$$\varphi_m(r,\phi) = \frac{1}{\sqrt{2\pi|m|!2^{|m|}}l_B}\left(\frac{r}{l_B}\right)^{|m|} e^{im\phi}e^{-r^2/4l_B^2}\,. \tag{5.25}$$

It is assumed that in the state $m = 0$, which is completely localized, an electron creates a surface dimple $\xi(r)$ and gains some additional energy due to the holding electric field. This kind of medium polarization cloud just removes the degeneracy of the ground state under a magnetic field. According to Cheng and Platzman [184], this should affect the position of the CR from surface electrons, because when the electron is excited to a higher Landau level, its average radius $\langle r\rangle$ changes, and this in turn changes its average height $\langle z \rangle$ inside the dimple, and therefore its interaction energy with the dimple.

Consider the nondegenerate gas (G) state of the electron system. Following [184], we note that the electron functions that are connected by dipole transitions are φ_0 and φ_{-1}. The charge distribution across the dimple for these two electron states is shown in Fig. 5.18. The dimple profile $\xi(r)$ is determined by the pressure balance equation, which yields the Fourier transforms

$$\xi_q^{(0)} = -\frac{e(E_\perp + E_q)}{\alpha(q^2 + \kappa^2)} |\varphi_0|_q^2 \,, \tag{5.26}$$

where $|\varphi_m|_q^2$ is the Fourier transform of $|\varphi_m(\boldsymbol{r})|^2$. The average interaction energy

$$\langle m|V_{\mathrm{int}}|m\rangle = \sum_{\boldsymbol{q}} V_q \xi_q^{(0)} \left\langle m \left| \mathrm{e}^{\mathrm{i}\boldsymbol{q}\cdot\boldsymbol{r}} \right| m \right\rangle \tag{5.27}$$

depends on m, causing a frequency shift $\Delta\omega_{\mathrm{G}}$ of the CR given by

$$\begin{aligned} \hbar\Delta\omega_{\mathrm{G}} &= \langle -1|V_{\mathrm{int}}|-1\rangle - \langle 0|V_{\mathrm{int}}|0\rangle \\ &= -\sum_{\boldsymbol{q}} e(E_\perp + E_q)\xi_q^{(0)} \left(|\varphi_0|_q^2 - |\varphi_{-1}|_q^2 \right) \,. \end{aligned} \tag{5.28}$$

Here we have taken into account the fact that the matrix elements $\langle m|\mathrm{e}^{\mathrm{i}\boldsymbol{q}\cdot\boldsymbol{r}}|m\rangle$ represent the Fourier transforms $|\varphi_m|_q^2$. Because $\xi_q^{(0)} < 0$, the shift of the CR is positive, which means that for the resonance condition $\omega_{\mathrm{c}} + \Delta\omega_{\mathrm{G}} = \omega$ to be fulfilled, one must apply a weaker magnetic field (downward shift).

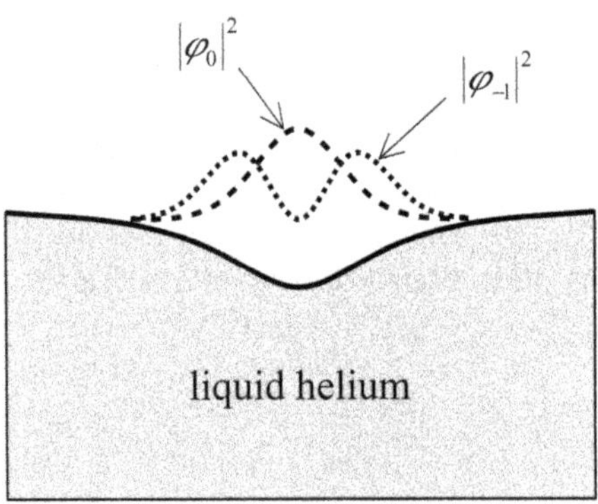

Fig. 5.18. Schematic view of the charge density for the angular momentum states $m = 0$ and $m = -1$ connected by dipole transitions

If the temperature is not ultra-low ($T \sim 0.5\,\mathrm{K}$), the polarization part of the electron–ripplon interaction can be substantially simplified: $E_q \simeq \Lambda\gamma q/3e$. In this case, the shift of the CR position can be obtained in the analytical form [192]

$$\Delta\omega_{\mathrm{G}} = \frac{e^2}{8\pi\alpha\hbar}\left(E_{\perp}^2 + \sqrt{\pi}E_{\perp}E_{\mathrm{G}} + E_{\mathrm{G}}^2\right) , \tag{5.29}$$

where $E_{\mathrm{G}} = \Lambda\gamma/3el_B$. Under the experimental conditions of Edel'man [183], the value of the effective holding field that comes from the polarization attraction is $E_{\mathrm{G}} \approx 139\,\mathrm{V/cm}$. In [184], the frequency shift was found numerically. The factor $1/8\pi$ of (5.29) corresponds exactly to the numerical factor 0.04 found by Cheng and Platzman.

In a highly correlated state of the electron liquid, the individual electron states of the form given in (5.25) are no longer useful for the CR absorption theory. For example, including the Coulomb electric field of other electrons in the definition of $\hbar\Delta\omega_{\mathrm{G}}$ from (5.27) would give a CR peak shift induced by internal forces which is impossible according to Kohn's theorem (see Sect. 3.4). In the Coulomb liquid, an AC electric field of high frequency can excite magnetoplasmons rather than single-electron states. For example, in the liquid state we have $\Omega_{\mathrm{mp}}(k) = \sqrt{\omega_{\mathrm{c}}^2 + \Omega_k^2}$, where Ω_k is the plasmon dispersion under zero magnetic field. Considering the CR under a uniform excitation field $(k \approx 0)$, we have no peak shift induced by Coulomb forces.

Now imagine that we have surface dimples $\xi^{(0)}(r)$ under each electron. For small electron displacements along the interface, we can consider the electron potential in the single dimple as an oscillatory potential with a characteristic frequency ω_{d} of electron oscillations. Then the Coulomb liquid magnetoplasmon dispersion will have the form

$$\Omega_{\mathrm{mp}}(k) = \sqrt{\omega_{\mathrm{c}}^2 + \omega_{\mathrm{d}}^2 + \Omega_k^2} . \tag{5.30}$$

Under conditions $k \approx 0$ and $\omega_{\mathrm{c}} \gg \omega_{\mathrm{d}}$, the peak shift can be defined as

$$\Delta\omega_{\mathrm{mp}} = \Omega_{\mathrm{mp}}(0) - \omega_{\mathrm{c}} \simeq \frac{\omega_{\mathrm{d}}^2}{2\omega_{\mathrm{c}}} , \tag{5.31}$$

which agrees qualitatively with the single-electron result of Cheng and Platzman.

If electrons form the Wigner solid, we have to take into consideration the spectrum of transverse phonons. In this instance, as we shall see in Chap. 7, the magnetic field mixes the transverse and longitudinal phonons of the electron crystal. As a result, the dispersion of the relevant magnetoplasmon mode $\Omega_+(k)$ is a combination of dispersions of longitudinal $\Omega_{\mathrm{l},k}$ and transverse $\Omega_{\mathrm{t},k}$ phonons, and the cyclotron frequency:

$$2\Omega_+(k) = \Omega_{\mathrm{l},k}^2 + \Omega_{\mathrm{t},k}^2 + \omega_{\mathrm{c}}^2 + \sqrt{(\Omega_{\mathrm{l},k}^2 + \Omega_{\mathrm{t},k}^2 + \omega_{\mathrm{c}}^2)^2 - 4\Omega_{\mathrm{l},k}^2\Omega_{\mathrm{t},k}^2} . \tag{5.32}$$

In the absence of surface dimples $(\omega_{\mathrm{d}} \to 0)$ and in the long-wavelength limit, the frequency of transverse phonons is much lower than the frequency of longitudinal phonons $\Omega_{\mathrm{t},k} \ll \Omega_{\mathrm{l},k}$. In this case, the mode $\Omega_+(k)$ obviously coincides with the usual magnetoplasmons of the 2D electron liquid $\Omega_+(k) \simeq$

$\sqrt{\omega_c^2 + \Omega_k^2}$. If surface dimples are present, in the long-wavelength limit there is no limiting case $\Omega_{t,k} \ll \Omega_{l,k}$ because the reconstructed longitudinal and transverse modes, i.e.,

$$\tilde{\Omega}_p(k) = \sqrt{\omega_d^2 + \Omega_p^2(k)} \,, \quad p = \mathrm{l}, \mathrm{t} \,, \tag{5.33}$$

have the same limiting frequency ω_d determined by the electron interaction with the motionless surface dimple. This phonon spectrum (5.33) should be employed in (5.32) for the magnetophonon mode $\Omega_+(k)$ instead of the phonon spectrum $\Omega_{p,k}$ above the flat surface. Thus, even at $k \approx 0$, the contribution of transverse phonons cannot be disregarded, and the CR peak shift

$$\Delta\omega_{\mathrm{mp}} = \left[\omega_d^2 + \omega_c^2/2 + (\omega_c^2/2)\sqrt{1 + 4\omega_d^2/\omega_c^2}\right]^{1/2} - \omega_c \simeq \omega_d^2/\omega_c \tag{5.34}$$

becomes two times greater.

In the limit of strong magnetic fields, the characteristic frequency ω_d can be defined as [192]

$$\omega_d^2 = \sum_{\boldsymbol{g}} \frac{n_s e^2 (E_\perp + E_G)^2}{2\alpha m} \exp\left(-g^2 \left\langle u_+^2 \right\rangle /2\right) \,, \tag{5.35}$$

where $\left\langle u_+^2 \right\rangle$ is the mean-square displacement of electrons in the mode $\Omega_+(k)$, and $\boldsymbol{g}$ is the reciprocal lattice vector. For $\left\langle u_+^2 \right\rangle \ll a^2 \sim 1/n_s$, the summation over $\boldsymbol{g}$ can be carried out using Poisson's formula. Assuming that the cyclotron frequency is much higher than the typical frequency of short-wavelength phonons of the WS, which results in $\left\langle u_+^2 \right\rangle \to l_B$, accurate evaluation yields the following shift in the CR position:

$$\Delta\omega_S = \frac{e^2}{4\pi\alpha\hbar}\left(E_\perp^2 + \sqrt{\pi}E_\perp E_S + E_S^2\right) \,, \tag{5.36}$$

where $E_S \simeq \sqrt{2}E_G$. Here the most important term, proportional to $E_\perp^2$, is exactly 2 times greater than the analogous term in (5.29) found for the gas state.

As follows from the simple considerations given above, when passing the WS transition range, the shift of the CR peak position as a function of $E_\perp^2$ should bend to the line which corresponds to a shift two times greater than that found for the gas state of SEs. This kind of curve bending was seen and even noted in Edel'man's early data presentation. These data are shown in Fig. 5.19. The low holding field data is well described by (5.29), which corresponds to the gas state of SEs. With a further increase in the holding field, above the region of $E_\perp \approx 400\,\mathrm{V/cm}$, which corresponds to $\Gamma^{(\mathrm{pl})} \approx 140$ (at $T \simeq 0.4$–$0.43\,\mathrm{K}$), the data decline and show a stronger holding field dependence. Unfortunately, the data were not systematic enough (they were obtained with a certain spread in external parameters such as temperature,

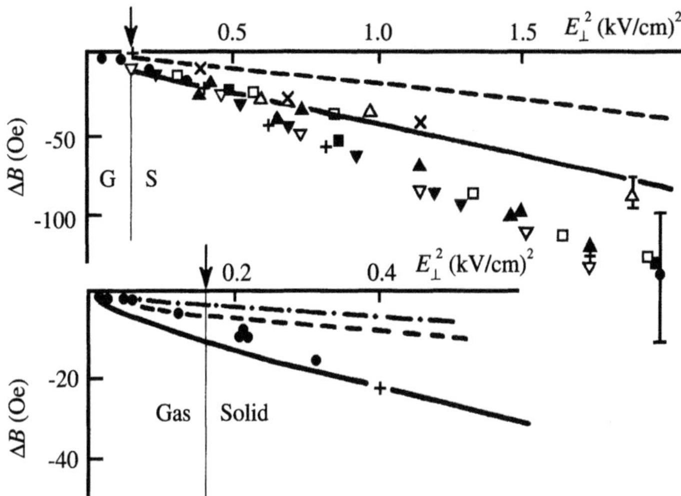

Fig. 5.19. CR peak shift vs. $E_\perp^2$. WS theory (*continuous curve*), single-electron evaluation according to (5.29) (*dashed curve*), the Cheng–Platzman theory (*dash-dotted curve*). *Symbols* represent Edel'man's data[9] found for the different experimental parameters discussed in the text

liquid helium thickness, etc.) to claim that the bending was caused by the appearance of the electron solid. The data measured under the condition $E_\perp \gg 2\pi e n_s$ are closer to the continuous curve. As for the data measured under the equilibrium condition $E_\perp = 2\pi e n_s$, they show an even stronger bending because, in this case, the characteristic frequency of the short-wavelength phonons contributing to $\langle u_+^2 \rangle$ is larger than ω_c, and therefore $\langle u_+^2 \rangle$ is smaller than l_B^2. In any case, the behavior of the CR data agrees with the magnetopolaron concept of the peak shift of the CR discussed above. Some of the ideas presented in this section will help us to understand the behavior of the Wigner solid on the surface of liquid helium, discussed in Chap. 7.

6 Interface Polarons

6.1 Relation to the General Polaron Problem

Polarons were invented by Landau and Pekar [193] in 1948 as electrons strongly coupled to medium excitations (phonons). Generally, a polaron is a self-sustained complex which consists of a localized electron and a medium excitation cloud induced by the electron. The electron is localized in the polaron center-of-mass reference frame. As a whole, the polaron can move freely with a strongly enhanced effective mass due to the medium excitation cloud. The theory was based on a variational procedure employing a Gaussian trial wave function for the localized electron. A later study of the strong coupling limit by Miyake in 1976 [194] showed that the Gaussian approximation actually gives a very accurate result. In typical solids under strong coupling conditions, the electron localization radius is about the size of the atomic cell. Therefore, the continuum theory or the Fröhlich Hamiltonian cannot describe electron properties in the strong coupling regime. This is the reason for the appearance of the so-called small polaron theory. In this sense, the system of SEs above the surface of liquid helium, or a helium film, represents an interesting case for the polaron problem: the strong electron–ripplon coupling limit is consistent with the continuum approach.

The particular problem of electron detrapping from the polaron to the quasi-free state has been attracting broad interest for a long while now [195–197]. An intriguing question that has been intensively discussed is whether electron detrapping occurs continuously or discontinuously when one varies the coupling constant α_{cc} characterizing the strength of electron interaction with medium vibrations. In other words, is there any critical behavior of the major electron properties at a certain, finite value of the coupling constant α_{c}^{*}, which can be considered as a sort of phase transition? In a naive picture, a discontinuity is associated physically with electron transition from a self-trapped state of finite localization radius L to a free state with $L = \infty$ at a certain value of the coupling constant.

In spite of the fact that self-trapping (or detrapping) transitions were claimed to exist in a large group of variational approaches, qualitative analysis based on the modern operator theory [198] shows no self-trapping transition for the standard Fröhlich system of the free optical polaron, when the phonon frequency satisfies the condition $\omega(q) \geq \omega > 0$. For acoustical polarons, some localization criteria were reported [199–201]. These criteria were formulated in terms of the stability index, which is a combination of the dimension index D, the medium dispersion index ν ($\omega_q \propto q^{\nu}$), and the force range index λ, where $g_q = \sqrt{\hbar/2\mu_q\omega_q}V_q \propto q^{-\lambda}$, V_q is the electron–medium coupling, and μ_q is an inertial quantity associated with medium vibrations. In this treatment, there is a marginal stability index value which separates self-trapped and free electron states. Because the stability index is usually a fixed intrinsic property of a system or a theoretical model (independent of α_c), self-trapping (or detrapping) cannot arise as a result of varying the coupling constant α_c, with the exception of the marginal stability index case.

The 2D electron system of interface electrons on the free surface of liquid helium represents a remarkable example for the polaron theory because the surface of a quantum liquid has no atomic cells, and the electron–ripplon interaction can be varied over a wide range by changing the holding electric field $E_{\perp}$ and the thickness d of the helium film. The self-consistent suppression of the helium surface and electron localization in this dimple was first discussed by Shikin [2, 37] in 1970–71, using the harmonic approximation. These interface polarons were initially called surface anions in analogy with bulk anions of liquid helium. As mentioned previously, the bulk liquid helium anion is a self-sustained 3D complex: an electron creates a bubble in order to reduce its interaction energy with the medium [202]. The radius of the bubble is determined by a variational procedure, namely, from the condition that the total energy of the complex is minimal. In this sense, the ripplonic polaron pictured as an electron plus the surface dimple represents a liquid helium anion raised up to the surface. Dynamic properties of ripplonic polarons were studied by Shikin and Monarkha [203]. Properties of the interface polarons above helium films were investigated by Monarkha [204] and Sander [205]. Important developments in the ripplonic polaron theory were conducted by Jackson and Platzman [206, 207], Hipòlito, Farias and Studart [208] and Saitoh [209] in the framework of the path integral formalism introduced by Feynman [210] in 1955.

From the general point of view, a very interesting result was obtained by Jackson and Platzman [206] when considering the ripplonic polaron above a liquid helium film. Using an acoustical model of the medium excitation spectrum $\omega_q = cq$ with a cutoff $q < q_c = \kappa$ (where κ is the capillary constant for surface excitations of the helium film), they found that the effective mass of the ripplonic polaron above the helium film undergoes an extremely rapid change (an almost step-like change) when the coupling constant α_{cc} approaches a certain critical value α_{cc}^{*} [206]. At the same time, it was shown that the polaron energy as a function of the variational parameter has no minimum for $\alpha_{cc} < \alpha_{cc}^{*}$. This transition was referred to as 'localization' because the drastic change in the effective mass was shown to occur over a very narrow range of the coupling constant ($< 10\%$).

In the variational analysis of [200], the acoustical polaron model of Jackson and Platzman ($D = 2$, $\nu = 1$ and $\lambda = -1/2$) was shown to correspond to the marginal critical value of the stability index

$$\delta = D - \nu - 2\lambda - 2 = 0 , \tag{6.1}$$

which yields a sort of 'critical' behavior of the polaron energy ($\mathcal{E}_{\mathrm{p}}$) considered as a function of the electron localization radius (L) and the coupling constant. The polaron energy

$$\mathcal{E}_{\mathrm{p}} \propto \frac{(\alpha_{\mathrm{c}}^{*} - \alpha_{\mathrm{c}})}{L^2} \tag{6.2}$$

changes sign at $\alpha_{\mathrm{c}} = \alpha_{\mathrm{c}}^{*}$. Still, at any value of the coupling constant, there is no minimum giving a finite localization radius L. Therefore the critical value α_{c}^{*} separates the free-electron ($L \to \infty$) and shrunken ($L \to 0$) states.

In another analysis based on symmetry arguments [201], the substantially different stability index

$$\sigma = \frac{D + 2 - 2\lambda}{\nu} \tag{6.3}$$

was introduced, and the ground state was shown to be delocalized for arbitrary coupling strength if $\sigma > 2$. There is a proof that the ground state is delocalized for $\sigma > 3$ [201, 211]. A sort of non-analyticity of the effective mass was found to exist approximately for $2 < \sigma < 3$. The ground state is localized for $\sigma < 2$ [201, 212]. Applying this criterion to the acoustical model of Jackson and Platzman, the authors of [201] reported that the corresponding value of the stability index was $\sigma = 5 > 2$, which precludes any 'localization' transition.

It should be noted that the above-mentioned analysis of [200, 201] does not fully correspond to the model used by Jackson and Platzman. The model employed in the ripplonic polaron treatment was actually a cutoff model, which assumes that there are no medium excitations if $q > q_{\mathrm{c}} = \kappa$. This wave-number cutoff is the most important assumption of the model, which actually introduces the 'localization' transition into the polaron treatment of [206], because the capillary constant κ combined with the localization radius of the strong coupling limit L_0 relates directly to the coupling constant $\alpha_{\mathrm{c}} = 2/\kappa^2 L_0^2$. The cutoff dispersion model of medium vibrations involved in the polaron cloud ($q \sim L_0$) thus becomes somewhat dependent on the coupling constant, which appears to be crucial for the self-trapping transition. The cutoff model of the medium dispersion is broadly used in studies of the polaron problem in the framework of the path integral method. We therefore pay some attention to this point in our discussion of interface polarons in Sect. 6.2.2.

Another important point is that the cutoff model of medium vibrations does not really reflect the actual excitation spectrum of the liquid helium film, which has no cutoff at $q \sim \kappa$:

$$\omega_q = \left[\frac{\alpha}{\rho}\left(q^2+\kappa^2\right) q \tanh(qd)\right]^{1/2} \simeq \omega_0 \tilde{q}\sqrt{1+\tilde{q}^2}\,, \tag{6.4}$$

where α is the surface tension, ρ is the liquid helium mass density, d is the film thickness, $\kappa = \sqrt{\rho G_d/\alpha}$ is the capillary constant, $\tilde{q} = q/\kappa$, and $\omega_0 = \sqrt{\alpha d/\rho\kappa^2}$. The second form of the ripplon dispersion represented in (6.4) implies that $qd \ll 1$. For $q \sim \kappa$, the medium dispersion curve just bends and the dispersion index changes smoothly from $\nu = 1$ ($q \ll \kappa$) to $\nu = 2$ ($q \gg \kappa$).

It is of obvious interest to investigate what happens for the actual form of the medium vibration spectrum given in (6.4). Will a variational approach using the exact dispersion form give a self-trapping transition at a finite value of α_c? If so, how will it relate to the stability criteria found previously in [200, 201]? The answers to these questions, given by Monarkha and Kono [213] quite recently, will be discussed in Sect. 6.2.2. We shall see that the detrapping transition ($L \to \infty$ as $\alpha_{cc} \to 1$) exists even for the actual ripplon spectrum, but that the polaron energy and the dimple mass are continuous and quite smooth functions of the coupling constant. The extreme sharpness or step-like behavior of the polaron mass reported previously [206] was shown to result from the very weak dependence of the polaron radius on the coupling constant in the cutoff approximation. Therefore, when passing through the region $\alpha_{cc} \simeq \alpha_{cc}^* = 1$, the cutoff model qualitatively agrees with the naive picture wherein detrapping occurs from some finite radius L to $L = \infty$. For the actual dispersion of (6.4), the electron localization radius depends strongly on the coupling constant $L \propto 1/\sqrt{\alpha_{cc}-1}$ and the detrapping point $\alpha_{cc} = 1$ is reached with zero binding energy and $L = \infty$. The detrapping transition is due to the fact that the dimensionless parameter $\tilde{q}^2$ becomes dependent on the coupling constant for the wave numbers $q \sim \sqrt{2}/L$ giving the major contribution to the polaron properties: $\tilde{q}^2 \sim \alpha_c/l^2$ (where $l = L/L_0$ is the normalized electron localization radius). Therefore, as the coupling constant α_c decreases and passes through the region $\alpha_c \sim 1$, the dispersion index ν of medium vibrations of (6.4) transforms from 2 to 1, crucially affecting the polaron stability index.

In our discussion of the interface polarons, we shall confine ourselves to the conventional adiabatic polaron treatment. The main reason is that, for the problem of ripplonic polarons, it reproduces well (even numerically) the results obtained in the more sophisticated path integral method. An additional advantage is that this transparent treatment offers an easy way to consider more complicated cases, e.g., using the actual dispersion of the medium excitations of the helium film. We shall start our discussion with the strong coupling approximation, introducing the main properties of the interface polarons (Sect. 6.2.1). The problem of electron detrapping is considered in Sect. 6.2.2. The behavior of the polaron effective mass near the detrapping transition is discussed in Sect. 6.3.1. In Sects. 6.3.2–6.3.4, we discuss the mobility of interface polarons in different regimes of the liquid substrate with regard to the bulk quasi-particle mean free path. Even though the evidence for single-electron interface polarons reported in a number of experiments cannot yet be considered as totally conclusive, the polaronic effect for interface electrons forming the Wigner solid has proven to be very important. Therefore, besides the instructive theoretical links with the general polaron problem, the discussions of interface polarons presented in this chapter will help us to understand the remarkable properties of the 2D electron solid formed on the free surface of liquid helium.

6.2 Ground-State Properties

6.2.1 Strong Coupling Theory

The essential starting point for the variational approach is that the electron wave function is assumed to be localized in the relative coordinates, which implies that the coupling constant α_{cc} is strong enough. The strict definition of the coupling constant α_{cc} will be given in the next section when we consider the weak coupling regime. The Hamiltonian of a single SE in the presence of ripplons can be written as

$$H = \frac{p^2}{2m_{\mathrm{e}}} + \frac{1}{2S_{\mathrm{A}}} \sum_{\boldsymbol{q}} \left(\frac{1}{\mu_q} \pi_{\boldsymbol{q}} \pi_{-\boldsymbol{q}} + \mu_q \omega_q^2 \xi_{\boldsymbol{q}} \xi_{-\boldsymbol{q}} \right) + \frac{1}{\sqrt{S_{\mathrm{A}}}} \sum_{\boldsymbol{q}} \xi_{\boldsymbol{q}} V_q \exp(\mathrm{i}\boldsymbol{q} \cdot \boldsymbol{r}) , \tag{6.5}$$

where μ_q and ω_q are the ripplon quantities introduced in Sect. 1.5.1. Following the conventional procedure of the strong coupling theory, we assume that the wave function of an electron in the polaron complex has the Gaussian form

$$\varphi(r) = \frac{1}{\sqrt{\pi} L} \exp\left(-\frac{r^2}{2L^2} \right) , \tag{6.6}$$

where L is the polaron radius. We then consider the expectation of the Hamiltonian, viz.,

$$\langle H \rangle_{\mathrm{e}} = \langle \varphi | H | \varphi \rangle , \tag{6.7}$$

which is the quantum average over the fast zero-point motion of the electron, as a function of the variational parameter L.

Taking into account the fact that

$$\langle \varphi | \mathrm{e}^{\mathrm{i}\boldsymbol{q}\cdot\boldsymbol{r}} | \varphi \rangle = \exp\left(-\frac{q^2 L^2}{4} \right) , \tag{6.8}$$

we find that

$$\begin{aligned} \langle H \rangle_{\mathrm{e}} = & \frac{\hbar^2}{2m_{\mathrm{e}} L^2} \\ & + \frac{1}{2S_{\mathrm{A}}} \sum_{\boldsymbol{q}} \left[\frac{1}{\mu_q} \pi_{\boldsymbol{q}} \pi_{-\boldsymbol{q}} + \mu_q \omega_q^2 \xi_{\boldsymbol{q}} \xi_{-\boldsymbol{q}} + \sqrt{S_{\mathrm{A}}} \left(\xi_{\boldsymbol{q}} + \xi_{-\boldsymbol{q}} \right) V_q \mathrm{e}^{-q^2 L^2/2} \right] , \end{aligned} \tag{6.9}$$

where the first term represents the zero-point kinetic energy of the localized electron. The ripplon Hamiltonian of (6.9) contains terms linear in $\xi_{\boldsymbol{q}}$, which means that the flat surface with $\xi = 0$ is not the proper ground state. We define new ripplonic variables by $\xi_{\boldsymbol{q}} = \xi_q^{(0)} + \xi'_{\boldsymbol{q}}$ and choose $\xi_q^{(0)}$ (the dimple profile) in order to eliminate the linear terms. We thus have

$$\xi_q^{(0)} = -\frac{V_q}{\mu_q \omega_q^2 \sqrt{S_{\mathrm{A}}}} \mathrm{e}^{-q^2 L^2/4} , \tag{6.10}$$

and

$$\langle H \rangle_{\mathrm{e}} = \frac{1}{2S_{\mathrm{A}}} \sum_{\boldsymbol{q}} \left[\frac{\pi_{\boldsymbol{q}} \pi_{-\boldsymbol{q}}}{\mu_q} + \mu_q \omega_q^2 \xi'_{\boldsymbol{q}} \xi'_{-\boldsymbol{q}} \right] + \frac{\hbar^2}{2m_{\mathrm{e}} L^2} - \sum_{\boldsymbol{q}} \frac{V_q^2}{2\mu_q \omega_q^2 S_{\mathrm{A}}} \mathrm{e}^{-q^2 L^2/2} , \tag{6.11}$$

Here the first term represents the energy of ripplon oscillations with respect to the new equilibrium surface profile. The third term is the energy gain due to the reconstruction of the surface. It is instructive to note that this term is negative, regardless of the actual sign of the interaction potential V_q. Therefore, the polaron cloud can actually be a deficiency or even a hole, as it is for helium anions.

The localization radius L is found from the condition that the total polaron energy $\mathcal{E}(L)$ at $T = 0$ [the last two terms of (6.11)] has a minimum. Polarons are energetically favorable if $\mathcal{E}(L_{\mathrm{min}}) < 0$. This gives

$$\frac{L}{2} \sum_{\boldsymbol{q}} \frac{q^2 V_q^2}{\mu_q \omega_q^2} \mathrm{e}^{-q^2 L^2/2} - \frac{\hbar^2}{m L^3} = 0 . \tag{6.12}$$

In most cases of the ripplonic polaron problem, one can substitute $V_q = eE_{\perp}^*$, where $E_{\perp}^* = E_{\perp} + \Lambda_{\mathrm{s}}/ed^2$ is the effective holding field due to the external electric field and polarization field of the solid substrate.

Recalling the expressions for μ_q and ω_q given in Sect. 1.5.1, in the limiting case $\kappa^2 L^2/2 \ll 1$, the localization radius can be obtained in the analytical form

$$L^2 = L_0^2 \equiv \frac{4\pi\alpha\hbar^2}{m(eE_{\perp}^*)^2} . \tag{6.13}$$

It is interesting to note that this localization radius, found by means of the variational procedure, is $\sqrt{2}$ times larger than the radius given by the oscillatory approximation [2, 37]. The respective normalized Gaussian wave functions are shown in Fig. 6.1 by dashed (variational solution) and dotted (oscillatory approximation) curves. The continuous curve in this figure represents the exact numerical solution for the self-consistent wave function obtained by Marques and Studart [214]. Although the exact wave function is not a Gaussian, it is very close to the Gaussian trial wave function of (6.6) (dashed curve) and it differs significantly from the wave function of the oscillatory approximation (dotted curve). We conclude therefore that the variational approach with the Gaussian trial wave function provides us with quite accurate results concerning the polaron binding energy and effective mass.

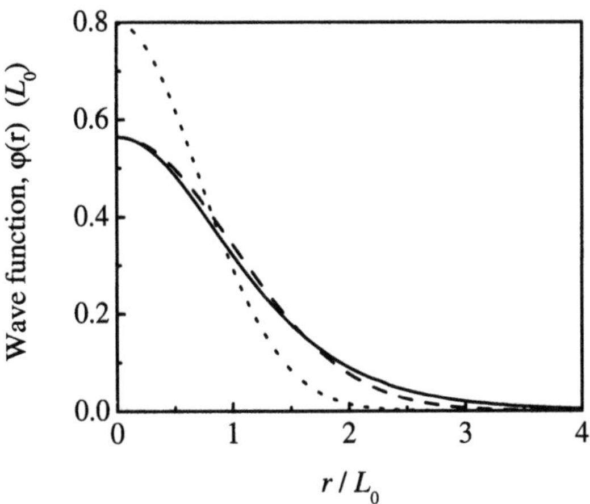

Fig. 6.1. Ground state wave function of the electron in the self-trapped state. *Continuous curve*: exact numerical evaluation of the Schrödinger equation given by Marques and Studart [214]. *Dashed curve*: Gaussian variational function with L defined by (6.13). *Dotted curve*: oscillatory approximation

In the strong coupling limit, the minimum value of the polaron energy $\mathcal{E}_0$ is [203, 204]

$$\mathcal{E}_0 \simeq -\frac{(eE_\perp^*)^2}{8\pi\alpha}\left[\ln\left(\frac{2}{\gamma_0\kappa^2L_0^2}\right)-1\right] , \tag{6.14}$$

where $\gamma_0 \simeq 1.78$ is the Euler constant. For bulk liquid helium and films covering substrates with a large dielectric constant, we have $\kappa L_0 \ll 1$. In this case, the polaron energy is certainly negative, which means that the formation of ripplonic polarons is energetically favorable. In the following section we shall see that the expression $\mathcal{E}_0(L_{\min})$ for the polaron energy, valid at arbitrary values of the parameter κL_0, is negative as soon as $\kappa L_0 < \sqrt{2}$.

Generally, one should distinguish the electron localization radius L_0 from the size of the surface dimple. The dimple usually has an additional length parameter κ^{-1} which cuts off the logarithmic tails of the surface profile according to

$$\xi^{(0)}(r) = -\frac{eE_\perp}{2\pi\alpha}\int_0^\infty \frac{qJ_0(qr)}{q^2+\kappa^2}\mathrm{e}^{-q^2L_0^2/4}\mathrm{d}q\,. \tag{6.15}$$

The existence of two characteristic lengths for the dimple profile is clearly seen in the numerical graph of Fig. 6.2. There is a distinct change in the surface profile at $r \sim L_0$. Still, the dimple tails (continuous) spread far beyond the electron density profile $\varphi^2(r)$ (dashed) up to $r \sim \kappa^{-1}$. In the strong coupling limit, the dimple depth is given by

$$\xi(0) \simeq -\frac{eE_{\perp}^{*}}{2\pi\alpha} \ln\left(\frac{2}{\kappa L_0 \sqrt{\gamma_0}}\right) .$$

For the holding fields $E_{\perp}$ usually used in experiments with SEs on liquid helium, $|\xi(0)|$ is less than 1.7×10^{-8} cm. In the range $r > \kappa^{-1}$, the dimple tails

$$\xi(r) \simeq -\frac{eE_{\perp}^{*}}{2\pi\alpha} K_0(\kappa r)$$

are cut off at an exponential rate.

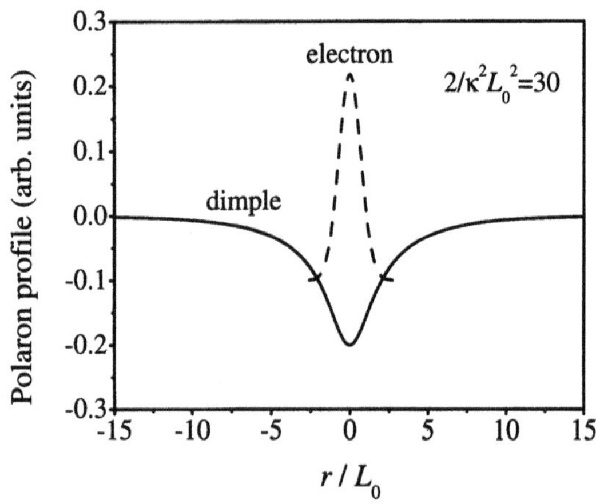

Fig. 6.2. Ripplonic polaron profile. *Continuous curve*: dimple profile. *Dashed curve*: electron charge density distribution profile. Evaluations were performed for $2/\kappa^2 L_0^2 = 30$

As the typical polaron binding energy is very small, the strongest holding field possible must be applied to reach the condition $|\mathcal{E}_0| > T$. For bulk liquid helium under conditions $n_s \ll E_{\perp}/2\pi e$, the charged surface becomes unstable at $E_{\perp}^{(\mathrm{crit})} = 2\sqrt{\pi}(\rho G\alpha)^{1/4} \simeq 2.9\,\mathrm{kV/cm}$ [9]. Moreover, one should take into account the fact that, at a finite electron density, the dimple tail cannot exceed half of the electron spacing $a \simeq 1/\sqrt{n_s}$. This can be done approximately by replacing $\kappa \to \sqrt{\kappa^2 + 4n_s}$ or introducing a long-wave-number cutoff $q_0 \simeq g_1$ in the sum over $\boldsymbol{q}$, where g_1 is the smallest reciprocal lattice wave number of the fictitious electron crystal for the given density. For $E_{\perp} = E_{\perp}^{(\mathrm{crit})}$, the polaron radius is $L \simeq 1.63 \times 10^{-5}$ cm, which satisfies the requirement $\kappa L \ll 1$. Thus, for $n_s = 5 \times 10^6\,\mathrm{cm}^{-2}$, the polaron binding energy cannot be larger than $|\mathcal{E}_0| \simeq 0.072\,\mathrm{K}$. It reaches this value at $E_{\perp} \to E_{\perp}^{(\mathrm{crit})}$. At $T \approx 0.072\,\mathrm{K}$, the plasma coupling parameter $\Gamma^{(\mathrm{pl})} \simeq 91$, which means that the condition $|\mathcal{E}_0| > T$ can be reached before the system undergoes the Wigner transition. It should be noted that, at a fixed holding electric field, the polaron binding energy above liquid ^{3}He is approximately 2.3 times larger than above liquid ^{4}He due to the smaller surface tension. Still, the maximum value of $E_\perp$ is reduced because $E_\perp^{(\mathrm{crit})} \propto (\rho\alpha)^{1/4}$.

In order to increase the polaron binding energy, Monarkha [204] and Sander [205] proposed to use a helium film of thickness $d \approx 100\,\text{Å}$ covering a dielectric substrate with a large dielectric constant. This makes the effective holding field $E_\perp^* = eE_\perp + \Lambda_s/d^2$ substantially stronger and increases $|\mathcal{E}_0|$ to values $\approx 6\,\mathrm{K}$. Unfortunately, in practice, it is rather difficult to make a dielectric substrate with the necessary flatness to avoid electron localization above surface irregularities [69, 215]. Still, there are experimental data [216] which show that the electron mobility above films with $d > 300\,\text{Å}$ is substantially lower than the free-electron mobility and close to the mobility of a ripplonic polaron, in spite of the rather small binding energies.

Concerning other methods used to describe interface polarons, an interesting version of the Feynman polaron approach applied to SEs on helium films at finite temperatures was proposed by Saitoh [209]. The form of the ground state energy obtained in this theory transforms into the result of (6.14) at zero temperature and differs from that found by Jackson and Platzman by a logarithmic factor. The difference between the results of these two path integral approaches arises because they use different models of the medium excitation dispersion at $q > \kappa$.

6.2.2 Detrapping Transition

As usual, there is a certain freedom in the definition of the polaron coupling constant α_{cc}. We define it from the condition that the average phonon number of the polaron cloud $\langle N \rangle = \alpha_{\mathrm{cc}}$ in the weak coupling regime. Then the value $\alpha_{\mathrm{cc}} = 1$ represents the transition regime from weak to strong coupling.

Following the standard perturbation procedure of quantum theory, we find that

$$\langle N \rangle = \sum_{\boldsymbol{q}} \frac{V_q^2 Q_q^2}{\left(\varepsilon_{\boldsymbol{k}} - \varepsilon_{\boldsymbol{k}-\boldsymbol{q}} - \hbar\omega_q\right)^2}\,, \tag{6.16}$$

where $\varepsilon_{\boldsymbol{k}}$ is the free electron spectrum. Using the actual expressions for ω_q and Q_q gives

$$\langle N \rangle = \alpha_{\mathrm{cc}} \int_0^\infty \sqrt{\frac{\tanh(x\lambda\kappa d)}{(1+\lambda^2 x^2)x\lambda\kappa d}}\,\frac{\mathrm{d}x}{(1+x)^2}\,, \tag{6.17}$$

where we have introduced the dimensionless parameters

$$\alpha_{\mathrm{cc}} = \frac{(eE_\perp^*)^2}{4\pi\alpha}\frac{2m_{\mathrm{e}}}{\hbar^2\kappa^2}\,, \qquad \lambda = \frac{2m_{\mathrm{e}}}{\hbar}\sqrt{\frac{\alpha d}{\rho}}\,. \tag{6.18}$$

The parameter λ does not depend on the properties of the solid substrate and it has a rather weak dependence on the helium film thickness. At $d \simeq 10^{-6}$ cm estimates show that the parameter λ is very small $\lambda \simeq 2.7 \times 10^{-3}$. For typical helium films, the parameter κd is not large (at least it is smaller than $1/\lambda$). Under these conditions $\tanh(x\lambda\kappa d) \simeq x\lambda\kappa d$ and the integral of (6.17) is equal to unity with high accuracy. Therefore, in the weak coupling limit, the average ripplon number $\langle N \rangle$ is equal to α_{cc}.

The relation $\langle N \rangle = \alpha_{cc}$ indicates that α_{cc} introduced above can serve as a convenient choice for the polaron coupling constant. Recalling the expression for L_0 given in (6.13), we find that the coupling constant chosen above relates in a remarkable way to the parameter $\kappa^2 L_0^2/2$ used previously:

$$\alpha_{cc} = \frac{2}{\kappa^2 L_0^2} \,. \tag{6.19}$$

In the original paper of Jackson and Platzman [206, 207], the coupling constant was defined as $\alpha_{cc}/2$. The advantage of the notation of (6.18) and (6.19) is that, in addition to the simple relation $\langle N \rangle = \alpha_{cc}$, the detrapping transition which we discuss below occurs exactly at $\alpha_{cc} \to 1$.

Taking into account the fact that $E_d \propto d^{-2}$ and $\kappa^2 \propto d^{-4}$, we conclude that as $E_\perp \to 0$ the coupling constant α_{cc} is practically independent of d and (in contrast to λ) is mainly determined by the properties of the substrate material. For typical substrate materials with a large dielectric constant, such as metal or glass, estimates show that $\kappa L_0 \approx 0.1$ which means that the coupling constant α_{cc} is large, although for a variety of substrates, $0.02 \leq \alpha_{cc} \leq 20$ according to [206].

The detrapping transition of Jackson and Platzman was obtained using the cutoff model of the medium excitation spectrum shown by the dashed curve in Fig. 6.3. It is clear that this dispersion differs substantially from the actual dispersion of ripplons in the helium film, shown there by the continuous curve. Before proceeding with the exact dispersion form, we would like to show that, for the cutoff model, the method employed here reproduces the detrapping transition found previously by means of the path integral method.

Firstly, we note that for the pure acoustical spectrum $\omega_q \propto q$ without the cutoff, the last two terms of (6.11) reproduce the result of [200]: the normalized polaron energy $\tilde{\mathcal{E}}_\text{p} = 2mL_0^2\mathcal{E}_\text{p}/\hbar^2 = (1-\alpha_{cc})/l^2$ changes its sign when the coupling constant α_c passes unity, but there is no minimum to fix the normalized localization radius $l = L/L_0$ when $\alpha_{cc} > 1$. The situation changes a great deal for the cutoff model with $q_\text{c} = \kappa$. In this case, the summation over $|\boldsymbol{q}| < \kappa$ yields a different polaron energy equation:

$$\tilde{\mathcal{E}}_\text{p} = \frac{1}{l^2}\left[1 - \alpha_{cc}\left(1 - \mathrm{e}^{-l^2/\alpha_\text{c}}\right)\right] \,. \tag{6.20}$$

Here the term $-\exp(-l^2/\alpha_\text{c})$ introduced by the cutoff prevents the electron from shrinking to $l = 0$ for $\alpha_\text{c} > 1$. For $\alpha_\text{c} > 1$, the polaron energy has a

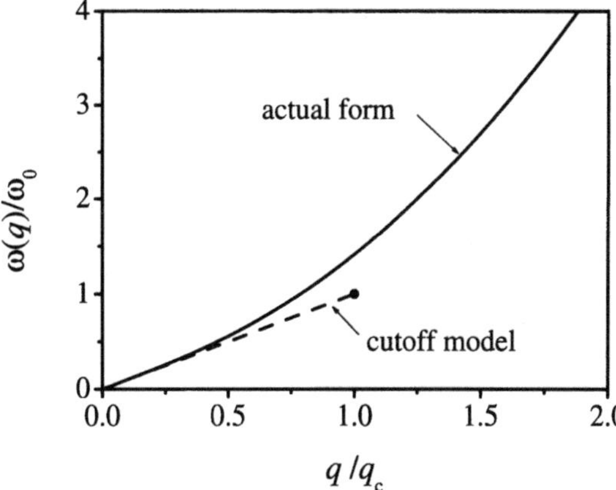

Fig. 6.3. The actual ripplon dispersion for the liquid helium film (*continuous curve*) and the cutoff model of Jackson and Platzman (*dashed curve*)

minimum at a finite l, while for $\alpha_{\text{cc}} < 1$, there is no minimum. The polaron energy as a function of the electron localization radius is shown in Fig. 6.4a for different values of the coupling constant. One can see that the position of the minima (dashed line) does not change very much when the coupling constant approaches unity. This is an important property of the cutoff model, which (as we shall see in Sect. 6.3.1) explains the sharpness of the detrapping transition in the model of Jackson and Platzman. Thus, even in the simple variational method of the adiabatic theory, we obtained the polaron detrapping transition of Jackson and Platzman at the same value of the coupling constant. This inspires us to apply this method to the actual form of the ripplon dispersion.

In the general case, for the ripplon spectrum of (6.4), the normalized polaron energy can be written as

$$\tilde{\mathcal{E}}_{\text{p}}(l) = \frac{1}{l^2} - \int_0^\infty \frac{\mathrm{e}^{-x}\mathrm{d}x}{x + l^2/\alpha_{\text{cc}}} \,. \tag{6.21}$$

Typical dependencies $\tilde{\mathcal{E}}_{\text{p}}(l)$ for different values of the coupling constant are shown in Fig. 6.4b. Comparing Figs. 6.4a and b, one can conclude that, for the actual dispersion, the polaron radius shown by the dashed line increases with $\alpha_{\text{cc}} \to 1$ much faster than for the cutoff model (Fig. 6.4a).

In order to find the the polaron radius as a function of the coupling constant, one should solve (6.12). It is convenient to rewrite this equation in the form

$$\frac{1}{l^2} = \int_0^\infty \frac{x}{x + l^2/\alpha_{\text{cc}}} \mathrm{e}^{-x}\mathrm{d}x \,. \tag{6.22}$$

At $\alpha_{\text{cc}} \gg 1$, the solution of (6.22) is evident: $l = 1$, in accordance with (6.13). This limiting case describes the situation where the capillary constant

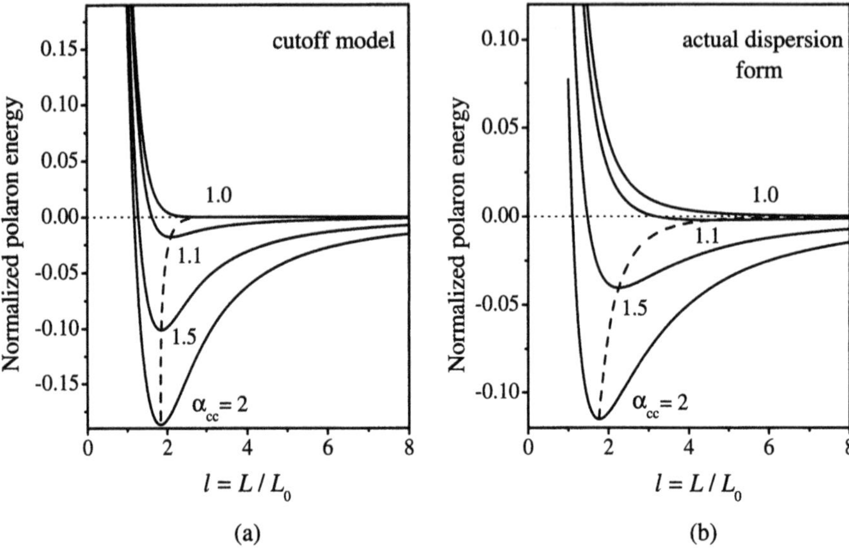

Fig. 6.4. Normalized polaron energy $\tilde{\mathcal{E}}_\mathrm{p}$ as a function of the electron localization radius for different values of the coupling constant: (**a**) The cutoff model, (**b**) the actual dispersion form. The positions of the minima are shown by *dashed curves*

κ can be disregarded in the ripplon dispersion as compared to q^2, and it corresponds to the following stability indexes introduced in (6.1) and (6.3): $\delta = -2 < 0$ and $\sigma = 2$. In the opposite limit $\alpha_\mathrm{c} \ll 1$, there is no solution for any finite l. The only solution $l = \infty$ describes uncoupled free electron states. The respective stability indexes are different: $\delta = 0$ and $\sigma = 5 > 2$. Therefore, for the actual dispersion, the marginal value $\delta = 0$ discussed in [200] is achieved only in the limiting case $l^2/\alpha_\mathrm{c} \gg 1$. The stability index thus becomes dependent on the coupling constant and reaches the marginal value when $\alpha_\mathrm{c} \to \alpha_\mathrm{c}^*$, which causes electron detrapping.

In terms of the dimensionless parameter $x_q = q^2L^2/2$, the ripplon dispersion of a helium film with $qd \ll 1$ can be rewritten as

$$\frac{\omega_q}{\omega_0} = \frac{\alpha_{\mathrm{cc}}}{l^2}\sqrt{x_q\left(x_q + \frac{l^2}{\alpha_{\mathrm{cc}}}\right)}\,. \tag{6.23}$$

The major contribution in the sum over $\boldsymbol{q}$ comes from the range $x_q \sim 1$. It is therefore clear that the dispersion (ν), range (λ) and stability (σ) indexes become strongly dependent on the coupling constant. The stability index σ increases substantially when the coupling constant passes through the region $\alpha_{\mathrm{cc}} \sim 1$. Therefore, important changes in the solution of (6.22) are expected in the intermediate coupling regime, which we now analyze in more detail.

By changing the integration variable, (6.22) can be rearranged in the following way:

$$l^2 = 1 + \frac{l^4}{\alpha_{cc}} \int_0^\infty \frac{e^{-x}}{x + l^2/\alpha_{cc}} dx$$
$$\equiv 1 + \frac{l^4}{\alpha_{cc}} e^{l^2/\alpha_{cc}} \left[-\mathrm{Ei}\left(-\frac{l^2}{\alpha_{cc}} \right) \right] , \qquad (6.24)$$

convenient for iteration and for analysis of the intermediate coupling regime. The first line of (6.24) shows that l increases when the coupling constant α_{cc} decreases. Anticipating critical behavior of the localization radius ($l \to \infty$ when $\alpha_{cc} \to \alpha_{cc}^*$), we can use the asymptotic form

$$\mathrm{Ei}(-x) \simeq -\frac{e^{-x}}{x} \left(1 - \frac{1!}{x} + \frac{2!}{x^2} \right) \qquad (6.25)$$

in the second line of (6.24). This yields the analytical solution

$$l^2 = \frac{2\alpha_{cc}^2}{\alpha_{cc} - 1} \simeq \frac{2}{\alpha_{cc} - 1} , \qquad (6.26)$$

valid when $\alpha_{cc} - 1 \ll 1$. The localization radius $l \to \infty$ and the parameter $l^2/\alpha_{cc} \to \infty$ if the coupling constant α_{cc} approaches unity. This result indicates that the stability index δ reaches the marginal value 0 when $\alpha_{cc} \to 1$.

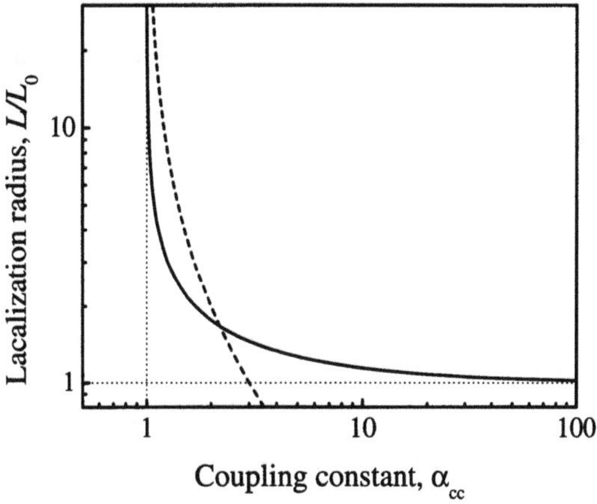

Fig. 6.5. Electron wave function localization radius vs. the coupling constant α_{cc}. *Continuous curve*: numerical solution of (6.22). *Dashed curve*: analytical form valid near $\alpha_{cc} = 1$ [213]

The numerical solution of (6.22) shown in Fig. 6.5 by the continuous curve agrees with the asymptotic behavior of (6.26) (dashed curve). The rapid increase in the electron localization radius near the critical value of the coupling constant agrees with the simple physical picture of the detrapping transition. The important difference between the result found above and the naive picture of electron detrapping from a finite value of L to $L = \infty$ is that the electron localization radius increases smoothly when $\alpha_{cc} \to 1$, and the detrapping point ($\alpha_{cc} = 1$) is reached when $L = \infty$ and $\mathcal{E}_0 = 0$.

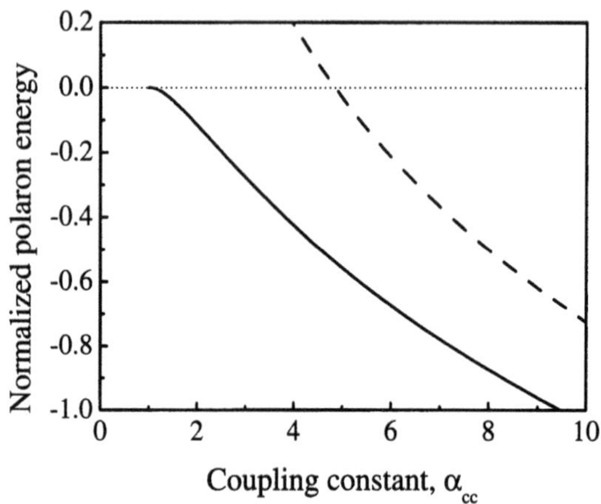

Fig. 6.6. Normalized polaron energy $\tilde{\mathcal{E}}_0$ vs. the coupling constant. *Continuous curve*: numerical evaluation of (6.27). *Dashed curve*: strong coupling theory (6.14) [213]

In the general case, the ground state polaron energy in units $\hbar^2/2m_e L_0^2 \equiv (eE_\perp^*)^2/8\pi\alpha$ can be found as

$$\begin{aligned}\tilde{\mathcal{E}}_0 &= -\int_0^\infty \frac{(1-x)}{x + l^2/\alpha_{cc}} \mathrm{e}^{-x} \mathrm{d}x \\ &= \left(1 + \frac{l^2}{\alpha_{cc}}\right) \mathrm{e}^{l^2/\alpha_{cc}} \mathrm{Ei}\left(-\frac{l^2}{\alpha_{cc}}\right) + 1\,. \end{aligned} \qquad (6.27)$$

Here the zero-point energy of the electron $1/l^2$ is transformed according to the equation (6.22) for the polaron radius. In the limiting case $\alpha_{cc} \gg 1$, (6.27) agrees with the result found previously for the strong coupling theory because $\mathrm{Ei}(-x) \simeq -\ln(1/\gamma_0 x)$ if $x \ll 1$. From the numerical graph of Fig. 6.6, one can conclude that, for arbitrary values of the coupling constant ($\alpha_{cc} > 1$), $\mathcal{E}_0$ (continuous curve) is always negative [unlike the asymptotic form of (6.14) (dashed curve)]. Still, it falls rapidly to zero as $\alpha_{cc} \to 1$. There is no non-analyticity of the ground state energy as $\alpha_{cc} \to 1$ (with the exception of the fact that there is no minimum for $\alpha_{cc} < 1$), which can be seen from the asymptotic behavior of (6.27) when $\alpha_{cc} - 1 \ll 1$:

$$\tilde{\mathcal{E}}_0 \simeq -\frac{(\alpha_{cc} - 1)^2}{4}\,. \qquad (6.28)$$

This equation is valid only in the range $\alpha_{cc} \geq 1$. The asymptotic form of (6.28) also indicates that the polaron binding energy falls rapidly to zero when the coupling constant approaches unity. Due to this strong decrease in the polaron binding energy $|\mathcal{E}_0|$, one has to cool the system down to $T \ll (eE_{\perp}^*)^2/8\pi\alpha$ in order to get closer to the region where the electron localization radius increases fast, which is consistent with the finite temperature results of Jackson and Platzman [207] and Peeters and Jackson [217].

To conclude, the detrapping transition of ripplonic polarons on a liquid helium film [$l \propto \sqrt{2/(\alpha_c - 1)}$ as $\alpha_c \to 1$] originates from the unusual dispersion form of the long-wavelength medium vibrations, which represents a very interesting case of the polaron problem. The dispersion and stability indexes for the medium vibrations involved in the polaron cloud become dependent on the coupling constant, so that the stability index reaches the marginal value when $\alpha_c \to 1$. Such a situation has not been discussed previously in the general stability analysis of the polaron problem [198, 201]. The cutoff model of the medium excitation spectrum often used in theoretical studies [206, 217] somehow reflects this feature of the ripplonic polaron. However, for polaron properties depending strongly on the electron localization radius, this model cannot be considered as totally accurate. As we shall see in the following section, at zero temperature, it introduces an over-sharp behavior of the polaron mass near the transition point. The actual form of the ripplon dispersion results in large but smooth and continuous changes in the polaron energy and mass, which agrees with the general analysis of [201].

6.3 Transport Along the Interface

Coupling to the medium excitation cloud drastically changes electron transport properties. When an electron moves slowly along the interface, the surface dimple contributes to the polaron effective mass. In the strong coupling regime, this contribution dominates and the interface polaron has a huge effective mass. Regarding polaron mobility, it depends on the conditions of the liquid substrate. At relatively high temperatures, the polaron mobility above liquid ^{4}He is limited by the viscous friction acting on surface dimples. At low temperatures $T < 1\,\mathrm{K}$, the system enters the long mean-free-path regime for bulk excitations of liquid ^{4}He, and dimples become more mobile than electrons scattered by ripplons. A different situation occurs for SEs above liquid ^{3}He, which represents a highly viscous Fermi liquid. Even in the long mean-free-path regime, a surface dimple is very slow [54] because of ^{3}He quasi-particle reflection from the surface dimple. In the following sections, we shall discuss polaron transport along the interface for different coupling regimes and different conditions of the liquid substrate.

6.3.1 Effective Mass

The effective mass of an interface polaron is the sum of the free electron mass m_e and the mass of the surface dimple M_d. Because typical dimple lengths (L and κ) are long enough, M_{d} can be found in the usual framework of hydrodynamics.

When the dimple moves along the surface with a constant velocity $\boldsymbol{V}_0$, its profile induces a velocity field in the bulk liquid $\boldsymbol{v}(\boldsymbol{r}, z)$. In the laboratory reference frame, the moving profile is described by $\xi^{(0)}(\boldsymbol{r} - \boldsymbol{V}_0 t)$. We assume that the profile is smooth $\nabla\xi \ll 1$. Then, using the relation $v_z(\boldsymbol{r}, 0) = \partial\xi/\partial t$, one can find the boundary condition on the liquid surface:

$$v_z = -\boldsymbol{\nabla}\xi \cdot \boldsymbol{V}_0 \,, \tag{6.29}$$

which determines the velocity field in the bulk liquid: $v_z(\boldsymbol{r}, 0) < 0$ on the front side of the dimple and $v_z(\boldsymbol{r}, 0) > 0$ on the rear side.

Consider the infinitesimal element $\mathrm{d}S_{\mathrm{A}}$ of the dimple surface. The outward normal of this element is a function of the in-plane coordinates:

$$\boldsymbol{n}(x, y) = \frac{\hat{\boldsymbol{z}} - \boldsymbol{\nabla}\xi}{\sqrt{1 + (\nabla\xi)^2}} \,, \tag{6.30}$$

where $\hat{\boldsymbol{z}}$ is the unit vector directed along the z-coordinate axis. For smooth profiles, $n_z \simeq 1$, $n_r \simeq -\partial\xi/\partial r$, and $v_z \simeq \boldsymbol{v} \cdot \boldsymbol{n}$. Then, in the moving reference frame ($\boldsymbol{v}' = \boldsymbol{v} - \boldsymbol{V}_0$), (6.29) agrees with the rigid boundary condition $\boldsymbol{v}' \cdot \boldsymbol{n} = 0$ at the liquid surface.

From standard textbooks, we know that the kinetic energy of the fluid motion and the associated mass caused by the velocity field $\boldsymbol{v}(\boldsymbol{r}, z)$ can be expressed as a surface integral

$$M_{\mathrm{d}} = \frac{\rho}{V_0^2} \int (\boldsymbol{v} \cdot \boldsymbol{n}) \Phi \, \mathrm{d}S_{\mathrm{A}} \,, \tag{6.31}$$

where $\Phi(\boldsymbol{r}, z)$ is the hydrodynamic potential $[\boldsymbol{v} = \boldsymbol{\nabla}\Phi]$. The integral is taken over the free liquid boundary, where $\boldsymbol{v} \cdot \boldsymbol{n} \neq 0$. The first factor of the integrand $(\boldsymbol{v} \cdot \boldsymbol{n})$ is known because of the boundary condition (6.29). In order to find the second factor Φ, we have to solve the equation $\Delta\Phi = 0$ for the incompressible liquid with boundary condition (6.29) at the liquid surface and with boundary condition

$$\frac{\partial\Phi}{\partial z} = 0 \tag{6.32}$$

at the solid substrate ($z = -d$). Direct evaluation yields

$$\Phi(\boldsymbol{r}, 0) = -\mathrm{i} \sum_{\boldsymbol{q}} \frac{\boldsymbol{q} \cdot \boldsymbol{V}_0}{q} \xi_q^{(0)} \coth(qd) \mathrm{e}^{\mathrm{i}\boldsymbol{q} \cdot \boldsymbol{r}} \,. \tag{6.33}$$

Combining (6.29) and (6.33) in (6.31), the associated mass of the surface dimple can be found in the convenient form

$$M_{\mathrm{d}} = \frac{\rho}{2} \sum_{\boldsymbol{q}} (\xi_q^{(0)})^2 q \coth(qd) \,, \tag{6.34}$$

valid for arbitrary smooth profiles.

For the particular dimple profile defined in (6.15) and the actual ripplon dispersion of the helium film, the dimple mass is

$$M_{\mathrm{d}} = M_{\mathrm{d}}^{(0)} l^2 \int_0^{\infty} \frac{\mathrm{e}^{-x} \mathrm{d}x}{(x + l^2/\alpha_{\mathrm{cc}})^2} , \tag{6.35}$$

where

$$M_{\mathrm{d}}^{(0)} = \frac{\hbar^2 \rho}{4 m_{\mathrm{e}} \alpha d} . \tag{6.36}$$

Thus, in the strong coupling limit, we have [204]

$$M_{\mathrm{d}} \simeq \alpha_{\mathrm{cc}} M_{\mathrm{d}}^{(0)} = \frac{\rho \left(e E_{\perp}^{*} \right)^2}{8 \pi \alpha^2 \kappa^2 d} . \tag{6.37}$$

It is remarkable that this expression for the polaron mass was also reproduced in the framework of the Feynman path integral approach by Saitoh [209] and by Peeters and Jackson [217]. The cause of the agreement between the results of these different approaches is that the dimple mass of the strong coupling limit does not depend on the electron localization radius l. In the opposite limit of small values of the parameter α_{cc}/l^2, the associated dimple mass

$$M_{\mathrm{d}} \simeq \alpha_{\mathrm{cc}}^2 M_{\mathrm{d}}^{(0)} \frac{1}{l^2} \tag{6.38}$$

depends strongly on the electron localization radius.

For the purposes of comparison, it is instructive to analyze the cutoff model as well. Using the acoustical spectrum with cutoff $q_{\mathrm{c}} = \kappa$ in (6.34), we find that

$$M_{\mathrm{d}} = \frac{M_{\mathrm{d}}^{(0)}}{l^2} \alpha_{\mathrm{cc}}^2 \left(1 - \mathrm{e}^{-l^2/\alpha_{\mathrm{cc}}} \right) . \tag{6.39}$$

Therefore in the strong coupling limit $l^2/\alpha_{\mathrm{cc}} \ll 1$, we have $M_{\mathrm{d}} \simeq \alpha_{\mathrm{cc}} M_{\mathrm{d}}^{(0)}$, which agrees remarkably with the result found for the actual dispersion [see (6.37)]. Moreover, in the opposite limit $\alpha_{\mathrm{cc}}/l^2 \ll 1$, the dimple mass $M_{\mathrm{d}} \simeq \alpha_{\mathrm{cc}}^2 M_{\mathrm{d}}^{(0)}/l^2$ has the same dependence on the electron localization radius as that found for the actual dispersion. Therefore, for the two different models of the medium excitation spectrum discussed, one can only expect different behavior of the polaron mass as a function of the coupling constant in the regime of intermediate coupling $\alpha_{\mathrm{cc}} \sim 1$, because of the different behavior of $l(\alpha_{\mathrm{cc}})$ shown previously in Fig. 6.4 by dashed curves.

For the actual dispersion form at $\alpha_{\mathrm{cc}} \to 1$ ($l \simeq 2/\sqrt{\alpha_{\mathrm{cc}} - 1} \to \infty$), the dimple mass vanishes smoothly:

$$M_{\mathrm{d}} \simeq M_{\mathrm{d}}^{(0)} \frac{\alpha_{\mathrm{cc}} - 1}{2} , \tag{6.40}$$

because of the steady increase in the localization radius. The numerical graph of $M_{\mathrm{d}}(\alpha_{\mathrm{c}})$ evaluated according to (6.35) is shown by the continuous curve in

Fig. 6.7. For $d = 100$ Å, estimates give $M_{\rm d}^{(0)}/m_{\rm e} \simeq 1.35 \times 10^5$. Therefore, the polaron mass does indeed change rapidly from a value of the order of 10^5–$10^6 m_{\rm e}$ to the free electron value in qualitative agreement with the result of Jackson and Platzman found in the framework of the path integral method. Still, in contrast with the sharp fall of the polaron mass found in [206], the continuous curve of Fig. 6.7 is smooth.

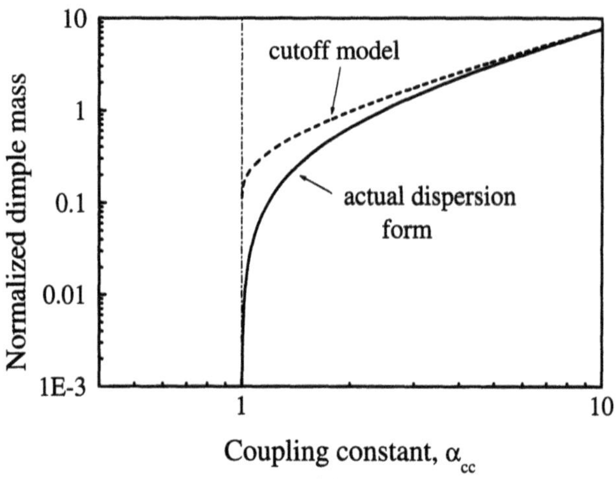

Fig. 6.7. Normalized dimple mass $M_{\rm d}/M_{\rm d}^{(0)}$ as a function of the coupling constant. *Continuous curve*: numerical evaluation of (6.35) for the actual dispersion form. *Short-dashed curve*: cutoff model result discussed in the text [213]

The sharpness of the fall in the polaron mass obtained in [206] can be ascribed to the cutoff approximation. This conclusion follows from the analysis of the cutoff model in the framework of the adiabatic variational method [see (6.20)]. In this case, the localization radius l only increases logarithmically as $\alpha_{\rm c} \to 1$. This means that the dimple mass $M_{\rm d} \simeq \alpha_{\rm c}^2 M_{\rm d}^{(0)} l^{-2}$ depends weakly on the difference $\alpha_{\rm c} - 1$ when the coupling constant approaches its critical value. Because the polaron energy has no minimum for $\alpha_{\rm c} < 1$ and, for the cutoff model, $M_{\rm d}(\alpha_{\rm c})$ is nearly constant when $\alpha_{\rm c} > 1$, the mass fall shown in Fig. 6.7 by the short-dashed curve really looks like a step function. The cutoff model thus leads to a result which is somewhat similar to the naive picture of the electron detrapping transition from a finite value of L to $L = \infty$. The strong dependence $L(\alpha_{\rm c})$ obtained for the actual dispersion eliminates the sharpness of the detrapping transition for interface polarons above a helium film.

6.3.2 Viscosity Drag of Self-Trapped Electrons

The velocity field induced by the dimple motion causes dissipation in the bulk liquid which can under certain conditions be described in terms of a viscosity η. In this case, the velocity of the dimple and mobility are found by balancing the work performed by the driving electric field and the energy dissipated in the bulk liquid. Such an approach is applicable for ^{3}He in the hydrodynamic regime and for ^{4}He films at rather high temperatures.

The viscosity drag leads to the polaron mobility equation [204]

$$\mu = \frac{e}{\eta \left(\mathcal{I}_1 + \mathcal{I}_2 \right)} , \tag{6.41}$$

where

$$\mathcal{I}_1 = \frac{1}{2\pi} \int_0^\infty \xi_q^2 \left(1 - \mathrm{e}^{-2qd} \right) q^4 \mathrm{d}q , \tag{6.42}$$

$$\mathcal{I}_2 = \frac{1}{2\pi} \int_0^\infty \xi_q^2 \frac{\left[1 + (2qd)^2 \right]}{\sinh(2qd) - 2qd} q^4 \mathrm{d}q . \tag{6.43}$$

We have not entered into the details of the evaluations resulting in the above formula. Some of them will be revealed in Chap. 8 which deals with transport properties of the Wigner solid.

For massive liquid helium ($d = \infty$), $\mathcal{I}_2 \ll \mathcal{I}_1$ and the dimple mobility becomes independent of the capillary constant (κ):

$$\mu = \frac{2\sqrt{2\pi} e \alpha^2 L}{\eta \left(e E_\perp^* \right)^2} . \tag{6.44}$$

In the opposite limiting case and under conditions $L \gg (\kappa d)^2 d$, one can disregard $\mathcal{I}_1$ in comparison with $\mathcal{I}_2$, whence

$$\mu \simeq \frac{4\pi \alpha^2 e (\kappa d)^2 d}{3\eta \left(e E_\perp^* \right)^2} . \tag{6.45}$$

It is very important that the dimple mobility of (6.45) depends on the dimple tail radius κ^{-1}, which is much larger than the electron localization radius L of the strong coupling limit. Therefore, at a finite electron density, one has to take into account the cutoff of the dimple tails because of the electron spacing which we discussed in Sect. 6.2.1.

In spite of evident experimental difficulties in conducting measurements of electron mobility above helium films, very interesting results were obtained by Mende, Kovdrya and Nikolaenko [215] in 1984 for helium films with depth $d > 200$ Å covering a pyroceramic substrate. Their data are shown in Fig. 6.8 together with theoretical curves evaluated by Marques and Studart [214] according to (6.34) and (6.41). The agreement between experiment and theory is very good, if the electron mobility is considered as a function of d. Still, the temperature dependence of the experimental data reported is not really consistent with the temperature dependence of the liquid helium viscosity. It should be noted that a number of experiments with SEs on helium films were interpreted in terms of interface polaron formation [216, 218–220]. However, at the present time, evidence for single-electron polaron formation cannot be considered as totally conclusive.

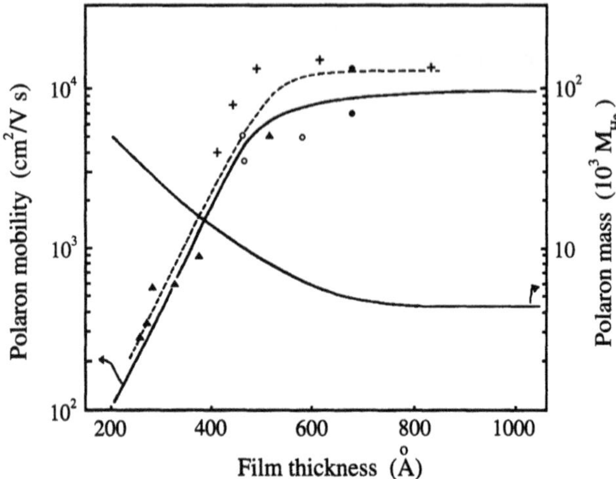

Fig. 6.8. Self-consistent mobility and effective mass of ripplonic polarons as a function of the film thickness for helium films on pyroceramic under a holding electric field $E_{\perp} = 1800\,\mathrm{V/cm}$ (corresponding to $n_s = 10^9\,\mathrm{cm}^{-2}$). Experimental data (*symbols*) were obtained by Mende, Kovdrya and Nikolaenko [215]. Theoretical results evaluated according to (6.41) and (6.34) (*continuous curves*) are taken from Marques and Studart [214]. The *dashed curve* is the best fit to the experimental data

6.3.3 Long Mean-Free-Path Regime

At low temperatures the mean-free-path of bulk excitations of liquid helium becomes longer than the typical dimple size. In this instance, the drag force acting on a moving dimple can be evaluated as the quasi-particle momentum absorbed by the dimple.

The momentum transfer between a quasi-particle and the surface element due to elastic reflection is

$$\Delta \boldsymbol{p} = 2(\boldsymbol{p} \cdot \boldsymbol{n})\boldsymbol{n} ,$$

where $\boldsymbol{p}$ is the 3D momentum of the bulk quasi-particle. $\Delta \boldsymbol{p}$ is directed along the outward normal $\boldsymbol{n}$ and its value depends on the angle between $\boldsymbol{p}$ and $\boldsymbol{n}$. The infinitesimal flux of bulk quasi-particles reflected by a surface element $\mathrm{d}S_{\mathrm{A}}$ can be written as

$$\delta \Phi_{\mathrm{f}} = \left(\boldsymbol{n} \cdot \frac{\partial E_{\boldsymbol{p},\sigma}}{\partial \boldsymbol{p}} \right) f_{\boldsymbol{p},\sigma} \mathrm{d}S_{\mathrm{A}} , \tag{6.46}$$

where $E_{\boldsymbol{p},\sigma}$ and $f_{\boldsymbol{p},\sigma}$ are the energy spectrum and distribution function of bulk helium quasi-particles and σ is the spin variable.

In order to find the drag force, we have to sum $\Delta \boldsymbol{p} \delta \Phi_{\mathrm{f}}$ over σ and over the momentum half-space (dependent on $\boldsymbol{n}$) from which quasi-particles can reach the element $\mathrm{d}S_{\mathrm{A}}$, and then take the integral over the free surface of liquid helium [54]:

$$\boldsymbol{F} = \int \mathrm{d}S_{\mathrm{A}} \sum_{\sigma} \sum_{\boldsymbol{p}(\boldsymbol{n})}{}' 2\boldsymbol{n} \frac{(\boldsymbol{n} \cdot \boldsymbol{p})|\boldsymbol{n} \cdot \boldsymbol{p}|}{p} \left| \frac{\partial E_{p,\sigma}}{\partial p} \right| f_{\boldsymbol{p},\sigma} , \tag{6.47}$$

where $\sum_{\boldsymbol{p}(\boldsymbol{n})}$ indicates the summation range mentioned above. In the integrand, the sign $|\ldots|$ is introduced so that this equation is also applicable to excitations which have their momentum and velocity pointing in opposite directions. It is clear that, for the equilibrium distribution function $f_0(E_{p,\sigma})$, the total drag along the surface is zero. For a moving dimple, the quasi-particle distribution function considered in the moving frame is shifted: $f_{\boldsymbol{p},\sigma} = f_0(E_{p,\sigma} - \boldsymbol{p} \cdot \boldsymbol{V}_0)$. Therefore, in (6.47), instead of $f_{\boldsymbol{p},\sigma}$, one can use the linear correction $\delta f_{\boldsymbol{p},\sigma} = (-\partial f_0/\partial E_{p,\sigma}) p V_0 \cos(\boldsymbol{p} \wedge \boldsymbol{V}_0)$, where $(\boldsymbol{p} \wedge \boldsymbol{V}_0)$ is the angle between $\boldsymbol{p}$ and $\boldsymbol{V}_0$.

We assume that the x-axis is directed along the velocity $\boldsymbol{V}_0$. When taking the integral over the momentum half-space, it is convenient to use cylindrical coordinates with θ_p and φ_p being the azimuth and polar angle of the quasi-particle momentum ($\boldsymbol{n} \cdot \boldsymbol{p} = np \cos\theta_p$). Noting that

$$\frac{1}{2\pi} \int_0^{2\pi} \cos(\boldsymbol{p} \wedge \boldsymbol{V}_0) \mathrm{d}\varphi_p = n_x \cos\theta_p ,$$

the friction acting on the surface dimple can be found as

$$F_x = \frac{V_0}{(2\pi)^2 \hbar^3} \left(\int n_x^2 \mathrm{d}S_{\mathrm{A}} \right) \frac{1}{2} \sum_{\sigma} \int_0^{\infty} p^4 \left| \frac{\partial E_{p,\sigma}}{\partial p} \right| \left(-\frac{\partial f_0}{\partial \varepsilon_{p,\sigma}} \right) \mathrm{d}p . \tag{6.48}$$

This equation can be applied to any kind of bulk excitation whose reflection from the surface is specular.

In the case of liquid ^{3}He, for the isotropic quasi-particle spectrum with a small gap Δ,

$$E_{p,\sigma} = \sqrt{(E_{p,\sigma}^{(0)})^2 + \Delta^2} , \tag{6.49}$$

the low temperature mobility equation was obtained by Monarkha and Kono [54]:

$$\frac{e}{\mu} = \frac{\hbar (k_{\mathrm{F}}^{(\mathrm{qp})})^4}{4\pi^2} \left(\int n_x^2 \mathrm{d}S_{\mathrm{A}} \right) 2 f_0(\Delta) . \tag{6.50}$$

Here $\hbar k_{\mathrm{F}}^{(\mathrm{qp})}$ is the Fermi momentum of liquid ^{3}He quasi-particles. This result is of a rather general form, which can be applied to any boundary shape. For instance, if we assume that the dimple is hemispherical ($\int n_x^2 \mathrm{d}S = 2\pi R_0^2/3$) and that the quasi-particle cross-section does not change in the superfluid phase, then (6.50) will yield half the value found for an electron bubble in liquid ^{3}He [221, 222]. In the case of the polaron dimple on the surface of ^{3}He, we have

$$n_x \simeq -\frac{\partial \xi}{\partial x} ,$$

and (6.50) gives the polaron mobility in the long mean-free-path regime.

The effective collision frequency of the polaron dimple defined by the relation $e\mu^{-1} = m_e \nu_d^*$ can thus be written as

$$\nu_d^* = \frac{\hbar (k_F^{(qp)})^4}{8\pi^2 m_e S_A} 2 f_0(\Delta) \sum_q q^2 \left| \xi_q^{(0)} \right|^2 . \tag{6.51}$$

In the normal phase of liquid ^{3}He $[\Delta(T) = 0,\ 2f_0(0) = 1]$, just above the superfluid transition, the polaron conductivity $[\sigma = e^2 n_e/(m_e \nu_d^*)]$ is independent of temperature, in contrast to the strong temperature dependence found for the hydrodynamic regime: $\sigma(T) \propto 1/\eta(T) \propto T^2$.

6.3.4 Ripplon-Limited Mobility

In the case of SEs on liquid ^{4}He, when the temperature decreases, the liquid substrate rapidly falls out of the hydrodynamic description because the mean-free-path of bulk excitations becomes longer than the dimple radius. At the same time the number of bulk excitations of the Bose liquid decreases fast with cooling. In this case, polaron scattering by surface excitations (ripplons) can be important.

Ripplon-limited mobility of polarons above liquid helium films was studied by Saitoh [209] for the strong coupling regime. Employing Feynman's path integral technique, he found an exponential increase in mobility with cooling:

$$\mu = \frac{e}{\pi T} \sqrt{\frac{\alpha}{\rho}} \exp \frac{(e E_\perp^*)^2}{16 \pi \alpha T} . \tag{6.52}$$

The prefactor μ is independent of the coupling constant α_{cc}. The origin of the exponential temperature dependence can be understood in the following way.

As discussed in Chap. 3, at $\omega = 0$ the effective collision frequency of electrons with ripplons is proportional to the electron DSF $S(q, -\omega_q)$. According to (2.22), for heavy particles with effective mass $m \simeq M_d$, the DSF contains the exponential factor

$$S(q, -\omega_q) \propto \exp \left[-\frac{(\varepsilon_q + \hbar \omega_q)^2}{4 \varepsilon_q T} \right] , \tag{6.53}$$

where

$$\varepsilon_q = \frac{\hbar^2 q^2}{2 M_d} . \tag{6.54}$$

Consider now the last term in the argument of the exponential function, viz.,

$$-\frac{\hbar^2\omega_q^2}{4\varepsilon_q T} = -\frac{M_{\rm d}\alpha\left(\kappa^2+q^2\right)d}{2T\rho}\,.$$

It shows that, for any relation between q and κ, the DSF contains the exponential factor

$$S(q,-\omega_q) \propto \exp\left(-\frac{M_{\rm d}\alpha\kappa^2 d}{2T\rho}\right)\,, \tag{6.55}$$

which does not depend on q and can be taken out of the integral over q in the equation for the effective collision frequency. Substituting here the definition of $M_{\rm d}$ given for the strong coupling regime [see (6.37)], one can see that the exponential decrease in $S(q,-\omega_q)$ with cooling and the corresponding decrease in the effective collision frequency agree accurately with the exponential factor in the polaron mobility found using the path integral formalism [see (6.52)].

The frequency and temperature dependencies of the mobility of ripplonic polarons were studied by Peeters and Jackson [217, 223] for an AC driving electric field. They found that, in the intermediate coupling regime, the mobility of self-trapped electrons has a complicated temperature dependence. Below 0.3 K, there is a temperature range where the mobility increases with falling T, but then drops sharply and becomes independent of temperature in the low temperature regime.

7 Wigner Solid. I.
Dynamics on Rigid and Soft Interfaces

7.1 Contemporary Practice of an Old Hypothesis

A degenerate Fermi gas is not the only possible ground state for a collection of interacting electrons. It was Wigner [224] who pointed out that electrons in a structureless positive background may form a lattice at a low enough density, when the kinetic energy of electrons plays only a secondary role as compared to their potential energy. Remarkably, this prediction was made in 1934 when the 'simple' theory of metal electrons disregarding electron–electron interactions had given satisfactory results for most questions. The electron crystal was called the Wigner solid (WS). Until 1971 the Wigner solid was studied theoretically as an intriguing hypothesis, even though there was no experimental system where it could exist. In 1967 it was suggested by Van Horn [225] that the degenerate positive ions of a white dwarf star might form a Wigner lattice.

Progress in studying 2D electron systems in semiconductor structures and on the free surface of superfluid helium gave new momentum to the search for the WS. In 1971, Crandall and Williams [3] suggested that nondegenerate electrons on liquid helium might form a Wigner lattice at low enough temperatures ($T < 0.5\,\mathrm{K}$). An analogous suggestion for semiconductor inversion layers was given by Chaplik [226]. The main advantage of these systems is the possibility of varying the electron density over a wide range by a simple change in the holding electric field or the electric potential applied to one of the electrodes. In the study of the Wigner transition, the SEs on liquid helium and 2D electrons in semiconductor structures represent complementary systems because of the big difference in the effective masses of charged carriers. For semiconductor electrons ($m^* \simeq 0.07\, m_{\mathrm{e}}$), the electron crystal melts predominantly because of quantum effects, while for SEs on liquid helium ($m^* \simeq m_{\mathrm{e}}$), the melting of the electron lattice occurs via thermodynamic instability with respect to the spontaneous formation of dislocations.

Because of the long-range nature of the Coulomb forces, the behavior of an electron system with varying density differs substantially from that of an ordinary system of neutral particles. In the ordinary system, the interaction potential energy of particles increases with density faster than the kinetic energy, and they eventually become localized at the lattice sites in order to reduce their energy. In an electron system, the situation is opposite. For example, the 2D electron gas of areal density $n_s = 1/\pi r_0^2$ (here r_0 represents the typical interelectron spacing) has the mean potential energy per electron

$$U_C = \frac{e^2}{r_0} = \frac{2}{r_s} \quad [\mathrm{Ry}] , \tag{7.1}$$

At the same time, the average kinetic energy of electrons with the simple energy spectrum $\varepsilon_k \propto k^2$ can be rewritten as

$$K_e^{(0)} = \frac{\varepsilon_F}{2} = \frac{\hbar^2}{2m_e r_0^2} = \frac{1}{r_s^2} \quad [\mathrm{Ry}] , \tag{7.2}$$

where we have used the following conventions: $r_s = r_0/a_B$ is the dimensionless measure of the average interelectron distance, $a_B = \hbar^2/me^2$ is the Bohr radius, and $1\,\mathrm{Ry} = me^4/2\hbar^2 = 13.6\,\mathrm{eV}$. Therefore, in the low density regime $r_s \gg 1$, the potential energy of electrons becomes much larger than their kinetic energy, and the electrons might crystallize. For semiconductor electrons, one should replace m_e by m^* and e by $e/\sqrt{\epsilon}$, where ϵ is the dielectric constant. This replacement decreases the Coulomb energy U_C and increases the quantum kinetic energy, making the ratio U_C/K_0 smaller by approximately two or three orders of magnitude.

The total energy (per electron) of the WS state with a 2D hexagonal lattice was estimated as [123]

$$\mathcal{E}_{WS} = -\frac{2.212}{r_s} + \frac{1.628}{r_s^{3/2}} \quad [\mathrm{Ry}] , \tag{7.3}$$

where the negative sign of the Coulomb term is caused by the interaction with the positive background, and the second term represents the zero-point vibrational energy. The square lattice, which has a very close interaction energy ($-2.2005/r_s\,\mathrm{Ry}$), appears to be unstable with regard to transverse displacements for certain directions of the wave vector $\boldsymbol{q}$ [227]. According to [228], the energy of the Wigner solid should also have the correction b/r_s^2 owing to anharmonicities in the crystal. For a 3D case, the constant b was estimated to be slightly less than unity. It is clear that in the low density limit, $\mathcal{E}_{WS}$ is lower than the total energy of the Fermi gas, which is the reason for the WS phase transition. In the opposite limit of small r_s, the kinetic energy dominates the potential energy and electrons are in the gas state which has the minimum possible kinetic energy.

It is sometimes convenient to describe the one-component electron plasma using another dimensionless parameter

$$\Gamma^{(\mathrm{pl})} = U_C/K_e , \tag{7.4}$$

which is the ratio of the mean interaction energy to the mean kinetic energy. We have already introduced this parameter in Chap. 2 when discussing the properties of the nondegenerate Coulomb liquid. In the zero temperature limit, combining (7.1) and (7.2), we find the relation

$$\Gamma^{(\mathrm{pl})} = 2r_{\mathrm{s}} , \tag{7.5}$$

which means that the parameters $\Gamma^{(\mathrm{pl})}$ and r_{s} are physically equivalent. In the low density limit, the Fermi energy $\varepsilon_{\mathrm{F}} = 2/r_{\mathrm{s}}^2$ (the Fermi wave number is $k_{\mathrm{F}} = \sqrt{2\pi n_{\mathrm{s}}} = \sqrt{2}/r_0$) eventually becomes much smaller than temperature T and the electrons represent a nondegenerate system. In this case, $K_{\mathrm{e}} = T$ and the plasma parameter $\Gamma^{(\mathrm{pl})}$ increases with cooling and has the opposite dependence on the parameter r_{s}:

$$\Gamma^{(\mathrm{pl})} = \frac{e^2}{r_0 T} = \frac{2}{r_{\mathrm{s}} T} \frac{m_{\mathrm{e}} e^4}{2\hbar^2} . \tag{7.6}$$

Therefore, the plasma parameter $\Gamma^{(\mathrm{pl})}$ is not generally equivalent to the parameter r_{s}. It seems reasonable to use $\Gamma^{(\mathrm{pl})}$ instead of r_{s} when describing the WS melting point: $\Gamma^{(\mathrm{pl})} = \Gamma_{\mathrm{c}}^{(\mathrm{pl})} = \mathrm{const}$.

The excitation spectrum $\Omega_{p,k}$ of the WS consists of the longitudinal ($p = \mathrm{l}$) and transverse ($p = \mathrm{t}$) phonon modes, if no magnetic field is applied. In the long-wavelength limit, $\Omega_{\mathrm{l},k}$ coincides with the spectrum of 2D plasmons [229]:

$$\Omega_{\mathrm{l},k} = \sqrt{\frac{2\pi n_{\mathrm{s}} e^2}{m} k} . \tag{7.7}$$

This square-root dependence on the wave number cannot be found in the usual elastic theory which describes solids constituted of neutral particles with short-range internal forces. The long-wavelength ($q \ll r_0^{-1}$) transverse phonons have the usual sound-like dispersion [123]

$$\Omega_{\mathrm{t},k} = c_{\mathrm{t}} k , \qquad c_{\mathrm{t}} = \sqrt{0.138 \frac{e^2}{m r_0}} , \tag{7.8}$$

where $r_0 = 1/\sqrt{\pi n_{\mathrm{s}}}$.

In the short-wavelength region, the dispersion of WS phonons deviates from the simple asymptotes of (7.7) and (7.8). For arbitrary wave vectors, the WS phonon spectrum was evaluated by Bonsall and Maradudin [123], using Ewald's summation method. Their results are shown in Fig. 7.1. The frequency ω_0 is defined by $\omega_0^2 = 8e^2/m_{\mathrm{e}} a^3$, where a is the electron spacing: $n_{\mathrm{s}} = 2/\sqrt{3}a^2$. It should be noted that, in spite of strong deviations of the dispersion curves from the long-wavelength asymptotic forms, the Debye approximation for the WS spectrum gives quite adequate results in most cases. For example, considering the Brillouin zone to be a circle of radius k_{m} determined by the condition $\sum_{\boldsymbol{k}} 1 = N_{\mathrm{e}}$ (which gives $k_{\mathrm{m}} = \sqrt{4\pi n_{\mathrm{s}}}$), one can find the zero-point vibrational energy of the Wigner lattice in the Debye approximation $\mathcal{E}_{\mathrm{WS}}^{(0)} \approx 2/r_{\mathrm{s}}^{3/2}$, which is not far away numerically from the accurate result shown in (7.3) by the second term. Later we shall also see that this approximation yields the self-consistent Debye–Waller factor of the WS which agrees well (even numerically) with experimental data at ultra-low temperatures.

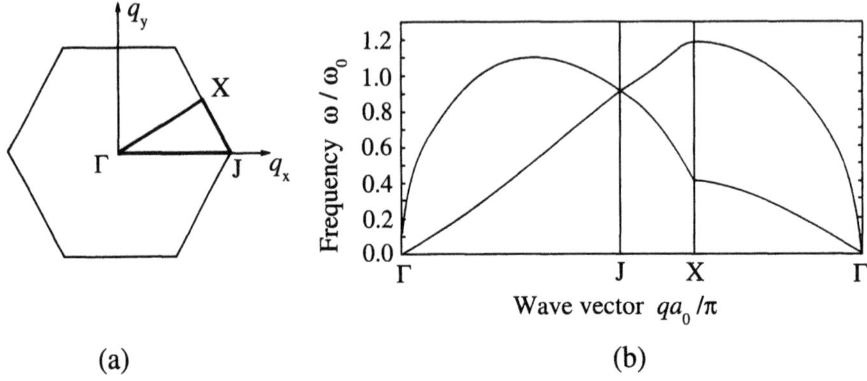

Fig. 7.1. First Brillouin zone for the 2D hexagonal lattice with the irreducible element outlined heavily (**a**) and the phonon dispersion curves for wave vectors along the irreducible element (**b**) [123]

The 2D Wigner solid was observed in the system of SEs on liquid helium in a remarkable experiment by Grimes and Adams [7]. The SEs were exposed to a nonuniform alternating electric field parallel to the surface. The power absorbed was measured vs. the frequency of the input signal. With falling temperature, the smooth absorption line burst suddenly into a peculiar curve with several resonances. These resonances occurred when the frequency of the electric field was close to the frequency of the capillary wave whose wave vector $\boldsymbol{q}$ equals one of the reciprocal lattice vectors of the 2D hexagonal lattice $\boldsymbol{g}$. Hence, because of the unique properties of the liquid helium substrate, it was possible in the same experiment to observe the WS transition and determine the structure of the 2D electron lattice. This experiment and the corresponding theoretical analysis of Fisher, Halperin and Platzman [230] will be discussed in detail in Sect. 7.4.2.

Soon after the observation of Grimes and Adams, the WS transition was also tracked by anomalies in both the holding field dependence [231] and the temperature dependence [50, 232, 233] of SE mobility. Research into the WS in the interface electron system resulted in an interesting application. Electrons localized in the lattice sites create a sublattice of surface dimples. The simple structure of the electron and dimple lattices allows one to use the 2D Wigner solid as a powerful tool for studying bulk and surface properties of superfluid ^{3}He [234, 235].

This chapter is mainly concerned with the non-dissipative dynamics of the 2D Wigner solid on rigid (flat) and soft interfaces. Section 7.2 is about the properties of the liquid–solid boundary of the 2D electron system. Section 7.3 describes the quantization of WS vibrations under a normal magnetic field. Because the lattice is composed of charged particles, the magnetic field introduces some new aspects into the lattice dynamics. For example, even in the long-wavelength limit, the magnetic field mixes longitudinal and transverse phonons. In this case, the structure of the electron Hamiltonian requires an extension of the conventional canonical transformations used to quantize the particle (electron) displacement from a lattice site $\boldsymbol{u}_l$. In typical solids, this transformation is just a rotation in the normal coordinate space. For the Wigner solid under a magnetic field, the transformation involves normal momenta. As a result, the expression for the mean-square displacement differs substantially from that found for typical neutral solids, and this is very important for obtaining the phase diagram of the 2D WS subject to a magnetic field.

Coupling of the Wigner solid with medium vibrations (ripplons) strongly affects the dynamics and excitation spectrum of the WS in the low frequency range. The electron crystal and dimple sublattice form a strongly coupled system whose excitation spectrum can be described by means of a self-consistent treatment. This treatment and relevant aspects of the WS dynamics on a soft interface are discussed in Sect. 7.4. Section 7.5 is concerned with the properties of the dynamical structure factor (DSF) of the 2D electron solid under different conditions. This quantity is crucial for understanding the transport properties of the WS, which we shall discuss in the following chapter. Important experiments on the Wigner solid of interface electrons, such as transverse sound and specific heat measurements [158, 248], are discussed in Sect. 7.6. Section 7.7 discusses properties of the bilayer crystal of charged particles which can be created in semiconductor heterostructures and in helium films. In contrast with the usual 2D Wigner solid with a fixed lattice structure, a number of structural phase transitions are expected in the bilayer electron crystal [35, 236].

7.2 Phase Diagram

The Wigner transition is expected to occur at some $r_s = r_s^{(c)} \gg 1$ or at $\Gamma^{(pl)} = \Gamma_c^{(pl)} \gg 1$. Regarding the exact value of $r_s^{(c)}$ for the 3D electron system, there were so many different and contradictory estimates, varying from numbers less than 10 to several hundred, that one could introduce a most probable estimate, close to 20 [237–239]. Still, Monte Carlo simulations of the classical one-component plasma showed that long-range order appeared at $\Gamma_c^{(pl)} = 125 \pm 15$ [240].

The same story can be told about the Wigner solid transition in two dimensions. Identifying the instability of the transverse phonon mode appearing in the self-consistent lattice vibration theory as the main mechanism for WS melting, Platzman and Fukuyama [241] found that $r_s^{(c)} \simeq 5$ in the quantum regime and $\Gamma_c^{(pl)} \simeq 3$ in the classical case. Computer simulations using molecular dynamics for a system of 10^4 electrons performed by Hockney and Brown [242] gave $\Gamma_c^{(pl)} \simeq 95 \pm 2$. The explanation of this discrepancy was given by the dislocation theory of melting in two dimensions (Kosterlitz and Thouless [243]) which, when applied to the Wigner lattice, gives $\Gamma_c^{(pl)} \simeq 78.7$ [244] in the harmonic approximation.

7.2.1 Boundary Shape

In order to analyze the shape of the liquid–solid phase boundary, Platzman and Fukuyama [241] introduced a very instructive physical picture. This picture was based on the assumption that the plasma parameter is constant along the melting curve $\Gamma^{(pl)}(n,T) = \Gamma_c^{(pl)}$. For example, in the classical regime, we find that the melting curve given by

$$n_s^{(c)}(T) = \left(\frac{\Gamma_c^{(pl)}}{\sqrt{\pi} e^2} \right)^2 T^2 \tag{7.9}$$

is actually a simple parabola. On the other hand, at high densities, the WS melts due to quantum effects, which restricts the upper boundary of the solid phase:

$$n_s < n_c \equiv \frac{4}{\pi} \left(a_B \Gamma_{(c)}^{(pl)} \right)^{-2} , \tag{7.10}$$

where we have used $K_e^{(0)}$ from (7.2) as the average kinetic energy.

In the general case, it is possible to find parametric equations for the melting boundary using $z = \exp(-\mu/T)$ as the parameter [241]. The chemical potential μ is defined by the condition

$$\pi n_s = \int_0^\infty f(\varepsilon_k) k dk = \frac{m_e T}{\hbar^2} \ln \left(1 + \frac{1}{z} \right) , \tag{7.11}$$

where $f(\varepsilon)$ is the Fermi distribution function. The kinetic energy per electron K_e entering the plasma parameter can be found as

$$K_e = \frac{1}{\pi n_s} \int_0^\infty \varepsilon_k f(\varepsilon_k) k dk = \frac{m T^2}{\pi n_s \hbar^2} F(z) , \tag{7.12}$$

where

$$F(z) = \int_0^\infty \frac{dx}{1 + z e^x} \tag{7.13}$$

is an auxiliary function.

Substituting (7.11) and (7.12) into the definition (7.4) of the plasma parameter, one can find the parametric equations for the liquid–solid boundary:

$$\frac{n_s(z)}{n_c} = \frac{[\ln(1+1/z)]^4}{4F^2(z)} , \tag{7.14}$$

$$\frac{T(z)}{T^*} = \frac{[\ln(1+1/z)]^3}{2F^2(z)} . \tag{7.15}$$

The characteristic temperature T^* is defined by

$$T^* = \frac{2me^4}{\hbar^2 \left(\Gamma_c^{(\mathrm{pl})}\right)^2} . \tag{7.16}$$

Figure 7.2 shows the phase diagram determined by (7.14) and (7.15). The important point is that the melting curve reaches both low temperature ends with zero derivative. For the classical regime, it follows directly from (7.9). In the high density regime, it can be seen from the low temperature asymptote of (7.12):

$$K_e \simeq K_e^{(0)} \left[1 + \frac{\pi^2}{3}\left(\frac{T}{\varepsilon_F}\right)^2\right] . \tag{7.17}$$

Using this asymptote in (7.14) and (7.15), we find

$$n_s(T) \simeq n_c \left[1 - \frac{2\pi^2}{3}\left(\frac{T}{\varepsilon_F}\right)^2\right] , \tag{7.18}$$

in agreement with the numerical plot of Fig. 7.2.

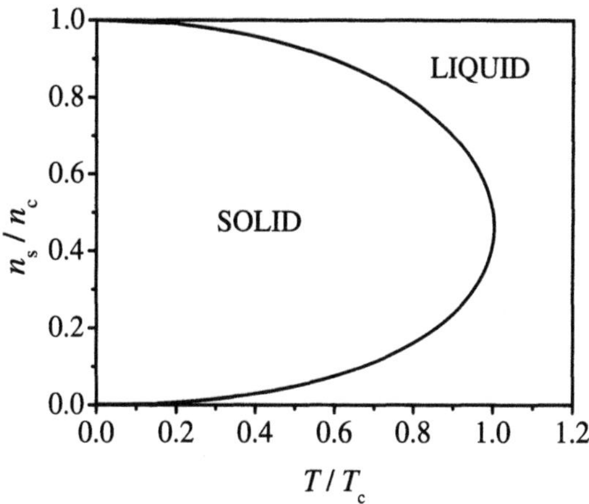

Fig. 7.2. Parametrized phase diagram of the 2D Coulombic system according to Platzman and Fukuyama [241] ($T_c \simeq 0.39\,T^*$)

The liquid–solid boundary shape of Fig. 7.2 follows from the assumption that the plasma parameter $\Gamma^{(\mathrm{pl})}$ does not change along the melting curve. In a real case, this may not be true, especially if quantum and classical melting are described by different mechanisms. This was not actually true even for the same melting mechanism discussed in [241], where $\Gamma_{\rm c}^{(\rm pl)}$ had different numbers in the quantum and classical cases. Nevertheless, the parametrized phase diagram of Platzman and Fukuyama is very transparent and applicable to other melting mechanisms.

7.2.2 Dislocation Melting in Two Dimensions

Consider now the dislocation theory of melting of 2D solids proposed by Kosterlitz and Thouless [243], for which the critical value $\Gamma_{\rm c}^{(\rm pl)}$ agrees well with available experimental data. This theory is based on the fact that in two dimensions a dislocation is associated with a point rather than with a line. Therefore, the elastic energy U of an isolated dislocation is not proportional to the size of the system, and it can be comparable with the entropy term in the free energy equation $F = U - TS$. At low temperatures the elastic energy dominates the free energy and $F > 0$, which means that thermally activated dislocations are quite rare. The melting transition occurs at a certain point $T = T_{\rm m}$, when the entropy term takes over and F changes sign. At $T > T_{\rm m}$ the free energy of a single dislocation is negative ($F < 0$) and dislocations appear spontaneously, destroying the solid phase.

In the elastic 2D theory, the energy of a dislocation U and the entropy S are

$$U = \frac{\mu_{\rm t} b^2 (1 + \tau_{\rm P})}{8\pi} \ln N_{\rm e} \,, \qquad S = \ln N_{\rm e} \,, \tag{7.19}$$

where $\mu_{\rm t}$ is the shear modulus, $\tau_{\rm P}$ is the Poisson ratio and b is the magnitude of the Burgers vector. Because both terms U and TS are proportional to $\ln N_{\rm e}$, isolated dislocations cannot appear in a large system below the melting point and thermally excited dislocations are mainly present as a collection of bound pairs of dislocations with equal and opposite Burgers vectors. Such pairs have finite energy comparable with $T_{\rm m}$ and may occur at $T < T_{\rm m}$ due to thermal excitation.

The above discussion is applicable to a typical system of neutral particles with a short-range interaction potential. In the case of the WS, there is an important question: in what sense can the elastic theory results of (7.19) be applied to a lattice constituted of charged particles? The right answer can be found by representing $\mu_{\rm t}$ and $\tau_{\rm P}$ of the 2D elastic theory in terms of the sound velocities $c_{\rm t}$ and $c_{\rm l}$:

$$\mu_{\rm t} = m n_{\rm s} c_{\rm t}^2 \,, \qquad \tau_{\rm P} = 1 - 2\frac{c_{\rm t}^2}{c_{\rm l}^2} \,. \tag{7.20}$$

The longitudinal velocity $c_{\rm l} = \partial \Omega_{{\rm l},k}/\partial k \propto 1/\sqrt{k}$ of the WS is infinite in the long-wavelength limit. For $ka \ll 1$, the transverse velocity $c_{\rm t} \ll c_{\rm l}$, and the Wigner solid can therefore be regarded as incompressible with Poisson ratio

$\tau_{\mathrm{P}} = 1$. Then, substituting (7.19) into the equation $U - TS = 0$, we find the melting temperature

$$T_{\mathrm{m}} = \frac{n_{\mathrm{s}} a^2 m c_{\mathrm{t}}^2}{4\pi} , \tag{7.21}$$

where a is the lattice spacing. (For a hexagonal lattice $a^2 = 2/\sqrt{3} n_{\mathrm{s}}$.)

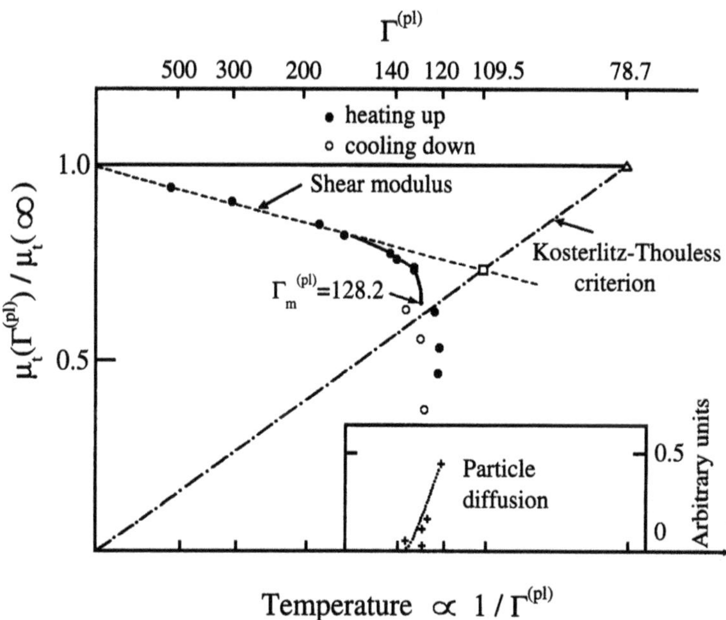

Fig. 7.3. Temperature dependence of the shear modulus $\mu \propto c_{\mathrm{t}}^2$ and particle diffusion [245]. *Circles* represent molecular dynamics results for $\mu(\Gamma^{(\mathrm{pl})})/\mu(\infty)$ as a function of $1/\Gamma^{(\mathrm{pl})}$. The particle diffusion is shown in the *inset*. The *continuous curve* ending at $\Gamma_{\mathrm{m}}^{(\mathrm{pl})} = 128.2$ displays the renormalization group results. The Thouless value $\Gamma^{(\mathrm{pl})} = 78.7$ is shown by a *triangle*

For the transverse sound velocity given in (7.8), the Kosterlitz–Thouless criterion yields $\Gamma_{\mathrm{c}}^{(\mathrm{pl})} \simeq 78.7$, which is slightly smaller the value $\Gamma_{\mathrm{c}}^{(\mathrm{pl})} = 137 \pm 15$ observed for SEs on liquid helium in the experiment of Grimes and Adams [7]. Using a molecular dynamics simulation, Morf [245] studied the temperature dependence of the shear modulus and found that it decreases with increasing T as

$$c_{\mathrm{t}}^2(T) = c_{\mathrm{t}}^2(0) \left(1 - 30.8 \frac{T}{e^2 \sqrt{\pi n_{\mathrm{s}}}}\right) , \tag{7.22}$$

up to $T \approx 0.9\, T_{\mathrm{m}}$. Then at higher T it falls rapidly as T approaches T_{m}, as shown in Fig. 7.3. Replacing c_{t}^2 in (7.21) by (7.22), Morf obtained $\Gamma_{\mathrm{c}}^{(\mathrm{pl})} \simeq 110$.

Moreover, including the effects of a finite density of dislocation pairs gave $\Gamma_c^{(pl)} \simeq 128$, which agrees well with experiment.

A portion of the phase diagram determined in experiments with SEs on liquid helium by Grimes and Adams [7] is shown in Fig. 7.4 by the $\sqrt{n_s}$ vs. T_m plot. This figure shows that the melting temperature is proportional to $\sqrt{n_s}$, which is consistent with the phase boundary defined by the equation $\Gamma^{(pl)}(n_s, T) = \Gamma_c^{(pl)}$. The WS transition was determined by the appearance of the resonance absorption due to excitation of standing capillary waves with $\boldsymbol{q}$ equal to a reciprocal lattice vector $\boldsymbol{g}$ [7], which will be discussed later. The same behavior of the liquid–solid boundary was also observed in experiments [101, 246], where the melting temperature was determined from the electron mobility anomalies.

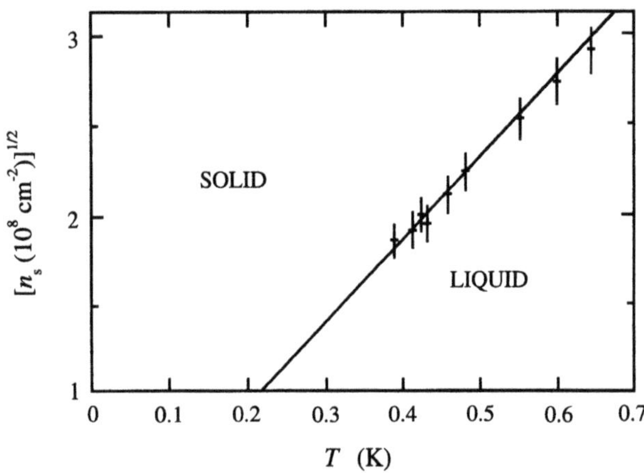

Fig. 7.4. Portion of the solid–liquid phase boundary for the 2D electron system on the surface of liquid helium [7]. Along the line, the plasma parameter $\Gamma^{(pl)} = 137$

The transverse velocity c_t and the shear modulus of the WS were measured in [247, 248] from the position of the hybrid phonon mode excited in the presence of the magnetic field applied normally to the system (see also Sect. 7.3.1). These experiments will be discussed in Sect. 7.6. Here we would like to note that the observed temperature dependence $\mu_t = mn_s c_t^2$ agrees well with the Morf formula of (7.22).

7.2.3 Quantum Melting Regime

Regarding the quantum melting regime, there is no guarantee that the upper phase boundary of Fig. 7.2 will still be described by the dislocation melting theory. The empirical melting criterion of Lindemann, often used for typical solids, states that a crystal melts when the ratio of the mean-square displacement of the particles from their lattice positions to a^2 or to $r_0^2 = 1/\pi n_s$

reaches a certain critical value. In the literature, there are two Lindemann parameters δ_0 and δ describing the liquid–solid boundary through the equations

$$\frac{\langle u^2 \rangle}{r_0^2} = \delta^2 , \qquad \frac{\langle u^2 \rangle}{a^2} = \delta_0^2 . \tag{7.23}$$

In three dimensions, the empirical value is $\delta = 1/4$ and consequently $\delta_0 \sim 0.1$. The exact value of δ_0 depends on the lattice structure. In two dimensions for the hexagonal lattice, these parameters are related by $\delta_0 = 3^{1/4}/\sqrt{2\pi}\delta$, and therefore $\delta_0 = 0.131$ if $\delta = 1/4$.

The mean-square displacement $\langle u^2 \rangle$ can be evaluated from the well known relation

$$\langle u^2 \rangle = \frac{\hbar}{2mN_\mathrm{e}} \sum_{p,\boldsymbol{k}} \frac{1}{\Omega_{p,k}} \coth \frac{\hbar \Omega_{p,k}}{2T} , \tag{7.24}$$

if no magnetic field is applied. For an infinite 2D system, it is well known that (7.24) results in $\langle u^2 \rangle = \infty$: at any nonzero temperature, there are transverse phonons $\hbar\Omega_{\mathrm{t},k} \ll 2T$ which make the integral $\int \mathrm{d}kk/\Omega_{\mathrm{t},k}^2$ logarithmically divergent at the lower bound ($k = 0$). For a finite system, the wave number is restricted by the size of the system L: $k \geq k_0 = 2\pi/L$. In this case, $\langle u^2 \rangle$ can be evaluated in the framework of the Debye approximation, which assumes that (7.7) and (7.8) describing the phonon spectrum of the WS are valid up to the maximum value of the wave number $k_\mathrm{m} = \sqrt{4\pi n_\mathrm{s}}$:

$$\langle u^2 \rangle = \langle u_0^2 \rangle + \frac{T}{2\pi m n_\mathrm{s} c_\mathrm{t}^2} \ln \frac{1 - \exp(-\hbar c_\mathrm{t} k_\mathrm{m}/T)}{1 - \exp(-\hbar c_\mathrm{t} k_0/T)} . \tag{7.25}$$

Here $\langle u_0^2 \rangle$ is the contribution from zero-point vibrations of both the longitudinal and transverse modes:

$$\frac{\langle u_0^2 \rangle}{a^2} = \frac{\hbar}{m c_\mathrm{t} k_\mathrm{m} a^2} \left(1 + \frac{2\sqrt{0.138}}{3} \right) = \frac{A}{\sqrt{r_\mathrm{s}}} , \tag{7.26}$$

and the proportionality factor is $A = (3/\sqrt{0.138} + 2)/(4\pi\sqrt{3}) \simeq 0.463$. In the second term of (7.25), describing thermal vibrations, the contribution from the longitudinal mode is disregarded because $\Omega_{\mathrm{l},k} \gg \Omega_{\mathrm{t},k}$.

At finite temperatures the logarithmic dependence of $\langle u^2 \rangle$ on the size of the system $[\langle u^2 \rangle - \langle u_0^2 \rangle \propto T \ln(T/\hbar c_\mathrm{t} k_0)]$ makes the Lindemann criterion doubtful, at least in its original form. In any case, one can consider the pure quantum melting regime by setting $T = 0$. Then (7.26) gives [249]

$$r_\mathrm{s}^{(\mathrm{c})} = \frac{A^2}{\delta_0^4} , \qquad \Gamma_\mathrm{c}^{(\mathrm{pl})} = 2 r_\mathrm{s}^{(\mathrm{c})} . \tag{7.27}$$

The critical density is determined as $n_\mathrm{c} = [\pi a_\mathrm{B}^2 (r_\mathrm{s}^{(\mathrm{c})})^2]^{-1}$. For $\delta_0 = 0.131$ resulting from the 3D empirical parameter $\delta = 1/4$, (7.27) gives $r_\mathrm{s}^{(\mathrm{c})} \simeq 722$.

This estimate means that either the melting curve of the phase diagram in the quantum and classical regimes is described by different critical values of the plasma parameter $\Gamma_c^{(pl)}$ or, in the 2D case, the Lindemann parameter δ is substantially larger than that established for 3D classical systems. A possible numerical error can be attributed in part to the Debye approximation.

To include thermal vibrations in the melting criterion, Bedanov et al. [250] considered displacement fluctuations between nearest neighbors and proposed to replace $\langle u^2 \rangle$ by

$$\langle |\boldsymbol{u}(\boldsymbol{R}+\boldsymbol{a}) - \boldsymbol{u}(\boldsymbol{R})|^2 \rangle \tag{7.28}$$

in the Lindemann criterion. Then the Monte Carlo simulations give $\delta_0' \simeq 0.3$ [251], which is substantially larger than the usual Lindeman parameter δ_0. Using (7.28) cuts off the contribution of long-wavelength phonons for which $\boldsymbol{u}(\boldsymbol{R}+\boldsymbol{a}) \simeq \boldsymbol{u}(\boldsymbol{R})$. In contrast, for uncorrelated motion of neighboring electrons due to short-wavelength phonons, (7.28) is equal to $2\langle u^2 \rangle$. This allows one to estimate $\delta_0 = \delta_0'/\sqrt{2} \simeq 0.212$, which gives substantially smaller values $r_s^{(c)} \simeq 106$ and $\Gamma_c^{(pl)} \simeq 212$ than those found above for $\delta_0 = 0.131$. Taking into account possible errors arising due to the Debye approximation, this estimate of $\Gamma_c^{(pl)}$ can be considered to be quite close to that found in the classical regime for the dislocation melting criterion.

Qualitatively, thermal vibrations can be included in the Lindemann criterion by replacing the weakly varying factor $\ln(T/\hbar c_t k_0)$ by a number C of the order of unity. Then, introducing $n = n_s/n_c$ and $t = T/T^*$, with $T^* = 0.138 \times 8\pi\delta_0^6/(\sqrt{3}A^2)\,\mathrm{Ry}$, the equation describing the liquid–solid boundary can be written in the form

$$n^{1/4} - 1 + \frac{t}{\sqrt{n}} C = 0\,. \tag{7.29}$$

Remarkably, in the classical regime ($n \ll 1$), (7.29) results in the same relation between T and n_s for the melting curve $t = \sqrt{n}/C$ as that found for the dislocation melting criterion. At the same time, the critical value of the plasma parameter, viz.,

$$\Gamma_c^{(pl)} = \frac{\sqrt{3}C}{0.138 \times 4\pi\delta_0^2}\,, \tag{7.30}$$

agrees with experiment if $C \simeq 2.35$ for the conventional value $\delta_0 = 0.131$ of the 3D case, or $C \simeq 6.2$ for the above estimate $\delta_0 = 0.212$ obtained for the 2D electron system.

In the quantum regime ($n \sim 1$), at low temperatures $t \ll 1$, the Lindemann criterion (7.29) results in a melting curve which decreases linearly with temperature $n(t) \simeq 1 - tC$. This differs qualitatively from the boundary shape found assuming $\Gamma^{(pl)} = \text{const.}$ along the melting curve [see (7.18) and Fig. 7.2].

A magnetic field applied normally to the system mainly affects the quantum region of the phase diagram. In the limit of strong magnetic fields, even a free electron can be described by a completely localized wave function, using the symmetrical gauge. As we shall see in the next section, $\langle u_0^2 \rangle$ decreases strongly with B while the contribution from thermal vibrations $\langle u^2 \rangle - \langle u_0^2 \rangle$ remains practically the same. In the framework of the dislocation melting theory, even a strong increase in the magnetic field does not much change the elastic energy of a dislocation [see (7.19)] and the melting temperature [see (7.21)], because the electron system becomes equivalent to a collection of point charged particles. On the other hand, quantum melting of the WS is strongly suppressed by the magnetic field, which is very important for semiconductor 2D electrons with a small effective mass. Thus, in spite of strong quantum effects, it is quite likely that the Wigner solid can be induced in the semiconductor 2D electron system by means of a high magnetic field oriented normally.

7.3 Normal Modes and Quantization Under a Magnetic Field

Generally, electron displacements from the lattice sites and the excitation spectrum of the Wigner crystal are strongly affected by a magnetic field directed normally to the system. It is interesting that it appears to be a much easier task to find the excitation spectrum of the WS in the presence of the magnetic field B than to describe the behavior (even qualitatively) of the mean-square displacement $\langle u^2 \rangle$ as a function of B [226, 252, 253]. It should be noted first that, even in the long-wavelength limit, the magnetic field mixes longitudinal and transverse phonons. As a result, one of the two new phonon modes has a frequency $\Omega_{-,k}(B)$ which decreases with the magnetic field as $\Omega_{-,k} \simeq \Omega_{1,k}\Omega_{t,k}/\omega_c \propto 1/B$, if $\omega_c \gg \Omega_{1,k}$. In this case, direct use of the conventional form (7.24) for the mean-square displacement leads to magnetic-field-induced melting of the WS because of the softening of the $\Omega_{-,k}$ mode. Remarkably, this conclusion is opposite to the correct answer first derived by Fukuyama [249] using the Green's function method.

The mean-square displacement and the phase diagram of the 2D WS under a magnetic field was also analyzed by Ulinich and Usov [254]. The conventional procedure for quantizing the phonon modes of a 2D lattice was extended by Monarkha and Sokolov [255] to apply to the electron crystal under a magnetic field. The extended class of canonical transformations, needed to represent the Hamiltonian of the WS as the Hamiltonian of a simple collection of independent phonons, mixes normal coordinates $Q_{1,\boldsymbol{k}}$ and momenta $\mathcal{P}_{t,-\boldsymbol{k}}$. In this section, we give a brief description of this procedure, formulate the relationship between the electron displacement operator $\boldsymbol{u}_l$ and the creation and destruction operators of the phonon modes under a magnetic field, and discuss the behavior of the mean-square displacement as a function of B.

The lattice dynamics is described by the electron displacement operator

$$\boldsymbol{u}_l = N_{\mathrm{e}}^{-1/2} \sum_{\boldsymbol{k}} \boldsymbol{u}_{\boldsymbol{k}} \exp(\mathrm{i}\boldsymbol{k} \cdot \boldsymbol{R}_l) , \tag{7.31}$$

introduced for a lattice site $\boldsymbol{R}_l$, where $\boldsymbol{k}$ is the wave-vector from the first Brillouin zone. The potential energy of the lattice is usually written as

$$U = \frac{1}{2} \sum_{l,l',\alpha,\beta} \mathcal{D}_{l,l'}^{\alpha,\beta} u_l^{(\alpha)} u_{l'}^{(\beta)} = \frac{m}{2} \sum_{\boldsymbol{k},\alpha,\beta} D_{\boldsymbol{k}}^{\alpha\beta} u_{\boldsymbol{k}}^{(\alpha)} u_{-\boldsymbol{k}}^{(\beta)} , \tag{7.32}$$

where $\mathcal{D}_{l,l'}^{\alpha,\beta}$ is the force-constant tensor

$$D_{\boldsymbol{k}}^{\alpha\beta} = \frac{1}{m} \sum_{n'} \mathcal{D}_{n,n'}^{\alpha\beta} \exp\left(-\mathrm{i}\boldsymbol{k} \cdot \boldsymbol{R}_{n,n'}\right) , \tag{7.33}$$

and $\boldsymbol{R}_{n,n'} = \boldsymbol{R}_n - \boldsymbol{R}_{n'}$. The magnetic field affects the kinetic energy of the crystal,

$$K = \frac{1}{2m} \sum_{n} \left[\boldsymbol{p}_n + \frac{e}{c}\boldsymbol{A}(\boldsymbol{r}_n)\right]^2 , \tag{7.34}$$

where $\boldsymbol{r}_n = \boldsymbol{R}_n + \boldsymbol{u}_n$.

If no magnetic field is applied, the two solutions for the phonon dispersion can be written as

$$\Omega_{1,2}^2(\boldsymbol{k}) = \frac{1}{2}\left[D_{\boldsymbol{k}}^{xx} + D_{\boldsymbol{k}}^{yy} \pm \sqrt{(D_{\boldsymbol{k}}^{xx} - D_{\boldsymbol{k}}^{yy})^2 + 4(D_{\boldsymbol{k}}^{xy})^2}\right] , \tag{7.35}$$

where the signs + and − standing before the square root correspond to the longitudinal and transverse phonons in the long-wavelength limit $ka \ll 1$. The dynamic matrix $D_{\boldsymbol{k}}^{\alpha\beta}$ and the phonon dispersion at arbitrary values of ka (Fig. 7.1) were evaluated in [123] using Ewald's summation method. In the long-wavelength limit, the dynamic matrix can be rewritten in terms of the longitudinal and transverse dispersions:

$$D_{\boldsymbol{k}}^{\alpha\beta} = \Omega_{\mathrm{t},\boldsymbol{k}}^2 \delta_{\alpha\beta} + \left(\Omega_{\mathrm{l},\boldsymbol{k}}^2 - \Omega_{\mathrm{t},\boldsymbol{k}}^2\right) \frac{k_\alpha k_\beta}{k^2} , \tag{7.36}$$

which is just an inversion of (7.35).

The phonon spectrum $\Omega_{\pm,k}$ of the WS under a magnetic field can be found from the equation of motion [226]:

$$\Omega_{\pm}^2 = \frac{\Omega_{\mathrm{l}}^2 + \Omega_{\mathrm{t}}^2 + \omega_{\mathrm{c}}^2}{2} \pm \sqrt{\left(\frac{\Omega_{\mathrm{l}}^2 + \Omega_{\mathrm{t}}^2 + \omega_{\mathrm{c}}^2}{2}\right)^2 - \Omega_{\mathrm{l}}^2 \Omega_{\mathrm{t}}^2} . \tag{7.37}$$

Here we use the notation Ω_{l} and Ω_{t} for the two modes of the WS under zero magnetic field [see (7.35)] which transform into the longitudinal and

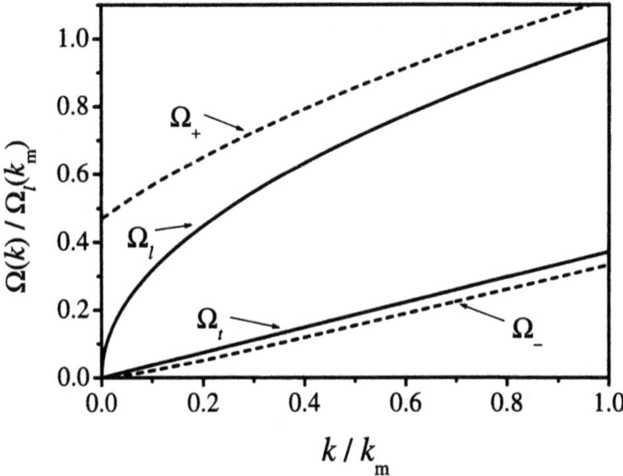

Fig. 7.5. Typical changes in the phonon spectrum of the WS induced by a normal magnetic field $[\omega_c/\Omega_l(k_m) = 0.468]$. The modes $p = l$ and $p = t$ are shown in the Debye approximation

transverse modes for $ka \ll 1$. Typical dispersion curves $\Omega_\pm(k)$ are shown in Fig. 7.5. In the limit of high magnetic fields, (7.37) describes widely spaced magnetophonon modes $\Omega_+ \simeq \omega_c \gg \Omega_- \simeq \Omega_l\Omega_t/\omega_c$.

The conventional form (7.24) for the mean-square displacement $\langle u^2 \rangle$, frequently used in the literature, is valid only for a quite limited class of transformations from the displacements $u_{\boldsymbol{k}}^{(\alpha)}$ to the normal coordinates $Q_{p,\boldsymbol{k}}$, which represent a sort of rotation:

$$\boldsymbol{u}_{\boldsymbol{k}} = \frac{1}{\sqrt{m}} \sum_p \boldsymbol{e}_{p,\boldsymbol{k}} Q_{p,\boldsymbol{k}} \,, \tag{7.38}$$

where $\boldsymbol{e}_{p,\boldsymbol{k}}$ are the orthonormal polarization vectors of the normal modes

$$e_{l,\boldsymbol{k}}^{(x)} = e_{t,\boldsymbol{k}}^{(y)} = \cos\phi \,, \qquad e_{l,\boldsymbol{k}}^{(y)} = -e_{t,\boldsymbol{k}}^{(x)} = \sin\phi \,, \tag{7.39}$$

$$\tan\phi = \frac{D_{\boldsymbol{k}}^{yx}}{\Omega_{l,\boldsymbol{k}}^2 - D_{\boldsymbol{k}}^{yy}} \,. \tag{7.40}$$

In this case, we have the simple relation

$$\langle u_l^2 \rangle = \sum_{\boldsymbol{k}} |\boldsymbol{u}_{\boldsymbol{k}}|^2 = \frac{1}{m} \sum_{p,\boldsymbol{k}} |Q_{p,\boldsymbol{k}}|^2 \,, \tag{7.41}$$

which leads straightforwardly to (7.24), using the standard expression

$$Q_{p,\boldsymbol{k}} = \sqrt{\frac{\hbar}{2\Omega_{p,\boldsymbol{k}}}} \left(a_{p,\boldsymbol{k}} + a_{p,-\boldsymbol{k}}^\dagger \right) \,. \tag{7.42}$$

The important point is that, in the presence of a magnetic field, the normal mode representation of the WS Hamiltonian $K + U$ cannot be obtained by means of any rotation [see (7.38)] because of the kinetic energy term.

7.3.1 Phonon Mode Mixing Induced by a Magnetic Field

In terms of the normal coordinates $Q_{p,\boldsymbol{k}}$ and velocities $\dot{Q}_{p,\boldsymbol{k}}$ the Lagrangian $\mathcal{L}_0$ and Hamiltonian H_0 of the WS under zero magnetic field can be expressed in forms typical of independent harmonic oscillators. The magnetic field term can be found from the corresponding correction to the Lagrange function:

$$\delta\mathcal{L} = \frac{e}{c}\sum_{\boldsymbol{n}} \boldsymbol{A}(\boldsymbol{r}_{\boldsymbol{n}})\frac{\mathrm{d}\boldsymbol{r}_{\boldsymbol{n}}}{\mathrm{d}t} = -m\omega_{\mathrm{c}}\frac{1}{2}\sum_{\boldsymbol{k}}\left(u_{\boldsymbol{k}}^{(x)}\dot{u}_{-\boldsymbol{k}}^{(y)} + u_{-\boldsymbol{k}}^{(x)}\dot{u}_{\boldsymbol{k}}^{(y)}\right) . \tag{7.43}$$

Here and below, we use the Landau gauge $\boldsymbol{A} = (0, Bx, 0)$ and disregard total derivatives with respect to time, which do not affect the equations of motion. Employing the form of $u_{\boldsymbol{k}}^{(\alpha)}$ given in (7.38) and disregarding total derivatives once again, we find

$$\delta\mathcal{L} = -\omega_{\mathrm{c}}\sum_{\boldsymbol{k}} Q_{\mathrm{l},\boldsymbol{k}}\dot{Q}_{\mathrm{t},-\boldsymbol{k}} . \tag{7.44}$$

It is now clear that the magnetic field term mixes longitudinal and transverse phonon modes.

The momentum $\mathcal{P}_{p,\boldsymbol{k}}$ canonically conjugate to the coordinate $Q_{p,\boldsymbol{k}}^{*}$ can be found according to the common rule

$$\mathcal{P}_{p,\boldsymbol{k}} = \frac{\partial\mathcal{L}}{\partial\dot{Q}_{p,\boldsymbol{k}}^{*}} . \tag{7.45}$$

For the Lagrange function $\mathcal{L} = \mathcal{L}_0 + \delta\mathcal{L}$, equation (7.45) implies

$$\dot{Q}_{\mathrm{l},\boldsymbol{k}} = \mathcal{P}_{\mathrm{l},\boldsymbol{k}} , \qquad \dot{Q}_{\mathrm{t},\boldsymbol{k}} = \mathcal{P}_{\mathrm{l},\boldsymbol{k}} + \omega_{\mathrm{c}}Q_{\mathrm{l},\boldsymbol{k}} . \tag{7.46}$$

The latter equation is reminiscent of the relation between the velocity and momentum of a single particle under a magnetic field.

In terms of the new variables ($Q_{p,-\boldsymbol{k}}$ and $\mathcal{P}_{p,\boldsymbol{k}}$), the Lagrangian $\mathcal{L}$ differs from the Lagrangian of independent oscillators $\mathcal{L}_0$ only by the replacement $\Omega_{\mathrm{l},k}^2 \to \Omega_{\mathrm{l},k}^2 + \omega_{\mathrm{c}}^2$, because the nondiagonal terms of $\delta\mathcal{L}$ are compensated by corresponding kinetic energy terms in $\mathcal{L}_0$ arising due to (7.46). In contrast, the Hamiltonian

$$H = \sum_{p,\boldsymbol{k}} \mathcal{P}_{p,\boldsymbol{k}}\dot{Q}_{p,-\boldsymbol{k}} - \mathcal{L} , \tag{7.47}$$

acquires nondiagonal terms

$$H = \sum_{k}\left\{\frac{1}{2}\left[|\mathcal{P}_{\mathrm{l},\boldsymbol{k}}|^2 + (\Omega_{\mathrm{l},\boldsymbol{k}}^2 + \omega_{\mathrm{c}}^2)\,|Q_{\mathrm{l},\boldsymbol{k}}|^2\right]\right.$$
$$\left. + \frac{1}{2}\left(|\mathcal{P}_{\mathrm{t},\boldsymbol{k}}|^2 + \Omega_{\mathrm{t},\boldsymbol{k}}^2\,|Q_{\mathrm{t},\boldsymbol{k}}|^2\right) + \omega_{\mathrm{c}}Q_{\mathrm{l},\boldsymbol{k}}\mathcal{P}_{\mathrm{t},-\boldsymbol{k}}\right\} . \tag{7.48}$$

The magnetic field thus affects the WS Hamiltonian in two ways:

- it increases the frequency of longitudinal phonons $\Omega_{\mathrm{l},\boldsymbol{k}} \to \sqrt{\Omega_{\mathrm{l},\boldsymbol{k}}^2 + \omega_{\mathrm{c}}^2}$ in accordance with the spectrum of 2D magnetoplasmons,
- it causes the appearance of nondiagonal terms $\omega_{\mathrm{c}} Q_{\mathrm{l},\boldsymbol{k}} \mathcal{P}_{\mathrm{t},-\boldsymbol{k}}$, which mix the longitudinal and transverse phonon modes.

It is clear that the Hamiltonian of (7.48) cannot be transformed to the normal mode form by any rotation or point transformation involving only normal coordinates. We have to use a canonical transformation involving $\mathcal{P}_{\mathrm{t},-\boldsymbol{k}}$ [254]. There is an elegant and instructive way of performing this transformation by separating it into two distinct steps [255]. Firstly, we use the canonical transformation which just renames momenta and coordinates of the transverse mode:

$$\Phi_{\mathrm{t},\boldsymbol{k}} = \frac{1}{\Omega_{\mathrm{t},\boldsymbol{k}}} \mathcal{P}_{\mathrm{t},\boldsymbol{k}} \, , \qquad \Pi_{\mathrm{t},\boldsymbol{k}} = -\Omega_{\mathrm{t},\boldsymbol{k}} Q_{\mathrm{t},\boldsymbol{k}} \, , \tag{7.49}$$

where $\Pi_{\mathrm{t},\boldsymbol{k}}$ is the new momentum, and leaves the variables of the longitudinal mode unchanged:

$$\Phi_{\mathrm{l},\boldsymbol{k}} = Q_{\mathrm{l},\boldsymbol{k}} \, , \qquad \Pi_{\mathrm{l},\boldsymbol{k}} = \mathcal{P}_{\mathrm{l},\boldsymbol{k}} \, . \tag{7.50}$$

This kind of canonical transformation is usually introduced as an instructive example showing the equivalence of the momentum and coordinate in the Hamiltonian formalism [256]. We take advantage of this well known example, noting that such a substitution immediately transforms the WS Hamiltonian into a typical form for neutral solids. Indeed, the nondiagonal terms of the transformed Hamiltonian now depend only on the new coordinates

$$\omega_{\mathrm{c}} \Omega_{\mathrm{t},\boldsymbol{k}} \Phi_{\mathrm{l},\boldsymbol{k}} \Phi_{\mathrm{t},\boldsymbol{k}}$$

and can be eliminated by the usual rotation in coordinate space.

The second step of the canonical transformation is obvious:

$$\Phi_{\mathrm{l},\boldsymbol{k}} = Q_{+,\boldsymbol{k}} \cos\lambda - Q_{-,\boldsymbol{k}} \sin\lambda = \sum_{p=+,-} p \zeta_p Q_{p,\boldsymbol{k}} \, , \tag{7.51}$$

$$\Phi_{\mathrm{t},\boldsymbol{k}} = Q_{+,\boldsymbol{k}} \sin\lambda + Q_{-,\boldsymbol{k}} \cos\lambda = \sum_{p=+,-} \zeta_{-p} Q_{p,\boldsymbol{k}} \, , \tag{7.52}$$

which makes the WS Hamiltonian diagonal in terms of the new coordinates $Q_{\pm,\boldsymbol{k}}$. Here we use the notation

$$\sin^2\lambda = \frac{\Omega_{\mathrm{t},\boldsymbol{k}}^2 - \Omega_{-,\boldsymbol{k}}^2}{\Omega_{+,\boldsymbol{k}}^2 - \Omega_{-,\boldsymbol{k}}^2} \, , \tag{7.53}$$

$\zeta_+ = \cos\lambda$, $\zeta_- = \sin\lambda$, the factor p in (7.51) is the sign of the corresponding term ($+$ or $-$) depending on the sign of p, and the subscript $-p$ denotes the opposite ($-$ or $+$), i.e., it is $-$ if $p = +$.

The variables $Q_{\pm,\boldsymbol{k}}$ and $\mathcal{P}_{\pm,\boldsymbol{k}}$ represent the new normal coordinates and momenta of the WS under a magnetic field. In terms of these variables, the WS Hamiltonian has the canonical form

$$H = \frac{1}{2}\sum_{\boldsymbol{k}}\sum_{p=+,-}\left(|\mathcal{P}_{p,\boldsymbol{k}}|^2 + \Omega_{p,\boldsymbol{k}}^2 |Q_{p,\boldsymbol{k}}|^2\right) . \tag{7.54}$$

When describing the new modes, we assume that the polarization subscript p runs over $+$ and $-$. Then (7.42), which establishes the relationship between the normal mode coordinate $Q_{p,\boldsymbol{k}}$ and the many-body creation and destruction operators, will also be applicable for $B \neq 0$. The analogous expression for the normal momenta is

$$\mathcal{P}_{p,\boldsymbol{k}} = \mathrm{i}\sqrt{\frac{\hbar\Omega_{p,\boldsymbol{k}}}{2}}\left(a_{p,-\boldsymbol{k}}^{\dagger} - a_{p,\boldsymbol{k}}\right) , \tag{7.55}$$

where $a_{p,\boldsymbol{k}}^{\dagger}$ and $a_{p,\boldsymbol{k}}$ are the creation and destruction operators of the new modes ($p = +, -$).

Summarizing the transformations used above, one can rewrite the Fourier transform of the displacement operator as

$$\begin{aligned} \boldsymbol{u}_{\boldsymbol{k}}(t) &= \frac{1}{\sqrt{m}}\left[\mathbf{e}_{\mathrm{l},\boldsymbol{k}}\Phi_{\mathrm{l},\boldsymbol{k}}(t) - \mathbf{e}_{\mathrm{t},\boldsymbol{k}}\frac{\Pi_{\mathrm{t},\boldsymbol{k}}(t)}{\Omega_{\mathrm{t},\boldsymbol{k}}}\right] \\ &= \frac{1}{\sqrt{m}}\sum_{p=\pm}\left[p\zeta_p \mathbf{e}_{\mathrm{l},\boldsymbol{k}} Q_{p,\boldsymbol{k}}(t) - \frac{\zeta_{-p}}{\Omega_{\mathrm{t},\boldsymbol{k}}}\mathbf{e}_{\mathrm{t},\boldsymbol{k}}\mathcal{P}_{p,\boldsymbol{k}}(t)\right] . \end{aligned} \tag{7.56}$$

Taking into account the definitions (7.42) and (7.55), the electron displacement operator can be expressed in terms of operators $a_{\pm,\boldsymbol{k}}$ and $a_{\pm,\boldsymbol{k}}^{\dagger}$, viz.,

$$\boldsymbol{u}_n = \sum_{\boldsymbol{k}}\sum_{p=+,-}\sqrt{\frac{\hbar}{2N_{\mathrm{e}}m\Omega_{p,\boldsymbol{k}}}}\left(\boldsymbol{E}_{p,\boldsymbol{k}} a_{p,\boldsymbol{k}} \mathrm{e}^{\mathrm{i}\boldsymbol{k}\cdot\boldsymbol{R}_n} + \boldsymbol{E}_{p,\boldsymbol{k}}^{*} a_{p,\boldsymbol{k}}^{\dagger} \mathrm{e}^{-\mathrm{i}\boldsymbol{k}\cdot\boldsymbol{R}_n}\right) , \tag{7.57}$$

with the new polarization vectors

$$\boldsymbol{E}_{p,\boldsymbol{k}} = p\zeta_p \mathbf{e}_{\mathrm{l},\boldsymbol{k}} + \mathrm{i}\zeta_{-p}\frac{\Omega_{p,\boldsymbol{k}}}{\Omega_{\mathrm{t},\boldsymbol{k}}}\mathbf{e}_{\mathrm{t},\boldsymbol{k}} . \tag{7.58}$$

In the presence of the magnetic field, $\boldsymbol{E}_{\pm,\boldsymbol{k}}$ are complex vectors and they are no longer orthonormalized:

$$|\boldsymbol{E}_{p,\boldsymbol{k}}|^2 = \frac{\left|2\Omega_{p,\boldsymbol{k}}^2 - \Omega_{\mathrm{l},\boldsymbol{k}}^2 - \Omega_{\mathrm{t},\boldsymbol{k}}^2\right|}{\Omega_{+,\boldsymbol{k}}^2 - \Omega_{-,\boldsymbol{k}}^2} \neq 1 , \tag{7.59}$$

$$\boldsymbol{E}_{+,\boldsymbol{k}}\cdot\boldsymbol{E}_{-,\boldsymbol{k}}^{*} = \omega_{\mathrm{c}}\frac{\Omega_{\mathrm{l},\boldsymbol{k}} - \Omega_{\mathrm{t},\boldsymbol{k}}}{\Omega_{+,\boldsymbol{k}}^2 - \Omega_{-,\boldsymbol{k}}^2} \neq 0 . \tag{7.60}$$

Here we used (7.53) and the property $\Omega_{+,\boldsymbol{k}}\Omega_{-,\boldsymbol{k}} = \Omega_{\mathrm{l},\boldsymbol{k}}\Omega_{\mathrm{t},\boldsymbol{k}}$. A sort of normalization remains for $|\boldsymbol{E}_{+,\boldsymbol{k}}|^2 + |\boldsymbol{E}_{-,\boldsymbol{k}}|^2 = 2$. In the limit of high magnetic fields, we find the asymptotes $|\boldsymbol{E}_{-,\boldsymbol{k}}|^2 \simeq (\Omega_{\mathrm{l},\boldsymbol{k}}^2 + \Omega_{\mathrm{t},\boldsymbol{k}}^2)/\omega_{\mathrm{c}}^2 \ll 1$ and $|\boldsymbol{E}_{+,\boldsymbol{k}}|^2 \simeq 2$.

The relations (7.57)–(7.60) allow one to perform standard quantum-mechanical evaluations for the Wigner solid subject to a magnetic field. As an important example, we consider the mean-square displacement.

7.3.2 Mean-Square Displacement

The extension of the conventional equation for the mean-square displacement, applicable to a WS subject to a magnetic field, can be found using the representation of the electron displacement operator given in (7.57):

$$\langle u^2 \rangle = \frac{\hbar}{2mN_{\mathrm{e}}} \sum_{p=+,-} \sum_{\boldsymbol{k}} \frac{|\boldsymbol{E}_{p,\boldsymbol{k}}|^2}{\Omega_{p,\boldsymbol{k}}} \coth \frac{\hbar \Omega_{p,\boldsymbol{k}}}{2T} , \tag{7.61}$$

where $|\boldsymbol{E}_{p,\boldsymbol{k}}|^2$ is defined by (7.59). This equation is similar to the conventional form of (7.24), and in the limit $B \to 0$, the two equations become equivalent because $|\boldsymbol{E}_{p,\boldsymbol{k}}|^2 \to 1$.

The important point is that, under a magnetic field, $|\boldsymbol{E}_{p,\boldsymbol{k}}|^2 \neq 1$. This originates from the canonical transformation $Q_{\mathrm{t},\boldsymbol{k}} = -\Pi_{\mathrm{t},\boldsymbol{k}}/\Omega_{\mathrm{t},\boldsymbol{k}}$ of (7.49) and from the fact that

$$\langle \mathcal{P}_{p,\boldsymbol{k}} \mathcal{P}_{p,-\boldsymbol{k}} \rangle = \frac{\hbar \Omega_{p,\boldsymbol{k}}}{2} (2n_{p,\boldsymbol{k}} + 1) \tag{7.62}$$

differs from

$$\langle \mathcal{Q}_{p,\boldsymbol{k}} \mathcal{Q}_{p,-\boldsymbol{k}} \rangle = \frac{\hbar}{2\Omega_{p,\boldsymbol{k}}} (2n_{p,\boldsymbol{k}} + 1) , \tag{7.63}$$

where $n_{p,\boldsymbol{k}}$ is the Bose distribution function of phonons with polarization $p = +, -$. Employing the explicit form of $|\boldsymbol{E}_{p,\boldsymbol{k}}|^2$ given by (7.59), the equation for the mean-square displacement can be transformed to

$$\langle u^2 \rangle = \frac{\hbar}{2mN_{\mathrm{e}}} \sum_{p=+,-} \sum_{\boldsymbol{k}} \frac{2\Omega_{p,\boldsymbol{k}}^2 - \Omega_{\mathrm{l},\boldsymbol{k}}^2 - \Omega_{\mathrm{t},\boldsymbol{k}}^2}{\Omega_{p,\boldsymbol{k}}^4 - \Omega_{\mathrm{l},\boldsymbol{k}}^2 \Omega_{\mathrm{t},\boldsymbol{k}}^2} \Omega_{p,\boldsymbol{k}} \coth \frac{\hbar \Omega_{p,\boldsymbol{k}}}{2T} , \tag{7.64}$$

In this representation, the equivalence of (7.61) and the formula found in [249, 254] becomes evident.

Regarding the properties of $\langle u^2 \rangle$ as a function of B, we note first that the softening of the phonon mode induced by the magnetic field, viz.,

$$\Omega_{-,\boldsymbol{k}} \to \Omega_{\mathrm{l},\boldsymbol{k}}\Omega_{\mathrm{t},\boldsymbol{k}}/\omega_{\mathrm{c}} ,$$

cannot increase $\langle u^2 \rangle$ because of the factor

$$|\boldsymbol{E}_{-,\boldsymbol{k}}|^2 \to (\Omega_{\mathrm{l},\boldsymbol{k}}^2 + \Omega_{\mathrm{t},\boldsymbol{k}}^2)/\omega_{\mathrm{c}}^2 \, .$$

Thus, in the limit of high magnetic fields $\omega_{\mathrm{c}} \gg \Omega_{\mathrm{l}}$ and under the condition $\hbar\Omega_{-} \ll 2T$, the contribution of thermal vibrations of the soft mode

$$\langle u_{-}^2 \rangle = \frac{T}{mN_{\mathrm{e}}} \sum_{\boldsymbol{k}} \left(\frac{1}{\Omega_{\mathrm{t},\boldsymbol{k}}^2} + \frac{1}{\Omega_{\mathrm{l},\boldsymbol{k}}^2} \right) \tag{7.65}$$

coincides with the contribution of both longitudinal and transverse phonon modes at $B = 0$ and $2T \gg \hbar\Omega_{\mathrm{l}}(k_{\mathrm{m}})$. At the same time, the zero-point vibration contribution is proportional to

$$\frac{|\boldsymbol{E}_{-,\boldsymbol{k}}|^2}{\Omega_{-,k}} \to \frac{\Omega_{\mathrm{l},k}}{\Omega_{\mathrm{t},k}\omega_{\mathrm{c}}} \propto \frac{1}{B} \, , \tag{7.66}$$

and decreases with the magnetic field, which affects the stability region of the WS in the quantum regime.

In the Debye model with the cutoff wave number $k_{\mathrm{m}} = \sqrt{4\pi n_{\mathrm{s}}}$ and as $B \to \infty$, (7.61) results in the zero temperature fluctuations

$$\langle u_{+}^2 \rangle = l_B^2 \, , \qquad \langle u_{-}^2 \rangle \simeq 1.94 \, l_B^2 \, , \tag{7.67}$$

where $l_B = \sqrt{\hbar/m\omega_{\mathrm{c}}}$ is the electron magnetic length. We have already used the first relation of (7.67) in Sect. 5.4 when describing the shift of the CR position for the Wigner solid state of SEs. The above formulae can also be used for the Lindemann melting criterion in the presence of the magnetic field. In the limit of high fields, we have

$$r_{\mathrm{s}}^{(\mathrm{c})} \simeq 0.9 \frac{l_B}{a_{\mathrm{B}}\delta_0} \, , \tag{7.68}$$

which shifts the solid–liquid boundary upward: $n_{\mathrm{s}}^{(\mathrm{c})} \simeq 0.393 \, \delta_0^2 / l_B^2 \propto B$.

7.4 Coupling with Medium Vibrations

As discussed in Chap. 6, in the 2D electron system formed on the 'soft' substrate of liquid helium, the interaction with surface displacements may lead to the existence of polaronic states, where an electron becomes self-trapped, creating a surface dimple. In the WS state, electrons are already localized at lattice sites owing to the Coulomb forces, and they therefore automatically create a lattice of surface dimples $\xi_0(\boldsymbol{r})$ because of the electron–ripplon interaction. The first analysis of this polaronic effect induced by the Wigner solid transition was given by Monarkha and Shikin [257]. Because the interaction energy $eE_{\perp}\xi_0$ is very small compared with the average Coulomb energy, the appearance of the dimple lattice has practically no effect on the

WS transition point. The static deformation of the helium surface $\xi_0(\boldsymbol{r})$ just accompanies the WS state. On the other hand, as shown in further studies [7, 230], the dimple lattice drastically affects the dynamics of the WS and its excitation spectrum at low frequencies. In a remote analogy with the ripplonic polaron problem, the renormalization of the WS excitation spectrum can be found by means of a self-consistent procedure [258, 259], which we discuss in Sect. 7.4.3.

Qualitatively, changes in the WS spectrum induced by the dimple lattice can be described by recalling the general structure of the medium response to the alternating motion of electrons given in Chap. 3. Firstly, we note that the medium response depends crucially on the relation between the frequency of electron motion ω and the typical frequency of medium excitations (ripplons) ω_{g_1}, where g_1 is the smallest (nonzero) reciprocal lattice vector. At low frequencies $\omega \ll \omega_{g_1}$, the dimples follow electrons, contributing to their effective mass: $M = m_{\mathrm{e}} + M_{\mathrm{d}}$. In this instance, the electron and dimple lattices move in phase, causing the renormalization of phonon frequencies $\Omega_p^{(\mathrm{s})} \simeq \sqrt{m_{\mathrm{e}}/M}\,\Omega_p(k)$, where $p = \mathrm{l}, \mathrm{t}$. At high frequencies $\omega \gg \omega_{g_1}$, dimples can be considered as motionless, and the WS excitation spectrum becomes similar to the spectrum of optical modes:

$$\Omega_p^{(\mathrm{f})}(k) = \sqrt{\omega_{\mathrm{f}}^2 + \Omega_p^2(k)}\,, \tag{7.69}$$

where ω_{f} is the frequency of oscillations of a single electron in the potential created by a fixed surface dimple.

If the electron–medium coupling is strong, as it is for SEs on liquid helium, then ω_{f} is much larger than ω_{g_1}, and the associated mass of the dimple is huge, i.e., $M_{\mathrm{d}} \gg m_{\mathrm{e}}$. We therefore conclude that the WS phonon spectrum over the soft substrate is split into two separate branches which we shall call fast (f) and slow (s) modes. The dimple size and the limiting frequency ω_{f} depend on the mean-square displacement $\langle u_{\mathrm{f}}^2 \rangle$ of electrons in the fast modes. On the other hand, the contribution of thermal vibrations to $\langle u_{\mathrm{f}}^2 \rangle$ depends on the limiting frequency ω_{f}, because of the specific behavior of two-dimensional systems in the long-wavelength limit. This is the reason for using the self-consistent procedure describing $\langle u_{\mathrm{f}}^2 \rangle$ and ω_{f}.

7.4.1 Dimple Lattice and Medium Response Force

Consider now the WS dynamics on the soft substrate in more detail. In the presence of interface displacements $\xi(\boldsymbol{r})$, the interaction Hamiltonian of the electron crystal formed on liquid helium can be written as

$$V_{\mathrm{int}} = \frac{1}{\sqrt{S_{\mathrm{A}}}} \sum_{\boldsymbol{q}} V_q \xi_{\boldsymbol{q}} \sum_{\boldsymbol{n}} \exp\left(\mathrm{i}\boldsymbol{q} \cdot \boldsymbol{R}_{\boldsymbol{n}} + \mathrm{i}\boldsymbol{q} \cdot \boldsymbol{u}_{\boldsymbol{n}}\right)\,, \tag{7.70}$$

where $V_q = e(E_{\perp} + E_q)$ is the electron–ripplon coupling defined previously in (1.119). The Lagrangian of the electron–ripplon system with interaction is

$\mathcal{L} = \mathcal{L}_0 - V_{\rm int}$. Then, using (7.70) and the properties of interface excitations given in Sect. 1.5.1, we find Lagrange's equation of motion for ξ_q:

$$\ddot{\xi}_{\boldsymbol{q}} + \omega_q^2 \xi_{\boldsymbol{q}} = -\frac{V_q}{\sqrt{S_{\rm A}}\mu_q} \sum_{\mathbf{n}} \exp\left[-\mathrm{i}\boldsymbol{q} \cdot \boldsymbol{R_n} - \mathrm{i}\boldsymbol{q} \cdot \boldsymbol{u_n}(t)\right] , \qquad (7.71)$$

where, according to (1.53), $\mu_q = \rho/q \tanh(qd)$. Ripplon frequencies $\omega_q = \sqrt{\alpha/\rho} q^{3/2}$ are usually much lower than typical frequencies Ω_p of Wigner solid phonons.

When describing the low-frequency motion of the electron lattice ($\omega \leq \omega_g \ll \omega_{\rm f}$), one can average the right-hand side of (7.71) over the vibrations of the fast mode $\Omega_p^{(\rm f)}$, which also includes the zero-point vibrations. This procedure is equivalent to averaging the interaction Hamiltonian $V_{\rm int}$ defined by (7.70) and it causes a reduction in the electron–ripplon coupling

$$V_q \to V_q \exp(-W_q) , \qquad W_q = q^2 \left\langle u_{\rm f}^2 \right\rangle /4 , \qquad (7.72)$$

to be used in (7.71) for the slow modes. Here and below, the lattice displacement operator for the fast mode is denoted by $\boldsymbol{u}_{\rm f}(\boldsymbol{n}, t)$. To find (7.72), we used Bloch's identity, given previously in (1.154), and the relation

$$\left\langle \left[u_{\rm f}^{(q)}\right]^2 \right\rangle = \left\langle u_{\rm f}^2 \right\rangle /2 , \qquad (7.73)$$

valid in two dimensions. The superscript (q) denotes the projection onto the direction of the wave vector $\boldsymbol{q}$. Here $\left\langle u_{\rm f}^2 \right\rangle$ is considered as a given quantity. It will be determined later in Sect. 7.4.3, using the self-consistent procedure.

If there are no slow electron displacements $\boldsymbol{u}_{\rm s}(\boldsymbol{n}, t)$ from lattice sites, the solution of (7.71) represents the static dimple profile (dimple sublattice)

$$\xi_0(\boldsymbol{r}) = -\sum_{\boldsymbol{g}} \frac{n_{\rm s} V_g}{\mu_g \omega_g^2} \mathrm{e}^{-W_g} \cos\left(\boldsymbol{g} \cdot \boldsymbol{r}\right) , \qquad (7.74)$$

which is similar to the dimple profile of the ripplonic polaron discussed in Chap. 6 [see (6.15)]. The important difference is that, instead of summing over all possible wave vectors $\boldsymbol{q}$, we now sum over all the reciprocal lattice vectors, and $\sqrt{\langle u_{\rm f}^2 \rangle}$ appears instead of the electron localization radius L.

The next instructive instance to be considered here is the model in which all electrons have the same slow displacement $\boldsymbol{u}_{\rm s}(t)$ caused, for example, by a uniform driving electric field. Then the Fourier transform of the dimple profile which follows from the equation of motion can be obtained in the form

$$\xi_{\boldsymbol{g}}(t) = -\frac{N_{\rm e} g V_g}{\rho \omega_g \sqrt{S_{\rm A}}} \mathrm{e}^{-W_g} \int_{-\infty}^{t} \sin\left[\omega_g (t - t')\right] \mathrm{e}^{-\mathrm{i}\boldsymbol{g} \cdot \boldsymbol{u}_{\rm s}(t') + \delta t'} \mathrm{d}t' . \qquad (7.75)$$

Here we have added the infinitesimal quantity $\delta > 0$ in the exponent in order to arrive at the proper ground state solution [see (7.74)] at $\boldsymbol{u}_{\rm s} = 0$, which implies an adiabatic switching-on of the electron–ripplon interaction. The analogous procedure in quantum many-particle physics is an important feature of the Feynman diagram technique.

Consider now the force acting on the moving WS owing to the interaction with medium vibrations. This in-plane force can be found from the equation $\boldsymbol{F} = (m/\hbar e)\mathrm{i}\left[\boldsymbol{J}, V_{\text{int}}\right]$, where $\boldsymbol{J} = -e\sum_e \boldsymbol{v}^{(e)}$ is the current operator, or directly from $\boldsymbol{F} = -\sum_e \partial V_{\text{int}}/\partial \boldsymbol{r}_e$. For the particular form (7.70) of the interaction Hamiltonian, we have

$$\boldsymbol{F} = -\frac{\mathrm{i}}{\sqrt{S_{\mathrm{A}}}}\sum_{\boldsymbol{q}} \boldsymbol{q} V_q \xi_q n_{-\boldsymbol{q}} \, , \tag{7.76}$$

where $n_{\boldsymbol{q}} = \sum_e \exp(-\mathrm{i}\boldsymbol{q}\cdot\boldsymbol{r}_e)$ is the electron density operator. Assuming that the uniform motion of the electron solid is slow, resulting in surface displacements of (7.75), and that the interaction Hamiltonian is averaged over the fast mode vibrations

$$n_{-\boldsymbol{q}} \to \langle n_{-\boldsymbol{q}}\rangle^{(\mathrm{f})} = N_{\mathrm{e}} \exp\left(-W_q + \mathrm{i}\boldsymbol{q}\cdot\boldsymbol{u}_{\mathrm{s}}\right)\delta_{\boldsymbol{q},\boldsymbol{g}} \, , \tag{7.77}$$

the force acting on the WS can be written as

$$\boldsymbol{F} = -N_{\mathrm{e}}\sum_{\boldsymbol{g}} \boldsymbol{g}\frac{n_{\mathrm{s}} g V_g^2}{\rho\omega_g}\mathrm{e}^{-2W_g}\int_{-\infty}^{t} \sin\left[\omega_g(t-t')\right]\mathrm{e}^{\delta t'}\sin\left\{\boldsymbol{g}\cdot\left[\boldsymbol{u}_{\mathrm{s}}(t) - \boldsymbol{u}_{\mathrm{s}}(t')\right]\right\}\mathrm{d}t' \, . \tag{7.78}$$

Generally, this force has a nonlinear dependence on the electron displacements.

For small displacements $\boldsymbol{u}_{\mathrm{s}}$, the sine function in (7.78) can be expanded in $\boldsymbol{g}\cdot[\boldsymbol{u}_{\mathrm{s}}(t) - \boldsymbol{u}_{\mathrm{s}}(t')]$ giving the linear reaction force. Therefore, assuming $\boldsymbol{u}_{\mathrm{s}}(t) \propto \exp(-\mathrm{i}\omega t)$, we have

$$\frac{\boldsymbol{F}(t)}{N_{\mathrm{e}}} = m_{\mathrm{e}}\omega\left[w(\omega) + \mathrm{i}\nu(\omega)\right]\boldsymbol{u}_{\mathrm{s}}(t) \, , \tag{7.79}$$

where

$$w(\omega) = \frac{1}{2\omega}\sum_{\boldsymbol{g}} v_g^2 \mathrm{e}^{-2W_g}\frac{\omega^2}{\omega_g^2 - \omega^2} \, , \tag{7.80}$$

$$\nu(\omega) = \frac{\pi}{4}\sum_{\boldsymbol{g}} v_g^2 \mathrm{e}^{-2W_g}\delta(\omega - \omega_g) \, . \tag{7.81}$$

We have introduced a new coupling parameter $v_q = (n_{\mathrm{s}}/m\alpha)^{1/2}V_q$ which has units of frequency. The imaginary term in (7.79), proportional to $\mathrm{i}\nu(\omega)$, arises due to the Weierstrass (or Sokhotsky–Weierstrass) relation when putting $\delta \to +0$. It represents kinetic friction of the Wigner lattice, which is antiparallel to the velocity of the lattice $\boldsymbol{v}_{\mathrm{s}}(t) = -\mathrm{i}\omega\boldsymbol{u}_{\mathrm{s}}(t)$.

Equation (7.79) agrees with the general phenomenological treatment of the medium response force discussed in Chap. 3. According to this, the function $w(\omega)$ is the real part of the memory function. At the same time, the function $\lambda(\omega)$, defined previously by the relation $\lambda(\omega) - \lambda(0) = \omega w(\omega)$, determines the medium reaction to the WS motion. It can be written as

$$\lambda(\omega) = \frac{1}{2} \sum_{g} v_g^2 \mathrm{e}^{-2W_g} \frac{\omega_g^2}{\omega_g^2 - \omega^2} \, . \tag{7.82}$$

We used this structure of $\lambda(\omega)$ in Sect. 5.4 when discussing the shift in position of the cyclotron resonance for SEs on liquid helium.

The delta-function structure of the effective collision rate $\nu(\omega)$ is due to resonant emission of surface excitations on the liquid helium by the moving electron lattice under the condition that WS phonons are not involved in the scattering events. Later we shall see that, for scattering events accompanied by creation or annihilation of WS phonons, the effective collision frequency is a rather smooth function of ω.

It is instructive to note that the same expressions for $w(\omega)$ and $\nu(\omega)$ can be found by means of the memory function approach discussed in Sect. 3.4. Indeed, employing the Kramers–Kronig relations, which follow from the spectral representation of (3.50), and the equation for $\langle F_x(t) F_x \rangle_\omega$ found in Sect. 3.4 [see (3.56)], we obtain

$$\begin{aligned} \mathrm{Re}\, G^{(\mathrm{R})}_{F_x, F_x}(\omega) = n_{\mathrm{s}} \sum_{\boldsymbol{q}} q_x^2 V_q^2 Q_q^2 \int_{-\infty}^{\infty} \frac{\mathrm{d}\omega'}{2\pi\hbar} \frac{1 - \mathrm{e}^{-\hbar\omega'/T}}{\omega - \omega'} \\ \times \left[(N_q + 1) S(-\boldsymbol{q}, \omega' - \omega_q) + N_q S(\boldsymbol{q}, \omega' + \omega_q) \right] . \end{aligned} \tag{7.83}$$

As we shall see in the next section, the simplest approximation for the dynamical structure factor of the 2D electron lattice to be employed in the conductivity equations of the strong coupling regime can be written as

$$S(q, \omega) \simeq 2\pi N_{\mathrm{e}} \mathrm{e}^{-2W_q} \delta(\omega) \sum_{g} \delta_{\boldsymbol{q},\boldsymbol{g}} \, , \tag{7.84}$$

where e^{-2W_q} is the self-consistent Debye–Waller factor, as discussed above. Substituting this expression into (7.83), we have

$$\mathrm{Re}\, G^{(\mathrm{R})}_{F_x, F_x}(\omega) \simeq \frac{m_{\mathrm{e}} N_{\mathrm{e}}}{2} \sum_{g} v_g^2 \mathrm{e}^{-2W_g} \frac{\omega_g^2}{\omega^2 - \omega_g^2} \, . \tag{7.85}$$

Recalling the representation for the memory function $M(\omega) \simeq M_{\mathrm{F}}(\omega)$ given in (3.43), we straightforwardly arrive at $\mathrm{Re}\, M(\omega) = w(\omega)$ with $w(\omega)$ defined by (7.80). In Sect. 8.2.1, we shall see that the same approximation for $S(q, \omega)$ and the quantum transport framework equation for the effective collision frequency yield the proper form of $\nu(\omega)$ given above in (7.81) if $\hbar\omega \ll T$.

In the limiting case $\omega \ll \omega_g$, the force acting on a single electron by the medium is $\boldsymbol{F}/N_{\mathrm{e}} \propto \omega^2$, and the real part of the memory function is proportional to the frequency of the WS motion: $w(\omega) \propto \omega$. This means that electron motion is affected by an additional inertia via the associated dimple mass

$$M_{\mathrm{d}} = \frac{m_{\mathrm{e}}}{2} \sum_{\boldsymbol{g}} \left(\frac{v_g}{\omega_g} \right)^2 \mathrm{e}^{-2W_g} . \tag{7.86}$$

Simple estimates show that, for SEs on liquid helium, $M_{\mathrm{d}} \gg m_{\mathrm{e}}$, which indicates that the electron–ripplon system is in the strong coupling regime.

In the opposite limit of high frequencies $\omega \gg \omega_g$, the medium reaction has the oscillatory form $\boldsymbol{F}/N_{\mathrm{e}} \simeq -m\omega_{\mathrm{f}}^2 \boldsymbol{u}_{\mathrm{s}}$, where

$$\omega_{\mathrm{f}}^2 = \frac{1}{2} \sum_{\boldsymbol{g}} v_g^2 \mathrm{e}^{-2W_g} . \tag{7.87}$$

This is consistent with the treatment of surface dimples as motionless objects. Even though (7.79) was found for slow displacements with $\omega \ll \omega_{\mathrm{f}}$, the above equation for ω_{f} remains valid because the medium reaction force becomes independent of frequency when $\omega_g^2 \ll \omega^2 \ll \omega_{\mathrm{f}}^2$ under the strong coupling conditions $\omega_{\mathrm{f}} \gg \omega_{g_1}$.

7.4.2 Coupled Modes

The medium response force (7.79) acting on moving electrons has a resonance structure when ω is close to the frequencies of ripplons whose wave vectors coincide with the reciprocal lattice vectors $\omega = \omega_g$. The alternating current of the 2D electron solid excites capillary waves whose wavelength is commensurable with the period of the electron lattice. According to the transport framework discussed in Chap. 3, the resonance structure of the medium response force straightforwardly leads to the resonance structure of the electron conductivity and power absorption in an alternating driving field. The possibility of resonance excitation of capillary waves by the WS in the presence of an alternating electric field $\delta E_{\perp} \propto \mathrm{e}^{-i\omega t}$ applied normally to the surface was proposed in [260] and analyzed in detail in [257]. The resonance excitation of ripplons by the WS moving parallel to the surface, realized experimentally by Grimes and Adams [7], appeared to be more effective than that calculated for a perpendicular alternating field. The reason is that the amplitude of motion of interface electrons in the vertical direction is very small because of the strong binding to the interface. The alternating electric field $E'_{\perp}$ is much smaller than the holding electric field $E_{\perp}$, so that the time-dependent pressure $eE'_{\perp}(t)n_{\mathrm{s}}$ acting on the surface is very small. For a parallel alternating field $E'_{\parallel}$, the time-dependent pressure $eE_{\perp}n_{\mathrm{s}}i\boldsymbol{g} \cdot \boldsymbol{u}_n(t)$ which follows from (7.71) is much stronger because it is proportional to the holding electric field.

In the experiment of Grimes and Adams [7], the driving electric field was nonuniform. It was induced in a cylindrical cell and had axial symmetry. The power absorbed by the WS under this field is shown in Fig. 7.6. A small decrease in temperature from $T = 0.46\,\mathrm{K}$ to $T = 0.44\,\mathrm{K}$ results in the sudden appearance of the resonance structure. The theoretical explanation of these resonances was given by Fisher, Halperin and Platzman [230], who showed that the ripplon coupling with the WS phonons with wave numbers $k \sim 1/R_{\mathrm{el}}$ determined by the size of the electron pool R_{el} is very important. Here we begin our discussion of coupled phonon–ripplon modes with analysis of the simpler case of a uniform excitation field ($k = 0$). We then extend our results so that they can be applied to the case of a nonuniform excitation field.

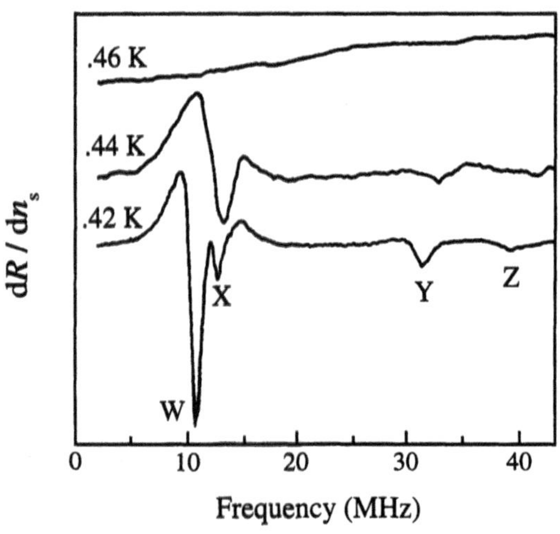

Fig. 7.6. Experimental traces displaying the sudden appearance of coupled plasmon–ripplon resonances with decreasing temperature [7]. The resonances only appear below 0.457 K, where the sheet of electrons has crystallized into a triangular lattice

We shall number the absolute values of the reciprocal lattice vectors g and the corresponding ripplon frequencies ω_g by an integer subscript ($n = 1, 2, \ldots$) as g_n and ω_n. The sequence of ripplon frequencies ω_n depends on the lattice structure and, for the hexagonal lattice, it is given by

$$\omega_n = \sqrt{\frac{\alpha}{\rho}} g_n^{3/2} , \qquad g_n = G\sqrt{n_s p_n} , \tag{7.88}$$

where

$$G = \frac{2^{3/2}\pi}{3^{1/4}} \simeq 6.75 , \qquad p_n = l^2 + m^2 + ml , \tag{7.89}$$

and $m, l = 0, \pm 1, \pm 2, \ldots$, so that $p_n = 1, 3, 4, 7, 9, \ldots$. Thus, observing the sequence of resonances at $\omega = \omega_n$, one can make out what kind of lattice structure is realized in the electron system.

When studying electron–ripplon resonances, it is convenient to introduce the dimensionless coupling parameter

$$C_n = \frac{1}{2}\left(\frac{v_n}{\omega_n}\right)^2 \mathrm{e}^{-2W_n} \sum_{|g|=g_n} 1 , \tag{7.90}$$

where $W_n = W_g$ and $v_n = v_g$, both taken at $g = g_n$. For example, $C_1 = 3\,(v_1/\omega_1)^2 \exp(-2W_1)$, but $C_4 = 6\,(v_4/\omega_4)^2 \exp(-2W_2)$ because there are 6 reciprocal lattice vectors for $n = 1, 2, 3, 5, \ldots$ ($p_n = 1, 3, 4, \ldots$) and 12 reciprocal lattice vectors for $n = 4, 7, 9, 10, 13$ ($p_n = 7, 13, 19, 21, 28$). In the region near the melting temperature $T \sim T_{\mathrm{m}}$, which corresponds to the conditions of Fig. 7.6, the coupling parameter C_n decreases fast with n due to the high-frequency Debye–Waller factor (HFDWF) $\exp(-g_n^2 \langle u_{\mathrm{f}}^2\rangle /2)$. Under the conditions of the experiment of Grimes and Adams ($n_{\mathrm{s}} = 4.5 \times 10^8\,\mathrm{cm}^{-2}$ and $T = 0.4\,\mathrm{K}$), estimates based on the self-consistent theory for $\langle u_{\mathrm{f}}^2\rangle$ to be discussed in the next section yield $C_1 \sim 10^2$ and $C_2 \sim 1 \ll C_1$. At ultra-low temperatures, the HFDWF is close to unity and the electron–ripplon coupling with $g_n > g_1$ is also important.

As mentioned above, the quantity $w(\omega)$ defined in (7.80) represents the real part of the memory function $M(\omega)$. The imaginary part of the memory function is the effective collision frequency $\mathrm{Im}\, M(\omega) \equiv \nu(\omega)$. Here we focus on the the real part, which affects the dynamics of the electron lattice. The imaginary part, responsible for dissipation, will be discussed in the next chapter. The real part enters the conductivity equation (3.48) in the combination $\omega + \mathrm{Re}\, M(\omega) = \omega + w(\omega)$. This can be conveniently rewritten as $\omega \mathcal{Z}(\omega)$, where we introduce a dimensionless response function

$$\mathcal{Z}(\omega) = 1 + \sum_{n=1}^{\infty} C_n \frac{\omega_n^2}{\omega_n^2 - \omega^2} , \tag{7.91}$$

which becomes unity in the absence of coupling with medium vibrations ($C_n \to 0$). Then, according to the memory function formalism, the conductivity of the WS can be written as

$$\sigma(\omega) = \frac{e^2 n_{\mathrm{s}}}{m} \frac{1}{\nu - \mathrm{i}\omega \mathcal{Z}(\omega)} . \tag{7.92}$$

This relation indicates that, under the conditions $\nu \ll \omega_n(1 + C_1)$, the power absorption proportional to $\mathrm{Re}\, \sigma(\omega)$ has resonances when $\mathcal{Z}(\omega) = 0$. Because the system of SEs is under the strong coupling regime ($C_1 \gg 1$), according to (7.91), there are no conductivity resonances near the primary ripplon frequency ω_1: $|\mathcal{Z}(\omega)| \gg 1$. Even if the imaginary part $\nu(\omega)$ of the memory function has a resonance structure at $\omega \to \omega_1$, it will be compensated by the corresponding antiresonance: $\mathcal{Z}(\omega) \to \infty$ at $\omega \to \omega_1$. As a matter of fact the absence of an antiresonance at $\omega \to \omega_1$ in the experiment of Grimes and Adams indicates that there is a resonance of $\nu(\omega)$ which compensates the antiresonance. The frequency dependence of the effective collision frequency $\nu(\omega)$ will be discussed in the next chapter (Sect. 8.2). Thus, in a uniform driving field, the resonance absorption occurs only at frequencies $\omega_2, \omega_3, \omega_4$ and so on [261]. This distinguishes the parallel field resonances from the resonances predicted in [260] for the perpendicular field excitation. In the latter case, resonance absorption occurs at all ripplon frequencies ω_n defined by (7.88).

In order to obtain the excitation spectrum of the coupled phonon–ripplon modes, let us assume that the driving electric field has the long-wavelength structure $E_{\parallel}^{(x)}(t) = E_0 \exp(\mathrm{i}kx - \mathrm{i}\omega t)$. If the magnetic field is weak enough not to spoil the separation of the WS phonon spectrum into the slow and fast modes, the WS conductivity tensor has the form

$$\sigma_{xx}(\omega, k) = \frac{e^2 n_{\mathrm{s}}}{m} \frac{\nu - \mathrm{i}\Phi_{\mathrm{t},k}/\omega}{\omega_{\mathrm{c}}^2 + (\nu - \mathrm{i}\Phi_{\mathrm{l},k}/\omega)(\nu - \mathrm{i}\Phi_{\mathrm{t},k}/\omega)} , \tag{7.93}$$

$$\sigma_{yx}(\omega, k) = \frac{\omega_{\mathrm{c}}}{\nu - \mathrm{i}\Phi_{\mathrm{t},k}/\omega} \sigma_{xx}(\omega, k) , \tag{7.94}$$

where

$$\Phi_{p,k}(\omega) = \omega^2 \mathcal{Z}(\omega) - \Omega_{p,k}^2 , \tag{7.95}$$

with $p = \mathrm{l}, \mathrm{t}$. These relations are easily obtained from the WS equations of motion in the absence of the electron–ripplon coupling by changing ω^2 into $\omega^2 \mathcal{Z}(\omega)$ in the corresponding expression for $\Phi_{p,k}(\omega)$. This includes the inertia of the dimple lattice in the electron equations. In the limiting case of a uniform driving electric field ($k \to 0$), (7.93) and (7.94) agree with the general structure of the conductivity tensor discussed in Chap. 3, which deals with the quantum transport framework [see (3.8)].

Under zero magnetic field ($\omega_{\mathrm{c}} \to 0$), the longitudinal driving force cannot excite the transverse mode [$(\nu - \mathrm{i}\Phi_{\mathrm{t},k}/\omega)$ drops out of (7.93)] and we have

$$\mathrm{Re}\, \sigma_{xx}(\omega) = \frac{e^2 n_{\mathrm{s}}}{m} \frac{\nu}{\nu^2 + \left[\omega^2 \mathcal{Z}(\omega) - \Omega_{\mathrm{l},k}^2\right]^2 / \omega^2} . \tag{7.96}$$

The power absorption thus has a resonance maximum at ω which satisfies the equation

$$\omega^2 - \Omega_{\mathrm{l},k}^2 + \omega^2 \sum_{n=1}^{\infty} C_n \frac{\omega_n^2}{\omega_n^2 - \omega^2} = 0 . \tag{7.97}$$

This is the secular equation of Fisher, Halperin and Platzman [230] for coupled phonon–ripplon modes. For uniform motion ($\Omega_{\mathrm{l},k} \to 0$), it transforms into the secular equation $\mathcal{Z}(\omega) = 0$ discussed above.

The presence of the plasma frequency in (7.97) crucially affects the resonance structure of $\mathrm{Re}\,\sigma(\omega)$ in the low frequency range $\omega < \omega_1$. If the wave number k of plasmons excited in the experimental cell is determined by the size of the system, $\Omega_{p,k}^2$ and $C_1\omega_1^2$ are of the same order of magnitude. In this case, the secular equation has a solution at $\omega < \omega_1$. The position of this resonance depends on the holding electric field and temperature, because of the coupling parameter $C_1(E_\perp, T) \propto V_q^2 \exp(-q^2 \langle u_{\mathrm{f}}^2 \rangle /2)$. This resonance corresponds to the excitation of the slow $(\omega < \omega_1)$ coupled phonon–ripplon mode whose spectrum transforms into $\sqrt{m_{\mathrm{e}}/M}\,\Omega_{1,k}$ in the low-frequency limit $\omega \ll \omega_1$. The schematic view of the dispersion of the longitudinal coupled modes is shown in Fig. 7.7. The excitation of the slow plasmon–ripplon modes with the two smallest wave numbers k_1 and k_2, shown in this figure by vertical lines, explains the appearance of the resonances W and X in the experiment of Grimes and Adams.

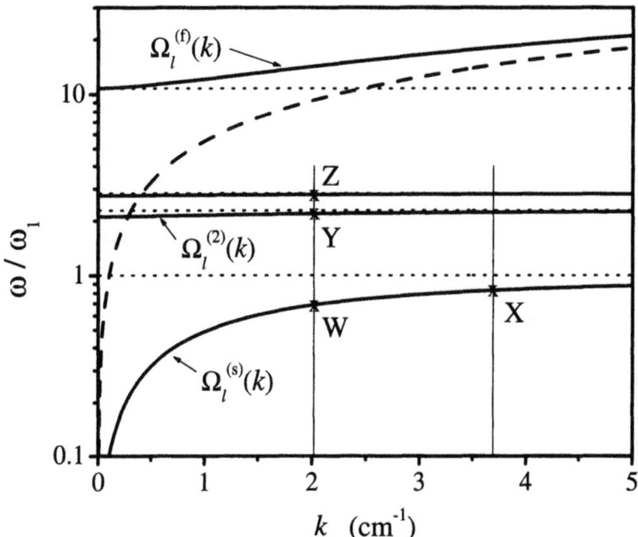

Fig. 7.7. Schematic view of the dispersion relation of the longitudinal coupled modes (*continuous curve*) after Fisher, Halperin and Platzman [230]. The *dashed curve* shows the uncoupled mode spectra and *vertical lines* represent the wave vectors excited in the experiment. Resonances are labeled as in Fig. 7.6

In the region near the melting temperature $T < T_{\mathrm{m}}$, the coupling parameter $C_2 \ll C_1$. In this case, disregarding coupling with $n > 1$, we find the analytical solution for slow mode dispersion $(\omega < \omega_1)$:

$$\Omega_{p,k}^{(\mathrm{s})} = \frac{\omega_1 \Omega_{p,k}}{\sqrt{C_1\omega_1^2 + \Omega_{p,k}^2}} . \tag{7.98}$$

Thus, in the long-wavelength limit $\Omega_{p,k} \ll \sqrt{C_1}\omega_1$, we have the simple renormalization of the phonon modes $\Omega_{p,k}^{(\mathrm{s})} \simeq \Omega_{p,k}/\sqrt{C_1}$. With an increase in the wave number k, when $\Omega_{p,k}$ becomes much larger than $\sqrt{C_1}\omega_1$, the coupled mode approaches the limiting frequency ω_1.

As shown in [255], in the presence of surface dimples, the Hamiltonian of the strongly coupled electron–ripplon system can be rearranged to a diagonal form by means of the transformation

$$Q_{p,\boldsymbol{k}} = M_{p,\boldsymbol{k}}^{(\mathrm{s})} S_{p,\boldsymbol{k}} + M_{p,\boldsymbol{k}}^{(\mathrm{f})} F_{p,\boldsymbol{k}} \,, \tag{7.99}$$

where

$$M_{p,\boldsymbol{k}}^{(\mu)} = \frac{\left|\omega_1^2 - \left(\Omega_{p,\boldsymbol{k}}^{(\mu)}\right)^2\right|}{\sqrt{\omega_{\mathrm{f}}^2\omega_1^2 + \left[\omega_1^2 - \left(\Omega_{p,\boldsymbol{k}}^{(\mu)}\right)^2\right]^2}} \,, \tag{7.100}$$

$\mu = (\mathrm{s}, \mathrm{f})$ labels modes, and $S_{p,\boldsymbol{k}}$ and $F_{p,\boldsymbol{k}}$ are normal coordinates for the slow and fast modes. An important consequence of this is that when $\omega_{\mathrm{f}} \gg \omega_1$ the contribution of slow modes to the electron mean-square displacement is proportional to $\omega_1^2 - \left(\Omega_{p,k}^{(\mathrm{s})}\right)^2$:

$$\langle u_{\mathrm{s}}^2 \rangle = \frac{\hbar}{2m_{\mathrm{e}}N_{\mathrm{e}}} \sum_{p,\boldsymbol{k}} \frac{\left[\omega_1^2 - \left(\Omega_{p,k}^{(\mathrm{s})}\right)^2\right]^2}{\left\{\left[\omega_1^2 - \left(\Omega_{p,k}^{(\mathrm{s})}\right)^2\right]^2 + \omega_{\mathrm{f}}^2\omega_1^2\right\} \Omega_{p,k}^{(\mathrm{s})}} \coth \frac{\hbar\Omega_{p,k}^{(\mathrm{s})}}{2T} \,, \tag{7.101}$$

and it therefore decreases quickly in the short-wavelength range, where the slow mode transforms into the pure ripplon mode $\omega_{\boldsymbol{g}+\boldsymbol{k}}$ (in our case, $k \ll g$). It is instructive to note that, even under zero magnetic field, (7.101) differs from the conventional form of (7.24) due to the coupling with medium vibrations. Only for long-wavelength phonons (when $\Omega_{p,k}^{(\mathrm{s})} \ll \omega_1$) does (7.101) acquire the conventional form of the mean-square displacements of the lattice constituted of charged particles with effective mass $M \simeq m_{\mathrm{e}}\omega_{\mathrm{f}}^2/\omega_1^2 \simeq C_1 m_{\mathrm{e}}$.

The secular equation (7.97) also has solutions near higher frequencies ω_n with $n \geq 2$. The dispersion is given by

$$\Omega_{p,k}^{(n)} = \omega_n \sqrt{1 - \frac{C_n\omega_n^2}{C_1\omega_1^2 + \Omega_{p,k}^2}} \,. \tag{7.102}$$

These modes are positioned closely below the characteristic frequencies ω_n due to the condition $C_n \ll C_1$. Because these modes are close to ω_n their contribution to electron vibrations is very small and they represent mainly ripplon vibrations with $\omega_{\boldsymbol{g}+\boldsymbol{k}}$. As shown in [261, 262], the inclusion of damping of ripplons γ_g can eliminate coupled modes close to ω_n with $n \geq 2$. In this case, electron–ripplon resonances with $n \geq 2$ are due to the resonance structure of $\nu(\omega)$ (see Sects. 8.2.1 and 8.2.3).

7.4.3 Self-Consistent Debye–Waller Factor

In the original phonon–ripplon coupling theory of Fisher et al. [230], the separation of electron displacements into a 'slow part' and a 'fast part' was made by introducing a small wave vector cutoff k_c for the fast modes. This parameter was adjusted to provide the right values of the high-frequency Debye–Waller (HFDWF) factor $F_g = \exp(-2W_g)$ and the frequencies $\Omega_1^{(s)}(k_1)$ and $\Omega_1^{(s)}(k_2)$ fitting the positions of the phonon–ripplon resonances W and X in the experiment of Grimes and Adams. Namaizawa [258] proposed a shear-mode self-consistency for the HFDWF based on the self-consistent field formalism. The set of three self-consistent equations found there gave a numerical solution for the HFDWF which agreed (at least qualitatively) with the experiment of Grimes and Adams.

An elaborate theory of WS vibration modes coupled to ripplons which gives a systematic treatment of the above effects was proposed by Eguiluz et al. [42]. They found that the quantitative explanation of the experimental data of Grimes and Adams does not follow from the harmonic phonon–ripplon Hamiltonian employed. A simple version of the self-consistent theory for the HFDWF with the reduced set of self-consistent equations was proposed by Monarkha and Shikin [259]. This theory gave simple analytical solutions for ω_f and $\langle u_f^2 \rangle$ which appear to be in good (even numerical) agreement with the large body of experimental data obtained to date. In this section we give a brief discussion of these approaches.

Equation (7.79) for the force acting on moving electrons is found for $\omega \ll \omega_f$ after averaging over the fast modes. At higher frequencies $\omega_1 \ll \omega \ll \omega_f$, it becomes independent of ω. It is therefore reasonable to assume that this expression also holds for the fast modes, with ω_f and $\langle u_f^2 \rangle$ to be determined in a self-consistent way. This is also supported by the approximation of the motionless dimple lattice: for fast modes $\Omega_f > \omega_f$, the dimples are still, if the system is in the strong coupling regime $\omega_f \gg \omega_1$. In turn, the static dimple profile $\xi_0(\boldsymbol{r})$ depends on $\langle u_f^2 \rangle$. This is reminiscent of the adiabatic approximation for a single polaron discussed in Chap. 6, where $\xi_0(\boldsymbol{r})$ was determined by the electron localization radius L. Following this analogy, one can expect a substantial difference between the mean-square displacement $\langle u_f^2 \rangle$ found in the harmonic approximation and that obtained in the self-consistent way.

Equation (7.87) for ω_f as a function of $\langle u_f^2 \rangle$ can be considered as the first in a set of two self-consistent equations. Another equation for $\langle u_f^2 \rangle$ as a function of ω_f can be found from a relation similar to (7.101) [255]:

$$\langle u_f^2 \rangle = \frac{\hbar}{2m_e N_e} \sum_{p,\boldsymbol{k}} \frac{\left[\left(\Omega_{p,\boldsymbol{k}}^{(f)} \right)^2 - \omega_1^2 \right]^2}{\left\{ \left[\left(\Omega_{p,\boldsymbol{k}}^{(f)} \right)^2 - \omega_1^2 \right]^2 + \omega_f^2 \omega_1^2 \right\} \Omega_{p,\boldsymbol{k}}^{(f)}} \coth \frac{\hbar \Omega_{p,\boldsymbol{k}}^{(f)}}{2T} , \quad (7.103)$$

which transforms into the conventional form (7.24) for $\omega_{\rm f} \gg \omega_1$. When $T < \hbar\Omega_{\rm t}(k_{\rm m})$, using $\Omega_{p,k}^{({\rm f})} = \sqrt{\omega_{\rm f}^2 + \Omega_{p,k}^2}$, we find

$$\langle u_{\rm f}^2 \rangle = \langle u_0^2 \rangle + u_T^2 \ln \frac{1}{1 - \exp(-\hbar\omega_{\rm f}/T)} , \tag{7.104}$$

where $\langle u_0^2 \rangle$ is the mean-square displacement for the WS above a flat surface at $T = 0$ and

$$u_T^2 = \frac{T}{2\pi m_{\rm e} n_{\rm s} c_{\rm t}^2} . \tag{7.105}$$

Equation (7.104) represents the second equation which completes the set of self-consistent equations. Here the transverse velocity $c_{\rm t}$ arises from the WS excitation spectrum over the flat surface. Therefore, the effect of phonon anharmonicity can be included as the Morf correction [245] presented in (7.22).

If the temperature is not ultra-low ($\hbar\omega_{\rm f} \ll T$), disregarding phonon coupling with ripplons of $q = g > g_1$, the solution of the self-consistent set of equations (7.87) and (7.104) can be found in the analytical form [259]

$$\omega_{\rm f}^2 \simeq 3v_1^2 \exp\left(-g_1^2 \langle u_{\rm f}^2 \rangle /2\right) \equiv \omega_{\rm f,a}^2 , \tag{7.106}$$

$$\langle u_{\rm f}^2 \rangle \simeq \frac{\langle u_0^2 \rangle + u_T^2 \ln\left(T/\sqrt{3}\hbar v_1\right)}{1 - g_1^2 u_T^2/4} \equiv \langle u_{\rm f,a}^2 \rangle , \tag{7.107}$$

where $v_1 = (n_{\rm s}/m\alpha)^{1/2} V_1$ is as introduced in (7.80).

It is interesting to compare the result given in (7.106) and (7.107) with experiment and other theories. We note first that, under the conditions of the experiment of Grimes and Adams, $n_{\rm s} = 4.55 \times 10^8\,{\rm cm}^{-2}$ and $T = 0.42\,{\rm K}$, according to [230], the perfect fit to the resonance position data would occur if the parameters $W_1 = g_1^2 \langle u_{\rm f}^2 \rangle /4$ and $F_1 = \exp(-2W_1)$ had the values $W_1 = 0.735$ and $F_1 = 0.23$. The approach of Fisher et al. [230] does not allow one to evaluate these numbers from first principles. The elaborate theory of Eguiluz et al. [42] results in the coupled mode dispersion which agrees formally with the one given by Fisher et al. [230] if the Debye–Waller function entering the dispersion equations is

$$W_g = \frac{1}{4} g^2 \zeta^2 , \qquad \zeta^2 = \frac{T}{2\pi m n_{\rm s} c_{\rm t}^2} \ln(2b) , \tag{7.108}$$

where b is a constant of the order of unity. (The analysis given in [42] suggests that $0.62 \le b \le 1.28$.) In spite of the different physical meaning, the quantity ζ^2 is reminiscent of the mean-square displacement due to thermal transverse phonons [see (7.72)]. There is also a logarithmic factor in (7.108), but its value is much smaller than the one given by the high frequency mode [see (7.107)]. The fitting value of $\ln(2b)$ should be about 8.45 in order to give $F_1 = 0.23$. Instead the theory gives $\ln(2b) \le 0.93$, which is too small for a quantitative agreement. At the values of $n_{\rm s}$ and T mentioned above, the theory of [42] gives $W_1 \leq 0.1$ and $F_1 \geq 0.815$, which disagrees with the 'perfect fit' values cited by Fisher et al. [230]. One may speculate that this strong discrepancy is caused by the use of the harmonic approximation. From Chap. 6, we know that for a single-electron polaron the self-consistent procedure results in a substantially larger value of the electron localization radius L than that given by the harmonic oscillator model [2, 37]. Therefore one may expect the self-consistent procedure for the WS coupled with surface dimples to yield larger values for $\langle u_f^2 \rangle$ and W_1 and a smaller value for F_1.

The self-consistent theory of Namaizawa [258] gives the estimates $W_1 = 0.44$ and $F_1 = 0.41$ which are indeed closer to the fitting values. The analytical solution of the self-consistent theory of [259] presented in (7.106) and (7.107) [taking into account the correction to $c_t^2(T)$ linear in T] gives the values $W_1 = 0.557$ and $F_1 = 0.328$. It should be noted that the values of T and n_s used for these estimates are close to the WS melting point and hence that the transverse sound velocity can be even smaller than the one given by (7.22). This would make F_1 closer to the 'perfect fit' number. Therefore the results given by the self-consistent approach can be considered as quite satisfactory, even for quantitative explanation of the experiment of Grimes and Adams.

Later experimental study of the slow coupled phonon–ripplon modes conducted by Sivokon' et al. [263] showed that the self-consistent DWF and $\langle u_f^2 \rangle$ defined by (7.107) give a good description of experimental data in the temperature range $0.2 \leq T \leq 0.6\,\mathrm{K}$ without any adjusting parameter. The temperature dependencies of the positions W and X of the two lowest resonant frequencies, corresponding to the excitation wave numbers $k = k_1$ and $k = k_2$, respectively, are shown in Fig. 7.8. The continuous curves were evaluated using (7.98), obtained for the slow coupled modes and valid in the limiting case $C_1 \gg C_2$. The HFDWF entering the coupling parameter C_1 was taken according to the self-consistent theory resulting in (7.106) and (7.107). Experimental signals of very small amplitude were used to avoid nonlinear effects on the positions of the electron–ripplon resonances. Figure 7.8 shows good agreement between experiment and the self-consistent approach for the high-frequency Debye–Waller factor over a broad temperature range. The deviation of data from the continuous curves observed at ultra-low temperatures $T < 0.2\,\mathrm{K}$ are reportedly caused by the reduction in the surface tension for liquid ^{4}He in the presence of a very small amount of ^{3}He. The temperature dependence of the surface tension of such a solution with concentration 5×10^{-7} was used to plot the dashed curves fitting the data at $T < 0.2\,\mathrm{K}$. Another reason for this deviation is the breakdown of the approximation $C_1 \gg C_2$ which occurs at ultra-low temperatures.

In recent experiments with the WS on the surface of liquid ^{3}He performed by Kiricheck et al. [235], the temperature dependence of the HFDWF was measured by studying the position of the high-frequency optical resonances down to ultra-low temperatures (see also the discussion below). Before dis-

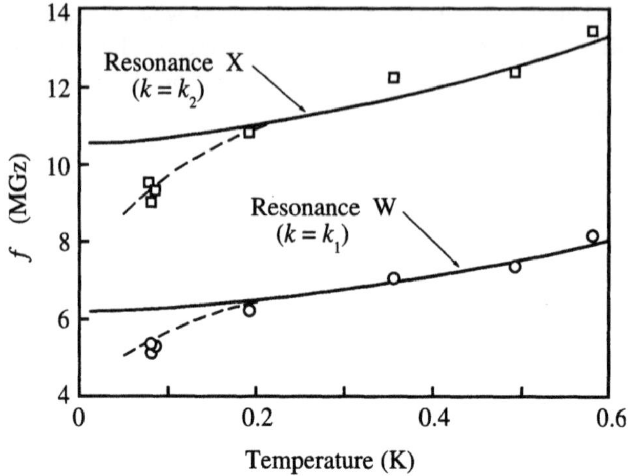

Fig. 7.8. Temperature dependence of the resonant frequencies for resonances W and X corresponding to wave numbers $k = k_1$ and $k = k_2$, respectively [263]. *Circles* and *squares*: experimental data. *Continuous curves*: theory based on the self-consistent DWF and using the inequality $C_2 \ll C_1$. The presence of a small amount of ^{3}He with concentration 5.5×10^{-7} is taken into account by the *dashed curves*

cussing these data, we need to extend the solution of the self-consistent equations for it to be applicable when $T < \hbar\omega_\mathrm{f}$.

At ultra-low temperatures $T < \hbar\omega_\mathrm{f}$, one cannot expand (7.104) in $\hbar\omega_\mathrm{f}/T$. In this case, a more accurate solution for the mean-square displacement of the fast mode normalized to the analytical solution of (7.107) $x = \langle u_\mathrm{f}^2 \rangle / \left\langle u_\mathrm{f,a}^2 \right\rangle$ can be found by rearranging the self-consistent equations in the following way [54]:

$$x = 1 - \frac{u_T^2}{\left\langle u_\mathrm{f,a}^2 \right\rangle (1 - g_1^2 u_T^2/4)} \ln \left[\frac{1 - \exp(-\hbar\omega_\mathrm{f}/T)}{\hbar\omega_\mathrm{f}/T} \sqrt{S_\mathrm{f}(x)} \right] , \tag{7.109}$$

$$\omega_\mathrm{f}(x) = \sqrt{3} v_1 \exp\left(- \frac{g_1^2 \left\langle u_\mathrm{f,a}^2 \right\rangle x}{4} \right) \sqrt{S_\mathrm{f}(x)} , \tag{7.110}$$

where $S_\mathrm{f}(x)$ is determined as the dimensionless sum

$$S_\mathrm{f}(x) = \sum_{n=1}^{\infty} \tau_n \left(\frac{V_n}{V_1} \right)^2 \exp\left[- \left(g_n^2 - g_1^2 \right) \left\langle u_\mathrm{f,a}^2 \right\rangle \frac{x}{2} \right] , \tag{7.111}$$

and $\tau_n = 1, 1, 1, 2, 1, \ldots$ is the degeneracy of $|\boldsymbol{g}| = g_n$ divided by 6. From (7.109), (7.110) and (7.111), it is easy to see that $S_\mathrm{f}(x) \to 1$ and $x \to 1$ if the temperature is high, i.e., $T \gg \hbar\omega_\mathrm{f}$, and the phonon–ripplon coupling with

$n > 1$ can be disregarded. On the other hand, at ultra-low temperatures, the second term in (7.109) is very small and we still have $x \simeq 1$, in spite of the fact that $S_{\mathrm{f}}(x)$ is much larger than unity due to the contribution from terms with $n > 1$.

The numerical solution of (7.109) and (7.110) is shown in Fig. 7.9. According to this, $\langle u_{\mathrm{f}}^2 \rangle$ is quite close to the approximate analytical result of (7.107) ($x \approx 1$ to an accuracy of 10%), which agrees with the above qualitative conclusions. The analytical solution thus has the right asymptotic behavior, not only at high temperatures but also at ultra-low temperatures, where $\left\langle u_{\mathrm{f,a}}^2 \right\rangle \simeq \langle u_0^2 \rangle$. Therefore, for simple evaluations, one may use the analytical solution $\langle u_{\mathrm{f}}^2 \rangle \simeq \left\langle u_{\mathrm{f,a}}^2 \right\rangle$ in the self-consistent HFDWF at practically all temperatures. Still, one should keep in mind that the HFDWF is close to unity at low temperatures and in the sums over n the terms with $n > 1$ cannot be disregarded. For example, the numerical solution for the characteristic frequency ω_{f} is substantially larger than the approximate solution because of the factor $\sqrt{S_{\mathrm{f}}(x)}$ which deviates from unity at ultra-low temperatures, as shown in Fig. 7.9.

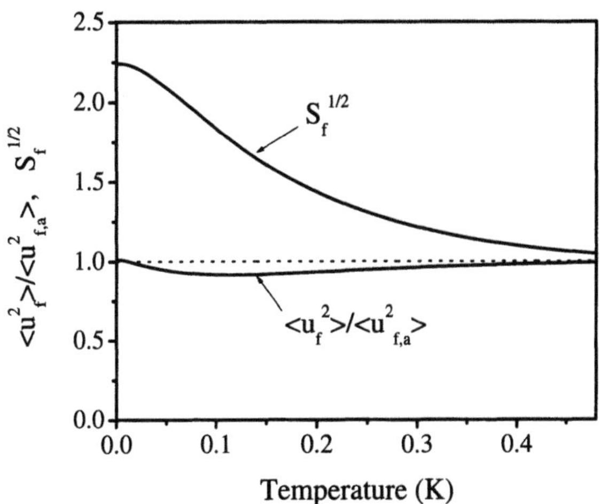

Fig. 7.9. Temperature dependence of the mean-square displacement of an electron in the fast mode, normalized by $\langle u_{\mathrm{f,a}}^2 \rangle$, and the function $\sqrt{S_{\mathrm{f}}}$ which determines the frequency ω_{f} of electron oscillations in a dimple [54]

The resonance excitation of the optical coupled phonon–ripplon mode under a weak magnetic field was studied by Kirichek et al. [235] for SEs above heavily viscous liquid ^{3}He. In this case, the position of the measured resonance is determined by (7.37) for the $\Omega_{-,k}$ mode in which the frequencies $\Omega_{p,k}$ are to be replaced by $\sqrt{\omega_{\mathrm{f}}^2 + \Omega_{p,k}^2}$. The continuous curve of Fig. 7.10

represents the exact numerical solution of the self-consistent equations (7.109) and (7.110). The dashed curve describes the optical mode frequency taking into account only the phonon–ripplon coupling with $n = 1$. Figure 7.10 shows that terms with $n \gg 1$ dominate the sum for $S_{\mathrm{f}}(x)$ and that the temperature dependence of the self-consistent HFDWF agrees well with the experimental data without any adjusting parameter. Thus the same self-consistent Debye–Waller factor gives a good description of both the low frequency and high frequency coupled phonon–ripplon modes of the Wigner solid above the liquid helium surface.

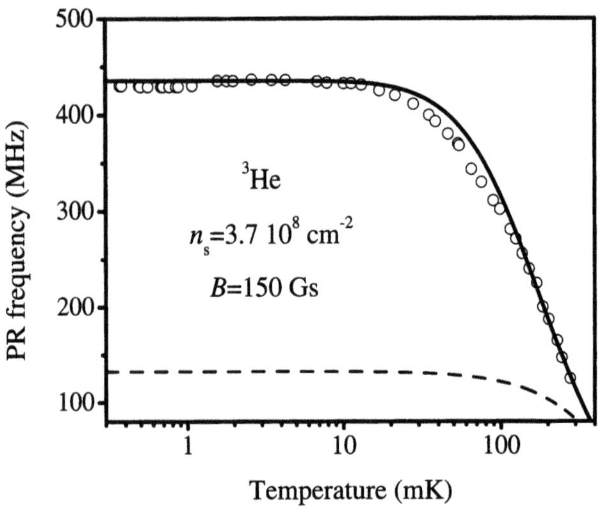

Fig. 7.10. Optical plasmon resonance frequency of the WS above liquid ^{3}He vs. temperature. *Circles*: experimental data. *Continuous curve*: theory. *Dashed curve*: contribution from the smallest reciprocal lattice vectors $\boldsymbol{g}_1$ [149]. The holding electric field is $E_{\perp} = 392\,\mathrm{V/cm} > 2\pi e n_{\mathrm{s}}$

7.4.4 Coupling Under a Magnetic Field

Consider now the phonon–ripplon coupling of the Wigner solid subject to a normal magnetic field. Firstly, we note that the long-wavelength part of the WS phonon spectrum is strongly affected even by a very weak magnetic field. As mentioned above, the fast mode changes when $\omega_{\mathrm{c}} \sim \omega_{\mathrm{f}}$, which gives $\omega_{\mathrm{c}} \sim \sqrt{C_1}\omega_1$ and $B \sim B_{\mathrm{f}} \equiv (mc/e)\sqrt{C_1}\omega_1 \sim 47\,\mathrm{Gs}$, under the conditions of the experiment of Grimes and Adams. Another characteristic magnetic field B_{s} can be introduced for the slow mode, which changes when the renormalized cyclotron frequency $\omega_{\mathrm{c}}^* \simeq \omega_{\mathrm{c}}/C_1$ becomes of the order of $\Omega_{1,k}^{(\mathrm{s})} \sim \Omega_1/\sqrt{C_1}$. Under the experimental conditions mentioned above, viz., $\Omega_{1,k} \sim C_1\omega_1$, we have $B_s = (mc/e)C_1\omega_1 \sim 447\,\mathrm{Gs}$, which is approximately one order of magnitude larger than B_f, due to the fact that $\sqrt{C_1} \approx 9.5$. Substantially higher magnetic fields $B \sim B_D = (mc/e)\Omega_p(k_m) \sim 0.5$–$1.3\,\mathrm{T}$ affect the short range part of the phonon spectrum, where phonon–ripplon coupling is unimportant. We thus have the strong inequalities $B_f \ll B_s \ll B_D$.

If the magnetic field is rather weak $B \leq B_s$, its influence on the HFDWF can be approximately disregarded. This conclusion follows from the fact that the contribution of the zero-point vibrations to $\langle u^2 \rangle$ is not affected by the low B, while the thermal vibration contribution does not change much even under high magnetic fields, according to (7.65). In addition, we note that the magnetic field $B \sim B_s$ affects the fast modes only at small k, while the main contribution of thermal vibrations to $\langle u_f^2 \rangle$ comes from phonons with $k \sim \omega_f/c_t \sim 700\,\mathrm{cm}^{-1}$, whose spectrum is practically unchanged under such weak magnetic fields.

The spectrum of the fast phonon modes can be found from (7.37) by replacing $\Omega_{p,k}$ for the WS on a flat surface by $\sqrt{\omega_f^2 + \Omega_{p,k}^2}$. Typical changes in the WS fast phonon modes induced by the magnetic field are shown in Fig. 7.11. In the limit $\omega_c \gg \omega_f$, the spectrum of the lowest new mode, viz.,

$$(\Omega_{-,k}^{(f)})^2 = \frac{\left(\omega_f^2 + \Omega_{l,k}^2\right)\left(\omega_f^2 + \Omega_{t,k}^2\right)}{\omega_c^2 + \Omega_{l,k}^2}\,, \tag{7.112}$$

decreases with B. This eventually spoils the separation of electron vibrations into the slow and fast modes, when $\Omega_{-,k} \sim \omega_1$, unless the temperature is low enough to make $\langle u_f^2 \rangle \simeq \langle u_0^2 \rangle$. In the limit of extremely high magnetic fields, one can define $\Omega_{+,k}^{(f)}$ as the only fast mode.

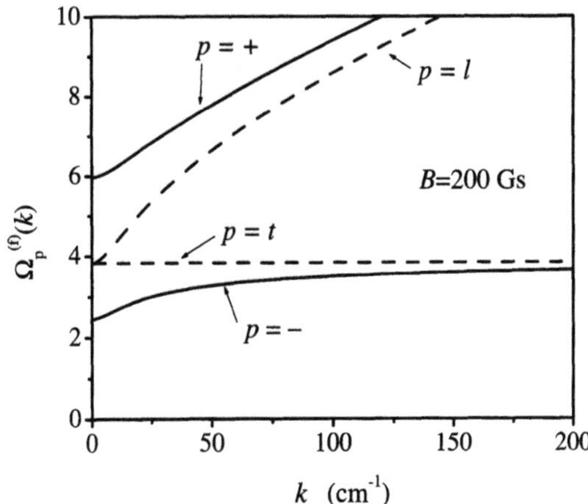

Fig. 7.11. Typical changes induced by a magnetic field in the fast mode dispersions in the long-wavelength range

Regarding the slow phonon modes, one can easily describe their spectrum only in the zero temperature limit. The dispersion equation for these modes can be found from the denominator of (7.93):

$$\left[\mathcal{Z}(\omega)\omega^2 - \Omega_{1,k}^2\right]\left[\mathcal{Z}(\omega)\omega^2 - \Omega_{t,k}^2\right] = \omega^2\omega_c^2 \,. \qquad (7.113)$$

The qualitative behavior of the slow phonon spectrum can be understood by disregarding phonon–ripplon coupling with $n > 1$. In this case, the secular equation can be rewritten in the form

$$\left[\omega^2 - \left(\Omega_{1,k}^{(s)}\right)^2\right]\left[\omega^2 - \left(\Omega_{t,k}^{(s)}\right)^2\right] = \omega^2\omega_c^2\left[\frac{(\omega_1^2 - \omega^2)\Omega_{1,k}^{(s)}\Omega_{t,k}^{(s)}}{\omega_1^2\Omega_{1,k}^2\Omega_{t,k}^2}\right]^2 . \qquad (7.114)$$

This equation shows that for $\omega \ll \omega_1$, when $\Omega_{p,k}^{(s)} \simeq \Omega_{p,k}/\sqrt{C_1}$, the magnetic field affects the WS phonon spectrum in the same way as it does on a flat surface. The important difference is that the right-hand side of (7.114) becomes $\omega^2(\omega_c^*)^2$, where $\omega_c^* = \omega_c/C_1 = eB/Mc$ is the renormalized cyclotron frequency. As a result, the polaronic effect which reduces the spectrum of the longitudinal and transverse phonons by the factor $\sqrt{C_1}$ does not significantly reduce the hybrid mode $\Omega_{-,k} \simeq \Omega_{1,k}\Omega_{t,k}/\omega_c$ because in this case the renormalizations $\Omega_{p,k} \to \Omega_{p,k}/\sqrt{C_1}$ ($p = 1, t$) and $\omega_c \to \omega_c/C_1$ compensate each other. In the limiting case $C_1 \gg 1$, we find

$$\Omega_{-,k}^{(s)} = \frac{\Omega_{1,k}\Omega_{t,k}}{\sqrt{\omega_c^2 + C_1\Omega_{1,k}^2}} \,, \qquad (7.115)$$

which shows that substantially stronger magnetic fields are needed to affect the slow transverse mode ($\omega_c \sim \sqrt{C_1}\Omega_{1,k}$) than the transverse mode on a flat surface ($\omega_c \sim \Omega_{1,k}$).

The right-hand side of (7.114) decreases significantly when $\omega \to \omega_1$ due to the factor $(\omega_1^2 - \omega^2)^2$. This means that the magnetic field cannot affect the part of the phonon–ripplon spectrum close to ω_1. This agrees with the above conclusion that, as $\omega \to \omega_n$, the coupled phonon–ripplon mode transforms into the pure ripplon mode. In the limit of strong magnetic fields, the dispersion of the mode $\Omega_{-,k}^{(s)}$ coupled to surface vibrations is close to the unperturbed spectrum $\Omega_{-,k} \simeq \Omega_{1,k}\Omega_{t,k}/\omega_c$, with the exception of the regions near $\Omega_{-,k} = \omega_n$, where it deviates strongly and transforms into the pure ripplon modes.

7.5 Dynamical Structure Factor

Following van Hove [264], the dynamical structure factor (DSF) $S(q,\omega)$ is usually introduced when studying the cross-section for the scattering of particle fluxes or X rays from liquids or solids. For example, the application of thermal neutron scattering has played a very important role in understanding the physical properties of matter from about 1950. In this book we have frequently emphasized the importance of the electron DSF for accurate description of quantum transport properties in highly correlated Coulomb liquids and solids. According to the quantum transport framework discussed in Chap. 3, for highly correlated electrons, the effective collision frequency ν as a function of ω and the CR linewidth can be expressed in terms of $S(q,\omega)$. Equations (3.55) and (3.57) represent typical examples of this relation. Therefore, in order to understand the transport properties of the Wigner solid, we have to pay more attention to its fundamental correlation function.

According to the definitions $S(q,\omega) = N_{\mathrm{e}}^{-1} \langle n_{\boldsymbol{q}}(t) n_{-\boldsymbol{q}}(0) \rangle$ and $n_{\boldsymbol{q}} = \sum_m \exp(-\mathrm{i}\boldsymbol{q}\cdot\boldsymbol{r}_m)$, the DSF of the 2D WS can be written as

$$S(q,\omega) = \frac{1}{N_{\mathrm{e}}} \sum_{\boldsymbol{l},\boldsymbol{m}} \mathrm{e}^{\mathrm{i}\boldsymbol{q}\cdot(\boldsymbol{R}_m - \boldsymbol{R}_l)} \int_{-\infty}^{\infty} \mathrm{e}^{\mathrm{i}\omega t} \left\langle \mathrm{e}^{-\mathrm{i}\boldsymbol{q}\cdot\boldsymbol{u}_l(t)} \mathrm{e}^{\mathrm{i}\boldsymbol{q}\cdot\boldsymbol{u}_m(0)} \right\rangle \mathrm{d}t\,, \qquad (7.116)$$

where we have used the notation $\boldsymbol{r}_m(t) = \boldsymbol{R}_m + \boldsymbol{u}_m(t)$, $\boldsymbol{R}_m$ is the 2D lattice site vector, and $\boldsymbol{u}_m$ is the displacement operator, which is a linear function of the creation and destruction operators. The conventional way of evaluating (7.116) is to use the identity of (2.29) established for non-commuting operators A and B. In the case considered here, $A = -\mathrm{i}qu_l^{(q)}(t)$ and $B = \mathrm{i}qu^{(q)}(0)$, where the superscript (q) denotes projection onto the direction of the vector $\boldsymbol{q}$. We can now use the Bloch identity. This procedure is analogous to evaluating the self-correlation function of an electron in the harmonic oscillatory potential given in Sect. 2.1.3 [see (2.33) and (2.35)].

Summarizing the transformations used, we can rewrite the correlation function entering the definition of the DSF in the following way:

$$\left\langle \mathrm{e}^{-\mathrm{i}\boldsymbol{q}\cdot\boldsymbol{u}_l(t)} \mathrm{e}^{\mathrm{i}\boldsymbol{q}\cdot\boldsymbol{u}_m(0)} \right\rangle = \exp\left[-q^2 \left\langle \left(u_l^{(q)} \right)^2 \right\rangle + q^2 \left\langle u_l^{(q)}(t) u_m^{(q)}(0) \right\rangle \right], \qquad (7.117)$$

which is similar to the relation (2.31) and (2.32). The first term in the exponent represents the conventional Debye–Waller factor (DWF). In two dimensions $\langle (u_l^{(q)})^2 \rangle = \langle u_l^2 \rangle /2$. Still, in this case, one should not rush into extracting the DWF, because of the divergence in the mean-square displacement.

For the displacement operator of (7.57), the DSF can be written in the form

$$S(q,\omega) = \sum_{\boldsymbol{l}} \mathrm{e}^{-\mathrm{i}\boldsymbol{q}\cdot\boldsymbol{R}_l} \int_{-\infty}^{\infty} \exp\left[\mathrm{i}\omega t - h_q(\boldsymbol{l},t)\right] \mathrm{d}t\,, \qquad (7.118)$$

where

$$h_q(\boldsymbol{l},t) = 2q^2 \sum_p \left[w_p(0,0) - w_p(\boldsymbol{l},t) \right], \qquad (7.119)$$

$$w_p(\boldsymbol{l},t) = \frac{\hbar}{4N_\mathrm{e}m}\sum_{\boldsymbol{k}} \frac{\left|E^{(q)}_{p,\boldsymbol{k}}\right|^2}{\Omega_{p,k}} \left[(n_{p,k}+1)\mathrm{e}^{\mathrm{i}(\boldsymbol{k}\cdot\boldsymbol{R}_l - \Omega_{p,k}t)} + n_{p,k}\mathrm{e}^{-\mathrm{i}(\boldsymbol{k}\cdot\boldsymbol{R}_l - \Omega_{p,k}t)}\right] ,$$

$n_{p,k}$ is the WS phonon distribution function and the index p can be l or t (or + or −, if the system is under a magnetic field). This equation shows no unusual singularity in the DSF of an infinite 2D solid because the thermal vibration contribution to $w_p(0,0) - w_p(\boldsymbol{l},t)$ contains the factor $1-\cos(\boldsymbol{k}\cdot\boldsymbol{R}_l - \Omega_{p,k}t)$, which cuts off the divergence. As we shall see in the following, the properties of the correlation functions in two dimensions nevertheless strongly affect the Wigner solid DSF.

In order to simplify evaluations, it is a convention to use the isotropic Debye model with $c_l = c_\mathrm{t}$. In this case $\sum_p \left|E^{(q)}_{p,\boldsymbol{k}}\right|^2 = 1$. For the Wigner solid, one has to pay more attention to this point. Note first that the 2D WS has only one acoustical mode ($p = \mathrm{t}$). Longitudinal phonons with $\Omega_{p,k} \propto \sqrt{k}$ do not contribute to the divergence of the DSF and can usually be disregarded, with the exception of the zero-point vibration term. Even for the electron solid with screened Coulomb interaction, c_l is much larger than the transverse sound velocity c_t, and the contribution from thermal longitudinal phonons is very small. Nevertheless, in most cases of interest, one can replace $\left|E^{(q)}_{p,\boldsymbol{k}}\right|^2$ in (7.119) by its angular average $\left|\boldsymbol{E}_{p,\boldsymbol{k}}\right|^2/2$ when $k \ll g$. In the presence of a strong magnetic field, one has to take into account the fact that the polarization vectors $\boldsymbol{E}_{p,\boldsymbol{k}}$ are not orthonormalized and depend on B, as discussed in Sect. 7.3.

7.5.1 Conventional Approximations

Two important approximations are commonly used in neutron scattering theory. The high temperature approximation assumes that for $T > T_\mathrm{D}$, where T_D is the WS Debye temperature, the function $2q^2 w_p(0,0) \gg 1$ and therefore that the main contribution to the integral of (7.118) comes from the region near $t = 0$, where the difference $[w_p(0,0) - w_p(\boldsymbol{l},t)]$ is small. One can therefore expand $w_p(\boldsymbol{l},t)$ in powers of $\Omega_{p,k}t$ up to quadratic terms and use the incoherent approximation ($\boldsymbol{l} = 0$), which gives

$$S(q,\omega) = \int_{-\infty}^{\infty} \mathrm{d}t \exp\left[\mathrm{i}\frac{(\hbar\omega - \varepsilon_q)}{\hbar}t - \frac{\varepsilon_q K_\mathrm{e}}{\hbar^2}t^2\right] , \qquad (7.120)$$

where

$$K_\mathrm{e} = (2N_\mathrm{e})^{-1}\sum_{p,\boldsymbol{k}} \hbar\Omega_{p,k}\left(n_{p,k} + 1/2\right) \qquad (7.121)$$

is the mean kinetic energy per electron and ε_q is the single-electron spectrum in the absence of the magnetic field. The evaluation of the integral of (7.120) yields a Gaussian form similar to the one for nondegenerate 2D electrons given in (2.22). The difference is that the electron temperature T_e is replaced by K_e, which approaches T_e in the limit $T_e \gg T_D$. Thus at high temperatures and under zero magnetic field, the WS and nondegenerate electron gas have the same DSF.

The low temperature approximation for the DSF of a 3D solid assumes that $2q^2 w_p(0,0)$ is small. In this case, the Debye–Waller factor

$$\exp\left[-2q^2 \sum_p w_p(0,0)\right]$$

is usually left as it is, while $\exp[2q^2 \sum_p w_p(\boldsymbol{l},t)]$ is expanded to give elastic terms, one-phonon terms, two-phonon terms, and so on [80]:

$$S = S^{(\mathrm{elas})} + S^{(\mathrm{1ph})} + S^{(\mathrm{2ph})} + \cdots .$$

For example, in the case considered here, we find

$$\begin{aligned} S(\boldsymbol{q},\omega) &= 2\pi \mathrm{e}^{-2W_q}\delta(\omega) N_e \sum_{\boldsymbol{g}} \delta_{\boldsymbol{q},\boldsymbol{g}} \\ &+ \frac{\pi\hbar}{m_e}\mathrm{e}^{-2W_q} \sum_{p,\boldsymbol{k}} \frac{|\boldsymbol{q}\cdot\boldsymbol{E}_{p,\boldsymbol{k}}|^2}{\Omega_{p,k}} \left\{ (n_{p,k}+1)\delta(\omega-\Omega_{p,k}) \sum_{\boldsymbol{g}} \delta_{\boldsymbol{q}-\boldsymbol{k},\boldsymbol{g}} \right. \\ &\left. + n_{p,k}\delta(\omega+\Omega_{p,k}) \sum_{\boldsymbol{g}} \delta_{\boldsymbol{q}+\boldsymbol{k},\boldsymbol{g}} \right\} + \cdots , \end{aligned} \quad (7.122)$$

where $W_q = q^2 \sum_p w_p(0,0)$. The first term of this expansion is usually called the elastic term, which implies that no phonons of the target are excited (in our case the WS lattice represents the moving target). At the same time, the scattering process can be inelastic because of excitation or absorption of medium vibration quanta according to the general form (3.57) of the effective collision frequency. The second term of (7.122) describes one-phonon processes of emission and absorption of the target excitations, and it contains the energy conservation functions $\delta(\omega \mp \Omega_{p,k})$.

7.5.2 Correlations in Two Dimensions

In two dimensions, the low temperature approach encounters a problem because of the logarithmic divergence in the mean-square displacement $\langle u_{\boldsymbol{n}}^2 \rangle$. As noted above, at a fixed $\boldsymbol{R}_l$, the difference $w_p(0,0) - w_p(\boldsymbol{l},t)$ entering (7.118) has no logarithmic dependence on the size of the system, in accordance with [42], while the separate parts [such as $w_p(0,0)$] do so, due to thermal vibrations. Still, this difference depends on $\boldsymbol{R}_l$ in such a way that it crucially affects the DSF, which is the sum over all lattice sites. The problem thus concerns the proper treatment of low frequency excitations of the 2D solid.

For interface electrons, we must treat them differently in the regimes of weak and strong coupling with medium vibrations.

In the weak coupling regime, the influence of medium vibrations on the electron DSF can be disregarded. Following Janccovici [265], and Imry and Gunther [266], let us first analyze the behavior of the static structure factor defined by

$$\tilde{S}(\boldsymbol{q}, t=0) = N_{\mathrm{e}}^{-1} \langle n_{\boldsymbol{q}} n_{-\boldsymbol{q}} \rangle = \sum_{\boldsymbol{l}} \mathrm{e}^{-\mathrm{i}\boldsymbol{q}\cdot\boldsymbol{R}_l - h_{\boldsymbol{q}}(\boldsymbol{l},0)} , \tag{7.123}$$

with $h_{\boldsymbol{q}}(\boldsymbol{l}, t)$ as defined in (7.119). In three dimensions, this quantity has sharp peaks at $\boldsymbol{q} = \boldsymbol{g}$, proportional to N_{e}. The same holds in two dimensions if $T = 0$. At finite temperatures, the low-frequency oscillations affect these peaks. Consider first $\tilde{S}(\boldsymbol{q}, t=0)$ at $\boldsymbol{q} = \boldsymbol{g}$. In this case, for the isotropic Debye model, the contribution of thermal vibrations can be evaluated directly:

$$h_g(\boldsymbol{l}, 0) \simeq \frac{2T}{T_g} \int_0^{\omega_{\mathrm{m}}} \frac{\mathrm{d}\omega}{\omega} \left[1 - J_0(\omega R_l / c_{\mathrm{t}})\right] , \tag{7.124}$$

where

$$T_g = \frac{8\pi n_{\mathrm{s}} m c_{\mathrm{t}}^2}{g^2} \tag{7.125}$$

is the characteristic temperature depending strongly on the reciprocal lattice vector, $\omega_{\mathrm{m}} = \min(c_{\mathrm{t}} k_{\mathrm{m}}, T/\hbar)$, and $k_{\mathrm{m}} = \sqrt{4\pi n_{\mathrm{s}}}$. We have taken into account the fact that, for the 2D Wigner solid, there is only one acoustical mode, which makes the characteristic temperature T_g two times larger than that found in [265, 266] for typical 2D solids.

Integrating (7.124) by parts, we find the following asymptote valid at large distances ($R_l \to \infty$):

$$h_g(l, 0) \simeq \frac{2T}{T_g} \ln \frac{\gamma_0 \omega_{\mathrm{m}} R_l}{2 c_{\mathrm{t}}} , \tag{7.126}$$

where $\gamma_0 \simeq 1.781$ is Euler's constant. The argument of the logarithmic function contains the ratio of upper (ω_{m}) and lower (c_{t}/R_l) cutoffs of the integral $\int \mathrm{d}\omega/\omega$, which are quite clear from (7.124). Substituting this asymptote into the structure factor gives

$$\tilde{S}(\boldsymbol{g}, t=0) \simeq \left(\frac{2c_{\mathrm{t}}}{\gamma_0 \omega_{\mathrm{m}}}\right)^{2T/T_g} \sum_{l} \frac{1}{R_l^{2T/T_g}} . \tag{7.127}$$

This sum diverges when $T < T_g$ and $N_{\mathrm{e}} \to \infty$, which is the reason for introducing the characteristic temperature T_g. We then replace the sum $\sum_l$ by an integral $n_{\mathrm{s}} \int \mathrm{d}^2 \boldsymbol{R}_l$, which yields [266]

$$\tilde{S}(\boldsymbol{g}, t=0) \simeq \gamma_0^{-2T/T_g} \frac{N_{\mathrm{e}}^{1-T/T_g} - 1}{1 - T/T_g} . \tag{7.128}$$

It is clear that at $T = 0$ we find $\tilde{S}(\boldsymbol{g}, t = 0) = N_e$. At finite temperatures, Bragg peaks are lower. The characteristic temperature T_g is a critical point which depends strongly on the reciprocal lattice vector g. Comparing T_g given by (7.125) and the melting temperature T_m according to the dislocation melting criterion (7.21), we conclude that the singularity at $T \to T_g$ is not really important for the Wigner solid because $T_{g_1} = 6\,T_m$. Still, T_g decreases rapidly with g.

Consider now the broadening of Bragg peaks induced by long-wavelength fluctuations. For this purpose, we assume that $\boldsymbol{q}$ is close to one of the reciprocal lattice vectors $\boldsymbol{g}$. Then the thermal vibration contribution to $h_q(\boldsymbol{l}, t)$ can be written as

$$h_q(\boldsymbol{l}, t) \simeq \frac{2T}{T_q}\frac{1}{2\pi}\int_0^{2\pi} \mathrm{d}\varphi \int_0^{\omega_m/c_t} \frac{\mathrm{d}k}{k}\left[1 - \cos(kR_l\cos\varphi - c_t k t)\right] . \quad (7.129)$$

In the static limit ($t = 0$), the integral of this equation can be estimated as the logarithm of the ratio of the upper and lower cutoffs, similar to (7.126). When the main contribution to $\tilde{S}(\boldsymbol{q}, t = 0)$ comes from large R_l ($|\boldsymbol{q} - \boldsymbol{g}| \ll g$), we have

$$\tilde{S}(\boldsymbol{q}, t = 0) \sim 2\pi n_s \sum_{\boldsymbol{g}} \left(\frac{\omega_m}{c_t}\right)^{-\alpha_g(T)} \int_0^{R_A} R^{1-\alpha_g(T)} J_0(|\boldsymbol{q} - \boldsymbol{g}|\, R)\mathrm{d}R , \quad (7.130)$$

where $\alpha_g(T) = 2T/T_g$ and $R_A = \sqrt{S_A/\pi}$. Obviously, we have $\tilde{S}(\boldsymbol{q}, t = 0) = N_e \sum_{\boldsymbol{g}} \delta_{\boldsymbol{q},\boldsymbol{g}}$ if $T = 0$:

$$\tilde{S}(\boldsymbol{q}, t = 0) \sim 2N_e \sum_{\boldsymbol{g}} \frac{J_1(|\boldsymbol{q} - \boldsymbol{g}|\, R_A)}{|\boldsymbol{q} - \boldsymbol{g}|\, R_A} , \quad (7.131)$$

because $J(x) \simeq x/2$ at small values of its argument. For $|\boldsymbol{q} - \boldsymbol{g}|\, R_A \gg 1$, the ratio $J_1(x)/x$ of (7.131), proportional to

$$\frac{1}{(|\boldsymbol{q} - \boldsymbol{g}|\, R_A)^{3/2}} \cos\left(|\boldsymbol{q} - \boldsymbol{g}|\, R_A - 3\pi/4\right) , \quad (7.132)$$

becomes very small.

When $T/T_g > 1/4$ ($\alpha_g > 1/2$), one can replace the upper limit of the integral by infinity, obtaining

$$\tilde{S}(\boldsymbol{q}, t = 0) \propto \sum_{\boldsymbol{g}} \frac{1}{|\boldsymbol{q} - \boldsymbol{g}|^{2-\alpha_g(T)}} . \quad (7.133)$$

This smearing of the Bragg peaks is consistent with the well known power-law decrease in the correlation function:

$$\langle n(r)n(0)\rangle - n_s^2 \propto \frac{1}{r^{\alpha_g(T)}} \cos(\boldsymbol{g} \cdot \boldsymbol{r}) , \quad (7.134)$$

where g corresponds to the smallest value of $\alpha_g(T)$. (In the case considered here, it is the smallest reciprocal lattice vector.)

When $T/T_g < 1/4$ ($\alpha_g < 1/2$), which is more suitable for the Wigner solid, the integral in (7.130) diverges as $R_A \to \infty$. Considering the asymptotic behavior of $J_0(x)$ at large values of its argument, one can see that the structure factor contains an oscillating factor $\cos(|\boldsymbol{q}-\boldsymbol{g}|\, R_A - 3\pi/4)$ [266], similar to the one found for $T = 0$ in (7.132).

To describe the related smearing of the $\delta(\omega)$-shaped spikes in the elastic term of the low temperature asymptote for the DSF (7.122), Dykman and Rubo [267] proposed the approximation

$$h_q(\boldsymbol{l},t) \approx \frac{2T}{T_q} \ln\left(\omega_{\mathrm{m}}\sqrt{t^2 + bR_l^2/c_{\mathrm{t}}^2}\right) + q^2\left\langle u_0^2\right\rangle/2\,, \qquad (7.135)$$

where the last term is the contribution from zero-point vibrations and b is ~ 1. This form of $h_q(\boldsymbol{l},t)$ follows qualitatively from (7.129). For this approximation, the corresponding spikes have a non-Lorentzian shape. We shall discuss these results in Sect. 8.2.1, which deals with the resonance structure of the electron collision rate.

7.5.3 Strong Coupling and Consistency Requirements

The consideration of density fluctuations in the 2D electron solid given above is valid for the regime of weak coupling with medium excitations (ripplons). It is a much more complicated task to find the electron DSF in the strong coupling regime, which is actually realized for SEs on liquid helium. Firstly, it should be noted that we are interested in the electron correlation function which describes electron conductivity of the WS strongly coupled to interface excitations. We then recall that the real and imaginary parts of the conductivity relaxation kernel [$w(\omega)$ and $\nu(\omega)$] are not independent, but are related via the analytical properties of the memory function. The latter can be seen even from the simple expressions for $w(\omega)$ and $\nu(\omega)$ given in (7.80) and (7.81). The Debye–Waller factors $\exp(-2W_g)$ entering the equations for $w(\omega)$ and $\nu(\omega)$ should be the same quantity. This means that the density correlation function $S(\boldsymbol{q},\omega)$ to be used in the corresponding equation for the effective collision frequency should be consistent with the real part of the memory function $w(\omega)$, which determines the secular equation for coupled phonon–ripplon modes [see (7.97)].

In the last section [see (7.85)], we found that the real and imaginary parts of the conductivity relaxation kernel which follows from the memory function formalism coincide with the result of direct evaluation of the medium response force, if the WS dynamical structure factor is approximated by the first term of the low temperature expansion of (7.122) with the self-consistent high-frequency Debye–Waller factor $\exp(-q^2\left\langle u_{\mathrm{f}}^2\right\rangle/2)$ standing for $\exp(-2W_q)$. Thus, in the framework of the conductivity treatment consistent with the secular equation for the coupled modes given above, one should exclude the slow coupled modes from the WS DWF.

The physical reason for the exclusion of the slow modes is that the most important part of the electron interaction with medium excitations is included in the Hamiltonian of the slow coupled modes and in the secular equation which caused the renormalization of the WS phonon spectrum. As a result, the ripplon scattering of slow WS phonons should be reduced. Therefore, the density correlation function discussed in the following and still denoted by $S(q,\omega)$ is not the whole DSF of the Wigner solid but only a part of it to be used in the equation for the conductivity relaxation kernel of the strong coupling theory. The self-consistent treatment also implies that the replacement

$$\exp(-2W_q) \rightarrow \exp(-q^2 \langle u_{\mathrm{f}}^2 \rangle /2)$$

should be made in all terms of the low temperature expansion of (7.122). This is equally important for elastic, one-phonon and multi-phonon terms. For renormalized modes $[\mu = (\mathrm{s},\mathrm{f})]$, the polarization vector has the coupling factor:

$$\boldsymbol{E}_{p,\boldsymbol{k}}^{(\mu)} = M_{p,\boldsymbol{k}}^{(\mu)} \boldsymbol{e}_{p,\boldsymbol{k}} \,, \tag{7.136}$$

where $\boldsymbol{e}_{p,\boldsymbol{k}}$ are the polarization vectors of longitudinal ($p = \mathrm{l}$) and transverse ($p = \mathrm{t}$) modes of the WS solid on the rigid, flat substrate. The factor $M_{p,\boldsymbol{k}}^{(\mu)}$ was defined previously in (7.100). Luckily, for fast modes ($\mu = \mathrm{f}$) under strong coupling conditions, this factor is very close to unity.

In the self-consistent treatment discussed above, the breakdown of the low temperature approximation for the electron DSF given in (7.122) can be caused by an increase in $\langle u_{\mathrm{f}}^2 \rangle$ with temperature, which means that multi-phonon processes involving two or more WS phonons become important. According to [80], this can be taken into account in the framework of the incoherent approximation, since for multi-phonon terms the conservation of energy and crystal momentum are not as restrictive as for one-phonon terms. In the static approximation, which disregards terms $\sum_s \Omega_s$ in the energy conservation δ-functions, the phonon series is heavily compensated and a reasonable low-temperature approximation in two dimensions is therefore obtained by replacing $\langle u^2 \rangle$ entering the DWF of (7.122) by the zero-point term $\langle u_0^2 \rangle$. In different ways, this approximation was actually used in [268, 269] to study WS phonon damping. The replacement stated above also extends (7.122) into a higher temperature range, where the one-phonon terms of (7.122) with the usual DWF die out due to the thermal vibration contribution. Still, these results have no correct asymptotic behavior at high energies and one should use a more sophisticated method for the evaluation of multi-phonon terms.

Another way of treating the multi-phonon terms, proposed for neutron scattering [270], is to represent them in a form which has a proper asymptotic behavior at high energies. Because the incoherent approximation is poor for one-phonon terms, following [80], we apply it only to multi-phonon terms. Then we can write

$$S(q,t) = \sum_{\mathbf{m}} \mathrm{e}^{-\mathrm{i}\boldsymbol{q}\cdot\boldsymbol{r}_m - 2W_q} \left[1 + 2q^2 w(\boldsymbol{m},t)\right] + \mathrm{e}^{-2W_q} \sum_{n=2}^{\infty} \frac{[2q^2 w(0,t)]^n}{n!} , \tag{7.137}$$

where $w(\boldsymbol{m},t) = \sum_p w_p(\boldsymbol{m},t)$. Expanding terms with $n \geq 2$ in powers of $\Omega_{p,\boldsymbol{k}}t$ as described above, we finally obtain the multi-phonon correction to the electron DSF in the form

$$\delta S(q,\omega) \simeq \tau_0 \sqrt{\frac{2\pi}{\Delta}} \mathrm{e}^{-2W_q} \sum_{n=2}^{\infty} \frac{(2W_q)^n}{n!\sqrt{n}} \exp\left[-\frac{(\omega\tau_0 - n)^2}{2n\Delta}\right] , \tag{7.138}$$

where we use the notation

$$\Delta = \frac{2K_{\mathrm{e}}\tau_0}{\hbar} - 1 , \qquad \tau_0 = \frac{m\left\langle u_{\mathrm{f}}^2\right\rangle}{\hbar} . \tag{7.139}$$

The multi-phonon terms are represented as Gaussian functions of ω. We shall use this multi-phonon correction to the electron DSF in the following chapter when studying the high-frequency conductivity of the WS.

7.5.4 High Magnetic Field Case

In the presence of a high magnetic field, when the phonon modes of the WS are widely spaced from each other on the frequency axis ($\Omega_{+,\boldsymbol{k}} \gg \Omega_{-,\boldsymbol{k}}$), it is possible to combine the two extreme approximations discussed above and to avoid the divergence of $w_p(0,0)$. For the low frequency mode $\Omega_{-,\boldsymbol{k}}$, one can use the high temperature approximation ($T \gg \hbar\Omega_{-,\boldsymbol{k}}$) expanding $w_-(0,t)$ in powers of $\Omega_{-,\boldsymbol{k}}t$, while for the high frequency mode $\Omega_{+,\boldsymbol{k}} > \omega_{\mathrm{c}} \gg T/\hbar$, it is possible to use the low temperature approximation (the Debye–Waller factor of the $\Omega_{+,\boldsymbol{k}}$ mode has no divergence owing to the limiting frequency ω_{c}) and expand the factor $\exp[2q^2 w_+(\boldsymbol{n},t)]$ in powers of $2q^2 w_+(\boldsymbol{n},t)$. The latter expansion would give energy conservation functions $\delta(\omega - \Sigma\Omega_{+,\boldsymbol{k}}) \simeq \delta(\omega - N\omega_{\mathrm{c}})$ for emission of N high frequency WS phonons, if the $\Omega_{-,\boldsymbol{k}}$ mode is disregarded.

In the limiting case considered here, the low frequency mode provides the following terms in the exponent of (7.118):

$$2q^2\left[w_-(0,t) - w_-(0,0)\right] \simeq -x_q \frac{\Gamma_{\mathrm{C}}^2}{4\hbar^2}\left(t^2 + \frac{\mathrm{i}t\hbar}{T}\right) , \tag{7.140}$$

$$\Gamma_{\mathrm{C}}^2 = \frac{\hbar\omega_{\mathrm{c}}}{N_{\mathrm{e}}} \sum_{\boldsymbol{k}} \left|\boldsymbol{E}_{-,\boldsymbol{k}}\right|^2 \hbar\Omega_{-,\boldsymbol{k}} \left(2n_{-,\boldsymbol{k}} + 1\right) , \tag{7.141}$$

where $x_q = q^2 l^2/2$ is a dimensionless parameter, and we have used the isotropic model approximation. For $\omega_{\mathrm{c}} \gg \Omega_{\mathrm{l},\boldsymbol{k}}$, we have $\Gamma_{\mathrm{C}} = \sqrt{2} e E_{\mathrm{f}}^{(0)} l$, with

$$eE_{\mathrm{f}}^{(0)} = \sqrt{\frac{mT}{N_{\mathrm{e}}} \sum_{\boldsymbol{k}} \left(\Omega_{\mathrm{l},k}^2 + \Omega_{\mathrm{t},k}^2 \right)} \ . \tag{7.142}$$

The quantity $E_{\mathrm{f}}^{(0)}$ is a measure of the average fluctuational electric field acting on an electron. The most important physics comes from the first term of (7.140), proportional to t^2. Because of this term, affecting the time integral of (7.118), the conventional energy conservation function $\delta(\omega - N\omega_{\mathrm{c}})$ is transformed into a heavily broadened Gaussian function.

For the high-frequency mode, one can assume $n_{+,k} \ll 1$ and $|\boldsymbol{E}_{+,\boldsymbol{k}}|^2 \simeq 2$, which gives

$$2q^2 w_+(0,t) \simeq \frac{x_q}{N_{\mathrm{e}}} \sum_k \frac{\omega_{\mathrm{c}}}{\Omega_{+,k}} \mathrm{e}^{-\mathrm{i}\Omega_{+,k} t} \ , \tag{7.143}$$

and $2q^2 w_+(0,0) \simeq x_q$, where $\Omega_{+,k} \simeq \omega_{\mathrm{c}}$. Therefore, the high-frequency Debye–Waller factor $\exp(-x_q)$ can be kept as it is, while the DSF can be expanded in $2q^2 w_+(0,t)$, similarly to the conventional low-temperature expansion. In this way we obtain the multi-phonon emission terms for the high-frequency mode. The scattering events described by terms proportional to $\mathrm{e}^{-\mathrm{i}N\omega_{\mathrm{c}} t}$ are usually forbidden at low temperatures $T \ll \hbar\omega_{\mathrm{c}}$ (except for the term with $N = 1$), because ripplons and vapor atoms do not have enough energy in the laboratory frame for such emission. In our case, the multi-phonon emission of high-frequency phonons is possible due to the strong broadening introduced by the low-frequency mode [see (7.140)]. Such emission is accompanied by the creation and destruction of a great many low-frequency phonons in the $\Omega_{-,k}$ mode.

The straightforward evaluation of all terms of the expansion in $2q^2 w_+(0,t)$ yields

$$S(q,\omega) = \frac{2\sqrt{\pi}\hbar}{\Gamma_{\mathrm{C}}} \sum_{N=0}^{\infty} \frac{x_q^{N-1/2}}{N!} \exp\left[-x_q - \frac{\hbar^2 (\omega - N\omega_{\mathrm{c}} - x_q \Gamma_{\mathrm{C}}^2 / 4T\hbar)^2}{x_q \Gamma_{\mathrm{C}}^2} \right] . \tag{7.144}$$

The small frequency shift $x_q \Gamma_{\mathrm{C}}^2/4T\hbar$ originates from the second term of (7.140). Under the condition $\Gamma_{\mathrm{C}} < T$, when the fluctuational electric field can be considered as quasi-uniform, it can be disregarded. A similar approximation was used by Dykman [86] when studying the CR from the WS. The difference is that here, in addition to the resonance term ($N = 1$) obtained in [86], the WS DSF involves all quanta of cyclotron motion ($N\omega_{\mathrm{c}}$), which is important for fundamental reasons and for applications.

From (7.144), one can see that the DSF of the Wigner solid subject to a strong normal magnetic field is very similar to the DSF of a nondegenerate 2D electron gas found in Chap. 2 [see (2.38)]. It has a series of maxima positioned at the free electron excitation spectrum $\omega = N\omega_{\mathrm{c}}$ (with the small correction $x_q \Gamma_{\mathrm{C}}^2/4T\hbar$) and broadened due to the Coulomb interaction. On the other hand, the DSF of the WS has a singularity as $q \to 0$ and $\omega = 0$, which is reminiscent of what is found for the semi-classical nondegenerate electron gas [see (2.22)]. The similarity of (7.144) and (2.38) allows one to assume that the single-electron properties somehow remain, even for the Wigner solid state, if the magnetic field is strong enough.

Surprisingly, (7.144) evaluated for the electron solid state coincides with the DSF of the Coulomb liquid discussed in Chap. 2 [see (2.52)] in the limiting case of strong fluctuational fields $T > \Gamma_C \gg \Gamma_N$, when collision broadening can be disregarded. The quantity Γ_C can be called the Coulomb broadening of the electron DSF, although it bears no relation to the Landau level broadening. Moreover, according to [76, 89], the quasi-uniform fluctuational electric field reduces the collision broadening of Landau levels. The equivalence of the dynamic structure factors of the Wigner solid and of the Coulomb liquid mentioned above explains many experimental observations (reported also in [185]) that the WS transition does not affect the CR linewidth.

7.6 Shear Mode Excitation and Specific Heat Measurements

The detection of the electron–ripplon resonances at frequencies ω_2 and ω_3, corresponding to the simple hexagonal electron lattice, achieved by Grimes and Adams [7], was the first experimental observation of the Wigner solid. Considerable research has been carried out in order to observe the transverse sound mode, which is a distinctive feature of the crystalline state of the electron system, and to measure the shear modulus. Apparently, the low frequency transverse mode cannot be excited in the usual experiment with SEs because its frequency is lower than the electron collision rate. In contrast, the optical (fast) transverse mode has a high enough limiting frequency ω_f to be detected by a resonance method. As shown in Fig. 7.11, the excitation frequency of the mode $\Omega_{t,k}^{(f)}$ is practically independent of k for $n \sim 1/R_{el}$. The first observation of the fast transverse mode of the WS was reported by Gallet et al. [247] for small excitation wave numbers k. The lower electrode was arranged with a meander line producing a transverse excitation electric field in the electron layer. The position of the optical frequency was shown to increase with cooling, which reflects the increase in the HFDWF discussed above. The shear mode resonance appears sharply with decreasing temperature, at practically the same value of the plasma parameter $\Gamma^{(pl)} = 140 \pm 10$ as that reported by Grimes and Adams.

The same kind of experiment was conducted for a much shorter wavelength ($k = 520\,\mathrm{cm}^{-1}$) of the excitation field by Deville et al. [248]. The important point is that, in this range of wave numbers, the fast transverse mode is close to the transverse sound mode of the WS above a flat surface. This allows one to detect the elastic modulus of the electron solid with a sufficient accuracy. The experimental data of [248] are shown in Fig. 7.12, which is reminiscent of Fig. 7.3, obtained by Morf using computer simulations. The

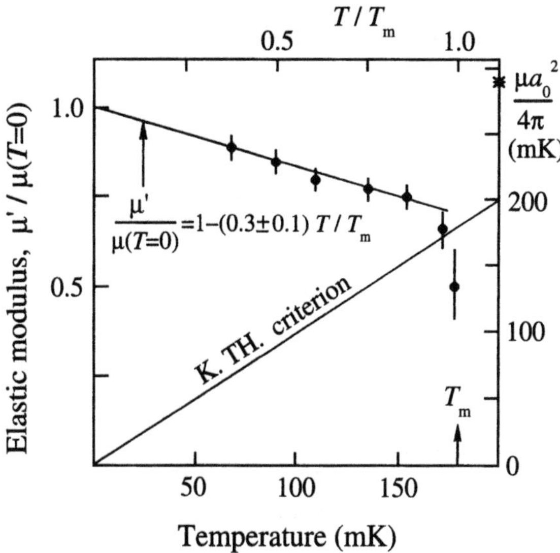

Fig. 7.12. Normalized elastic shear modulus, plotted as a function of temperature [248]. The *asterisk* indicates the calculated classical zero temperature value for the experimentally measured density

measured shear modulus data show a remarkable agreement with the results of molecular dynamics simulations [245].

The optical frequency $\omega_{\rm f}$ of the fast WS modes depends strongly on the electron temperature. Therefore the position of $\omega_{\rm f}(T)$ can serve as a thermometer for the electron layer in order to measure the WS specific heat [158]. Glattli, Andrei and Williams took advantage of the fact that the energy relaxation rate from SEs into liquid helium is very low, as discussed in Sect. 3.6. They found it easy to heat the layer with a pulse of power P for a time $\tau_{\rm e\text{-}e} \ll \Delta t \ll \tilde{\nu}_{\rm eff}^{-1}$ and make its temperature $T_{\rm e}$ substantially higher than the liquid helium temperature. The specific heat was determined directly by monitoring the electron temperature variation $\Delta T_{\rm e}$:

$$C = P\Delta t/\Delta T_{\rm e} \, ,$$

without interference with the sea of $\sim 10^{22}$ liquid helium atoms. The specific heat data measured in [158] are shown in Fig. 7.13. The data behave in accordance with the phonon contribution to C (continuous and dashed curves).

A precise study of the time evolution of the electron temperature performed by Glattli, Andrei and Williams [158] did not show any discontinuity in the entropy or its derivatives at the WS transition. This allows one to conclude that the melting of the 2D electron solid is not of the first order. A first

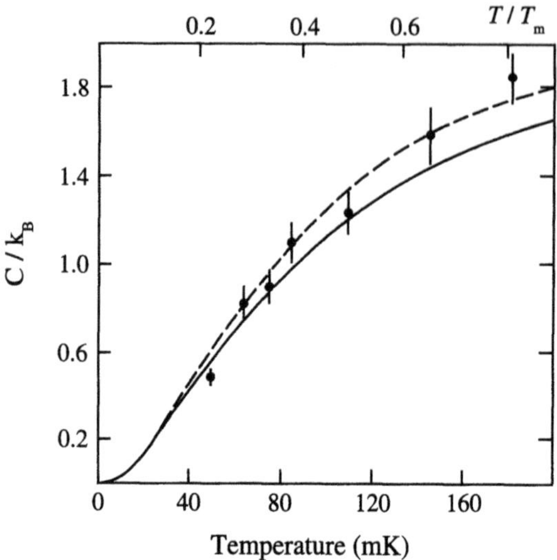

Fig. 7.13. Specific heat per electron in solid phase as a function of reduced temperature T/T_{m} [158]. Experimental values are compared with the phonon specific heat: $c_{\mathrm{t}}(T) = c_{\mathrm{t}}(0)$ (*continuous curve*) and $c_{\mathrm{t}}(T) = c_{\mathrm{t}}(0)(1 - 0.3T/T_{\mathrm{m}})$ (*dashed curve*)

order transition is expected in Chui's melting model [271], which assumes that dislocations group themselves into grain boundaries.

7.7 Bilayer Electron Crystals

A single-layer Wigner solid has the simplest (hexagonal) lattice structure. There are no structural phase transitions in such a solid. Vil'k and Monarkha [35] pointed out that an interesting ordering is expected to occur for charged bilayer systems. A change in the electron lattice structure for the bilayer electron solid can be induced by varying the spatial separation d of the layers, or by cooling it in the strong coupling regime ($d \ll a$). As shown in more recent studies by Goldoni and Peeters [236], for certain separations between two electron layers, even the structure with square electron sublattices can be energetically favorable.

Two interacting parallel 2D electron gases are realized in semiconductor heterostructures. Interesting 'drag' measurements have been conducted to probe the interlayer interaction [272, 273]. Under typical conditions the semiconductor bilayer electron systems are far from the Wigner solid transition point because of the small mass of charge carriers. Therefore the bilayer electron gas represents an interesting object for studying the quantum Hall effect [274, 275]. On the other hand, a strong magnetic field directed perpendicular to the layer quenches the kinetic energy, which may induce the Wigner solid in such quantum systems. The interaction between two electron layers can be used for a direct indication of the crystalline ordering in the magnetic freeze-out regime [276]. A classical version of the bilayer charged system can be realized on two neighboring interfaces of a helium film [35]: the electron crystal formed on the free surface of the liquid interferes with the crystal of negative ions collected on the liquid–solid boundary.

The origin of the structural phase transition is easily understood by considering two identical parallel 2D Wigner solids in the weak coupling regime ($d \gg a$, where a is the electron spacing in each layer). In this case, the strong intralayer interaction $V(r) = e^2/r$ forms 2D hexagonal lattices in each layer. The weak interlayer interaction $V(r) = e^2/\sqrt{r^2 + d^2}$ just fixes the relative positions of these lattices in order to minimize their interaction energies. The energetically favored configuration of the two electron sublattices is shown in Fig. 7.14a: the lattice sites of the upper layer are located just above the centers of the triangles formed by the lattice sites of the lower layer.

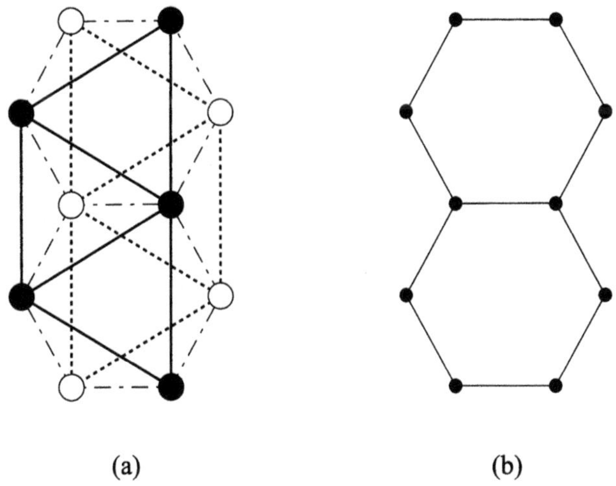

Fig. 7.14. Lattice structure of the bilayer electron crystal in the limit of weak interlayer coupling. (**a**) Top view of the staggered hexagonal lattice, distinguishing sublattices of different layers by *black* and *white circles*. (**b**) Composite hexagonal lattice

The two sublattices form a composite lattice whose unitary cell contains two 'atoms'. As a result, the bilayer crystal has optical modes with the limiting frequency ω_0 determined by the mutual Coulomb interaction between layers. The easiest way of finding this limiting frequency is to consider sublattices formed of charges with substantially different masses, as for the system consisting of electrons and ions in the helium film. In this case, the optical modes represent phonon modes of the first layer in the field of frozen (fixed) charges of the second layer. For arbitrary values of the dimensionless parameter $y = d/a$, the limiting frequency ω_0 was evaluated in [35] using the Ewald summation method:

$$\omega_0^2 = \frac{6b^{5/2}e^2}{\sqrt{\pi}m_e a^3}\left[\psi_{-1/2}(by^2, b) + \frac{1}{3}\psi_{3/2}\left(0, by^2 + \frac{b}{3}\right) - \frac{1}{b}\psi_{1/2}\left(0, by^2 + \frac{b}{3}\right)\right], \tag{7.145}$$

where $b = 2\pi/\sqrt{3}$ and the function $\psi_\nu(x, y)$ was defined previously in (2.91). For weak interlayer coupling ($y = d/a > 1$), we have the very simple form

$$\omega_0^2 \simeq \frac{3\pi e^2 n g_1}{m_e}\exp(-g_1 d), \tag{7.146}$$

where g_1 is the smallest (nonzero) reciprocal lattice vector and n is the charge density in each layer. For identical layers with the same mass of charge carriers m_e, electrons in the first layer and electrons in the second layer oscillate out of phase. In this case, the limiting frequency of optical modes is found from (7.145) and (7.146) by replacing the electron mass m_e by the reduced mass $m = m_e/2$, which makes ω_0^2 two times larger. The presence of the medium should be taken into account by means of the replacement $e^2 \to e^2/\epsilon$.

For a semiconductor bilayer system, we should take into account the fact that crystallization of electron layers is induced by a strong magnetic field applied normally. This field affects the phonon modes and the limiting frequency in a similar way to that found for the fast modes of the single-layer Wigner solid. In that case, the composite lattice was formed by electron and dimple sublattices. For layers with different carrier masses, we can use the result given by (7.112). Therefore, in the limit of high magnetic fields, we have

$$\omega_0(B) = \Omega_-^{(\mathrm{opt})}(k = 0) \simeq \frac{\omega_0^2(0)}{\omega_c}. \tag{7.147}$$

For identical layers, the formula found for $\omega_0(B)$ in [276] agrees with (7.147) if the electron mass m_e is replaced by the reduced mass $m = m_e/2$ in the expression for $\omega_0^2(0)$ given in (7.146).

In order to excite optical vibrations in the double-layer electron solid, it is necessary to break the symmetry between layers 1 and 2. For this reason, it was proposed to pin one of the layers (say, 2) so that the external electric field can affect only the other sublattice [276]. In this case, the situation becomes similar to that described above by the model with $m_2 \gg m_1$.

The bilayer electron solid also has acoustical modes. The important point is that the transverse phonon mode of the crystalline structure shown in Fig. 7.14b is unstable: $c_t^2 < 0$, where c_t is defined by $\omega_t(k) = c_t k$. This means that, when reducing the interlayer spacing, at a certain value of $y = d/a$, the staggered hexagonal lattice should be replaced by another lattice structure. Evaluations performed in [35] indicate that $c_t^2(y)$ of the staggered hexagonal lattice goes to zero as $y \to y_c \simeq 0.62$. Comparing the static energy per particle of different lattice structures, Goldoni and Peeters [236] found that the staggered hexagonal lattice (called phase V) should be replaced by a staggered rhombic lattice (phase IV) at somewhat larger values of the parameter $y \simeq 0.681$ ($\eta = d\sqrt{n_e} \simeq 0.732$, where n_e is the electron density in each layer).

In the limiting case of strong interlayer coupling $d \ll a$, electrons of different layers should be combined in a single hexagonal lattice. Therefore in this limit the bilayer crystal can be considered as a 2D binary alloy with a triangular lattice and mixing energy [35]

$$V(r) = 2e^2 \left(\frac{1}{r} - \frac{1}{\sqrt{r^2 + d^2}} \right) .$$

The concentration c of this alloy is defined as the fraction $c = N_1/(N_1 + N_2)$. For identical layers ($N_1 = N_2$) $c = 1/2$. It is convenient to introduce the random function $\Delta(\boldsymbol{l})$, which equals unity if the particle $\boldsymbol{l}$ is located in the upper layer (1), and zero if the particle $\boldsymbol{l}$ is located in the lower layer (2). Then, the distribution of particles in the binary alloy is described by $n(\boldsymbol{r_l}) = \langle \Delta(\boldsymbol{l}) \rangle$, which is the probability that the particle placed in the lattice site $\boldsymbol{l}$ belongs to the upper layer. In the disordered phase, $n(\boldsymbol{r_l})$ is independent of $\boldsymbol{l}$ and equal to c.

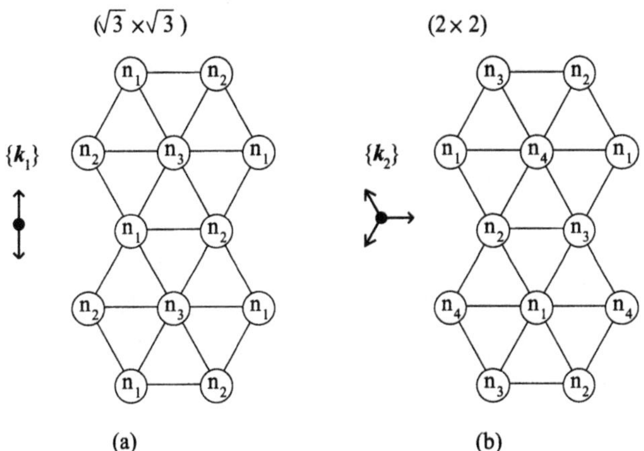

Fig. 7.15. Sublattice structures describing the bilayer electron crystal in the strong coupling regime ($d \ll a$). Two-dimensional binary alloys on a triangular Ising lattice corresponding to the stars (**a**) $\{\boldsymbol{k_1}\}$ and (**b**) $\{\boldsymbol{k_2}\}$

With falling temperature, the function $n(\boldsymbol{r_l})$ starts to take on several values and the original lattice decomposes into several sublattices for each of which $n(\boldsymbol{r_l}) = \text{const.} = n_\alpha$, where $\alpha = 1, 2, \ldots, M$ and M is the number of sublattices. In the self-consistent field approach, the function $n(\boldsymbol{r_l})$ is usually considered as a superposition of static concentration waves [277]:

$$n(\boldsymbol{r}_l) = c + \sum_{\boldsymbol{k}} n_{\boldsymbol{k}} \exp(\mathrm{i}\boldsymbol{k} \cdot \boldsymbol{r}_l) \, , \tag{7.148}$$

where $\boldsymbol{k}$ is a nonzero wave vector from the first Brillouin zone. The maximal temperature at which a nontrivial solution of the form (7.148) appears for the self-consistent field equation is [277]

$$T_{\mathrm{A}}^{(1)} = -c(1-c) \min V(\boldsymbol{k}) \, . \tag{7.149}$$

This is the temperature where the disordered phase becomes absolutely unstable. In (7.148), $\boldsymbol{k}$ is assumed to correspond to the collection (star) of wave vectors $\{\boldsymbol{k}_i\}$ for which $V(k)$ reaches its minimum. For the triangular Ising lattice considered here, these collections of wave vectors are the so-called Lifshitz stars $\{\boldsymbol{k}_1\}$ and $\{\boldsymbol{k}_2\}$ shown in Fig. 7.15 together with their sublattice structures. The star $\{\boldsymbol{k}_1\}$ corresponds to the structure $(\sqrt{3} \times \sqrt{3})$ in which n_α takes on three values $M = M_1 = 3$, as indicated in Fig. 7.15a. Another star $\{\boldsymbol{k}_2\}$ corresponds to the structure (2×2) with $M = M_2 = 4$ (Fig. 7.15b).

The analysis of the configuration energy at $T = 0$ shows that the structure (2×2) is energetically most favorable for $1/2 \le c < c_-$ and $c_+ < c < 1$, where $c_- \simeq 0.5453$ and $c_- \simeq 0.731$. In the first region, the most favorable distribution is $(n_1, n_2, n_3, n_4) = (1, 1, 0, 4c - 2)$, so that for identical layers $(c = 1/2)$ the distribution function n_α takes on only two different values (1,1,0,0). In the range $c_- < c < c_+$, the $(\sqrt{3} \times \sqrt{3})$ structure is more favorable.

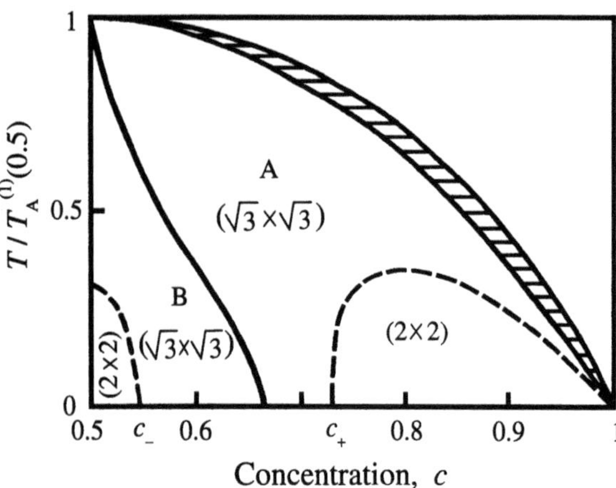

Fig. 7.16. Phase diagram of the bilayer electron crystal in the strong coupling regime $(d \ll a)$ for an arbitrary relation between numbers of charges N_1 and N_2 in the layers defined by the fraction $c = N_1/(N_1 + N_2)$ [35]. The system is described as a two-dimensional binary alloy on a triangular Ising lattice with mixing energy $V(r) \propto r^{-3}$. The *shaded region* corresponds to $T_{\mathrm{A}}^{(1)} < T < T_{\mathrm{A}}^{(2)}$

Regardless of the configurational energy, when reducing the temperature from the disordered phase $[n(\boldsymbol{r}_l) = c]$, the $(\sqrt{3} \times \sqrt{3})$ structure must appear first, even for $c = 1/2$, because $V(k_1) \simeq -2.33\, e^2d^2/a^3$ is smaller than $V(k_2) \simeq -1.88\, e^2d^2/a^3$. Then, at lower temperatures, one may expect an order–order phase transition $[(\sqrt{3} \times \sqrt{3}) \to (2 \times 2)]$ for certain values of c, as shown in the phase diagram of Fig. 7.16 taken from [35]. In this figure the boundary of the (2×2) phase (dashed curves) is shown schematically because this phase is unstable due to the formation of antiphase domains. The $(\sqrt{3} \times \sqrt{3})$ structure has two phases: phase A with $n_2 = n_3 = n$, and phase B where all n_α are different. The continuous curves in Fig. 7.16 show the results of numerical evaluations. The parameter $T_A^{(2)}$ represents the temperature above which the ordered phase A is absolutely unstable.

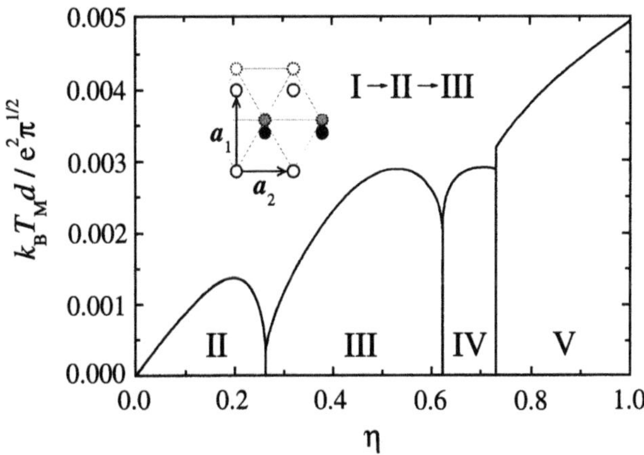

Fig. 7.17. Melting temperature of a bilayer crystal as a function of $\eta = d\sqrt{n_e}$, after Goldoni and Peeters [236]. The *insert* shows the lattice transformation from phase I [the superstructure (2×2) with $n_1 = n_2 = 1$ and $n_3 = n_4 = 0$] to phase III (staggered square lattice). Phase I is only favorable in a narrow region: $\eta < 0.006$. Phase IV involves rotation, so that the angle between $\boldsymbol{a}_1$ and $\boldsymbol{a}_2$ becomes smaller than $\pi/2$. Phase V corresponds to weakly coupled 2D Wigner solids, as shown in Fig. 7.14

Consider now the particular case $N_1 = N_2$ (c=1/2). When the (2×2) phase with the configuration $(1, 1, 0, 0)$ (phase I of [236]) is reached, a small increase in the parameter $\eta = d\sqrt{n_e}$ above 0.006 can lead to a structural transition to a staggered rectangular lattice (phase II), and then to a staggered square lattice (phase III), as indicated in Fig. 7.17. In this case, the initial triangular Wigner lattice is deformed. The insert of Fig. 7.17 shows these transformations as a result of a relative change in the absolute values of lattice vectors $\boldsymbol{a}_1$ and $\boldsymbol{a}_2$. The structure defined as phase IV involves rotation of $\boldsymbol{a}_1$ relative to $\boldsymbol{a}_2$. The melting temperature in Fig. 7.17 was determined from the modified Lindemann criterion discussed in Sect. 7.2.3 [see (7.28)]. It should be noted that, for $d \ll a$, the deformations of the initial triangular Wigner lattice discussed here are not expected in the disordered phase $[n(\boldsymbol{r}_l) = c]$ and in the $(\sqrt{3} \times \sqrt{3})$ phase which appears from the disordered phase when cooling.

8 Wigner Solid. II. Transport Properties

8.1 Solid Current

The surface of superfluid helium supporting the 2D Wigner solid does not impose an impurity potential or a regular potential on the electron lattice. At a finite frequency of the driving electric field, this solid is a quite mobile object. Hence, the system of SEs on liquid helium provides us with a unique possibility for studying transport properties of the extremely correlated state of interface electrons. As discussed in the last chapter, under typical conditions, the electron crystal is strongly coupled to surface dimples. It is obvious that the conductivity of the WS depends crucially on the frequency of the signal ω. If $\omega < \omega_1$, the surface dimples of liquid ^{4}He are involved in motion of the electron lattice, which increases the effective mass and affects the effective collision frequency of electrons. Soon after the observation of the coupled phonon–ripplon modes reported by Grimes and Adams, mobility anomalies at the WS transition point were reported in [101, 232]. An interesting mobility minimum near the melting temperature of the 2D electron solid supported by the liquid helium surface was reported in [50, 233]. We have already seen this anomaly in Fig. 3.1 as a maximum of the inverse quantity $1/\mu$ vs. temperature. The narrow excess scattering observed was reportedly identical to the temperature-dependent losses associated with the superfluid transition in thin helium films. However, there was no conclusive proof that this peak is caused by dissociation of dislocation pairs due to the Kosterlitz–Thouless melting mechanism.

The Wigner solid represents a particular case of the charge density wave. It should therefore be pinned in the field of a random potential. Such an 'impurity' potential appears for electrons supported by a helium film covering a solid substrate. In this case, substrate roughness, which causes a spatial variation of the image potential of the SEs, is the cause of the 'impurity' potential. The electron lattice is pinned by these traps and the system undergoes a sort of metal–insulator transition at the WS transition point. The strength of the 'impurity' potential can be varied by changing the film thickness. This transition was observed by Jiang and Dahm [278] for SEs on a helium film covering a glass substrate with $\epsilon = 7.3$. Their resistivity vs. temperature data are shown in Fig. 8.1. The resistivity of the fluid phase is nearly temperature-independent. Below a certain temperature T_c, the resistivity increases as $\rho = \rho_0 \exp(-\gamma T/T_c)$, which corresponds to the pinned electron solid state, where γ is a function of the excitation field, its frequency ω, and the film thickness d. It increases as these parameters are reduced.

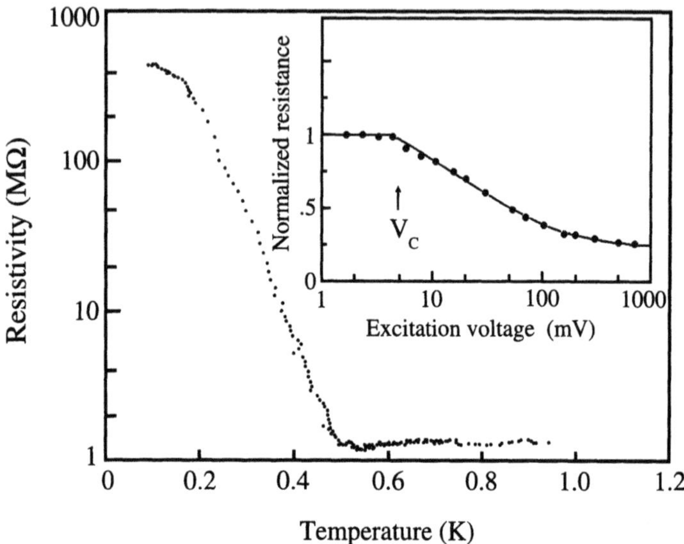

Fig. 8.1. Resistivity vs. temperature: $n_s = 2 \times 10^9\,\mathrm{cm}^{-2}$, $d = 380\,\text{Å}$. The excitation frequency and voltage were 100 Hz and 2 mV, respectively [246, 278]. The *insert* shows the normalized resistivity vs. excitation voltage: $n_s = 5 \times 10^9\,\mathrm{cm}^{-2}$, $d = 320\,\text{Å}$, $T = 690\,\mathrm{mK}$

With an increase in the amplitude of the driving electric field, the normalized resistance shown in the insert of Fig. 8.1 does not change until the excitation voltage V exceeds a critical value V_c which corresponds to a threshold field $E_c \simeq 5\,\mathrm{mV/cm}$. In the range $V > V_c$, it decreases with the driving field, and then, at a certain value of the electric field $E_N \sim 100\,\mathrm{mV}$, broadband noise sharply increases its amplitude. Such behavior of the WS is analogous to that of the sliding charge-density wave. It should be noted that the current and electric field profiles along the current direction were nonuniform in this experiment because of an intrinsic feature of the Sommer and Tanner method. The current has a maximum in the center of the sample and vanishes at the edges. This is the reason for the smooth depinning of the electron crystal caused by the driving electric field.

For the WS formed on the surface of massive liquid helium, there is no static pinning potential, as there is for electrons above helium films or supported by a solid cryogenic substrate. Nevertheless, the WS transport on the surface of liquid helium cannot be described as quasi-free solid motion because of the creation of the sublattice of surface dimples. By analogy with the single-electron self-trapping to a surface dimple discussed in Chap. 6, the creation of the dimple lattice and strong coupling to it can be considered as a self-pinning of the Wigner solid. This self-pinning does not change the WS transition point, because the coupling energy is much smaller than the mean Coulomb energy of interface electrons. Still, it crucially affects the WS transport, because the driving electric field is usually too weak to break the coupling between the electron and dimple sublattices. We recall here that some important properties of the Wigner solid of interface electrons, such as the electron confinement in the surface dimple described by $\sqrt{\langle u_{\mathrm{f}}^2 \rangle}$, and the limiting frequency ω_{f} of the fast phonon modes, are found by means of the self-consistent treatment, which also supports the analogy with interface polarons. This self-consistent treatment is very important, not only for obtaining the real part of the conductivity relaxation kernel $w(\omega) = \mathrm{Re}\, M(\omega)$ and the spectrum of the coupled phonon–ripplon modes. Because of the relation $\nu(\omega) = \mathrm{Im}\, M(\omega)$, the self-consistent approach also helps to describe the unusual transport properties of the electron lattice strongly coupled to the sublattice of heavy dimples, which is the subject matter of this chapter.

In Sect. 8.2, we discuss the effective collision frequency and phonon damping of the WS formed on the surface of liquid helium in the presence of a weak alternating driving electric field. The basic relations concerning WS transport are introduced in Sect. 8.2.1. We consider two different approaches to the description of the WS transport: the weak coupling theory and the self-consistent strong coupling treatment. In some cases, they arrive at substantially different results and conclusions. We discuss these differences and compare the final results with available experimental data.

The nondegenerate electron gas and Wigner solid have significantly different excitation spectra. For example, in the solid state there is the shear mode, and the electron conductivity of the WS reportedly depends strongly on the velocity c_{t} of transverse sound [268]. Because there is no such quantity in the gas state, it might be concluded that conductivities of the WS and noninteracting electrons should differ substantially. However, the opposite is true [149]: detailed analysis of the electron momentum relaxation time given in Sect. 8.2.1 indicates that, under weak coupling conditions, the conductivities of the Wigner solid and electron gas are remarkably close (even numerically) over wide ranges of temperatures and electron densities, in spite of the obvious differences in their energy excitation spectrums. It looks as though the Wigner solid state 'hides' itself in the conventional conductivity study. We have already mentioned the same property of the electron solid in Chap. 5, which dealt with cyclotron resonance absorption from interface electrons.

It appears that the strong coupling of the WS to the dimple sublattice makes the AC conductivity of the WS really different from that of the gas state. The most prominent differences occur when the frequency of the driving electric field is close to the characteristic frequencies of this coupled system. The limiting frequency of the fast (optical) phonon modes ω_{f} is an example. Besides plasmon resonance absorption at this frequency, there is a strong reduction in the electron collision rate and the WS phonon damping induced by the strong coupling of the WS to the dimple lattice. These properties of the WS conductivity and phonon damping are discussed in Sect. 8.2.2.

If the frequency of the driving electric field is close to a typical frequency of ripplons, whose wave vector $\boldsymbol{q}$ is equal to a reciprocal lattice vector $\boldsymbol{g}$ ($\omega \simeq \omega_n$, $n = 1, 2, 3, \ldots$), the electron conductivity $\sigma(k, \omega)$ exhibits the resonance anomaly because of the excitation of the coupled phonon–ripplon modes. In the last chapter, we mainly discussed the real part of the conductivity relaxation kernel describing the spectrum of the coupled phonon–ripplon modes. In Sect. 8.2.3, we discuss the transport aspects of this phenomenon which originate from the imaginary part of the conductivity relaxation kernel. It appears that the effective collision rate $\nu(\omega) = \mathrm{Im}\, M(\omega)$ exhibits a resonant structure itself, which interferes with the resonant terms of the real part of the conductivity relaxation kernel $w(\omega)$. The interplay between the peculiar frequency dependencies of $w(\omega)$ and $\nu(\omega)$ determines the absorption line shape of the electron–ripplon resonances.

Liquid ^{3}He supporting SEs represents a special case for WS transport limited by the slow motion of the dimple sublattice. At low temperatures this Fermi liquid has a huge viscosity which decreases fast with cooling. It undergoes the superfluid transition when $T \to T_c = 0.93\,\mathrm{K}$. The specific properties of this fluid affect the WS transport to an enormous extent. When the system is cooling down, the resistivity of SEs undergoes changes of several orders of magnitude in both directions (up and down) [54, 234]. The dimple lattice, reflecting bulk liquid quasi-particles, allows one to use the Wigner solid as a sensitive tool for studying the superfluid properties of this remarkable quantum liquid [234, 279]. The WS mobility over the surface of normal and superfluid ^{3}He is discussed in Sect. 8.3.

It is quite clear that a driving electric field of sufficient amplitude can cause depinning of the electron crystal from its self-pinned state. This sliding state of the Wigner solid on the surface of superfluid helium was reportedly observed by Shirahama and Kono [280], who showed that the transport of the WS subjected to a weak magnetic field is a nonlinear transport under typical experimental conditions. The nonlinear transport phenomena of the 2D electron solid caused by the Bragg–Cherenkov emission of surface excitations and by depinning from the dimple sublattice are discussed in Sect. 8.4.

8.2 AC Conductivity and Phonon Damping

8.2.1 Basic Relations

The general equations of the quantum transport framework given in Chap. 3 are applicable to the Wigner solid state as well. For typical electron densities realized on the free surface of liquid helium, the WS transition occurs at rather low temperatures, where electrons interact predominantly with capillary wave quanta (ripplons). The excitations involved belong to the long-wavelength part of the ripplon spectrum and one can therefore simplify $N_q^{(r)} + 1 \simeq N_q^{(r)} \simeq T/\hbar\omega_q$. Under this assumption the general relation between the effective collision frequency and the electron DSF [see (3.57)] can be transformed to the form

$$\nu(\omega) = \frac{1 - e^{-\hbar\omega/T}}{8\alpha m_e S_A} \frac{T}{\hbar\omega} \sum_{\boldsymbol{q}} V_q^2 \left[S_0(\boldsymbol{q}, \omega - \omega_q) + S_0(\boldsymbol{q}, \omega + \omega_q) \right] . \qquad (8.1)$$

Here we have taken into account the fact that the DSF of the Wigner solid generally depends on the direction of $\boldsymbol{q}$. One can see that in the quantum theory, the effective collision frequency (8.1) has a strong frequency dependence regardless of the actual dependence of $S_0(\boldsymbol{q}, \omega)$.

In order to obtain $\nu(\omega)$ and the conductivity of the WS, the right approximation for the electron DSF $S_0(\boldsymbol{q}, \omega)$ must be chosen. The easiest is the high temperature ($T > T_D$) approximation

$$S(q, \omega) = \hbar \sqrt{\frac{\pi}{\varepsilon_q K_e}} \exp \left[-\frac{(\varepsilon_q - \hbar\omega)^2}{4\varepsilon_q K_e} \right] , \qquad (8.2)$$

which follows from (7.120). This equation coincides with the DSF of the nondegenerate electron gas if $K_e \to T$. Therefore, we expect the WS and nondegenerate gas to have similar effective collision frequencies in this temperature regime. It is instructive to compare the AC conductivity of the WS with that of the electron gas on a broader temperature scale. For this reason, we shall extend the form of (8.2) to low temperatures and the corresponding results will be marked with the subscript NDG.

Regardless of the frequency of the driving electric field ω, the frequency of ripplons ω_q can be disregarded in the argument of the electron DSF chosen above. Then the effective collision frequency of the WS and nondegenerate gas has the form [149]

$$\nu_{NDG}(\omega) = \frac{(1 - e^{-\hbar\omega/T})(eE_\perp)^2 T}{2\sqrt{\pi}\alpha\hbar^2\omega} \int_0^\infty V_G^2(x) \exp \left[-\left(x - \frac{\hbar\omega}{4K_e x} \right)^2 \right] dx . \qquad (8.3)$$

Here, in accordance with (1.72), we use the notation

$$V_G(x) = 1 + \frac{2\Lambda k_T^2 x^2}{eE_\perp} w_C \left(\frac{k_T^2 x^2}{\gamma^2} \right) , \qquad (8.4)$$

where $w_C(x)$ is the electron–ripplon coupling function defined in (1.73), $x = 0.5\sqrt{\varepsilon_q/K_e}$ is a dimensionless parameter, γ^{-1} determines the localization length of the electron wave function in the perpendicular direction, and $\hbar k_T = \sqrt{2mK_e}$ is the thermal momentum of electrons.

A numerical evaluation shows that, at high densities, the presence of ω in the integrand of (8.3) has only a weak impact on the frequency dependence of the effective collision frequency. In the limiting case $V_{\mathrm{G}}(x) \simeq 1$, this is easily seen even analytically, because the integral $\int_0^\infty \exp[-(x-a/x)^2]\mathrm{d}x = \sqrt{\pi}/2$ does not depend on the parameter a. For the same reason, the effective collision frequency does not depend much on the electron kinetic energy K_{e}. Thus the relation between AC and DC effective collision frequencies can be approximately written as

$$\frac{\nu_{\mathrm{NDG}}(\omega)}{\nu_{\mathrm{NDG}}(0)} \simeq \frac{T(1-\mathrm{e}^{-\hbar\omega/T})}{\hbar\omega} . \tag{8.5}$$

According to this relation, $\nu_{\mathrm{NDG}}(\omega)$ decreases sharply as compared to the DC result and acquires the proportionality factor $T/\hbar\omega$, if the temperature is lower than $\hbar\omega$. The strong frequency dependence of the electron collision rate is a pure quantum effect which results from the fluctuation–dissipation theorem. It appears regardless of the particular state of the electron system. The temperature dependence, which follows from (8.5) because of the quantum condition $T < \hbar\omega$, was observed experimentally for SEs on liquid helium in the plasmon resonance experiment [235, 281, 282]. We shall discuss these data later, after a brief analysis of the low temperature regime for the WS state.

At ultra-low temperatures, one cannot replace $N_q^{(\mathrm{r})}$ by $T/\hbar\omega_q$. In this case, assuming $T \ll \hbar\omega/4$, the asymptote of the collision frequency can be found in the analytical form

$$\nu_{\mathrm{NDG}}(\omega) \simeq \frac{1-\mathrm{e}^{-\hbar\omega/T}}{8\sqrt{\alpha\rho}\hbar\omega} q_\omega^{3/2} V_{q_\omega}^2 \left(2N_{q_\omega}^{(\mathrm{r})}+1\right) , \tag{8.6}$$

where $q_\omega = \sqrt{2m\omega/\hbar}$. Thus, at ultra-low temperatures ($N_{q_\omega}^{(\mathrm{r})} \ll 1$), the effective collision frequency and the plasmon resonance linewidth become independent of temperature. It should be noted that the asymptote of (8.6) found for the limiting case $T \ll \hbar\omega/4$ is surprisingly close to the exact result, even for $T \sim \hbar\omega$. For example, the use of $N_q^{(\mathrm{r})} \simeq T/\hbar\omega_q$ and $V_q \simeq eE_\perp$ in (8.3) and (8.6) gives the same result.

At low enough temperatures the electron system enters the regime where the usual Fermi gas becomes degenerate. Even though the Coulomb interaction energy of electrons on liquid helium is much higher than their Fermi energy $\varepsilon_{\mathrm{F}}^{(\mathrm{e})}$, it is instructive to consider here the degenerate electron gas as well. In contrast with conventional degenerate electron systems where $\hbar\omega \ll \varepsilon_{\mathrm{F}}^{(\mathrm{e})}$, the Fermi energy of surface electrons is low and, under the conditions of the plasmon resonance experiment, we have $\hbar\omega > \varepsilon_{\mathrm{F}}^{(\mathrm{e})}$. In this case, the Fermi distribution function $f(\varepsilon_k+\hbar\omega)$ can be disregarded as compared to unity, which yields the approximation of (2.25) for the electron DSF. Then the effective collision frequency of the degenerate Fermi gas can be written as

$$\nu_{\mathrm{DG}}(\omega)=\frac{1}{16\pi^2\hbar\omega\sqrt{\alpha\rho}n_{\mathrm{s}}}\int\limits_{q_-}^{q_+} q^{1/2}\left|V_q\right|^2(2N_q^{(\mathrm{r})}+1)\sqrt{(q_+^2-q^2)(q^2-q_-^2)}\mathrm{d}q\,. \tag{8.7}$$

It is assumed here that $T \ll \hbar\omega$. The quantities q_+ and q_- introduced in (2.26) represent two extreme values of the electron momentum exchange for the limiting case $\hbar\omega > \varepsilon_{\mathrm{F}}^{(\mathrm{e})}$. The collision frequency of degenerate electrons is also proportional to the temperature, and over a wide range of parameters, it is very close to the AC collision frequency of nondegenerate electrons. The latter can even be seen analytically, if $q_\omega \gg k_{\mathrm{F}}$. In this limiting case, q_+ and q_- are very close and the smooth functions $q^{1/2}V_q$ and $2N_q^{(\mathrm{r})}+1$ can therefore be taken at $q \simeq q_\omega$ and moved out of the integrand. Then, recalling the relation $k_{\mathrm{F}}^2 = 2\pi n_{\mathrm{s}}$, one finds that (8.7) reproduces the low temperature asymptote ($T < \hbar\omega$) of a nondegenerate electron gas [see (8.6)]. The similar behavior of the effective collision rates of nondegenerate and degenerate electrons found above is due to the fact that, in the low temperature limit and for $\hbar\omega > \varepsilon_{\mathrm{F}}^{(\mathrm{e})}$, the corresponding dynamic structure factors (2.22) and (2.25) have a sharp maximum at nearly the same wave vectors $q \simeq q_\omega$, as shown previously in Fig. 2.1.

If electrons form the Wigner solid, the low temperature asymptote of their DSF [see (7.122)] differs substantially from the DSF of free electrons. Consider first the elastic term (WS phonons are not involved in scattering events)

$$S^{(\mathrm{elas})}(\boldsymbol{q},\omega)=2\pi\mathrm{e}^{-2W_g}\delta(\omega)N_{\mathrm{e}}\sum_{\boldsymbol{g}}\delta_{\boldsymbol{q},\boldsymbol{g}}\,. \tag{8.8}$$

Here we should agree about the definition of the Debye–Waller function W_g, which is divergent for an infinite 2D system. In the weak coupling theory, one can assume that the real system is finite and at low temperatures W_g is small enough for the expansion given in (7.122). Another important point is that, for an infinite system, thermal vibrations can be excluded from $W_g(T)$ because they induce multi-phonon terms which are heavily compensating. Therefore the replacement $W_g(T) \to W_g(0)$ is quite a good low temperature approximation. Later we shall investigate the accuracy of this approximation.

In the strong coupling theory, the relaxation kernel of the electron conductivity consistent with the secular equation for the coupled plasmon–ripplon modes contains $S(\boldsymbol{q},\omega)$ with the high-frequency Debye–Waller factor (HFDWF), as discussed in Sect. 7.5.3. One can also see this from the final expression for the WS collision frequency. The substitution of $S^{(\mathrm{elas})}(\boldsymbol{q},\omega)$ for the WS DSF yields the collision frequency

$$\nu_{\mathrm{WS}}^{(\mathrm{elas})}(\omega)=\frac{\pi n_{\mathrm{s}}\left(1-\mathrm{e}^{-\hbar\omega/T}\right)}{4\alpha m_{\mathrm{e}}}\frac{T}{\hbar\omega}\sum_{\boldsymbol{g}}V_g^2\mathrm{e}^{-2W_g}\delta(\omega-\omega_g)\,. \tag{8.9}$$

In the limiting case $\hbar\omega_g \ll T$, this equation accurately transforms into the result given previously in (7.81) for the frictional force acting on the rigid 2D electron lattice. Because the factor $\exp(-2W_g)$ should be the same in both functions $\nu(\omega)$ and $w(\omega)$, representing the imaginary and real parts of the memory function, consistency requires $2W_g \to g^2 \langle u_f^2 \rangle /2$.

The WS collision frequency of (8.9) contains sharp peaks at the positions $\omega = \omega_g$, which corresponds to the resonance excitation of ripplons by the rigid 2D electron lattice. The ripplon damping should broaden these resonances into the usual Lorentzians. At frequencies ω which do not satisfy the resonance condition, the elastic term in the effective collision frequency can be disregarded. In order to find a better approximation for $S(\boldsymbol{q},\omega)$, let us consider the one-phonon terms. It is clear that the main contribution will come from transverse phonons because of their slow velocity c_t. Then the one-phonon terms of $S(\boldsymbol{q},\omega)$ given in (7.122) can be rewritten in the following way:

$$S^{(1\text{-ph})}(\boldsymbol{q},\omega) = \frac{\pi\hbar q^2 \mathrm{e}^{-2W_q}}{2m_e|\omega|}\left\{ \left[n_B(\hbar|\omega|)+1\right]\sum_{\boldsymbol{k}}\delta(\omega-\Omega_{t,k}) + n_B(\hbar|\omega|)\sum_{\boldsymbol{k}}\delta(\omega+\Omega_{t,k})\right\}\sum_{\boldsymbol{g}}\delta_{\boldsymbol{q},\boldsymbol{g}+\boldsymbol{k}}\,, \tag{8.10}$$

where $n_B(\varepsilon)$ is the Bose distribution function.

In the weak coupling theory, the reconstruction of the WS phonon spectrum is not taken into account in the electron DSF: $\Omega_{t,k} = c_t k$. We note also that in the most interesting cases $k \ll g$. Then direct evaluation of (8.10) yields

$$S^{(1\text{-ph})}(\boldsymbol{q},\omega) = \frac{\hbar q^2 S_A \mathrm{e}^{-2W_q}}{4m_e c_t^2}\left\{ \left[n_B(\hbar|\omega|)+1\right]\theta(\omega) + n_B(\hbar|\omega|)\theta(-\omega)\right\}\sum_{\boldsymbol{g}}\delta_{\boldsymbol{q},\boldsymbol{g}}\,, \tag{8.11}$$

where $\theta(x)$ is the conventional unit step function. We should remember that here ω is not the frequency of the driving electric field, but just the argument of the electron DSF, which is actually $\omega \pm \omega_q$ and can be negative. This is why we keep the sign $|\dots|$ in the above equation.

If the absolute value of the frequency argument of $S(q,\omega)$ is small compared with $T/\hbar$, then $n_B(\hbar|\omega|) \simeq T/\hbar|\omega| \gg 1$ and the one-phonon term reduces to

$$S^{(1\text{-ph})}(\boldsymbol{q},\omega) \simeq \frac{\hbar q^2 S_A}{4m_e c_t^2}\mathrm{e}^{-2W_q}\frac{T}{\hbar|\omega|}\sum_{\boldsymbol{g}}\delta_{\boldsymbol{q},\boldsymbol{g}}\,. \tag{8.12}$$

Note the unusual frequency dependence in the denominator, which leads to important physical conclusions. Substituting this form into the effective collision frequency equation (8.1) yields

$$\nu_{WS}^{(1\text{-ph})}(\omega) = \frac{\pi n_s}{8\alpha m_e}\sum_{\boldsymbol{g}}\alpha_g(T)V_g^2\mathrm{e}^{-2W_g}\left[\frac{1}{|\omega-\omega_g|}+\frac{1}{\omega+\omega_g}\right], \tag{8.13}$$

where we have used the temperature-dependent dimensionless parameter

$$\alpha_g(T) = \frac{2T}{T_g} = \frac{g^2 T}{4\pi m_e c_t^2 n_s} . \tag{8.14}$$

The outstanding feature of the one-phonon terms in the weak coupling treatment is that the resonance structure of the collision rate has a non-Lorentzian shape. Such a shape was predicted by Dykman [283] using the approximation (7.135) for the function $h_q(\boldsymbol{l}, t)$ in the electron DSF and assuming that the phonon–ripplon coupling is weak.

Thermal fluctuations of the infinite 2D electron solid introduce the following improvements in the resonance term. The resonant denominator $|\omega - \omega_g|$ is changed to $|\omega - \omega_g|^{1-\alpha_g}$, the proportionality factor $\alpha_g(T)$ in (8.13) is changed to

$$\frac{2}{\pi} \sin\left(\frac{\pi \alpha_g}{2}\right) \frac{\Gamma(1-\alpha_g)}{(\omega_m)^{\alpha_g}} ,$$

and the Debye–Waller function $W_g(T)$ is set to $W_g(0)$. Here $\Gamma(z)$ is the gamma function and the characteristic frequency $\omega_m = \min(T/\hbar, c_t k_m)$ was defined previously in (7.124). Thus, in the limiting case $\alpha_g(T) \ll 1$, the long-wavelength fluctuations of the infinite 2D electron solid can be disregarded and the one-phonon terms of (8.13) give a quite accurate result, if $W_g(T)$ is set to $W_g(0)$. It should be noted that, for the smallest reciprocal lattice vector, below the WS melting point, the parameter $\alpha_g \leq 1/3$, although it increases fast with g.

If the coupling of WS phonons with medium excitations is strong, the electron conductivity relaxation kernel considered here involves only fast WS phonon modes, as discussed above. In the self-consistent treatment, the phonon spectrum $\Omega_{t,k}$ entering (8.10) for the DSF is replaced by the fast mode dispersion $\sqrt{\omega_f^2 + c_t^2 k^2}$. The limiting frequency $\omega_f \gg \omega_g$, and therefore the phonon absorption term in the electron DSF [see (8.10)] can be omitted. Then, assuming $k \ll g$ and rearranging $n_B(\hbar\omega)+1$, the one-phonon term can be rewritten as

$$S^{(1\text{-ph})}(\boldsymbol{q}, \omega) = \frac{\hbar q^2 S_A e^{-2W_q}}{4 m_e c_t^2 \left(1 - e^{-\hbar\omega/T}\right)} \theta(\omega - \omega_f) \sum_{\boldsymbol{g}} \delta_{\boldsymbol{q},\boldsymbol{g}} , \tag{8.15}$$

where $2W_q = q^2 \langle u_f^2 \rangle /2$.

When substituting the above form of the WS DSF into the equation for the effective collision frequency given in (8.10), we disregard ω_g in comparison with ω_f in all smooth functions. However, we keep it in the θ-function because ω_g can be comparable with $\omega - \omega_f$. Then straightforward evaluation yields [149]:

$$\nu_{\text{WS}}^{(1\text{-ph})}(\omega) = \frac{T}{32 \alpha m_e^2 \omega c_t^2} \sum_{\boldsymbol{g}} g^2 |V_g|^2 e^{-g^2 \langle u_f^2 \rangle /2}$$
$$\times \left[\theta(\omega - \omega_g - \omega_f) + \theta(\omega + \omega_g - \omega_f)\right] . \tag{8.16}$$

The origin of the two terms in the square brackets is clear from the corresponding arguments of the θ-functions. The first term reflects ripplon emission, while the second is due to ripplon absorption processes.

The electron–ripplon scattering is usually treated quasi-elastically: the ripplon frequency ω_g is disregarded in the argument of the θ-functions. In this instance, the two terms of (8.16) give the same contribution. If such an approximation is applied and the limiting frequency $\omega_{\mathrm{f}} \to 0$ (weak coupling regime), then (8.16) reproduces the low temperature asymptote for the PR linewidth found by Dykman [268] and by Makabe and Saitoh [269] using different treatments:

$$\nu_{\mathrm{WS}}^{(*)}(\omega) = \frac{T}{16\alpha m_{\mathrm{e}}^2 \omega c_{\mathrm{t}}^2} \sum_g g^2 \left| V_g \right|^2 \mathrm{e}^{-2W_g(0)} . \tag{8.17}$$

This result is actually a high-frequency approximation, $\omega \gg \omega_g$. At lower frequencies, one should keep $|\omega - \omega_g|$ in the argument of the Bose distribution function $n_{\mathrm{B}}(|\omega - \omega_g|)$ which restores the resonance structure of (8.13) when $\hbar|\omega - \omega_g| \ll T$.

The approximations (8.2) and (8.15) for the electron DSF of the WS represent two opposite limiting cases of high and low temperatures. In the medium regime, the Debye–Waller function $2g^2 \left\langle u_{\mathrm{f}}^2 \right\rangle /2$ increases and the contribution of one-phonon processes dies out. Because the main contribution to the effective collision frequency of (8.16) and (8.17) comes from quite large reciprocal lattice vectors, the substantial increase in $2g^2 \left\langle u_{\mathrm{f}}^2 \right\rangle /2$ should be taken into account for rather high temperatures. First we note that the decrease in the contribution from one-phonon scattering with increasing T does not mean a real decrease in the effective collision frequency, because the inclusion of multi-phonon terms and the corresponding replacement of (8.15) by (8.2) makes the scattering stronger.

According to the discussion in Sect. 7.5.1, the influence of multi-phonon processes on the collision rate can be taken into account qualitatively if $W_g(T)$ is replaced by $W_g(0)$. A more rigorous treatment requires us to include the multi-phonon correction $\delta S(q,\omega)$ from (7.138), which yields

$$\nu_{\mathrm{WS}}^{(\text{m-ph})} = \frac{(1 - \mathrm{e}^{-\hbar\omega/T}) T (eE_\perp)^2}{4\sqrt{2\pi\Delta}\hbar^2 \omega \alpha} \sum_{n=2}^{\infty} \frac{1}{\sqrt{n}} \exp\left[-\frac{(\omega\tau_0 - n)^2}{2n\Delta} \right] Y_n , \tag{8.18}$$

where

$$Y_n = \int_0^\infty \frac{x^n \mathrm{e}^{-x}}{n!} V_{\mathrm{S}}^2(x) \mathrm{d}x , \qquad V_{\mathrm{S}}(x) = 1 + \frac{\Lambda x}{eE_\perp l_{\mathrm{f}}^2} w_{\mathrm{C}} \left(\frac{x}{2\gamma^2 l_{\mathrm{f}}^2} \right) , \tag{8.19}$$

and $l_{\mathrm{f}} = \sqrt{\langle u_{\mathrm{f}}^2 \rangle}$. Thus the effective collision frequency $\nu_{\mathrm{WS}}^{(\text{1-ph})} + \nu_{\mathrm{WS}}^{(\text{m-ph})}$ of the WS differs generally from the low-temperature asymptote of (8.16) and (8.17).

At low temperatures, the Wigner solid DSF as a function of the wave vector $\boldsymbol{q}$ represents a set of peaks located at the reciprocal lattice positions $\boldsymbol{q} = \boldsymbol{g}$. The dependence on the magnitude g is smooth and a maximum occurs at $g \sim 1/\sqrt{\langle u_0^2 \rangle}$, irrespective of the frequency ω. This behavior of the WS DSF is contrary to that found for the gas state of the electron layer. Therefore one could expect a big difference in transport properties of the WS and free electron gas at $T \ll T_{\mathrm{D}}$. At first glance, (8.16) and (8.17) found for the Wigner solid confirm this guess. Indeed, (8.16) contains the physical quantity, namely the transverse sound velocity c_{t}, attributed to the solid state only. There is no such quantity in the gas state. Following [268], it might be concluded that the WS can be revealed even from a study of the longitudinal mode damping. Surprisingly, a more detailed analysis [149] leads to the opposite conclusion.

Searching for a low temperature asymptote of (8.16) and (8.17), one can replace $W_g(T)$ by $W_g(0)$ and $\sum_{\boldsymbol{g}}$ by $N_{\mathrm{e}}^{-1} \sum_{\boldsymbol{q}}$. For reasons of simplicity, we consider the limiting case of high holding fields: $V_q \simeq eE_{\perp}$. We then obtain

$$\nu_{\mathrm{WS}}^{(*)}(\omega) \simeq \frac{T(eE_{\perp})^2}{8\pi m^2 \omega a n_{\mathrm{s}} c_{\mathrm{t}}^2 \left[\langle u_0^2 \rangle\right]^2} , \tag{8.20}$$

where $\langle u_0^2 \rangle$ is the zero-point mean-square fluctuation. In the Debye approximation,

$$\langle u_0^2 \rangle \simeq 1.248 \frac{\hbar}{2m\sqrt{\pi n_{\mathrm{s}}} c_{\mathrm{t}}} . \tag{8.21}$$

It is easy to see that inserting (8.21) into (8.20) leads to cancellation of the sound velocity c_{t}. As a result, the low temperature ($T \ll \hbar\omega$) asymptote of $\nu_{\mathrm{WS}}^{(*)}(\omega)$ is actually the same as that found for the nondegenerate electron gas [see (8.3)] up to a small numerical difference (about 20%) in the proportionality factor. Comparing (8.3) and (8.16), we also find that the frequency dependence of the WS conductivity is not unique, as might have been concluded. It is actually the same as the frequency dependence of the quantum conductivity of nondegenerate free electrons.

Equation (8.20) is found for the low-temperature regime ($\langle u_{\mathrm{f}}^2 \rangle \simeq \langle u_0^2 \rangle$), approximating the coupling potential by $V_q \simeq eE_{\perp}$. Therefore, the polarization part of the interaction potential V_q, important at low electron densities, and also thermal vibrations should affect the conclusion stated above. In order to take these things into account, we have to carry out numerical evaluations. Consider first an initially flat surface without the dimple lattice, or conditions where the frequency of the experimental signal is away from the limiting frequency ω_{f}. In these instances, both ripplon emission and absorption contribute to the effective collision frequency. The results of numerical evaluations of the momentum relaxation time $\tau = \nu_{\mathrm{eff}}^{-1}$ for different approximations are gathered in Figs. 8.2a–d. These figures are plotted for different electron densities and frequencies of the signal. Here we find it instructive to show curves evaluated for the degenerate gas (DG) and the low temperature asymptote for the Wigner solid ($\mathrm{WS}^{(1)}$) even beyond their validity ranges.

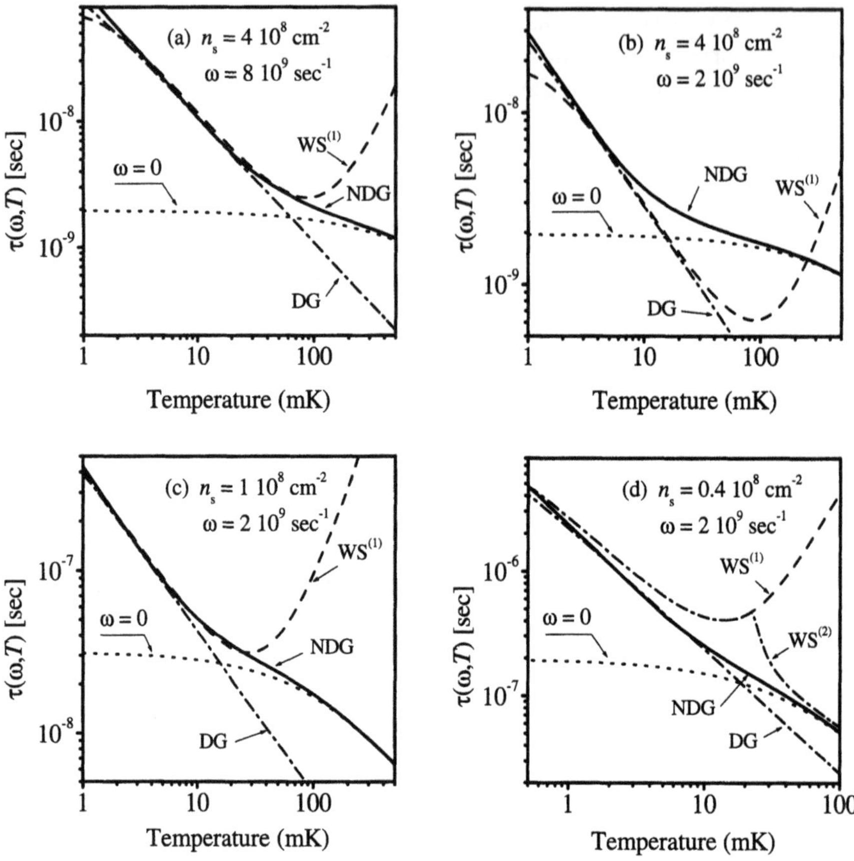

Fig. 8.2. Electron momentum relaxation time $\tau = \nu_{\text{eff}}^{-1}$ vs. temperature for different n_s and ω (**a**–**d**). *Continuous curve*: nondegenerate gas (NDG). *Dashed curve*: Wigner solid (one-phonon contribution to τ). *Dash-dotted curve*: degenerate Fermi gas (DG). The WS curve marked $\text{WS}^{(2)}$ in (**d**) takes into account the multi-phonon terms as discussed in the text. The *dotted curve* ($\omega = 0$) represents the DC relaxation time of nondegenerate electrons

The main conclusion which follows from these evaluations confirms the result of the simple analytical analysis given above. When varying electron density within an order-of-magnitude range, the low temperature curves representing the relaxation time of the WS (dashed line) and nondegenerate gas (continuous curve NDG) remain close. A substantial difference appears only for ultra-low temperatures, where the WS curve approaches the saturation regime faster than the gas curves. This difference appears because the reciprocal lattice vectors contributing to the effective collision frequency of (8.6) are substantially larger than q_ω of (8.6).

It should be noted that, in the low temperature range $T < \hbar\omega$, all AC curves deviate strongly from the DC approximation (dotted curve). Figures 8.2a and b show how a frequency change affects the relaxation time. It is clear that the frequency decrease changes the low temperature asymptotes in the same way (shifting leftward) for all approximations used: degenerate Fermi gas (DG), nondegenerate gas (NDG) and Wigner solid ($\mathrm{WS}^{(1)}$ marks the contribution from the one-phonon processes alone).

The next two figures (Figs. 8.2c and d) show the effect of a density change on the electron relaxation time at a fixed signal frequency (keeping the saturation condition $E_{\perp} = 2\pi e n_{\mathrm{s}}$). The density decrease appears to affect the low temperature asymptotes in the opposite way to a frequency decrease. In Figs. 8.2a–d, the frequency of the signal ω and the electron density n_{s} were varied independently. In a PR experiment, the resonant frequency increases with electron density because of the increase in the plasmon frequency and electron–ripplon coupling. As a result, the effect produced by a density change is usually compensated by the effect caused by a corresponding change in the resonant frequency.

Although the electron Fermi energy is rather small ($1 < \varepsilon_{\mathrm{F}}^{(\mathrm{e})} < 11\,\mathrm{mK}$) under the conditions of Figs. 8.2a–d, in the low temperature range, the curves (DG) calculated for the Fermi degenerate gas are also very close to the result found for nondegenerate electrons.

At medium temperatures, depending on n_{s} and ω, the relaxation time found for one-phonon processes (dashed curve $\mathrm{WS}^{(1)}$) bursts into a sharp increase due to the HFDWF. In this range, the electron DSF transforms from (8.15) to (8.2), which should actually lead to a decrease in τ. In order to see this, one should take into account multi-phonon processes. First, it should be noted that we cannot just carelessly add the contributions from (8.16) and (8.18), because the multi-phonon terms have poor asymptotic behavior at low temperatures if $K_{\mathrm{e}} > T$. The right asymptotic behavior at both high and low temperatures can be found just by replacing K_{e} by T. This can be considered as an upper bound for the electron relaxation time. For this case, the result of combining contributions from the one-phonon term and multi-phonon terms $\tau(\omega) = 1/(\nu_{\mathrm{WS}}^{(\text{1-ph})} + \nu_{\mathrm{WS}}^{(\text{m-ph})})$ is shown in Fig. 8.2d by the curve marked $\mathrm{WS}^{(2)}$. Although one cannot interpolate smoothly between different approximations (coherent and incoherent), the multi-phonon terms supposedly and correctly bend the WS curve down to the curve found for a nondegenerate electron gas.

As expected, the WS and nondegenerate electron gas have the same conductivity if T is larger than the WS Debye temperature T_{D}. According to Figs. 8.2a–d, the AC conductivities of the WS and nondegenerate electron gas appear to be very close in the opposite temperature regime ($T \ll T_{\mathrm{D}}$) as well. It is hard to imagine that there would be a substantial difference in the intermediate case $T \sim T_{\mathrm{D}}$. The latter is also supported by the inclusion of multi-phonon processes. Therefore we conclude that the high-frequency conductivities of the 2D WS and nondegenerate electron gas are very close and actually have the same dependencies on the main parameters of the system despite the big differences in their excitation spectra.

8.2.2 Spectrum-Splitting Reduction of Phonon Damping

The important assumption made when obtaining (8.17) [268, 269] is that the reconstruction of the WS phonon spectrum (in other words, the inclusion of the frequency ω_{f} of electron oscillations in a dimple) does not much affect the PR linewidth. The above statement is actually valid for the weak coupling regime (no dimple lattice present), or for excitation frequencies which are substantially higher than the position of the bottom of the optical modes.

The PR linewidth changes a great deal if the excitation frequency is very close to the limiting frequency of optical phonons ω_{f}, and the difference $\omega-\omega_{\mathrm{f}}$ becomes smaller than the ripplon frequency ω_g. This situation occurred in the plasmon resonance experiment [235, 281, 282], where the wavelength of excited optical transverse phonons was determined by the size of the experimental cell. In this case, the ripplon emission term of (8.16) becomes zero or negligibly small and the effective collision frequency reduces to

$$\nu_{\mathrm{WS}}^{(1\text{-ph})}(\omega) = \frac{T}{32\alpha m_{\mathrm{e}}^2 \omega c_{\mathrm{t}}^2} \sum_{\boldsymbol{g}} g^2 \left|V_g\right|^2 \mathrm{e}^{-g^2 \langle u_{\mathrm{f}}^2 \rangle /2} \theta(\omega - \omega_{\mathrm{f}} + \omega_g) \, , \qquad (8.22)$$

half of the result $\nu_{\mathrm{WS}}^{(*)}(\omega)$ found for the weak coupling regime [268, 269]. Under certain conditions, one or two reciprocal lattice vectors may contribute to ripplon emission, but at low temperatures the main contribution to the sum over the reciprocal lattice vectors comes from terms with $g \gg g_1$, and we still have the factor 1/2 for the effective collision frequency. If the fast phonon mode is excited by the external electric field in a plasmon resonance experiment, then the frequency of the signal satisfies $\omega \geq \omega_{\mathrm{f}}$ and the θ-function of (8.22) is obviously equal to unity.

Consequently, the spectrum-splitting of the WS phonons into fast and slow modes reduces $\nu_{\mathrm{WS}}^{(1\text{-ph})}(\omega)$ by a factor of one half if ω is close to the limiting frequency of the WS optical modes ω_{f}. Therefore, the WS curve, which is close to the curve of the nondegenerate electron gas at $T > T_{\mathrm{D}}$, should bend and follow the reduced asymptote of (8.22) with falling temperature rather than the asymptote $\nu_{\mathrm{WS}}^{(*)}$ found by disregarding the renormalization of the phonon spectrum. For the conditions of the experiment [235] ($n_{\mathrm{s}} = 1.4 \times 10^8\,\mathrm{cm}^{-2}$, $B = 0$), theoretical curves and data are shown in Fig. 8.3. It should be noted that, at high temperatures, the validity ranges of both low temperature WS asymptotes (continuous curves 1 and 2) are restricted at least by the dotted curve representing the DC relaxation time of the nondegenerate gas. One can see that at high temperatures, the experimental data follow the dashed (AC) curve of the nondegenerate electron gas.

Then, at lower temperatures, the data transform and follow the continuous curve 2 [the asymptote of (8.22)], which is substantially removed from the continuous curve 1 representing the case when the reconstruction of the WS spectrum is disregarded [see (8.17)]. In this way, the linewidth data indicate the appearance of the dimple lattice and favor the self-consistent treatment of the strong coupling regime.

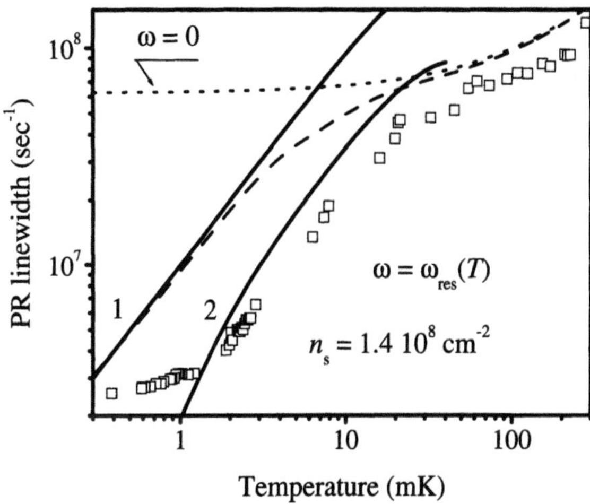

Fig. 8.3. Plasmon resonance linewidth vs. temperature under zero magnetic field: experimental data (*squares*), DC collision frequency of free electrons (*dotted curve*), AC collision frequency of the nondegenerate electron gas (*dashed curve*), one-phonon contribution to the effective collision frequency $\nu_{\mathrm{WS}}^{(*)}$ of the WS (*continuous curve* 1), WS 'collision frequency' $\nu_{\mathrm{WS}}^{(1\text{-ph})}$ affected by dimples (*continuous curve* 2) [149]

Another set of data was found for the WS subject to a weak magnetic field applied normally ($B = 150\,\mathrm{G}$) [235]. This field mainly affects the position of the resonance frequency, according to (7.37). In Fig. 8.4, we compare these data with the zero magnetic field theory, taking into account the above-mentioned change in the resonance frequency. Once again, the high temperature data follow the AC curve evaluated for the nondegenerate electron gas until it meets with the continuous curve 2. Then the data set bends and follows the continuous curve found for the reduced effective collision frequency [see (8.22)].

The origin of the temperature-independent regime appearing below 20 mK cannot be understood in the framework of the ripplon scattering. The wave vectors contributing to the effective collision frequency of the WS and electron gas are too small to make the ripplon distribution function $N_q^{(\mathrm{r})}$ less than unity at $T < 20\,\mathrm{mK}$. The temperature-independent contribution of the elastic term (7.122) is very small because of the factor $\exp(-g^2 \left\langle u_{\mathrm{f}}^2 \right\rangle /2)$: typical wave

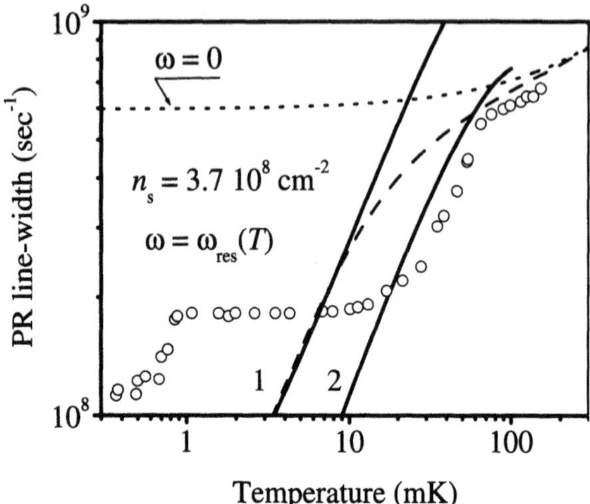

Fig. 8.4. Plasmon resonance linewidth vs. temperature under a weak magnetic field ($B = 150\,\mathrm{Gs}$). Experimental data (*circles*) [149]. The notation for theoretical curves is the same as in Fig. 8.3

numbers g^* defined by the equation $\omega = \omega_g$ are very large (for the conditions of Fig. 8.4, we estimate $g^* \simeq 1.6 \times 10^6\,\mathrm{cm}^{-1}$). Generally, (7.122) is also applicable for the surface sound mode of Fermi liquid ^{3}He [284, 285], although in this case the exact form of the interaction potential V_q is not known. The important point is that the velocity of the surface sound is expected to be high (of the order of the Fermi velocity of liquid ^{3}He), which significantly reduces $g^* \simeq 1.6 \times 10^5\,\mathrm{cm}^{-1}$ and increases the HFDWF, $\exp(-g^2 \langle u_{\mathrm{f}}^2 \rangle /2)$. The latter means that the direct decay of plasmons into a surface sound mode may be quite probable, leading to a temperature-independent PR linewidth. Quantitative analysis of this effect requires an additional study of the electron interaction with the surface sound mode. Another cause of the temperature-independent linewidth may appear owing to the small motion of the dimple lattice.

8.2.3 Resonance Structure of the Collision Rate

The available experimental information concerning the electron–ripplon resonances, which served as conclusive evidence of electron ordering in a 2D triangular Wigner lattice [7], can also be used to test theoretical models of WS transport: the weak coupling theory and the self-consistent strong-coupling approach.

According to the analysis in the last chapter, the WS conductivity depends quite specifically on the frequency of the driving electric field because of the excitation of the coupled phonon–ripplon modes. For example, the transport framework discussed in Chap. 3 results in the following form for the real part of the electron conductivity:

$$\mathrm{Re}\,\sigma(\omega) = \frac{e^2 n_{\mathrm{s}}}{m} \frac{\nu(\omega)}{\nu(\omega)^2 + [\omega + w(\omega)]^2} \equiv \frac{e^2 n_{\mathrm{s}}}{m} \frac{\nu(\omega)}{\nu(\omega)^2 + \omega^2 \mathcal{Z}^2(\omega)} \,. \tag{8.23}$$

Here we consider the case of a uniform excitation electric field. For nonuniform excitation, one has to include $\Omega_{1,k}^2$ in the denominator, as discussed in Sect. 7.4.2. The frequency dependencies of the functions $w(\omega)$ and $\mathcal{Z}(\omega)$ have the resonance structures and both are proportional to $\omega_g^2/(\omega_g^2 - \omega^2)$, when $\omega \to \omega_g$ [see (7.80) and (7.91)]. The conductivity resonances discussed in Sect. 7.4.2 occur when $\omega + w(\omega) = 0$, or, if the driving field is nonuniform, when ω coincides with the solution of the secular equation given in (7.97).

It should be noted that the effective collision frequency with smooth frequency dependence $\nu(\omega)$ cannot explain all observed features of the electron–ripplon resonances. Experimentally, if the coupling is strong, there is no conductivity anomaly exactly at $\omega = \omega_1$, but there is a resonance at a substantially lower, displaced frequency. Still, for a smooth dependence $\nu(\omega)$, the real part of the electron conductivity of (8.23) definitely has an antiresonance anomaly at $\omega \to \omega_1$: $\mathrm{Re}\,\sigma(\omega) \to 0$, owing to $\mathcal{Z}(\omega) \to \infty$. This does not happen in reality because the effective collision frequency, besides the smooth part, has the resonance term itself, as discussed above.

Consider first the weak coupling regime. In this approximation, $\mathcal{Z}(\omega)$ is close to unity ($C_n \ll 1$), and the conductivity resonances are due to the resonance structure of the effective collision frequency given in (8.13). Long-wavelength fluctuations of the infinite electron system affect this result in such a way that the line shape of the electron conductivity resonances acquires an additional and unusual broadening [283]:

$$\mathrm{Re}[\sigma(\omega)] \propto \frac{1}{|\omega - \omega_n|^{1-\alpha_g}} \,, \tag{8.24}$$

where the parameter $\alpha_g(T)$ is from (8.14). The proportionality factor which we have omitted in this relation decreases linearly with cooling. It is interesting that the condition $\alpha_g < 1$ required for the appearance of the electron–ripplon resonance in this instance depends strongly on g_n. The resonances can appear successively when the temperature is equal to the characteristic temperatures T_n^* defined by the equation $\alpha_{g_n}(T_n^*) = 1$. The succession of T_n^* is as follows: $T_1^* = 3T_{\mathrm{m}}$, $T_2^* = T_{\mathrm{m}}$, and $T_3^* = 3T_{\mathrm{m}}/4 < T_{\mathrm{m}}$, where $T_{\mathrm{m}} = n_{\mathrm{s}} a^2 m_{\mathrm{e}} c_{\mathrm{t}}^2 / 4\pi$ coincides with the WS melting temperature defined previously in (7.21). Therefore, at the melting point $T = T_{\mathrm{m}}$, according to the weak coupling theory, there should only be the main electron–ripplon resonance at $\omega = \omega_1$. For other resonance frequencies ($n > 1$), $\alpha_g \geq 1$, and the electron conductivity given in (8.24) does not have the resonance structure.

With falling temperature, according to the relation between T_3^* and T_{m} written above, the weak coupling theory results in a 25% delay in the appearance of the electron–ripplon resonance with $n = 3$, as compared to the resonance with $n = 2$. For example, resonance Z of the experiment of Grimes and Adams shown in Fig. 7.6 should not be observed before $T = (3/4) \times 0.46\,\mathrm{K} \simeq 0.345\,\mathrm{K}$. Still, it is already clearly seen at a substantially higher temperature, viz., $T = 0.42\,\mathrm{K}$.

A detailed study of the electron–ripplon resonances was also reported by Deville [286]. He observed the high-order resonances up to $\omega = \omega_{13}$. The succession of absolute values of the reciprocal lattice vectors is sometimes denoted by integers p_n defined by (7.89) or simply by the relation

$$g_n^2 = p_n g_1^2 \,. \tag{8.25}$$

For the 2D triangular lattice, the numbers relevant to the experiment [286] are $p_n = 1, 3, 4, 7, 9, 12, 13, 16, 19, 21, 25, 27, 28, 31, \ldots$. The experimental absorption trace from SEs on liquid helium is shown in Fig. 8.5, indicating electron–ripplon resonances with $1 \leq p_n \leq 28$. The high-order resonances appear successively at the progressively lower temperatures below T_m as their coupling constants C_n increase. It was reported that, at $T \simeq T_\mathrm{m}/5$, modes up to $p_{13} = 28$ can be detected for typical densities $n \simeq 2 \times 10^8$ to $12 \times 10^8\,\mathrm{cm}^{-2}$. Under these conditions, the characteristic temperature T_n^*, below which the resonance can be observed in the weak coupling theory, is substantially lower than T: $T_{13}^* = 3T_\mathrm{m}/28 \simeq 0.1\,T_\mathrm{m}$. The absorption trace with the lower expanded frequency scale present in Fig. 8.5 shows typical resonance shaping. Thus both experiments [7, 286] allow us to conclude that the weak coupling treatment of thermal fluctuations of the 2D WS overestimates the role of long-wavelength vibrations and does not actually help much in understanding the line shape and appearance of electron–ripplon resonances under typical experimental conditions.

At low temperatures, when $\alpha_g(T) \ll 1$, the non-Lorentzian structure of the collision rate $\nu(\omega) \propto 1/|\omega - \omega_n|$ discussed above is not caused by long-wavelength fluctuations of the infinite 2D electron crystal. It appears due to the inelastic scattering involving one WS phonon with a quite large wave number $k = |\omega - \omega_g|/c_\mathrm{t} \gg 1/L$. Long-wavelength fluctuations contribute to the temperature-dependent correction $\alpha_g(T)$ to the exponent -1. It is interesting that the same non-Lorentzian structure of the collision rate could be found if we included the slow coupled modes $\Omega_{\mathrm{t},\boldsymbol{k}}^{(\mathrm{s})}$ in the low temperature expansion for the WS DSF of the strong coupling regime. Still, as discussed in Sect. 7.5.3, this procedure is not consistent for the strong coupling treatment given in the last chapter, because the most important part of the electron–ripplon interaction proportional to $\xi_q i\boldsymbol{q} \cdot \boldsymbol{u}_l^{(\mathrm{s})}$ is already included in the Hamiltonian of the slow coupled phonon–ripplon modes strongly affecting the WS phonon spectrum, and it cannot lead to any scattering of excitations of the new modes.

If the slow coupled phonon–ripplon modes ($\omega < \omega_1$) are excluded from the DWF of the conductivity relaxation kernel, then at typical frequencies of the signal $\omega \sim \omega_g \ll \omega_\mathrm{f}$, the effective collision frequency $\nu(\omega)$ does not have

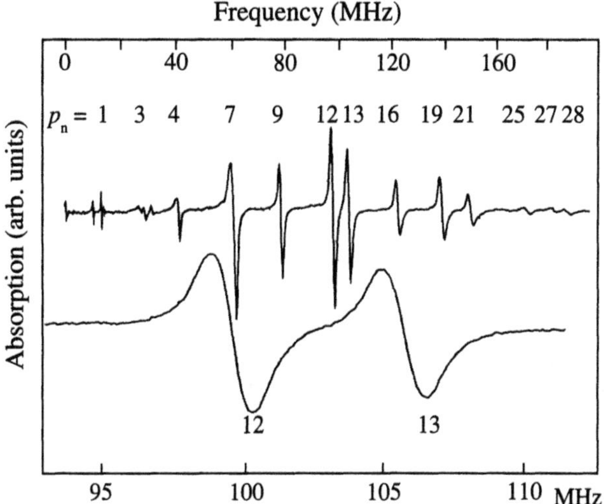

Fig. 8.5. Observed electron–ripplon resonances for $n_s = 5.25 \times 10^8\,\mathrm{cm}^{-2}$ and $T = 60\,\mathrm{mK}$ [286]. The phase-sensitive detector supplies a signal that is the derivative of the absorption signal with respect to the electron density. The order of p_n is indicated for each mode. Modes $p_6 = 12$ and $p_7 = 13$, drawn with the lower expanded frequency scale, show the typical linewidth of the resonances. The chosen conditions of potential modulation favor the detection of high-order resonances to the detriment of the $p_n = 1, 3$, and 4 modes, which could give a much larger signal under different conditions of modulation

the one-phonon terms [see (8.16)] because of the factor $\theta(\omega - \omega_{\mathrm{f}} \pm \omega_g)$. In this frequency region, only the multi-phonon terms contribute to $\nu(\omega)$. These terms, given in (8.18), have a smooth frequency dependence and can therefore serve as a broadening factor for the phonon–ripplon resonances observed when $\omega < \omega_1$. According to the numerical analysis of these terms shown in Fig. 8.2, in the high temperature range, they are close to the collision frequency of the nondegenerate electron gas. At least, the DC collision frequency of the nondegenerate electron gas can be taken as an upper bound for $\nu(\omega)$ entering the conductivity equation. Then the simple estimate implies that, under typical experimental conditions, ν, ω and ω_1 are of the same order. Still, the resonance excitation appears because the conductivity denominator in the nonuniform excitation field has the form

$$\nu^2 + \left[\omega \mathcal{Z}(\omega) - \Omega_{1,k}^2/\omega\right]^2 , \tag{8.26}$$

where $\mathcal{Z}(\omega) \propto C_1\omega_1^2/(\omega_1^2 - \omega^2) \gg 1$ and the plasmon excitation frequency over the flat surface $\Omega_{1,k} \gg \nu$. Both terms in the square brackets are much larger than ν, and therefore, when they compensate each other at the resonant frequency, the conductivity increases by a factor $(C_1\omega/\nu)^2 \sim C_1^2$.

Regarding the elimination of the conductivity antiresonance at $\omega = \omega_1$, we have to examine the elastic term in more detail. According to the definitions of $w(\omega)$ and $\nu(\omega)$ given in (7.80) and (7.81), the WS conductivity relaxation kernel $M(\omega) = w(\omega) + \mathrm{i}\nu(\omega)$ can be written as

$$M(\omega) = \frac{1}{2\omega} \sum_g v_g^2 \mathrm{e}^{-2W_g} \frac{\omega^2}{\omega_g^2 - \omega^2 - 2\mathrm{i}\omega\omega_g\delta} , \tag{8.27}$$

where $\delta > 0$ is the infinitesimal parameter. If energy dissipation in the medium is taken into account, the parameter $\omega_g\delta$ should be replaced by the damping of the capillary waves γ_g. In the case of superfluid ^{4}He with a small amount of ^{3}He impurity atoms absorbed in the Andreev surface states [55], the above replacement even constitutes a numerically accurate procedure [262]. For usual hydrodynamics, a more accurate treatment will be described in the next section.

With the damping of surface waves taken into account, the dimensionless response function $Z(\omega) = 1 + M(\omega)/\omega$ has the form

$$Z(\omega) = 1 + \sum_{n=1}^{\infty} C_n \frac{\omega_n^2}{\omega_n^2 - \omega^2 - 2\mathrm{i}\omega\gamma_n} , \tag{8.28}$$

where γ_n is γ_q with $q = g_n$. The real part of this function $\mathcal{Z}(\omega) = \mathrm{Re}\, Z(\omega)$ which should be used in the conductivity denominator and in the secular equation for the coupled modes is not infinite when $\omega \to \omega_1$:

$$\mathcal{Z}(\omega) \simeq \frac{1}{2} \frac{\omega_1(\omega_1 - \omega)}{(\omega_1 - \omega)^2 + \gamma_1^2} . \tag{8.29}$$

Then the imaginary part $\mathrm{Im}\, Z$, which determines the effective collision frequency $\nu^{(\mathrm{elas})}(\omega)$ in the conductivity equation, has the right resonance line shape

$$\nu^{(\mathrm{elas})}(\omega) = \omega\, \mathrm{Im}\, Z(\omega) \simeq \frac{1}{2} C_1 \frac{\gamma_1\omega_1}{(\omega_1 - \omega)^2 + \gamma_1^2} .$$

This form exactly compensates the resonance anomaly of $\mathcal{Z}^2(\omega)$ at $\omega \to \omega_1$ in the denominator of the WS conductivity formula.

In superfluid helium, the damping of ripplons is very small and therefore the broadening of the low-frequency resonances with $\omega < \omega_1$ (W and X resonances of [7]) is most probably determined by the multi-phonon terms of the electron collision frequency. Then the shape of the shifted phonon–ripplon resonances is close to the usual Lorentzian, which agrees with experimental observations. The origin of the electron–ripplon resonances at $\omega = \omega_n$ with high numbers n depends on the relation between the coupling constant C_n and the inverse dimensionless damping ω_n/γ_n. It is clear that, for $C_n \ll C_1$, even a small damping can eliminate the coupled modes which are close to ω_n with $n \geq 2$: owing to (8.29), the difference $\omega^2 \mathcal{Z}(\omega) - \Omega_{1,k}^2$ cannot be zero or substantially smaller than the main terms. In this instance, the principal cause of the resonant absorption at $\omega = \omega_n$ is the resonance structure of the effective collision rate $\nu^{(\mathrm{elas})}(\omega) = \omega \operatorname{Im} Z(\omega)$, which also has a Lorentzian shape.

To conclude this section, we note that the quantum transport framework for highly correlated electrons (Chap. 3) establishing the relation between the effective collision frequency and the equilibrium electron DSF appears to be very fruitful for the analysis of the AC conductivity and phonon damping of the 2D electron solid. By using different approximations for the electron DSF, one can reveal general effects associated with the 2D electron system, regardless of the actual state of the electron matter, as well as effects which exist only for the Wigner solid state.

8.3 Mobility over Normal and Superfluid ^{3}He

As noted above, the high-frequency conductivity of the electron system formed on a liquid helium surface does not change much at the WS transition. Moreover, the conductivities of the WS and electron gas appear to be close, even numerically. In the low-frequency case $\omega \ll \omega_g$, the dimple lattice accompanies the WS motion and one can expect a significant change in the electron transport. Above superfluid ^{4}He, dimples are very mobile (even more mobile than electrons) and one can expect the mobility of the WS to be of the same order of magnitude as that of the electron gas. Changes are observed in the electron conductivity as a function of T, such as a step-like decrease and an excess scattering near the melting point reported in [50]. A change in the holding field dependence of the electron mobility was also observed in [231]. Anyway, the situation changes a great deal if SEs are formed on the surface of Fermi liquid ^{3}He. In the case of bulk liquid ^{3}He with an initially flat surface, electrons become self-trapped in the heavily viscous Fermi liquid substrate because of the creation of the dimple lattice. The Fermi liquid properties of the substrate crucially affect the motion of the dimples and electron conductivity at low frequencies of the driving electric field [54, 234].

It is known that the liquid isotopes of helium, ^{3}He and ^{4}He, show prominent differences in bulk properties at low temperatures. The origin of the differences is understood on the basis of the different statistics of the isotopes. It should be emphasized that electron motion along the free surface itself cannot produce any substantial influence on the bulk liquid because of the very short penetration length of the electron wave function into the liquid. It is the motion of surface dimples that couples electron transport to the bulk properties of liquid helium ^{3}He.

The viscosity of normal ^{3}He increases with cooling as $\eta \propto 1/T^2$, and it becomes very high at ultra-low temperatures. The mobility of the electron crystal over liquid ^{3}He is therefore restricted by extremely slow surface dimples in the way discussed in Sect. 6.3. The high viscosity of Fermi liquid ^{3}He results in a high resistivity of the electron state and strong damping of the low-frequency coupled phonon–ripplon modes. With a typical electron spacing of $a \approx 10^{-4}$ cm, different temperature regimes can be expected for the WS transport above ^{3}He: hydrodynamic, long mean-free-path, and superfluid regimes. Each regime introduces a significant change in the WS resistivity as a function of temperature.

8.3.1 Viscous Drag of the Dimple Lattice

The hydrodynamic regime is valid if the quasi-particle mean free path $l_{\mathrm{f}} = 5\eta/(n_{\mathrm{He}}\hbar k_{\mathrm{F}})$ is shorter than the average dimple radius, which yields the temperature range $T > 20$ mK, where n_{He} is the density of liquid helium and $\hbar k_{\mathrm{F}}$ is the Fermi momentum of quasi-particles. In this regime, the WS mobility is determined by the viscosity of ^{3}He and, in order to find the medium response force, one has to include dissipation terms in the equation which determines $\xi_q(t)$.

The proper dissipative corrections can be found using the Navier–Stokes equation of an incompressible fluid [div$(\boldsymbol{v}) = 0$], viz.,

$$\rho \frac{\partial \boldsymbol{v}}{\partial t} + \nabla p = \eta \Delta \boldsymbol{v} \,, \tag{8.30}$$

with the boundary conditions at the free surface

$$\sigma_{zx}^{(\mathrm{str})} = \sigma_{zy}^{(\mathrm{str})} = 0 \,, \tag{8.31}$$

$$-p + 2\eta \frac{\partial v_z}{\partial z} = \alpha \Delta \xi' - \mathcal{P}'_{\mathrm{el}} \,, \tag{8.32}$$

where $\sigma_{ij}^{(\mathrm{str})}$ is the stress tensor, $\boldsymbol{v}$ is the fluid velocity field, p is the fluid pressure, ξ' is the time-dependent correction to the surface profile, and $\mathcal{P}'_{\mathrm{el}}$ is the linear approximation for the time-dependent electron pressure defined according to (7.71):

$$\mathcal{P}'_{\mathrm{el},\boldsymbol{g}} = -V_g n_{\mathrm{s}} \mathrm{e}^{-W_g} \mathrm{i} \boldsymbol{g} \cdot \boldsymbol{u}_{\mathrm{s}}(t) \,, \tag{8.33}$$

where $\boldsymbol{u}_{\mathrm{s}}(t)$ is the slow displacement of the electron lattice and $W_g \equiv g^2 w_{\mathrm{f}} = g^2 \langle u_{\mathrm{f}}^2 \rangle /4$ is the Debye–Waller function.

The solution whereby the velocity field is confined to the boundary $(\mathrm{Re}\sqrt{1 - \mathrm{i}\kappa} > 0,\ \kappa = \omega\rho/\eta g^2)$ can be written as

$$\xi'_{\boldsymbol{g}} = -\frac{g\mathcal{P}'_{\mathrm{el},\boldsymbol{g}}}{\rho \Delta_g} \,, \qquad \Delta_g = \omega_g^2 - \omega^2 - \delta_g^2(\omega) - \mathrm{i}2\omega\gamma_g(\omega) \,. \tag{8.34}$$

Functions $\delta_g^2(\omega)$ and $\gamma_g(\omega)$ represent the response of the viscous fluid to the time-dependent pressure:

$$\gamma_g(\omega) = \frac{\eta}{\rho} g^2 \phi(\kappa) \,, \qquad \delta_g^2 = \omega^2 \chi(\kappa) \,, \tag{8.35}$$

where

$$\phi(\kappa) = 2 - \frac{\sqrt{2}}{\kappa}\left(\sqrt{1+\kappa^2} - 1\right)^{1/2} , \tag{8.36}$$

$$\chi(\kappa) = \frac{4}{\kappa^2}\left[\frac{1}{\sqrt{2}}\left(\sqrt{1+\kappa^2} + 1\right)^{1/2} - 1\right] . \tag{8.37}$$

In the low viscosity regime ($\kappa \gg 1$), $\gamma_g \simeq 2\eta g^2/\rho$ is the damping of ripplons. In the opposite limiting case ($\kappa \ll 1$), $\gamma_g \simeq \eta g^2/\rho$ and $\delta_g^2 \simeq \omega^2/2$. It should be noted that, in the latter instance, γ_g does not represent the damping coefficient. Moreover, the solutions of the dispersion equation $\Delta_g = 0$ are purely imaginary in this regime, in accordance with the analysis of [287].

The surface profile $\xi_g = \xi_g^{(0)} + \xi_g'(t)$ should be substituted into (7.76) for the medium response force. Finally, the influence of the viscous drag can be described by replacing the dimensionless response function $\mathcal{Z}(\omega)$ of (7.91) by

$$Z(\omega) = 1 + \frac{1}{2}\sum_g \frac{v_g^2}{\omega_g^2} \mathrm{e}^{-2W_g} \frac{\omega_g^2(\omega^2 + \delta_g^2 + \mathrm{i}2\omega\gamma_g)}{\omega^2(\omega_g^2 - \omega^2 - \delta_g^2 - \mathrm{i}2\omega\gamma_g)} . \tag{8.38}$$

This function should be used in the secular equation for the coupled phonon–ripplon modes: $\omega^2 \mathrm{Re}\left[Z(\omega)\right] - \Omega_{p,k}$, as discussed in Sect. 7.4.2. For liquid ^{4}He, when $\omega \ll \omega_g$,

$$\mathrm{Re}\, Z(\omega) \simeq 1 + M_\mathrm{d}/m_\mathrm{e} \gg 1 , \tag{8.39}$$

where M_d is the effective mass of a single-electron dimple. In the case of very viscous ^{3}He, $\omega_g \ll 4\gamma_g$ and therefore $\mathrm{Re}\, Z(\omega) < 0$ (!). Moreover, the parameter $2\omega\gamma_g/\omega_g^2$ can be much larger than unity, even for such low frequencies as $\omega = 2\pi \times 10^5\,\mathrm{s}^{-1}$ typically used in experiments with surface electrons. Under the condition stated above, the response function has the oscillatory form

$$\mathrm{Re}\, Z(\omega) \simeq 1 - \omega_\mathrm{f}^2/\omega^2 < 0 , \tag{8.40}$$

which is typical for the high-frequency ($\omega \gg \omega_g$) WS vibrations above ^{4}He. Physically, this unusual behavior of the effective mass function means that dimples of the ^{3}He surface are nearly stiff, even for low frequencies ($\omega \ll \omega_g$).

Huge negative values of $\mathrm{Re}\, Z(\omega)$ lead to the important conclusion that the low-frequency solution of the dispersion equation of the coupled phonon–ripplon modes,

$$\omega^2 \mathrm{Re}\, Z(\omega) = \Omega_{p,k}^2 , \tag{8.41}$$

is purely imaginary ($\omega^2 < 0$). Here $\Omega_{p,k}$ is the longitudinal ($p = \mathrm{l}$) or transverse ($p = \mathrm{t}$) phonon frequency of the WS above a flat surface. This means that low-frequency ($\omega \ll \omega_\mathrm{f}$) resonances observed on ^{4}He by Grimes and Adams [7] cannot exist in the case of normal liquid ^{3}He at sufficiently low temperatures. For a typical electron density $n_\mathrm{e} = 2 \times 10^8\,\mathrm{cm}^{-2}$, the condition $\mathrm{Re}\, Z(0) < 0$ is realized at $T < 0.13\,\mathrm{K}$.

The conductivity of the WS is determined by (7.92) with $\mathcal{Z}(\omega)$ replaced by $Z(\omega)$ from (8.38). In the case of heavily viscous liquid ^{3}He, the effective collision frequency of electrons due to inelastic scattering involving WS phonons is much smaller than $\omega \mathrm{Im}\, Z$. We determine the mobility of the WS as $\mu = (\mathrm{Re}\,\sigma)/em_\mathrm{e}$. One can also introduce the effective collision frequency of the dimple lattice as

$$\mu = \frac{e}{m_\mathrm{e}\nu_\mathrm{d}} \,, \qquad \nu_\mathrm{d}(\omega) = \omega \frac{|Z(\omega)|^2}{\mathrm{Im} Z(\omega)} \,. \tag{8.42}$$

In the formal limit $\omega \to 0$ ($\mathrm{Im}\, Z \gg |\mathrm{Re}\, Z|$), we have

$$\nu_\mathrm{d}(0) = \frac{\eta}{m_\mathrm{e} n_\mathrm{s}} \sum_{\boldsymbol{g}} g^3 \left|\xi_g^{(0)}\right|^2 \,. \tag{8.43}$$

The same result can be found by means of the method discussed in Chap. 6 and used to evaluate the mobility of a 2D ripplonic polaron over a viscous liquid. Instead of an integral over wave vectors $\boldsymbol{q}$ contributing to a polaron dimple, we now have a sum over reciprocal lattice vectors $\boldsymbol{g}$. Just below the Wigner transition, the main contribution to ν_d comes from the terms with smallest $|\boldsymbol{g}|$. At lower temperatures, other terms also become important, contributing approximately 70% of the total value.

For a finite frequency ω, the increase in the viscosity $\eta \propto 1/T^2$ makes $2\omega\gamma_g \gg \omega_g^2$ ($|\mathrm{Re}\, Z| \gg \mathrm{Im}\, Z$), which substantially reduces $\mathrm{Im}\, Z(\omega)$. The same reduction might be expected for the effective collision frequency of (8.42). However, the real and imaginary parts of the dimensionless response function are combined in (8.42) in such a way that the final reduction in the effective collision frequency becomes rather small. Qualitatively, it can be seen from $\nu_\mathrm{d}(\omega) \sim \omega(\mathrm{Re}\, Z)^2/\mathrm{Im}\, Z \sim (\rho/m_\mathrm{e}n_\mathrm{e})g|\xi_g^{(0)}|^2\gamma_g \sim \nu_\mathrm{d}(0)$. At all frequencies, the maximum reduction appears to be smaller than 20%. It is much smaller than the change in $\nu_\mathrm{d}(0) \propto \eta(T)$ because of the increase in viscosity. The maximum reduction is independent of frequency, but it extends to higher temperatures with increasing ω. In practice, when $\omega \simeq 2\pi \times 10^5\,\mathrm{s}^{-1}$, the reduction is even less, because in the long mean-free-path limit, ν becomes independent of T at $T < 20\,\mathrm{mK}$.

We therefore conclude that the capacitive coupling technique used in experiments with electrons on liquid helium ^{3}He can reveal the DC transport properties of the WS, in spite of the high-frequency condition $\omega_g^2 \ll 2\omega\gamma_g$ which arises at low temperatures owing to the high viscosity. In this regime, the WS mobility decreases fast with cooling as $\mu \propto \eta^{-1} \propto T^2$.

8.3.2 Long Mean-Free-Path Regime

At $T < 20\,\mathrm{mK}$, the system is primarily in the long mean-free-path regime. In this case, transport properties of the WS are determined by the reflection of ballistic quasi-particles at the dimple lattice. At such low temperatures, the effective collision frequency $\nu_{\rm d}^*$ of the dimples is approximately two or three orders of magnitude higher than the collision frequency of SEs owing to surface excitations. Therefore, bulk quasi-particle scattering at the dimples is the main cause for the electron resistance.

Let us consider ordinary quasi-particle reflection at the free surface of liquid ^{3}He. We disregard Andreev reflection processes and coherent effects in the superfluid state. The later assumes that the electron spacing in the Wigner lattice is much larger than the coherence length. There are several reasons for considering the quasi-particle scattering at the dimples to be elastic. First, the momentum (along the surface) exchanged in a single collision is proportional to a very small factor $\nabla\xi \sim 10^{-5}$–10^{-6}. Secondly, the huge size of the dimples compared to the size of negative ions does not allow the dimple recoil in a single collision to be considered independently from other collisions. Additionally, dimples cannot be considered as independent objects, because they are bound by the strong Coulomb forces acting between electrons and by electron–ripplon coupling.

In the ballistic limit, the distribution function of quasi-particles moving toward the surface is formed far from the dimples. In the moving reference frame, it can be written as a drift-velocity-shifted Fermi function, viz., $f_0(E_{p,\sigma} - \boldsymbol{p}\cdot\boldsymbol{V}_{\rm dr})$, where $\boldsymbol{V}_{\rm dr}$ is the drift velocity and $E_{p,\sigma}$ is the quasi-particle energy, which is isotropic for the normal and superfluid-B phases of liquid ^{3}He.

In the presence of the dimples, the normal vector $\boldsymbol{n}(x,y)$ to the helium surface is proportional to $(\hat{\boldsymbol{z}} - \nabla\xi)$, where $\hat{\boldsymbol{z}}$ is the unit vector along the z-axis. The drag force and mobility of the single-electron dimple of the interface polaron was discussed in Sect. 6.3.3. Here we can use the main result (6.50), taking into account the fact that the surface displacement is formed by electrons localized at the lattice sites of the Wigner solid: $n_x \simeq -\partial\xi/\partial x$ with $\xi(\boldsymbol{r})$ defined by (7.74). Thus the effective collision frequency of the WS determined by the relation $e\mu^{-1} = m_{\rm e}\nu_{\rm d}^*$ can be written as

$$\nu_{\rm d}^* = \frac{\hbar (k_{\rm F}^{(\rm qp)})^4}{8\pi^2 m_{\rm e} n_{\rm s}} 2f_0(\Delta) \sum_{\boldsymbol{g}} g^2 \left|\xi_g^{(0)}\right|^2 . \tag{8.44}$$

In the normal phase [$\Delta(T) = 0$, $2f_0(0) = 1$], just above the superfluid transition, the WS conductivity [$\sigma = e^2 n_{\rm e}/(m_{\rm e}\nu_{\rm d}^*)$] is practically independent of temperature. The absolute value of the WS conductivity is very sensitive to the exact value of $k_{\rm F}$: $\sigma \propto 1/k_{\rm F}^4$. In the Landau theory, the Fermi momentum sticks to its free value $k_{\rm F}^{(0)}$. The alternative theory discussed in [288] and [289] assumes that half of the ^{3}He atoms are paired and form bosons, which yields $k_{\rm F} = 0.8k_{\rm F}^{(0)}$ [289]. This prediction can be checked experimentally, because the 20% reduction in $k_{\rm F}$ would increase WS conductivity by a factor of $\simeq 2.4$ in comparison with the result found according to the conventional theory of Fermi liquids.

The third important temperature regime occurs below the superfluid transition, $T < T_c \simeq 0.93\,\mathrm{mK}$, where the energy gap appears for the quasi-particle excitation spectrum $E_{p,\sigma} = \sqrt{(E^{(0)}_{p,\sigma})^2 + \Delta^2}$. The gap reduces the momentum absorbed by the moving dimples by a factor $2f(\Delta/k_BT)$. This gives an exponential decrease in the WS resistivity with cooling:

$$\frac{\sigma(T_c)}{\sigma(T)} = \frac{2}{\exp\left[\Delta(T)/k_BT\right] + 1} \, . \tag{8.45}$$

The temperature dependence of the energy gap can be interpolated as follows:

$$\Delta(T) = \Delta(0)\tanh\left[\frac{\Delta^*}{\Delta(0)}\sqrt{\frac{T_c}{T} - 1}\right] \, , \tag{8.46}$$

where $\Delta(0) = 1.74 \times k_BT_c$, and $\Delta^* \simeq 3.06 \times k_BT_c$.

At $T \approx 1\,\mathrm{mK}$, the mobility of the WS above liquid ^{3}He is far beyond the influence of the surface excitations and the basic properties of the WS, including the Debye–Waller factor, can be described in a strict way. Therefore, WS mobility measurements can be used for experimental study of the superfluid phases of liquid ^{3}He.

The WS resistance was measured by Shirahama, Kirichek and Kono [234] using the capacitive coupling method which we discussed in Sect. 4.2. A concentric copper electrode pair (the so-called Corbino disk) is set in the middle of the cell. The inner and outer electrode radius are 10 mm, and 15 mm respectively, while the gap between them is 0.1 mm. Liquid ^{3}He is introduced so as to set the free surface at 1 mm above the electrode. Surface electrons are generated by thermionic emission of a tungsten filament through a hole in the upper electrode at 600 mK. Electrons are confined on the helium surface by applying a positive voltage to the Corbino disk. The electron density is determined by the Wigner solid melting condition [7]: $e^2\sqrt{\pi n}/k_BT_c \simeq 137$.

The cell was mounted on a copper nuclear demagnetization refrigerator. The temperature was measured by a platinum NMR susceptibility thermometer mounted on the nuclear stage, which was calibrated by a ^{3}He melting curve thermometer. An AC voltage of 100 kHz was applied to the inner electrode. The conductivity was found by fitting the output current to the well known formula [see (4.14)]. The experimental data were taken during slow temperature sweeping by adiabatically demagnetizing or magnetizing the nuclear stage.

Experimental data [234] and theoretical results of (8.42) and (8.44) are shown in Fig. 8.6. At high temperatures ($> 20\,\mathrm{mK}$), the WS resistivity varies approximately as $\sigma^{-1} \propto T^{-2}$, in accordance with the result found for the hydrodynamic regime. Because it is impossible to give an accurate description of the transition into the long mean-free-path regime, we have made it sharp for each term in the sum over $|g|$. The corresponding transition temperatures depend on g_n according to the condition $g_n l_f(T) \simeq 1$. Therefore, the theoretical continuous line bends step-by-step to the long mean-free-path result with falling temperature. The experimental data show a smoother transition. In the long mean-free-path regime, the resistivity data are nearly temperature-independent and are in perfect agreement with the theory, which has no fitting parameters.

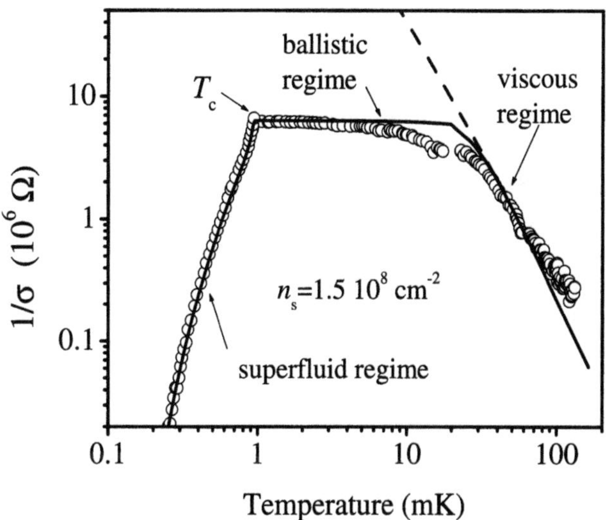

Fig. 8.6. WS resistivity over liquid ^{3}He vs. temperature for the saturation case, $E_\perp = 2\pi en$. *Continuous curve*: theoretical evaluations according to (8.42) and (8.44) without any fitting parameter. *Dashed curve*: viscous drag regime

As noted above, the WS resistivity is very sensitive to the absolute value of the Fermi momentum ($\sigma^{-1} \propto k_F^4$) which sticks to its free value $k_F^{(0)}$ in the Landau theory. Figure 8.6 does not show the reduction in k_F predicted in the alternative theories of normal ^{3}He [288, 289].

Below the superfluid transition a rapid decrease in resistivity is observed, in complete accordance with (8.45). The sharp drop in the resistivity at the superfluid transition is caused by the appearance of the energy gap for the quasi-particle spectrum. It should be pointed out that, from 0.93 mK to 0.23 mK, the WS resistivity decreases by more than three orders of magnitude. We conclude that, in the superfluid phase, WS transport is dominated by the scattering of Bogoliubov quasi-particles, whose distribution is controlled by the superfluid energy gap. The later proves that the WS can be developed as a sensitive quasi-particle detector.

The results discussed above are valid for the isotropic B-phase of the superfluid ^{3}He, which is characterized by an energy gap independent of the direction of the quasi-particle momentum. Superfluid ^{3}He also exists as an anisotropic liquid, in which excitations have an energy gap depending on the angle θ_l between $\boldsymbol{p} = \hbar\boldsymbol{k}$ and the angular-momentum quantization axis expressed by the vector $\boldsymbol{l}$:

$$\Delta(T, \theta_l) = \Delta(T) \sin \theta_l \, , \tag{8.47}$$

where $\Delta(T)$ is defined slightly differently from (8.46):

$$\Delta(T) = 2.031\, T_{\mathrm{c}} \tanh \left(1.687 \sqrt{\frac{T_{\mathrm{c}}}{T} - 1} \right) \, . \tag{8.48}$$

The configuration of $\boldsymbol{l}$ is an intrinsic property of superfluid ^{3}He called the texture, which is determined by chance at the instant of the phase transition. According to (8.48), the energy gap of ^{3}He-A excitations has point nodes in the direction of the texture, affecting their scattering at the dimples.

In order to describe scattering of excitations with angle-dependent energy gap, one has to take this into account when evaluating the integral of (6.47) over the momentum half-space. The final result can be expressed as an unconventional average of the ratio in (8.45):

$$\frac{R}{R_{\mathrm{N}}} = \frac{\sigma(T_{\mathrm{c}})}{\sigma(T)} = \left\langle \frac{2}{\exp\left[\Delta(T)/k_{\mathrm{B}}T\right] + 1} \right\rangle_{\theta_n} \, , \tag{8.49}$$

where the average $\langle \ldots \rangle_{\theta_n}$ is defined by [279]

$$\langle \ldots \rangle_{\theta_n} = \frac{2}{\pi} \int_0^{2\pi} \mathrm{d}\phi \int_0^{\pi/2} d\theta_n \sin(\theta_n) \cos^3(\theta_n) \ldots \, , \tag{8.50}$$

where θ_n is the angle between $\boldsymbol{p}$ and the normal. In general the angles θ_l and θ_n are different if $\boldsymbol{l}$ is not parallel to the surface normal. Therefore the WS conductivity along the surface can be a probe for the texture of superfluid ^{3}He-A near its surface.

It should be noted that, without a magnetic field the A-phase cannot have a free surface because it appears under some pressure above $\sim 2\,$Mpa. Under free surface conditions, the A-phase can be induced by a magnetic field of the order of 0.3 T. The direction of the magnetic field is very important for the WS excitation spectrum. It is convenient to choose it parallel to the surface, in order to reduce its effect on the WS properties and on the texture. Since spatial variation of the vector $\boldsymbol{l}$ costs energy, it tends to be uniform. At the boundary, however, it is energetically favorable to have the orientation perpendicular to the boundary. Near the free surface, $\boldsymbol{l}$ should be parallel to the normal vector. It is also known that $\boldsymbol{l}$ tends to be perpendicular to the magnetic field. It is obvious that the parallel orientation of the magnetic field is compatible with the expected orientation of $\boldsymbol{l}$.

In the case of normal orientation of $\boldsymbol{l}$, the angles θ_n and θ_l are the same. Then the nodes of the energy gap are directed to the surface and excitations heading to a surface dimple have smaller energy gaps, which must increase scattering and WS resistivity. The corresponding result is shown in Fig. 8.7

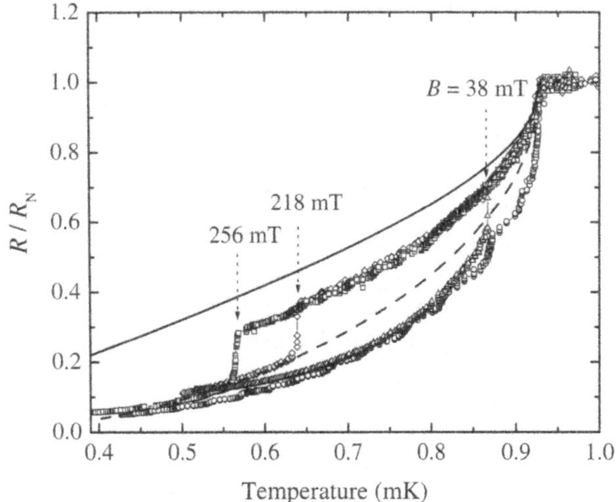

Fig. 8.7. WS resistivity under a magnetic field oriented parallel to the free surface [279]. *Continuous* and *dashed curves*: model calculations for the ^{3}He-A textures perpendicular and parallel to the surface, respectively. *Arrows* and magnetic field values indicate the second resistivity drops in experimental data. Conditions $n_s = 2.1 \times 10^8\,\mathrm{cm}^{-2}$, $E_\perp = 360\,\mathrm{V/cm}$. Measurements were conducted for $B = 0$ (*circles*), 38 (*triangles*), 218 (*diamonds*), and 256 mT (*squares*)

by the continuous curve. If the direction of $\boldsymbol{l}$ were parallel to the surface, the energy gap of excitations falling onto the surface would be near their maximum value and the result would be approximately similar to that found for the B-phase (dashed curve of Fig. 8.7). The measured WS resistivity appears to be somewhat less than the model calculation, as shown in Fig. 8.7. Still, at relatively high temperatures, the data are higher than expected for the parallel texture (dashed curve). With falling temperature, the data display a second drop (the first drop is associated with the transition from normal to superfluid ^{3}He-A), whose position depends on the magnitude of the magnetic field. The position of this drop nearly coincides with the phase transition temperature from ^{3}He-A to ^{3}He-B. This indicates that $\boldsymbol{l}$ is nearly perpendicular to the surface and that the gap nodes substantially increase the resistivity of the WS.

8.4 Nonlinear Transport

In the absence of a magnetic field directed normally, the conductivity measurement in the cylindrical geometry usually involves a driving electric field that excites electron motion and density fluctuations along the radius of the Corbino disk. (The electron velocity v_r is a function of the distance from the disk center r.) At low levels of the driving field, the conductivity measured is assumed to be independent of the signal strength, which corresponds to the linear transport regime. Switching on the magnetic field induces electron rotation around the symmetry axis: $v_\varphi \neq 0$. For the liquid state of the electron system, v_φ is just the Hall velocity v_{H}, and the system is still in the linear transport regime. Shirakhama and Kono [280, 290] observed a remarkable transformation of the electron transport after the electron system underwent the Wigner solid transition in the Corbino experiment with a weak magnetic field applied normally. In this experiment, under the input voltages V_{in} which correspond to the linear transport regime for the liquid phase, below the melting point, the electron transport becomes substantially nonlinear: σ_{xx} exhibits a very complicated dependence on the input signal V_{in}, as shown in Fig. 8.8.

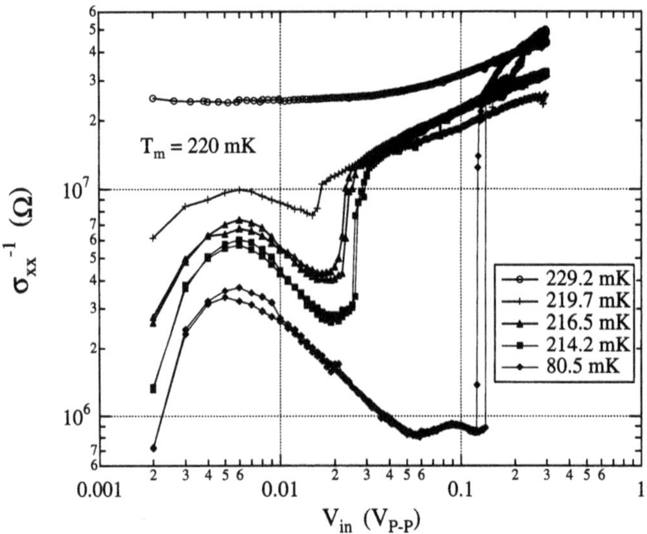

Fig. 8.8. Inverse conductivity as a function of the input voltage [280]: $B = 362\,\mathrm{G}$, $n_{\mathrm{s}} = 1.1 \times 10^8\,\mathrm{cm}^{-2}$, $\omega/2\pi = 100\,\mathrm{kHz}$, and $E_\perp = 92.5\,\mathrm{V/cm}$. Data sets were measured for the following succession of temperatures: 229.2, 219.7, 216.5, 214.2, 80.5 mK. The melting temperature was $T_{\mathrm{m}} = 220\,\mathrm{mK}$

From Fig. 8.8, one can distinguish three main regions, showing different behavior of σ_{xx}^{-1} as V_{in} increases. In the first region of small values of V_{in}, the inverse conductivity σ_{xx}^{-1} rises rapidly up to its maximum value. Under the experimental conditions, in this region, there is not even a sign of the linear transport regime. Then, in the second region of intermediate values of the input voltage, σ_{xx}^{-1} falls approximately as V_{in}^{-1}. In this range, the magnetoconductivity is characterized by the unusual magnetic field dependence $\sigma_{xx} \propto 1/B$. In the third region, at high V_{in}, the quantity σ_{xx}^{-1} suddenly jumps to a value close to that obtained for the fluid phase. When V_{in} is decreased, σ_{xx}^{-1} falls back to a smaller value of the input voltage, showing a hysteresis loop.

Qualitatively, the appearance of the strong non-Ohmic conductivity behavior when the system is cooled down to the Wigner solid state agrees with the analysis of the energy relaxation rate of the electron gas and solid given in Sect. 3.6. As shown in Fig. 3.3, in the WS state, the energy relaxation time of SEs due to two-ripplon emission processes is approximately one order of magnitude longer than in the gas state. Therefore, the same driving voltages that correspond to the linear transport regime for the 2D electron gas can cause the nonlinear response in the Wigner solid.

The nonlinear magnetoconductivity of the 2D Wigner solid in a Corbino geometry measured under different experimental conditions was later reported by Kristensen et al. [291]. These conductivity data vs. excitation voltage are shown in Fig. 8.9 for different frequencies of the signal. Once again this figure does not show any sign of the linear transport regime and the conductivity is not Ohmic down to extremely low values of the excitation voltage V_0. This linear increase in σ_{xx} with V_0 agrees qualitatively with the second region of the nonlinear conductivity measured by Shirakhama and Kono and presented in Fig. 8.8. The sharp decrease in σ_{xx} at larger V_0 in Fig 8.9 is qualitatively consistent with the sudden jump in σ_{xx}^{-1} seen in Fig 8.9. Still, these σ_{xx} data vary in a smoother way and there is no hysteresis loop.

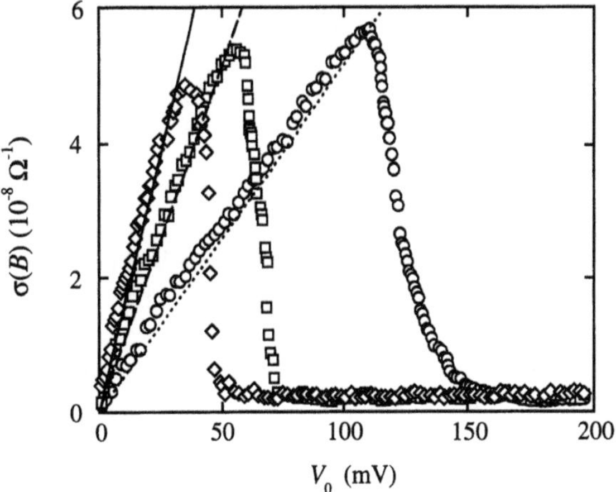

Fig. 8.9. Magnetoconductivity $\sigma_{xx}(B)$ vs. rms excitation voltage V_0 for $n_s = 1.13 \times 10^{12}\,\text{m}^{-2}$ at 0.07 K, $B = 0.2$ T, for $\omega/2\pi = 4$ (*circles*), 8 (*squares*), and 12 kHz (*diamonds*) [291]. *Curves* show the conductivities from (8.51) with $v_B = \omega_1/g_1 = 4.1\,\text{m/s}$

The explanation of the nonlinear conductivity given by Kristensen et al. [291] was based on the simple relation

$$\sigma_{xx} = \frac{J_x}{E_x} = \frac{J_x}{v_{\mathrm{H}} B} . \tag{8.51}$$

For the linear transport, both J_x and v_{H} are proportional to the excitation voltage V_0 and, naturally, $\sigma_{xx}(V_0)$ is constant. The authors assumed that the Hall velocity is fixed somehow at a certain constant value v_1 determined by the phase velocity of surface dimples. Then the relation of (8.51) would yield $\sigma_{xx} \propto V_0$. At the same time, the experiment was conducted under the condition that the current J_x is fixed and determined by the electrode geometry, the frequency ω and the driving voltage. In this case, (8.51) also yields the right magnetic field dependence as observed in the experiment $\sigma_{xx} \propto 1/B$.

In order to understand the differences in the nonlinear magnetoconductivity observed in these two experiments, one should first check whether one can rely on the transmission line model used to obtain σ_{xx} from the experimental signal. In this model, the equations for the distribution of the electrical current in the plane of the Corbino disk are found by assuming that the system is in the liquid state. In the Wigner solid state, there is a shear resistance affecting rotational transport. Let us consider the elasticity theory equations for a 2D body constituted of charged particles in the presence of a magnetic field. For radial and azimuthal components of the displacement field $\boldsymbol{u} = (u_r, u_\varphi)$, these equations can be written as

$$\omega^2 u_r + c_{\mathrm{l}}^2 \frac{\partial}{\partial r}\left[\frac{1}{r}\frac{\partial}{\partial r}(r u_r)\right] + \mathrm{i}\omega\omega_{\mathrm{c}} u_\varphi = \frac{eE_r(r)}{m_{\mathrm{e}}} , \tag{8.52}$$

$$\omega^2 u_\varphi + c_{\mathrm{t}}^2 \frac{\partial}{\partial r}\left[\frac{1}{r}\frac{\partial}{\partial r}(r u_\varphi)\right] - \mathrm{i}\omega\omega_{\mathrm{c}} u_r = 0 , \tag{8.53}$$

where c_{l} and c_{t} are the sound velocities of the 2D WS constituted of electrons with partly screened Coulomb interaction due to the metal electrodes. The longitudinal velocity c_{l} is the same for the liquid and solid states of the electron system. In the WS state, the equation for u_φ acquires a shear resistance term proportional to c_{t}^2 which substantially affects the electron motion. For example, in the limiting case $c_{\mathrm{t}}^2 \to \infty$, (8.53) has a solution $u_\varphi(r) \propto r$ which is just a solid rotation. In this instance, the solid rotation $v_\varphi(r) \propto r$ conflicts with the Hall motion $v_{\mathrm{H}}(r) = v_\varphi(r) = cE_r(r)/B$.

The strong coupling of the electron solid to the dimple lattice and the kinetic friction can be taken into account by means of the replacement

$$\omega^2 \to \left[\mathcal{Z}(\omega)\omega^2 - \mathrm{i}\omega\nu_{\mathrm{eff}}\right] \tag{8.54}$$

in the first terms of (8.52) and (8.53). This procedure is in accordance with the transport framework discussed previously: the real part of the memory function $w(\omega) = \omega[\mathcal{Z}(\omega) - 1]$ affects the inertia term of the equations of motion. The imaginary part (ν_{eff}) is included in the usual way.

It is clear that the shear term of the solid equations of motion given above can be disregarded only if

$$\left|\mathcal{Z}(\omega)\omega^2 - \mathrm{i}\omega\nu_{\text{eff}}\right| \gg c_{\mathrm{t}}^2/R^2 \,, \tag{8.55}$$

where R is the size of the Corbino disk. In this case, angular displacements of the WS are similar to the displacements in a gas state:

$$u_\phi \simeq -\frac{\omega_{\mathrm{c}}}{\nu_{\text{eff}} - \mathrm{i}\omega\mathcal{Z}} u_r \,. \tag{8.56}$$

The condition (8.55) means that rather high frequencies ω are required for the WS to respond like a gas in the Corbino geometry.

Then substitution of u_ϕ from (8.56) into (8.52) results in the following equation for u_r, similar to the corresponding equation in the transmission line model for the current density $j_r = -\mathrm{i}\omega e n_{\mathrm{s}} u_r$:

$$\frac{\partial^2 u_r}{\partial r^2} + \frac{1}{r}\frac{\partial u_r}{\partial r} + \left(\zeta^2 - \frac{1}{r^2}\right) u_r = \frac{eE_r}{m_{\mathrm{e}} c_{\mathrm{l}}^2} \,, \tag{8.57}$$

where

$$\zeta^2 = \frac{\mathcal{Z}\omega^2 - \mathrm{i}\omega\nu_{\text{eff}}}{c_{\mathrm{l}}^2} - \frac{\omega^2\omega_{\mathrm{c}}^2}{c_{\mathrm{l}}^2(\mathcal{Z}\omega^2 - \mathrm{i}\omega\nu_{\text{eff}})} \equiv \frac{\mathrm{i}\omega e^2 n_{\mathrm{s}}}{m_{\mathrm{e}} c_{\mathrm{l}}^2 \sigma_{xx}} \,, \tag{8.58}$$

$$\sigma_{xx} = \frac{e^2 n_{\mathrm{s}}}{m_{\mathrm{e}}} \frac{\nu_{\text{eff}} - \mathrm{i}\mathcal{Z}\omega}{\omega_{\mathrm{c}}^2 + (\nu_{\text{eff}} - \mathrm{i}\mathcal{Z}\omega)^2} \,. \tag{8.59}$$

It is easy to see that the quantity σ_{xx} coincides here with the electrical conductivity defined previously in (7.93) for a more general case. The relations (8.58) and (8.59) allow us to conclude that measurements based on the transmission line model can reveal the magnetoconductivity σ_{xx}, if the parameters of the system and external fields satisfy the condition (8.55). Additionally, we would like to emphasize the strong frequency dependence of σ_{xx} because of the coupling to the dimple lattice.

Under the conditions of the experiment of Shirakhama and Kono [280], $R = 3\,\mathrm{cm}$, $\omega/2\pi \sim 100\,\mathrm{kHz}$, $B \sim 360\,\mathrm{G}$ and $n_{\mathrm{s}} \sim 10^8\,\mathrm{cm}^{-2}$, the condition (8.55) required for the transmission line model to be applicable is fulfilled. The external excitation field is mainly localized in the gap between two Corbino electrodes $E_r(r) \propto \delta(r - r_1)$. The solution for the radial current density $j_\varphi(r)$ has a maximum at the gap position $r = r_1$ and decreases steadily for $r > r_1$ and $r < r_1$. According to (8.56), the Hall current $j_\varphi(r)$ is proportional to $j_r(r)$ as a function of r.

The experiment of Kristensen et al. [291] differs from that of Shirakhama and Kono [280] in two important respects:

- the radius of the Corbino disk used is very short ($R = 0.2\,\mathrm{cm}$),

- the frequency of the excitation voltage is low ($\omega/2\pi \sim 4\,\mathrm{kHz}$).

Under these conditions the requirement of (8.55) is strongly broken (electrons tend to rotate as a single body), and one cannot rely on the transmission line model unless an additional investigation of (8.52) and (8.53) with the proper edge conditions proves otherwise. As a qualitative agreement, we note that, in Fig. 8.9, an increase in the frequency of the excitation voltage makes the conductivity fall more sharply. If the WS rotates as a rigid body, then the angular current density increases towards the edge and, as a function of r, it is not proportional to the radial current. In this case, a small asymmetry in the guard electrode or a tilt can cause edge-pinning for the solid rotation because the guard voltage V_{G} is much higher than the excitation voltage: $V_{\mathrm{G}}/V_0 \sim 10^{-3}$.

We now consider some theoretical models for the nonlinear transport of the electron solid. In these models, the electron pool is assumed to be infinitely large and electrons can move steadily in the presence of static external fields. Such an idealization is also expected to reveal some important features of the WS motion in a finite geometry.

8.4.1 Bragg–Cherenkov Scattering

Consider an infinite electron solid moving in crossed electric and magnetic fields. The Hall velocity $v_{\mathrm{H}} \equiv v_y$ is assumed to be much larger than the velocity component v_x along the electric field. What happens to the dimple depth when v_{H} increases? In order to find the answer, we can use the solution for the interface displacements $\xi_{\boldsymbol{g}}(t)$ in the presence of slow electron displacements $\boldsymbol{u}_{\mathrm{s}}(t)$ given previously by (7.75). In the case considered here, $u_{\mathrm{s,y}}(t) = v_{\mathrm{H}}t$. Inserting this form of $u_{\mathrm{s},y}(t)$ into the exponent of the integrand of (7.75) and evaluating the time integral directly, we find

$$\xi_{\boldsymbol{g}}(v_{\mathrm{H}}, t) = \xi_{\boldsymbol{g}}^{(0)} \mathrm{e}^{-\mathrm{i}g_y v_{\mathrm{H}} t} \frac{1}{1 - g_y^2 v_{\mathrm{H}}^2/\omega_g^2 - \mathrm{i}\delta\, \mathrm{sgn}(g_y)} , \tag{8.60}$$

where $\xi_g^{(0)} = -(n_{\mathrm{s}} g V_g / \rho \omega_g^2) \mathrm{e}^{-W_g}$ is the the Fourier transform of the dimple profile for $v_{\mathrm{H}} = 0$, and $\delta > 0$ is an infinitesimal parameter defined previously. This kind of equation describing the dimple depth was found by Vinen [292] for a one-dimensional model.

Equation (8.60) shows that the depth of surface dimples increases in a resonant way when the Hall velocity approaches the phase velocity of the capillary waves with wave number g_y. The resonance condition is first met for the smallest reciprocal lattice vectors $\boldsymbol{g}_1$. This effect causes an increase in the associated mass and the drag force acting on the dimples. It is interesting that, for higher velocities $v_{\mathrm{H}} > v_{\boldsymbol{g}} = \omega_g/|g_y|$, the surface displacement ξ changes sign and the dimples turn to knobs. In the presence of damping, the infinitesimal parameter δ should be replaced by some finite quantity [292].

The medium response force acting on the moving WS can be found using (7.76), where the density fluctuation n_q is averaged over fast WS modes according to (7.76). Using the above dimple profiles and $u_{s,y}(t) = v_H t$, we arrive at the conclusion that the force does not depend on time and that, as a function of the drift velocity, it has a resonance structure

$$F_y(v_H) = -N_e \sum_{\boldsymbol{g}} \frac{\pi n_s g}{\rho \omega_g^2} V_g^2 e^{-2W_g} |g_y| \delta\left(1 - g_y^2 v_H^2/\omega_g^2\right)$$

$$= -N_e v_H \frac{\pi n_s}{\alpha} \sum_{\boldsymbol{g}} \left(\frac{g_y}{g}\right)^2 V_g^2 e^{-2W_g} \delta\left(\omega_g - g_y v_H\right) . \qquad (8.61)$$

In the presence of damping, the delta-functions are broadened accordingly and the friction force has a maximum magnitude, determined by the damping parameter, which arises from two effects: the natural damping of ripplons and the finite size of the area where the crystal can satisfy the condition for Bragg–Cherenkov scattering. The decrease in the friction force with an increase in the driving electric field can lead to a breakdown in the balance of forces in this system [292], which we shall discuss in the next section.

Here we would like to note that the medium response force given in (8.61) is caused by the emission of surface excitations. It has a sharp resonance structure if the 2D electron lattice is rigid and, formally, it is close to zero if $g_y v_H$ is substantially smaller than ω_g. The conventional kinetic friction $\boldsymbol{F}_{\text{fric}}$ acting on the moving dimples originates from the reflection of bulk liquid quasi-particles from the dimples. In both the hydrodynamic and the long mean-free-path regimes, according to (8.43) and (8.44),

$$\boldsymbol{F}_{\text{fric}} \propto -|\xi_g|^2 \boldsymbol{v} . \qquad (8.62)$$

This force is finite at $g_y v_H \ll \omega_g$, and it increases smoothly if the dimple depth $|\xi_g|$ increases with v_H. Behavior of the same kind can be obtained for the force given in (8.61), if we take into account the fact that bulk quasi-particle reflection introduces finite damping γ_q of surface waves. Then, at low velocities, the response force changes linearly with $\boldsymbol{v}$, it is proportional to γ_g, and it increases smoothly, when $g_y v_H$ becomes comparable with ω_g.

Equation (8.61) describes Bragg–Cherenkov scattering by the rigid 2D body: the plasmons of the WS solid are not involved in this process. Another approach to the problem of Bragg–Cherenkov scattering, proposed by Dykman and Rubo [267] in 1997, was based on the relation between the friction force and the electron DSF $S(\boldsymbol{q}, \omega)$ established previously for the nonlinear transport regime [84]. We discussed this relation in Sect. 3.3, which dealt with the force-balance method [see (3.14)]. Balancing the friction force $\boldsymbol{F}_{\text{scat}}$, acting on the WS because of electron–ripplon scattering, and the force due to the external fields $N_e e \boldsymbol{E}_{\parallel} + N_e m_e \omega_c (\boldsymbol{v} \times \hat{\boldsymbol{z}})$, we find the relation [267]

$$\sigma_{xx}(\boldsymbol{v}_H) \simeq \frac{n_s c^2}{B^2 v_H^2} \boldsymbol{v}_H \cdot \boldsymbol{F}_{\text{scat}}(\boldsymbol{v}_H) , \qquad (8.63)$$

valid for $v_{\mathrm{H}} = v_y \gg v_x$.

The electron DSF, defined in the electron center-of-mass reference frame, enters the equation for the kinetic friction in the form $S(\boldsymbol{q}, \omega_{\mathrm{q}} - \boldsymbol{q} \cdot \boldsymbol{v}_{\mathrm{H}})$. Under typical experimental conditions, $\hbar \boldsymbol{g} \cdot \boldsymbol{v}_{\mathrm{H}} \ll T$ and a strong nonlinearity appears in the electron transport owing to the Doppler correction in the argument of the DSF. In this case, the y-component of the frictional force $F_{\mathrm{scat}}^{(y)}$ can be represented as

$$F_{\mathrm{scat}}^{(y)} = -\sum_{\boldsymbol{q}} \frac{n_{\mathrm{s}} q V_q^2}{2\rho\omega_q^2} q_y^2 v_{\mathrm{H}} S(\boldsymbol{q}, \omega_q - \boldsymbol{q} \cdot \boldsymbol{v}_{\mathrm{H}}) . \tag{8.64}$$

The elastic terms of the low-temperature expansion of $S(\boldsymbol{q}, \omega_q - \boldsymbol{q} \cdot \boldsymbol{v}_{\mathrm{H}})$ given in (8.8) are proportional to $\delta(\omega_g - \boldsymbol{g} \cdot \boldsymbol{v}_{\mathrm{H}})$, which results in the singular resonance structure of $\boldsymbol{F}_{\mathrm{scat}}$. A direct evaluation of (8.64) based on the elastic approximation for the DSF yields a friction force which coincides with the second line of (8.61).

It is interesting that, for the weak coupling theory, the one-phonon terms of the WS DSF given in (8.10) lead to a different resonant structure of the kinetic friction. These terms contain the energy conservation δ-function $\delta(\omega_g - g_y v_{\mathrm{H}} \mp c_{\mathrm{t}} k)$. Evaluating the sum

$$\sum_{\boldsymbol{k}} \delta(x - c_{\mathrm{t}} k) = \frac{S_{\mathrm{A}} x}{2\pi c_{\mathrm{t}}^2} \theta(x) , \tag{8.65}$$

the frictional force of (8.64) corresponding to the one-phonon terms can be found in a form similar to that obtained for the electron–ripplon resonances [see (8.13)]:

$$F_{\mathrm{scat}}^{(y)}(v_{\mathrm{H}}) = -N_{\mathrm{e}} m_{\mathrm{e}} \nu(v_{\mathrm{H}}) v_{\mathrm{H}} ,$$

$$\nu(v_{\mathrm{H}}) = \frac{\pi n_{\mathrm{s}}}{2 m_{\mathrm{e}} \alpha} \sum_{\boldsymbol{g}} \left(\frac{g_y}{g}\right)^2 \frac{\alpha_g(T) V_g^2}{|\omega_g - g_y v_{\mathrm{H}}|} \mathrm{e}^{-2W_g} , \tag{8.66}$$

where the parameter $\alpha_g(T) = 2T/T_g$ was defined previously [see (8.14)]. If we assume $\alpha_g \ll 1$ and set $2W_g = g^2 \langle u_0^2 \rangle /2$, then this equation would coincide with the result of Dykman and Rubo [267], which was found by means of the approximation given in (7.135) of the last chapter. In the original theory, the frictional force increases as a power law in the reciprocal detuning $|\omega_g - g_y v_{\mathrm{H}}|^{-1}$, with the temperature dependent exponent $1 - \alpha_g(T)$. We note that, at the WS transition point $\alpha_{g_1}(T_{\mathrm{m}}) = 1/3$, and it is small enough to be disregarded at substantially lower temperatures. Compared with the result of the elastic approximation given in (8.61), the resonance structure of the one-phonon contribution to the frictional force has longer tails. If the Hall velocity is directed along $\boldsymbol{g}_1$, it should saturate as a function of the driving field at the value $v_1 = \omega_{g_1}/g_1$.

The Hall velocity v_{H} data of Kristensen [291] as a function of the driving voltage shown in Fig. 8.10 do indeed saturate to v_1 for small and intermediate

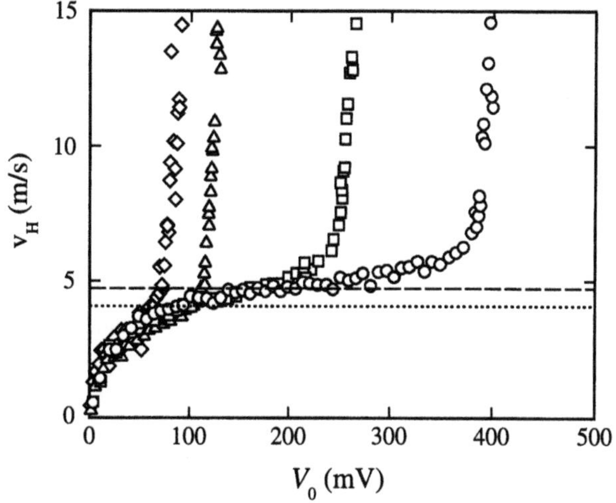

Fig. 8.10. Rms Hall velocity v_{H} vs. rms excitation voltage V_0 for $n_s = 1.06 \times 10^{12}$ (*diamonds*), 1.20×10^{12} (*triangles*), 1.57×10^{12} (*squares*), and $1.90 \times 10^{12}\,\mathrm{m}^{-2}$ (*circles*) at $T = 0.06\,\mathrm{K}$ [291]. Ripplon velocities are $v_1 = 4.09\,\mathrm{m/s}$ ($n_s = 1.06 \times 10^{12}\,\mathrm{m}^{-2}$ *dotted line*) and $v_1 = 4.74\,\mathrm{m/s}$ ($n_s = 1.90 \times 10^{12}\,\mathrm{m}^{-2}$ *dashed line*)

values of V_0. Then at large voltages, v_{H} increases in sharply. The saturation regime of these data is well described by Dykman and Rubo using the non-linear relation given in (8.63).

The above picture is valid for weak coupling of the WS to the dimple lattice. The real experimental situation is opposite. The phonon spectrum of the solid is split into two branches which we call the fast and slow modes. The slow modes are strongly renormalized because of the interaction, and they transform into the pure ripplon mode as $\Omega^{(s)}_{p,k} \to \omega_{g_1}$. Consistency requires these slow modes to be excluded from the Debye–Waller (DWF) factor of the conductivity relaxation kernel, as discussed in Sect. 7.5.3. In the self-consistent treatment, we have only the elastic term of (8.61), where the DWF should be set to the high-frequency DWF. Then the smooth saturation of the Hall velocity in the experiment with the nonuniform driving field can be due to the following two factors:

- the Bragg–Cherenkov condition is initially met only for a small part of the electrons because of the nonuniform conditions,
- the frictional force increases smoothly $[F_{\mathrm{fric}} \propto (\omega_g - g_y v_{\mathrm{H}})^{-2}]$ because of damping effects.

It should also be noted that in the experiments cited above, the driving electric field was an alternating field and the Hall velocity v_{H} was time-dependent.

8.4.2 Sliding Wigner Solid

The 2D electron solid on the free surface of liquid helium is strongly coupled to the dimple lattice induced by the Wigner phase transition. This coupling mainly affects the dynamic properties of electrons; the position of the melting point is not changed. It is nevertheless interesting to study experimentally the properties of the WS uncoupled from the dimple lattice. The possibility of such decoupling was already suggested by Fisher, Halperin and Platzman [230] in the first publication concerned with the phonon–ripplon coupling. They pointed out that the Wigner solid leaves the surface deformation if the WS moves faster than the ripplon phase velocity, or if the electron displacement is larger than the WS lattice constant. The latter case requires certain conditions because, as we saw in the last section, in the absence of damping, a steady, slow motion of the WS with displacement $u = vt$ cannot cause electron decoupling from the dimples. The same is true for low frequencies of the driving field. The frequency must be high enough. In any case, for frequencies $\omega > \omega_{g_1}$, large displacements $u \sim g_1^{-1}$ can be reached by electrons whose velocity is faster than the ripplon phase velocity.

It is instructive to introduce some numbers which characterize surface dimples and electron coupling. Under the conditions of experiments concerned with the nonlinear transport of the Wigner solid, $n_s \simeq 1 \times 10^8\,\mathrm{cm}^{-2}$ and $E_\perp = 2\pi e n_s$, the dimple depth defined approximately as $\left|\xi_g^{(0)}\right| = n_s g V_g / \rho \omega_g^2$ is estimated to be very short: $\left|\xi_{g_1}^{(0)}\right| \sim 10^{-11}\,\mathrm{cm}$. Then the binding energy of an electron to the surface dimple is extremely low: $V_{g_1}\xi_{g_1} \simeq eE_\perp \xi_{g_1} \sim 10^{-5}\,\mathrm{K} \ll T$ (!). Still, thermal motion which has much higher energies cannot decouple electrons from dimples because of the strong Coulomb interaction fixing electron positions at lattice sites. Thermal decoupling from surface dimples exists in the liquid state, and it is the main reason why there are no single-electron dimples in this state.

The surface dimples induced by the WS with lattice spacing $a \simeq 10^{-4}\,\mathrm{cm}$ are very smooth: $g_1\xi_1 \sim 6.5 \times 10^{-7}$. This restricts the maximum value of the restoring force in the deformation potential: $F_{\mathrm{res}}^{(\max)} \sim eE_\perp g_1 \xi_1$. Consider first the rigid potential model in which dimples are motionless. The estimates given above imply that, for a rigid deformation, an excitation voltage or driving electric field of very low amplitude $E_\parallel \sim 10^{-6} E_\perp$ can break the coupling and induce the sliding state of the WS. In other words, the periodic potential induced by the dimple lattice, which has local minima, is tilted in the presence of the driving electric field. When the local potential minima disappear at sufficiently strong electric fields, electrons can move freely and the WS slides. This sliding results in an abrupt jump in the electron conductivity. In the presence of a magnetic field, this sliding criterion $[E_\parallel > E_\perp \max(\nabla\xi)]$ holds similarly.

In experiments with the WS on superfluid helium, dimples are quite mobile and follow the electron motion. In this case, we have to qualify the sliding criterion. For example, consider the steady motion of the WS with a constant velocity. Then the driving force applied to the electrons is generally balanced by frictional forces acting on both electrons (e) and dimples (D): $e\boldsymbol{E}_\parallel = \boldsymbol{F}^{(\mathrm{e})}_{\mathrm{fric}} + \boldsymbol{F}_{\mathrm{res}}$, $\boldsymbol{F}_{\mathrm{res}} = \boldsymbol{F}^{(\mathrm{D})}_{\mathrm{fric}}$, where $\boldsymbol{F}_{\mathrm{res}}$ is the restoring force acting on electrons by dimples. It is clear that $\boldsymbol{F}_{\mathrm{res}}$ is also the strain of the fictitious spring that couples the electron and dimple, when the system moves steadily. If $eE_\parallel - F^{(\mathrm{e})}_{\mathrm{fric}}$ exceeds $F^{(\mathrm{max})}_{\mathrm{res}} = eE_\perp \max(\nabla\xi)$, the coupling spring is broken, causing sliding of the WS. In the presence of a magnetic field, we have to include the Lorentz force in the first equation given above, but this does not much change the subsequent analysis.

One should distinguish the following two extreme cases, shown schematically in the spring diagram of Fig. 8.11. The restoring force and the strain of the coupling spring is zero if the frictional force is applied only to electrons because of their scattering, and there is no frictional force applied to the dimples, as shown in the schematic diagram of Fig. 8.11a. In another extreme case shown in Fig. 8.11b, the frictional force is applied only to the dimple and it causes the stretching and strain of the fictitious spring binding the electron and dimple. In this case, $\boldsymbol{F}_{\mathrm{res}} = e\boldsymbol{E}_\parallel$. Generally, the frictional force is applied to both the electron and the dimple, and therefore $\boldsymbol{F}_{\mathrm{res}} < e\boldsymbol{E}_\parallel$. In any case, the restoring force F_{res} acting on electrons, which should actually be compared with the threshold value $F^{(\mathrm{max})}_{\mathrm{res}} \simeq eE_\perp \max(\nabla\xi)$, is equal to the frictional force of dimples. Because the frictional force acting on surface dimples is antiparallel to the current ($\boldsymbol{F}^{(\mathrm{D})}_{\mathrm{fric}} = -m_\mathrm{e}\nu_\mathrm{D}\boldsymbol{v}$), the sliding occurs

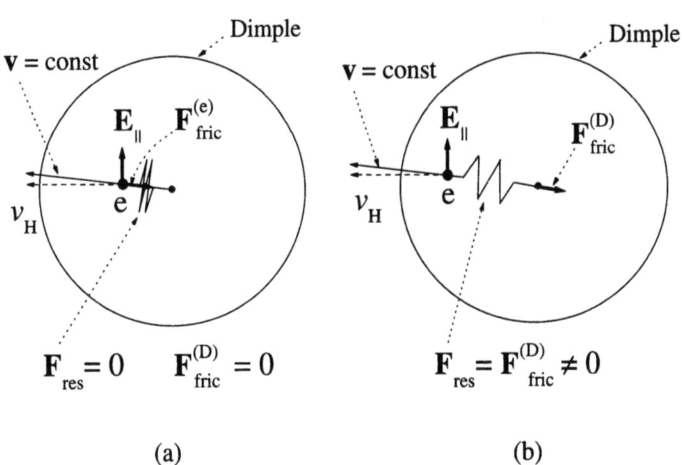

Fig. 8.11. Schematic spring diagram showing the average restoring force $\boldsymbol{F}_{\mathrm{res}}$ acting on an electron (e) by its dimple (D), moving with constant velocity, for two cases: (**a**) the frictional force is applied to the electron while the dimple frictional force is zero, (**b**) the frictional force is mainly applied to the dimple

in the direction of the average electron velocity $\boldsymbol{v}$, which is close to the Hall velocity $\boldsymbol{v}_{\mathrm{H}}$, if $\omega_c \gg \nu_e + \nu_D$.

For superfluid liquid ^{4}He at low temperatures, $\boldsymbol{F}_{\mathrm{fric}}^{(\mathrm{D})}$ is much smaller than the frictional force acting on the electrons from scatterers ($\nu_e \gg \nu_D$). In this case, $\boldsymbol{F}_{\mathrm{res}} = \boldsymbol{F}_{\mathrm{fric}}^{(\mathrm{D})} \ll eE_{\parallel}$ and, for a fixed depth of the moving dimple, the restoring force increases linearly with $\boldsymbol{v}$. When it reaches the threshold value $eE_{\perp}\max(\nabla\xi)$ at some critical velocity, the 'spring' is broken and the WS slides. The increase in the restoring force with electron velocity $\boldsymbol{v}$ can be accompanied by an increase in the dimple depth, $|\xi_g|$ because of Bragg–Cherenkov scattering when v_{H} becomes of the same order of magnitude as ω_{g_1}/g_1. Under conditions $\nu_e \gg \nu_D$, this does not spoil the sliding mechanism because the frictional force acting on the surface dimple increases even faster: $F_{\mathrm{fric}}^{(\mathrm{D})} \propto \xi_g^2$. [For example, see (8.43) found for the viscous transport regime.]

In the case of an alternating driving electric field of high enough frequency, one has to take inertia into account. This should reduce the threshold driving electric field because electrons and dimples have significantly different masses. In the extreme limit $\omega \gg \omega_{g_1}$, dimples are motionless and $F_{\mathrm{res}} = eE_{\parallel}$, regardless of the dimple drag force. When $\omega \ll \omega_{g_1}$, this effect can still be important, depending on the ratio ω/ω_{g_1}.

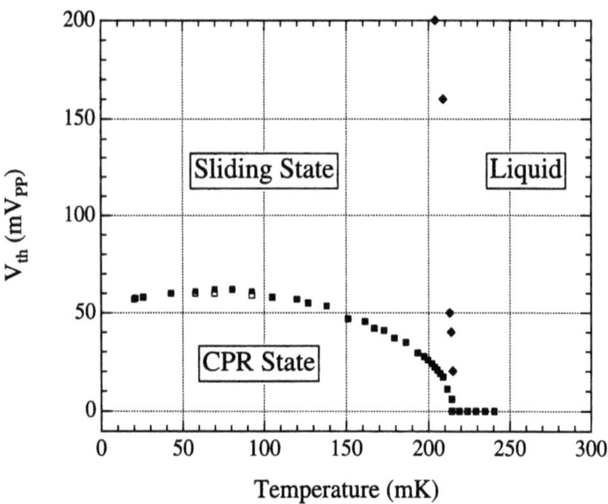

Fig. 8.12. Phase boundary in the V_{in}–T plane. *Squares* show the threshold voltage where σ_{xx}^{-1} jumps. *Lozenges* show the sliding solid–liquid transition. $n_s = 1.1 \times 10^8\,\mathrm{cm}^{-2}$, $B = 724\,\mathrm{G}$, $\omega/2\pi = 100\,\mathrm{kHz}$, and $E_{\perp} = 92.5\,\mathrm{V/cm}$

It seems reasonable to ascribe the sharp jump in $\sigma_{xx}^{-1}(V_{\mathrm{in}})$ at high input voltages in Fig. 8.8 to the sliding transition. Indeed, the threshold voltage for this jump was shown to increase linearly with the holding electric field $E_{\perp}$, which is consistent with the restoring force $eE_{\perp}\nabla\xi$ acting on an electron in the dimple, whose maximum value determines the threshold field. The dependence of V_{th} on this and other parameters of the system can be summarized as follows [280]:

$$V_{\mathrm{th}} \propto \frac{n_{\mathrm{s}}^{3/2} E_{\perp}}{\omega B^{0.8}} . \tag{8.67}$$

The temperature dependence of the threshold voltage is shown in Fig. 8.12 as a phase diagram in the V_{in}–T plane. The data marked with squares separate the coupled phonon–ripplon (CPR) state and the sliding state of the electron solid.

We have not discussed here the shape transformation of the dimple profile, when it moves with a finite restoring force $F_{\mathrm{res}} = F_{\mathrm{fric}}^{(\mathrm{D})}$. For the restoring force to appear, the average position of the electron should be shifted with respect to the bottom of the dimple towards the current direction. The displacement of the electron pressure affects the balance of forces in the z-direction and can change the shape of the dimple profile. This effect may also be a cause for the nonlinear transport of the WS.

Another possible explanation for the σ_{xx}^{-1} jump was proposed by Wilen and Gianetta [293]. They ascribe it to the shear melting of the WS by a large strain induced by the magnetic field. This model gives a good description of the B dependence of the threshold voltage. Still, the observed frequency dependence is unexplained in this case.

It should be emphasized that the above models correspond to steady motion of the WS in a uniform driving electric field. In experiments [280] and [291], the driving field was an alternating field. The real scenario for WS sliding depends on many parameters of the system, such as $E_{\|}/E_{\perp}\max(\nabla\xi)$, $\nu_{\mathrm{e}}/\nu_{\mathrm{D}}$, ω/ω_{g_1}, and v_{H}/v_{g_1}. At low frequencies ω, the WS shear is also involved in this interplay, as shown in the introductory part of this section. Therefore, the identification of the particular mechanism underlying nonlinear transport is very difficult, especially if we take into account the fact that the alternating driving field is nonuniform across the Corbino disk radius. We feel nevertheless that the models discussed in this section at least provide a qualitative understanding of the remarkable phenomenon of nonlinear WS transport.

[illegible] angle whose maximal value [illegible] [illegible] and other phenomena [illegible]
[illegible]

References

1. M.W. Cole and M.H. Cohen: Phys. Rev. Lett. **23**, 1238 (1969)
2. V.B. Shikin: Zh. Eksperim. Teor. Fiz. **58**, 1748 (1970) [Soviet Phys. JETP **31**, 936 (1970)]
3. R.S. Crandall and R. Williams: Phys. Lett. A **34**, 404 (1971)
4. E.P. Wigner: Phys. Rev. **46**, 1002 (1934)
5. K. von Klitzing, G. Dorda, and M. Pepper: Phys. Rev. Lett. **45**, 494 (1980)
6. D.C. Tsui, H.L. Stormer, and A.C. Gossard: Phys. Rev. Lett. **48**, 1559 (1982)
7. C.C. Grimes and G. Adams: Phys. Rev. Lett. **42**, 795 (1979)
8. L.P. Gor'kov and D.M. Chernikova: Pis'ma Zh. Eksp. Teor. Fiz. **18**, 119 (1973) [JETP lett. **18**, 68 (1973)]
9. V.S. Edel'man: Usp. Fiz. Nauk **130**, 675 (1980) [Sov. Phys. Usp. **23**, 227 (1980)]
10. T. Ando, A.B. Fowler, and F. Stern: Rev. Mod. Phys. **54**, 437 (1982)
11. G. Bastard, J.A. Brum, and R. Ferreira: Electronic States in Semiconductor Heterostructures. In: *Solid State Physics*, Vol. 44, ed. by H. Ehrenreich, D. Turnbull (Academic Press, Boston, San Diego, New York, London, Sydney, Tokyo, Toronto 1991) pp. 229–415
12. W.T. Sommer: Phys. Rev. Lett. **12**, 271 (1964)
13. C.C. Grimes, T.R. Brown, M.L. Burnes, and C.L. Zipfel: Phys. Rev. B **13**, 140 (1976)
14. T.M. Sanders and G. Weinreich: Phys. Rev. B **13**, 4810 (1976)
15. O. Hipólito, J.R.D. De Felicio, G.A. Farias: Sol. State Comm. **28**, 365 (1978)
16. E. Cheng, M.W. Cole, and M.H. Cohen: Phys. Rev. B **50**, 1136 (1994)
17. E. Fermi: Z. Physik **26**, 54 (1924)
18. R.S. Crandall and R. Williams: Phys. Rev. A **5**, 2183 (1972)
19. V.B. Shikin and Yu.P. Monarkha: Fiz. Nizk. Temp. **1**, 957 (1975) [Sov. J. Low Temp. Phys. **1**, 459 (1975)]
20. Yu.P. Monarkha and S.S. Sokolov: Fiz. Nizk. Temp. **4**, 685 (1978) [Sov. J. Low Temp. Phys. 6, 327 (1978)]
21. Yu.P. Monarkha: Fiz. Nizk. Temp. **5**, 994 (1979) [Sov. J. Low Temp. Phys. **5**, 470 (1979)]
22. M. Saitoh: J. Phys. Soc. Jpn. **42**, 201 (1977)
23. D.M. Chernikova: Zh. Eksp. Teor. Fiz. **68**, 249 (1975) [Sov. Phys. JETP **41**, 121 (1975)]
24. V.S. Edel'man: Pis'ma Zh. Eksp. Teor. Fiz. **25**, 422 (1977) [JETP Lett. **25**, 394 (1977)]
25. M. Saitoh and T. Aoki: J. Phys. Soc. Jpn. **44**, 71 (1978)
26. Yu.P. Monarkha and S.S. Sokolov: Fiz. Nizk. Temp. **5**, 1283 (1979) [Sov. J. Low Temp. Phys. **5**, 470 (1979)]

27. V.B. Shikin and Yu.P. Monarkha: J. Low Temp. Phys. **16**, 193 (1974)
28. K. Kajita: J. Phys. Soc. Jpn. **52**, 372 (1983)
29. W. Huang, J. Rudnick, and A. Dahm: J. Low Temp. Phys. **28**, 21 (1977)
30. Yu.P. Monarkha: Fiz. Nizk. Temp. **3**, 1459 (1977) [Sov. J. Low Temp. Phys. **3**, 702 (1977)]
31. F.M. Peeters and P.M. Platzman: Phys. Rev. Lett. **50**, 2021 (1983)
32. F.M. Peeters: Phys. Rev. B **30**, 159 (1984)
33. H. Etz, W. Gombert, W. Idstein, and P. Leiderer: Phys. Rev. Lett. **53**, 2567 (1984)
34. Yu.P. Monarkha and Yu.Z. Kovdrya: Fiz. Nizk. Temp. **8**, 215 (1982) [Sov. J. Low Temp. Phys. **8**, 107 (1982)]
35. Yu.M. Vil'k and Yu.P. Monarkha: Sov. J. Low Temp. Phys. **11**, 535 (1985) [Fiz. Nizk. Temp. **11**, 971 (1985)]
36. Yu.P. Monarkha, K. Shirahama, K. Kono, and F.M. Peeters: Phys. Rev. B **58**, 3762 (1998)
37. V.B. Shikin: Zh. Eksp. Teor. Phys. **60**, 713 (1971) [Sov. Phys. JETP **33**, 387 (1971)]
38. G.D. Mahan: *Many-Particle Physics*, 3rd edn., (Kluwer Academic/Plenum Publishers, New York, Boston, Dordrecht, London, Moscow 1990) pp. 65–106, pp. 499–576
39. M.W. Cole: Phys. Rev. B **2**, 4239 (1970)
40. G.D. Gaspari and F. Bridges: J. Low Temp. Phys. **21**, 535 (1975)
41. Yu.P. Monarkha: Fiz. Nizk. Temp. **3**, 587 (1977) [Sov. J. Low Temp. Phys. **3**, 282 (1977)]
42. A.G. Eguiluz, A.A. Maradudin, and R.J. Elliott: Phys. Rev. B **24**, 197 (1981)
43. Yu.P. Monarkha: Fiz. Nizk. Temp. **2**, 1232 (1976) [Sov. J. Low Temp. Phys. **2**, 600 (1976)]
44. Yu.P. Monarkha: Fiz. Nizk. Temp. **4**, 1093 (1978) [Sov. J. Low Temp. Phys. **4**, 515 (1978)]
45. B. Goodman: Phys. Rev. **110**, 888 (1958)
46. W.T. Sommer and D.J. Tanner: Phys. Rev. Lett. **27**, 1345 (1971)
47. C.C. Grimes and G. Adams: Phys. Rev. Lett. **36**, 145 (1976)
48. A.S. Rybalko, Yu.Z. Kovdrya, and B.N. Eselson: Zh. Eksp. Teor. Phys. Pis'ma Red. **22**, 569 (1975) [Sov. Phys. JETP Lett. **22**, 280 (1975)]
49. Y. Iye: J. Low Temp. Phys. **40**, 441 (1980)
50. R. Mehrotra, C.J. Guo, Y.Z. Ruan, D.B. Must, and A.J. Dahm: Phys. Rev. B **29**, 5239 (1984)
51. V.S. Edel'man: JETP Lett. **26**, 492 (1977) [Pis'ma Zh. Eksp. Teor. Fiz. **26**, 647 (1977)]
52. A.P. Volodin and V.S. Edel'man: JETP Lett. **30**, 633 (1979) [Pis'ma Zh. Eksp. Teor. Fiz. **30**, 668 (1979)]
53. K. Shirahama, S. Ito, H. Suto, and K. Kono: J. Low Temp. Phys. **101**, 439 (1995)
54. Yu.P. Monarkha and K. Kono: J. Phys. Soc. Jpn. **66**, 3901 (1997)
55. A.F. Andreev: Zh. Eksp. Teor. Fiz. **50**, 1415 (1966) [Sov. Phys. JETP **23**, 939 (1966)]
56. D.O. Edvards and W.F. Saam: The free surface of liquid helium. In: *Progress in Low Temperature Physics*, Vol. VII A, ed. by D.F. Brewer (North-Holland, Amsterdam 1978), pp. 283–369

57. Yu.P. Monarkha: Fiz. Nizk. Temp. **5**, 940 (1979) [Sov. J. Low Temp. Phys. **5**, 447 (1979)]
58. Yu.P. Monarkha and S.S. Sokolov: Fiz. Nizk. Temp. **7**, 45 (1981) [Sov. J. Low Temp. Phys. **7**, 22 (1981)]
59. F.M. Ellis, R.B. Hallock, M.D. Miller, and R.A. Guyer: Phys. Rev. Lett. **46**, 1461 (1981)
60. A.M. Troyanovskii and M.S. Khaikin: Zh. Eksp. Teor. Fiz. **81**, 398 (1981) [Sov. Phys. JETP **54**, 214 (1981)]
61. V.S. Edel'man and M.I. Faley: J. Low Temp. Phys. **52**, 301 (1983)
62. S.S. Sokolov, J.P. Rino, and N. Studart: Phys. Rev. B **51**, 11068 (1995)
63. P.W. Adams and M.A. Paalanen: Phys. Rev. Lett. **58**, 2106 (1987)
64. P.W. Adams: Surf. Sci. **263**, 663 (1991)
65. M.J. Stephen: Phys. Rev. B **36**, 5663 (1987)
66. K. Kono, U. Albrecht, and P. Leiderer: J. Low Temp. Phys. **82**, 279 (1991)
67. M.A. Paalanen and Y. Iye: Phys. Rev. Lett. **55**, 1761 (1985)
68. D. Cieslikowski, A.J. Dahm, and P. Leiderer: Phys. Rev. **58**, 1751 (1987)
69. Yu.P. Monarkha, U. Albrecht, K. Kono, and P. Leiderer: Phys. Rev. B **47**, 13812 (1993)
70. A.M. Troyanovskii, A.P. Volodin, and M.S. Khaikin: Pis'ma Zh. Eksp. Tepr. Phys. **29**, 421 (1979) [JETP Lett. **29**, 382 (1979)]
71. K. Kajita: Surf. Sci. **142**, 86 (1984)
72. K. Kajita: J. Phys. Soc. Jpn. **54**, 4092 (1985)
73. K. Kono, U. Albrecht, and P. Leiderer: J. Low Temp. Phys. **83**, 423 (1991)
74. T. Ando and Y. Uemura: J. Phys. Soc. Jpn. **36**, 959 (1974)
75. R.R. Gerhardts: Surf. Sci. **58**, 227 (1976)
76. Yu.P. Monarkha, E. Teske, and P. Wyder: Phys. Rev. B **62**, 2593 (2000)
77. T. Ando: J. Phys. Soc. Jpn. **37**, 622 (1974)
78. R. Kubo: J. Phys. Soc. Jpn. **12**, 570 (1957)
79. D.N. Zubarev: Usp. Fiz. Nauk **71**, 71 (1960) [Sov. Phys. Usp. **3**, 320 (1960)]
80. W. Marshall and S.W. Lovesey: *Theory of Thermal Neutron Scattering* (Clarendon Press, Oxford 1971)
81. R. Kubo, S.J. Miyake, and N. Hashitsume: Quantum theory of galvanomagnetic effects at extremely strong magnetic fields. In: *Solid State Physics* Vol. 17, ed. by F. Seitz, D. Turnbull (Academic Press, New York and London 1965) pp. 269–364
82. C. Kittel: *Quantum Theory of Solids* (Wiley, New York 1963)
83. D. Pines, and P. Nozières: *The Theory of Quantum Liquids* (W.A. Benjamin, New York 1966)
84. Yu.M. Vil'k and Yu.P. Monarkha: Sov. J. Low Temp. Phys. **15**, 131 (1989) [Fiz. Nizk. Temp. **15**, 235 (1989)]
85. M.I. Dykman and L.S. Khazan: Zh. Eksp. Teor. Fiz. **77**, 1488 (1979) [Sov. Phys. JETP **50**, 747 (1979)]
86. M.I. Dykman: J. Phys. C: Solid State Phys. **15**, 7397 (1982)
87. C. Fang-Yen, M.I. Dykman, and M.J. Lea: Phys. Rev. B **55**, 16272 (1997)
88. M.I. Dykman, C. Fang-Yen, and M.J. Lea: Phys. Rev. B **55**, 16249 (1997)
89. E. Teske, Yu.P. Monarkha, M. Seck, and P. Wyder: Phys. Rev. Lett. **82**, 2772 (1999)
90. Yu.P. Monarkha, E. Teske, and P. Wyder: Phys. Rev. B **59**, 14884 (1999)
91. R.H. Ritchie: Phys. Rev. **106**, 874 (1957)

92. R.A. Ferrell: Phys. Rev. **111**, 1214 (1958)
93. F. Stern: Phys. Rev. Lett. **18**, 546 (1967)
94. D.B. Mast, A.J. Dahm, and A.L. Fetter: Phys. Rev. Lett. **54**, 1706 (1985)
95. D.C. Glattli, E.Y. Andrei, G. Deville, J. Poitrenaud, and F.I.B. Williams: Phys. Rev. Lett. **54**, 1710 (1985)
96. M. Wassermeier, J. Oshinowo, J.P. Kotthaus, A.H. MacDonald, C.T. Foxon, and J.J. Harris: Phys. Rev. B **41**, 10287 (1990)
97. I. Grodnensky, D. Heitmann, and K. von Klitzing: Phys. Rev. Lett. **67**, 1019 (1991); N.B. Zhitenev, R.J. Haug, K. von Klitzing, and K. Eberl: Phys. Rev. Lett. **71**, 2292 (1993)
98. S. Ito, K. Shirahama, and K. Kono: J. Phys. Soc. Jpn. **66**, 533 (1997)
99. Yu.P. Monarkha, S. Ito, K. Shirahama, and K. Kono: Phys. Rev. Lett. **78**, 2445 (1997)
100. P.M. Platzman and N. Tzoar: Phys. Rev. B **13**, 3197 (1976)
101. B.N. Esel'son, A.S. Rybalko, and S.S. Sokolov: Fiz. Nizk. Temp. **6**, 1120 (1980) [Sov. J. Low Temp. Phys. **6**, 544 (1980)]
102. S. Giovanazzi, L. Pitaevskii, and S. Stringari: Phys. Rev. Lett. **72**, 3230 (1994)
103. M. Buttiker: Phys. Rev. B **38**, 9375 (1988)
104. V.A. Volkov and S.A. Mikhailov: Electrodynamics of two-dimensional electron systems in high magnetic fields. In: *Modern Problems in Condensed Matter Sciences*, Vol. 27.2, Chap. 15, ed. by V.M. Agranovich, A.A. Maradudin (North-Holland, Amsterdam 1991) pp. 855–907
105. O.I. Kirichek, P.K.H. Sommerfeld, Yu.P. Monarkha, P.J.M. Peters, Yu.Z. Kovdrya, P.P. Steijaert, R.W. van der Heijden, and A.T.A.M. de Waele: Phys. Rev. Lett. **74**, 1190 (1995)
106. Yu.P. Monarkha: Low Temp. Physics. **21**, 458 (1995)
107. V.B. Shikin: Pis'ma Zh. Eksp. Teor. Fiz. **47**, 471 (1988) [JETP Lett. **47**, 555 (1988)]
108. S.S. Nazin and V.B. Shikin: Zh. Eksp. Teor. Fiz. **94**, 133 (1988) [Sov. Phys. JETP **67**, 288 (1988)]
109. I.L. Aleiner and L.I. Glazman: Phys. Rev. Lett. **72**, 2935 (1994)
110. O.I. Kirichek, I.B. Berkutov, Yu.Z. Kovdrya, and V.N. Grigor'ev: J. Low Temp. Phys. **109**, 397 (1997)
111. Yu.P. Monarkha: Low frequency edge waves of the strongly correlated two-dimensional electron liquid and solid in a magnetic field. In: *Horizons in World Physics*, Vol. 236, ed. by O.I. Kirichek (Nova Science Publishers Inc., Huntington New York 2001) pp. 47–71
112. P.L. Elliot, C.I. Pakes, L. Skrbek, and W.F. Vinen: Phys. Rev. Lett. **75**, 3713 (1995)
113. Yu.P. Monarkha, F.M. Peeters, and S.S. Sokolov: J. Phys.: Condens. Matter **9**, 1537 (1997)
114. K. Kajita, Y. Iye, K. Kono, and W. Sasaki: Solid State Commun. **27**, 1379 (1978)
115. Y. Iye, K. Kono, K. Kajita, and W. Sasaki: J. Low Temp. Phys. **34**, 539 (1979)
116. K. Kono, K. Kajita, S. Kobayuashi, and W. Sasaki: Surf. Sci. **113**, 438 (1982)
117. S. Nagano, S. Ichimaru, H. Totsuji, and N. Itoh: Phys. Rev. B **19**, 2449 (1979)
118. Y. Iye, K. Kono, K. Kajita, and W. Sasaki: J. Low Temp. Phys. **38**, 293 (1980)
119. H. Totsuji: Phys. Rev. A **17**, 399 (1978)
120. N. Itoh, S. Ichimaru, and S. Nagano: Phys. Rev. B **17**, 2862 (1978)

121. N. Itoh and S. Ichimaru: Phys. Rev. B **22**, 1459 (1980)
122. Yu.M. Vil'k and Yu.P. Monarkha: Fiz. Nizk. Temp. **13**, 684 (1987) [Sov. J. Low Temp. Phys. **13**, 392 (1987)]
123. L. Bonsall and A.A. Maradudin: Phys. Rev. B **15**, 1959 (1977)
124. D.S. Fisher, B.I. Halperin, and R. Morf: Phys. Rev. B **20**, 4692 (1979)
125. E.Y. Andrei, S. Yücel, and L. Menna: Phys. Rev. Lett. **67**, 3704 (1991)
126. S. Yücel, L. Menna, and E.Y. Andrei: Phys. Rev. B **47**, 12672 (1993)
127. Yu.M. Vilk and A.E. Ruckenstein: Phys. Rev. B **48**, 11196 (1993)
128. V.A. Buntar', V.N. Grigor'ev, O.I. Kirichek, Yu.Z Kovdrya, and Yu.P. Monarkha: Fiz. Nizk. Temp. **12**, 796 (1986) [Sov. J. Low Temp. Phys. **12**, 450 (1986)]
129. D.K. Lambert and P.L. Richards: Phys. Rev. Lett. **44**, 1427 (1980)
130. C.L. Zipfel, T.R. Brown, and C.C. Grimes: Phys. Rev. Lett. **37**, 1760 (1976)
131. C. Kittel: *Introduction to Solid State Physics*, 4th edn. (Wiley, New York 1971) pp. 586–587
132. W. Götze and P. Wölfle: Phys. Rev. B **6**, 1226 (1972)
133. W. Götze and J. Hajdu: J. Phys. C **11**, 3993 (1978)
134. W. Götze and J. Hajdu: Solid State Commun. **29**, 89 (1979)
135. C.S. Ting, S.C. Ying, and J.J. Quinn: Phys. Rev. B **16**, 5394 (1977)
136. Y. Shiwa and A. Isihara: J. Phys. C **16**, 4853 (1983)
137. X.L. Lei and C.S. Ting: Phys. Rev. B **30**, 4809 (1984)
138. W. Cai, X.L. Lei, and C.S. Ting: Phys. Rev. B **31**, 4070 (1985)
139. M. Saitoh: Solid State Commun. **52**, 63 (1984)
140. M. Saitoh: J. Phys. Soc. Jpn. **56**, 706 (1987)
141. P.J.M. Peters, P. Scheuzger, M.J. Lea, Yu.P. Monarkha, P.K.H. Sommerfeld, and R.W. van der Heijden: Phys. Rev. B **50,** 11570 (1994)
142. B.I. Halperin: Phys. Rev. B **25**, 2185 (1982)
143. S.M. Girvin: In *Topological Aspects of Low Dimensional Systems*, ed. by A. Comtet, T. Jolicæur, S. Ouvry, F. David (EDP sciences and Springer-Verlag, Berlin 1998)
144. Zhao-bin Su and B. Sakita: Phys. Rev. B **40**, 9959 (1989)
145. D.R. Leadley, R.J. Nicholas, W. Xu, F.M. Peeters, J.T. Devreese, J. Singleton, J.A.A.J. Perenboom, L. van Bockstal, F. Herlach, C.T. Foxon, and J.J. Harris: Phys. Rev. B **48**, 5457 (1993)
146. W. Kohn: Phys. Rev. **123**, 1242 (1961)
147. H. Mori: Prog. Theor. Phys. **33**, 423 (1965)
148. Yu.P. Monarkha: Fiz. Nizk. Temp. **27**, 627 (2001) [Low Temp. Phys. **27**, 463 (2001)]
149. Yu.P. Monarkha and K. Kono: J. Phys. Soc. Jpn. **70**, 1617 (2001)
150. F.G. Bass and Yu.G. Gurevich: *Hot Electrons and Strong Electromagnetic Waves in Semiconductor and Gas-Discharge Plasma* (Nauka, Moscow 1975)
151. V.A. Buntar', Yu.Z. Kovdrya, V.N. Grigor'ev, Yu.P. Monarkha, and S.S. Sokolov: Fiz. Nizk. Temp. **13**, 789 (1987) [Sov. J. Low Temp. Phys. **13**, 451 (1987)]
152. V.A. Buntar', V.N. Grigor'ev, O.I. Kirichek, Yu.Z. Kovdrya, Yu.P. Monarkha, and S.S. Sokolov: J. Low Temp. Phys. **79**, 323 (1990)
153. I.N. Adamenko, A.V. Zhukov, and K.E. Nemchenko: Fiz. Nizk. Temp. **26**, 631 (2000) [Low Temp. Phys. **26**, 459 (2000)]
154. P.M. Platzman and G. Beni: Phys. Rev. Lett. **36**, 626 (1976)

155. P.M. Platzman, A.L. Simons, and N. Tzoar: Phys. Rev. B **16**, 2023 (1977)
156. Yu.P. Monarkha, V.B. Shikin, D.C. Glattli, and F.I.B. Williams: Phys. Lett. A **159**, 277 (1991)
157. Yu.P. Monarkha and V.B. Shikin: Fiz. Nizk. Temp. **17**, 934 (1991) [Sov. J. Low Temp. Phys. **17**, 487 (1991)]
158. D.C. Glattli, E.Y. Andrei, and F.I.B. Williams: Phys. Rev. Lett. **60**, 420 (1988)
159. D.C. Glattli: Thesis at the Université de Paris-Sud, Centre d'Orsay, Paris (1986)
160. R.E. Prange, S.M. Girvin (Eds.): *The Quantum Hall Effect*, 2nd edn. (Springer-Verlag, New York 1990)
161. D.R. Yennie: Rev. Mod. Phys. **59**, 781 (1987)
162. M. Büttiker: Phys. Rev. Lett. **57**, 1761 (1986)
163. X.G. Wen: Phys. Rev. B **43**, 11025 (1991)
164. B. Davydov and I. Pomeranchuk: J. Phys. (USSR) **2**, 147 (1940)
165. P.J.M. Peters, P. Scheuzger, M.J. Lea, Yu.P. Monarkha, P.K.H. Sommerfeld, and R.W. van der Heijden: Phys. Rev. B **50**, 11570 (1994)
166. R.W. van der Heijden, M.C.M. van de Sanden, J.H.G. Surewaard, A.T.A.M. de Waele, H.M. Gijsman, and F.M. Peeters: Europhys. Lett. **6**, 75 (1988)
167. J. Neuenschwander, P. Scheuzger, W. Joss, and P. Wyder: Physica **165–166**, 845 (1990)
168. P. Scheuzger, J. Neuenschwander, and P. Wyder: Helv. Phys. Acta. **64**, 170 (1991)
169. M.J. Lea, A.O. Stone, P. Fozooni, and J. Frost: J. Low Temp. Phys. **85**, 67 (1991)
170. Yu.P. Monarkha, E. Teske, and P. Wyder: Phys. Rep. **370**, No. 1, pp. 1–61 (2002)
171. R. Mehrotra and A.J. Dahm: J. Low Temp. Phys. **67**, 115 (1987)
172. L. Wilen and R. Giannetta: J. Low Temp. Phys. **72**, 353 (1988)
173. M.I. Dykman, M.J. Lea, P. Fozooni, and J. Frost: Phys. Rev. Lett. **70**, 3975 (1993)
174. M.J. Lea, P. Fozooni, P.J. Richardson, and A. Blackburn: Phys. Rev. Lett. **73**, 1142 (1994)
175. M.J. Lea, P. Fozooni, A. Kristensen, P.J. Richardson, K. Djerfi, M.I. Dykman, C. Fang-Yen, and A. Blackburn: Phys. Rev. B **55**, 16280 (1997)
176. J. Klier, I. Doicescu, and P. Leiderer: J. Low Temp. Phys. **121**, 603 (2000)
177. V.I. Ryzhii: Fiz. Tekh. Poluprovodn. **3**, 1704 (1969) [Sov. Phys. Semicond. **3**, 1432 (1969)]
178. H. Fukuyama, Y. Kuramoto, and P.M. Platzman: Phys. Rev. B **19**, 4980 (1979)
179. Yu.P. Monarkha and F.M. Peeters: Europhys. Lett. **34**, 611 (1996)
180. T.R. Brown and C.C. Grimes: Phys. Rev. Lett. **29**, 1233 (1972)
181. R.M. Ostermeier and K.W. Schwarz: Phys. Rev. Lett. **29**, 25 (1972)
182. V.S. Edel'man: Pis'ma Zh. Eksp. Teor. Fiz. **24**, 510 (1976) [JETP Lett. **24**, 468 (1976)]
183. V.S. Edel'man: Zh. Eksp. Teor. Fiz. **77**, 673 (1979) [Sov. Phys. JETP **50**, 338 (1979)]
184. A. Cheng and P.M. Platzman: Solid State Commun. **25**, 813 (1978)
185. L. Wilen and R. Giannetta: Phys. Rev. Lett. **60**, 231 (1988)

186. T.A. Kennedy, R.J. Wagner, B.D. McCombe, and D.C. Tsui: Solid State Commun. **22**, 459 (1977)
187. B.A. Wilson, S.J. Allen Jr., and D.C. Tsui: Phys. Rev. Lett. **44**, 479 (1980)
188. B.A. Wilson, S.J. Allen Jr., and D.C. Tsui: Phys. Rev. B **24**, 5887 (1981)
189. T. Ando: J. Phys. Soc. Jpn. **38**, 989 (1975)
190. J.P. Kotthaus, G. Abstreiter, and J.F. Koch: Solid State Commun. **15**, 517 (1974)
191. T. Aoki and M. Saitoh: J. Phys. Soc. Jpn. **48**, 1929 (1980)
192. Yu.P. Monarkha and V.B. Shikin: Sov. J. Low Temp. Phys. **14**, 439 (1988) [Fiz. Nizk. Temp. **14**, 798 (1988)]
193. L.D. Landau and S.I. Pekar: Zh. Eksp. Teor. Fiz. **18**, 419 (1948)
194. S.J. Miyake: J. Phys. Soc. Jpn. **41**, 747 (1976)
195. Y. Toyozawa: Prog. Theor. Phys. **26**, 29 (1961)
196. A. Sumi and Y. Toyozawa: J. Phys. Soc. Jpn. **35**, 137 (1973)
197. F.M. Peeters and J.T. Devreese: Phys. Rev. B **32**, 3515 (1985)
198. B. Gerlach and H. Löwen: Rev. Mod. Physics **63**, 63 (1991)
199. Y. Toyozawa and Y. Shinozuka: J. Phys. Soc. Jpn. **48**, 472 (1980)
200. S. de Sarma: Solid State Commun. **54**, 1067 (1985)
201. M.P.A. Fisher and W. Zwerger: Phys. Rev. B **34**, 5912 (1986)
202. G. Kareri, V. Fasoli, and F. Gaeta: Nuovo Cimento **15**, 774 (1960)
203. V.B. Shikin and Yu.P. Monarkha: Zh. Eksp. Teor. Fiz. **65**, 751 (1973) [Sov. Phys. JETP **38**, 373 (1973)]
204. Yu.P. Monarkha: Fiz. Nizk. Temp. **1**, 526 (1975) [Sov. J. Low Temp. Phys. **1**, 258 (1975)]
205. L. Sander: Phys. Rev. B **11**, 4350 (1975)
206. S.A. Jackson and P.M. Platzman: Phys. Rev. B **24**, 499 (1981)
207. S.A. Jackson and P.M. Platzman: Phys. Rev. B **25**, 4886 (1982)
208. O. Hipólito, G.A. Farias, and N. Studart: Surf. Sci. **113**, 394 (1982)
209. M. Saitoh: J. Phys. C: Solid State Phys. **16**, 6995 (1983)
210. R.P. Feynman: Phys. Rev. **97**, 660 (1955)
211. H. Spohn: J. Phys. A **19**, 533 (1986)
212. H. Spohn and R. Dümcke: J. Stat. Phys. **41**, 389 (1985)
213. Yu.P. Monarkha and K. Kono: Phys. Rev. B **65**, 212507-1 (2002)
214. G.E. Marques and N. Studart: Phys. Rev. B **39**, 4133 (1989)
215. F.F. Mende, Yu.Z. Kovdrya, and V.A. Nikolaenko: Sov. J. Low Temp. Phys. **11**, 355 (1985) [Fiz. Nizk. Temp. **11**, 646 (1985)]
216. O. Tress, Yu.P. Monarkha, F.C. Penning, H. Bluyssen, and P. Wyder: Phys. Rev. Lett. **77**, 2511 (1996)
217. F.M. Peeters and S.A. Jackson: Phys. Rev. B **34**, 1539 (1986)
218. E.Y. Andrei: Phys. Rev. Lett. **52**, 1984
219. B. Lehndorf and K. Dransfeld: J. Phys. (France) **50**, 2579 (1989)
220. B. Lehndorf, T. Vossloh, T. Ganzler, and K. Dransfeld: Surf. Sci. **263**, 674 (1992)
221. R.M. Bowley: J. Phys. C **9**, L151 (1976)
222. G. Baym, C.J. Pethick, and Salomaa: Phys. Rev. **38**, 845 (1977)
223. F.M. Peeters and S.A. Jackson: Phys. Rev. B **31**, 7098 (1985)
224. E.P. Wigner: Phys. Rev. **46**, 1002 (1934); Trans. Farad. Soc. **34**, 678 (1938)
225. H.M. Van Horn: Phys. Rev. **157**, 342 (1967)
226. A.V. Chaplik: Zh. Eksp. Teor. Fiz. **62**, 746 (1972) [Sov. Phys. JETP **35**, 395 (1972)]

227. H. Namaizawa and M. Voss: Phys. Rev. B **13**, 1370 (1976)
228. W.J. Carr: Phys. Rev. **122**, 1437 (1961)
229. R.S. Crandall: Phys. Rev. A **8**, 2136 (1973)
230. D.S. Fisher, B.I. Halperin, and P.M. Platzman: Phys. Rev. Lett. **42**, 798 (1979)
231. A.S. Rybalko, B.N. Esel'son, and Yu.Z. Kovdrya: Fiz. Nizk. Temp. **5**, 947 (1979) [Sov. J. Low Temp. Phys. **5**, 450 (1979)]
232. D. Marty, J. Pointernaud, and F.I.B. Williams: J. Phys. Lett. (France) **41**, L311 (1980)
233. R. Mehrotra, B.M. Guenin, and A.J. Dahm: Phys. Rev. Lett. **48**, 641 (1982)
234. K. Shirahama, O.I. Kirichek, and K. Kono: Phys. Rev. Lett. **79**, 4218 (1997)
235. O. Kirichek, M. Saitoh, K. Kono, and F.I.B. Williams: Phys. Rev. Lett. **86**, 4064 (2001)
236. G. Goldoni and F.M. Peeters: Phys. Rev. B **53**, 4591 (1996)
237. N.F. Mott: Philos. Mag. **6**, 287 (1961)
238. P. Nozières and D. Pines: Phys. Rev. **111**, 442 (1958)
239. A.A. Kugler: Ann. Phys. (N.Y.) **53**, 133 (1969)
240. R.C. Gann, S. Chakravarty, and G.V. Chester: Phys. Rev. B **20**, 326 (1979)
241. P.M. Platzman and H. Fukuyama: Phys. Rev. B **10**, 3150 (1974)
242. R.W. Hockney and T.R. Brown: J. Phys. C **8**, 1813 (1975)
243. J.M. Kosterlitz and D.J. Thouless: J. Phys. C **6**, 1181 (1973)
244. D.J. Thouless: J. Phys. C **11**, L189 (1978)
245. R.H. Morf: Phys. Rev. Lett. **43**, 931 (1979)
246. H.W. Jiang and A.J. Dahm: Surf. Sci. **229**, 352 (1990)
247. F. Gallet, G. Deville, A. Valdes, and F.I.B. Williams: Phys. Rev. Lett. **49**, 212 (1982)
248. G. Deville, A. Valdes, E.Y. Andrei, and F.I.B. Williams: Phys. Rev. Lett. **53**, 588 (1984)
249. H. Fukuyama: Solid State Comm. **19**, 551 (1976)
250. V.M. Bedanov, G.V. Gadiyak, and Y.E. Lozovik: Phys. Lett. A **109**, 289 (1985)
251. V.M. Bedanov, G.V. Gadiyak, and Y.E. Lozovik: Zh. Eksp. Teor. Fiz. **88**, 1622 (1985) [Sov. Phys. JETP **61**, 967 (1985)]
252. H. Fukuyama: Solid State Commun. **17**, 1323 (1975)
253. A.V. Chaplik: Zh. Eksp. Teor. Fiz. **72**, 1946 (1977) [Sov. Phys. JETP **45**, 1023 (1977)]
254. F.R. Ulinich and N.A. Usov: Zh. Eksp. Teor. Fiz. **76**, 288 (1979) [Sov. Phys. JETP **49**, 147 (1979)]
255. Yu.P. Monarkha and S.S. Sokolov: Fiz. Nizk. Temp. **8**, 350 (1982) [Sov. J. Low Temp. Phys. **8**, 173 (1982)]
256. L.D. Landau and E.M. Lifshitz: *Mechanics*, Course of Theoretical Physics, Vol. 1, 3rd edn. (Pergamon Press, Oxford, New York, Seoul, Tokyo 1976) pp. 143–145
257. Yu.P. Monarkha and V.B. Shikin: Zh. Eksp. Teor. Phys. **68**, 1423 (1975) [Sov. Phys. JETP **68**, 710 (1975)]
258. H. Namaizawa: Solid State Commun. 34, 607 (1980)
259. Yu.P. Monarkha and V.B. Shikin: Sov. J. Low Temp. Phys. **9**, 471 (1983) [Fiz. Nizkh. Temp. **9**, 913 (1983)]
260. V.B. Shikin: Pis'ma Zh. Eksp. Teor. Fiz. **19**, 657 (1974) [JETP Lett. **19**, 335 (1974)]

261. Yu.P. Monarkha: Fiz. Nizk. Temp. **6**, 685 (1980) [Sov. J. Low Temp. Phys. **6**, 331 (1980)]
262. Yu.P. Monarkha: Fiz. Nizk. Temp. **7**, 692 (1981) [Sov. J. Low Temp. Phys. **7**, 338 (1981)]
263. V.E. Sivokon', V.V. Dotsenko, Yu.Z. Kovdrya, and V.N. Grigor'ev: Fiz. Nizk. Temp. **22**, 1107 (1996) [Low Temp. Phys. **22**, 845 (1996)]
264. L. Van Hove: Phys. Rev. **95**, 249 (1954)
265. B. Jancovici: Phys. Rev. Lett. **19**, 20 (1967)
266. Y. Imry and L. Gunther: Phys. Rev. B **3**, 3939 (1971)
267. M.I. Dykman and Yu.G. Rubo: Phys. Rev. Lett. **78**, 4813 (1997)
268. M.I. Dykman: Zh. Eksp. Teor. Fiz. **82**, 1318 (1982) [Sov. Phys. JETP **55**, 766 (1982)]
269. H. Makabe and M. Saitoh: Physica E **6**, 876 (2000)
270. A. Sjolander: Ark. Fys. **14**, 315 (1958)
271. S.T. Chui: Phys. Rev. Lett. **48**, 933 (1982)
272. M.C. Bonsager, K. Flensberg, B.Y.-K. Hu, A.-P. Jauho: Phys. Rev. Lett. **77**, 1366 (1996) and references therein
273. H. Rubel, A. Fischer, W. Dietsche, K. von Klitzing, and K. Eberl: Phys. Rev. Lett. **78**, 1763 (1997)
274. Y.W. Suen, L.W. Engel, M.B. Santos, M. Shayegan, and D.C. Tsui: Phys. Rev. Lett. **68**, 1379 (1992)
275. J.P. Eisenstein, G.S. Boebinger, L.N. Pfeiffer, K.W. West, and S. He: Phys. Rev. Lett. **68**, 1383 (1992)
276. V.I. Falko: Phys. Rev. B **49**, 7774 (1994)
277. A.G. Khachaturyan: Zh. Eksp. Teor. Phys. **63**, 1421 (1972) [Sov. Phys. JETP **36**, 753 (1973)]
278. H.-W. Jiang and A.J. Dahm: Phys. Rev. Lett. **62**, 1396 (1989)
279. K. Kono: Physica B **280**, 112 (2000)
280. K. Shirahama and K. Kono: Phys. Rev. Lett. **74**, 781 (1995)
281. O.I. Kirichek, K. Shirahama, and K. Kono: J. Low Temp. Phys. **113**, 1103 (1998)
282. O.I. Kirichek, K. Shirahama, F.I.B. Williams, M. Saitoh, and K. Kono: Physica B **284–288**, 279 (2000)
283. M.I. Dykman: Fiz. Nizk. Temp. **10**, 453 (1984) [Sov. J. Low Temp. Phys. **10**, 233 (1984)]
284. I.A. Fomin: Zh. Eksp. Teor. Fiz. **61**, 2562 (1971) [Sov. Phys. JETP **34**, 1371 (1972)]
285. Yu.B. Ivanov: Zh. Eksp. Teor. Fiz. **79**, 1082 (1980) [Sov. Phys. JETP **52**, 549 (1980)]
286. G. Deville: J. Low Temp. Phys. **72**, 135 (1988)
287. L.D. Landau and E.M. Lifshitz: *Fluid Mechanics*, Course of Theoretical Physics, Vol. 6, 2nd edn. (Butterworth-Heinemann Ltd, Oxford 1987) pp. 93–94
288. J. Ranninger, S. Robaszkiewicz, A. Sulpice, and R. Tournier: Europhys. Lett. **3**, 347 (1987)
289. J.P. Bouchaud: New microscopic description of liquid helium 3. In: *Recent Progress in Many-Body Theories*, Vol. 2, ed. by Y. Avishai (Plenum Press, New York, 1990) pp. 331–335
290. K. Shirahama and K. Kono: J. Low Temp. Phys. **104**, 237 (1996)

291. A. Kristensen, K. Djerfi, P. Fozooni, M.J. Lea, P.J. Richardson, A. Santrich-Badal, A. Blackburn, and R.W. van der Heijden: Phys. Rev. Lett. **77**, 1350 (1996)
292. W.F. Vinen: J. Phys.: Condens. Matter **11**, 9709 (1999)
293. R. Giannetta and L. Wilen: Solid State Commun. **78**, 199 (1991)

Index

Springer Series in Solid-State Sciences

Editors: M. Cardona P. Fulde K. von Klitzing H.-J. Queisser

Springer Series in Solid-State Sciences

Editors: M. Cardona P. Fulde K. von Klitzing H.-J. Queisser

Springer Series in Solid-State Sciences

Editors: M. Cardona P. Fulde K. von Klitzing H.-J. Queisser

www.ingramcontent.com/pod-product-compliance
Ingram Content Group UK Ltd.
Pitfield, Milton Keynes, MK11 3LW, UK
UKHW021857190726
13853UKWH00003B/1306

* 9 7 8 3 6 6 2 1 0 6 4 0 2 *